Camella

W9-AJI-304

CONTENTS

Preface v

1 Building a Geometry Vocabulary 1
The Building Blocks of Geometry 2
Definitions and Postulates 6
Inductive Versus Deductive Reasoning 9
The IF . . . THEN . . . Sentence Structure 10
Review Exercises for Chapter 1 11

2 Measure and Congruence 15
Measuring Segments and Angles 16
Betweenness of Points and Rays 18
Congruence 21
Midpoint and Bisector 22
Diagrams and Drawing Conclusions 24
Properties of Equality and Congruence 25
Additional Properties of Equality 28
The Two-Column Proof Format 31
Review Exercises for Chapter 2 32

3 Angle Pairs and Perpendicular Lines 39
Supplementary and Complementary Angle Pairs 40
Adjacent and Vertical Angle Pairs 42
Theorems Relating to Complementary, Supplementary, and Vertical Angles 44
Definitions and Theorems Relating to Right Angles and Perpendiculars 48
A Word About the Format of a Proof 54
Review Exercises for Chapter 3 54

4 Parallel Lines 59
Planes and Lines 60
Properties of Parallel Lines 61
Converses and Methods of Proving Lines Parallel 70
The Parallel Postulate 73
Review Exercises for Chapter 4 75

5 Angles of a Polygon 81
The Anatomy of a Polygon 82
Angles of a Triangle 85
Exterior Angles of a Triangle 89
Angles of a Polygon 93
Review Exercises for Chapter 5 97

6 Proving Triangles Are Congruent 101
Correspondences and Congruent Triangles 102
Proving Triangles Congruent: SSS, SAS, and ASA Postulates 105
Proving Overlapping Triangles Congruent 110
Proving Triangles Congruent: AAS and Hy-Leg Methods 113
When Two Triangles Are NOT Congruent 118
Review Exercises for Chapter 6 119

7 Applying Congruent Triangles 125
Using Congruent Triangles to Prove Segments and Angles Congruent 126
Using Congruent Triangles to Prove Special Properties of Lines 129
Classifying Triangles and Special Segments 131
The Isosceles Triangle 135
Double Congruence Proofs 139
Review Exercises for Chapter 7 142

Cumulative Review Exercises: Chapters 1–7 149

8 Geometric Inequalities 153
Some Basic Properties of Inequalities 154
Inequality Relationships in a Triangle 155
The Indirect Method of Proof 158
Review Exercises for Chapter 8 162

9 Special Quadrilaterals 169
Classifying Quadrilaterals 170
Properties of a Parallelogram 171
Properties of Special Parallelograms 174
Proving a Quadrilateral Is a Parallelogram 178
Applications of Parallelograms 184
Properties of a Trapezoid 186
Review Exercises for Chapter 9 193

10 Ratio, Proportion, and Similarity 199
Ratio and Proportion 200
Proportions in a Triangle 207
When Are Polygons Similar? 211
Proving Triangles Similar 217
Proving Lengths of Sides of Similar Triangles in Proportion 218

Proving Products of Segment Lengths Equal 223
Review Exercises for Chapter 10 225

11 The Right Triangle 233
Proportions in a Right Triangle 234
The Pythagorean Theorem 237
Special Right-Triangle Relationships 243
Trigonometric Ratios 247
Indirect Measurement in a Right Triangle 252
Review Exercises for Chapter 11 256

Cumulative Review Exercises: Chapters 8–11 261

12 Circles and Angle Measurement 267
The Parts of a Circle 268
Arcs and Central Angles 273
Diameters and Chords 281
Tangents and Secants 284
Angle Measurement: Vertex on the Circle 288
Angle Measurement: Vertex in the Interior of the Circle 297
Angle Measurement: Vertex in the Exterior of the Circle 299
Using Angle-Measurement Theorems 304
Review Exercises for Chapter 12 307

13 Chord, Tangent, and Secant Segments 315
Equidistant Chords 316
Tangents and Circles 319
Similar Triangles and Circles 323
Tangent- and Secant-Segment Relationships 326
Circumference and Arc Length 331
Review Exercises for Chapter 13 335

14 Area and Volume 341
Areas of a Rectangle, Square, and Parallelogram 342
Areas of a Triangle and Trapezoid 346
Comparing Areas 353
Area of a Regular Polygon 356
Areas of a Circle, Sector, and Segment 361
Geometric Solids 364
Review Exercises for Chapter 14 369

15 Coordinate Geometry 377
The Coordinate Plane 378
Finding Area Using Coordinates 379
The Midpoint and Distance Formulas 382
Slope of a Line 388
Equation of a Line 394
Equation of a Circle 399
Proofs Using Coordinates 401
Review Exercises for Chapter 15 404

16 Locus and Constructions 409
Describing Points That Fit One Condition 410
Describing Points That Fit More Than One Condition 411
Locus and Coordinates 415
Basic Constructions 418
Review Exercises for Chapter 16 423

17 Transformation Geometry 427
Terms and Notation 428
Congruence Transformations 429
Classifying Isometries 434
Size Transformations 434
Types of Symmetry 436
Transformations in the Coordinate Plane 439
Composing Transformations 444
Review Exercises for Chapter 17 447

Cumulative Review Exercises: Chapters 12–17 452

Some Geometric Relationships and Formulas Worth Remembering 459

Glossary 465

Answers to Chapter Exercises 471

Solutions to Cumulative Review Exercises 494

Preface

Many geometry-help books tend to resemble factual handbooks that reduce the study of geometry to a list of formulas followed by a continuous stream of exercises. *E-Z Geometry* is different. This book "teaches" a typical introductory course in geometry, using a friendly writing style that incorporates many of the same strategies that an experienced classroom teacher would use.

This book has several special features that make it ideal for self-study and rapid learning:

- Most concepts are *not* presented in "finished form," but instead are carefully developed using a clear writing style with many helpful examples and diagrams. Emphasis is on *understanding* geometric principles as well as on knowing how to apply them in new problem situations.
- The discussions, numerous examples, and convenient summaries anticipate and answer the "why" types of questions students might ask if the material were being explained by a classroom teacher.
- Each chapter concludes with a comprehensive set of original exercises that are designed to clinch understanding of key ideas while building skill and confidence in solving problems. Also, three sets of cumulative review exercises, strategically placed in the book, serve as progress checkpoints and ensure long-term learning. Answers to all exercises are provided at the end of the book.
- Key geometric formulas and relationships have been organized in easy-to-read tables at the end of the book.
- Important geometric terms have been collected and placed in a glossary at the end of the book.

The latest edition of *E-Z Geometry* adds a chapter on Transformation Geometry, reorganizes some of the exercise sections, and includes a number of minor changes throughout the text designed to further improve readability.

LAWRENCE S. LEFF

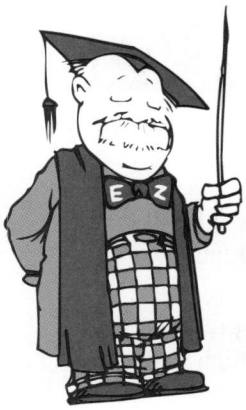

1

Building a Geometry Vocabulary

WHAT YOU WILL LEARN

Studying geometry is very different from studying elementary algebra. In geometry we are concerned with developing a logical structure in which mathematical relationships are *proved* as well as *applied*. In this chapter you will learn:

- some basic terms and assumptions of geometry;
- the nature of geometric proofs;
- the type of mathematical reasoning that forms the basis of this course.

SECTIONS IN THIS CHAPTER

- The Building Blocks of Geometry
- Definitions and Postulates
- Inductive Versus Deductive Reasoning
- The IF . . . THEN . . . Sentence Structure

The Building Blocks of Geometry

Studying geometry is, in a sense, like building a house. Geometry uses logical reasoning as the cement and the following types of statements as the basic building blocks:

- UNDEFINED TERMS. Some terms are so fundamental that they cannot be defined using simpler terms. *Point*, *line*, and *plane* are undefined terms in geometry. Although these terms cannot be defined, they can be described.
- DEFINED TERMS. New terms can be defined using undefined as well as previously defined terms, thereby creating an expanding "dictionary" of terms which makes it easier to describe geometric figures and relationships.
- POSTULATES. Some beginning principles in geometry, called *postulates,* are so basic that they cannot be arrived at using simpler facts. A **postulate** is a statement that is accepted without proof. For instance, the observation "Exactly one line can be drawn through two points" is a postulate.
- THEOREMS. Unlike a postulate, a theorem is a generalization that can be proved to be true. "Prove" simply means presenting a valid argument that uses a set of known facts and logical reasoning to show that a statement is true. The familiar fact that "The sum of the measures of the three angles of a triangle is 180" is a theorem.

Geometry is an example of a *postulational system* in which a beginning set of assumptions and undefined terms is used as a starting point in developing new relationships that are expressed as theorems. These theorems, together with the postulates and defined and undefined terms, are used to *prove* other theorems. Here, "to prove" simply means to use a logical chain of reasoning to show how undefined terms, definitions, postulates, and previously established theorems lead to a new generalization.

UNDEFINED TERMS

Table 1.1 lists some undefined terms.

TABLE 1.1

Undefined Term	Description	Notation
Point	A *point* indicates position; it has no length, width, or depth.	• A A point is named by a single capital letter.
Line	A *line* is a set of continuous points that extend indefinitely in either direction.	A B ℓ A line is identified by naming two points on the line and drawing a line over the letters: $\overleftrightarrow{AB}$

TABLE 1.1 *(continued)*

Undefined Term	Description	Notation
		Alternatively, a line may be named by using a single lowercase letter. line ℓ
Plane	A *plane* is a set of points that forms a flat surface that has no depth and that extends indefinitely in all directions.	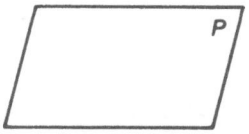 A plane is usually represented as a closed four-sided figure and is named by placing a capital letter at one of the corners.

Figure 1.1 illustrates that lines may lie in different planes or in the same plane. Line ℓ and line AB both lie in plane Q. Line k and line AB lie in plane P. Lines ℓ and k are contained in different planes, while line AB (the intersection of the two planes) is common to both planes. To simplify our discussions, we will always assume that we are working with figures that lie in the same plane. This branch of geometry takes a "flat," two-dimensional view of figures and is referred to as *plane geometry*. *Solid geometry* is concerned with figures and their spatial relationships as they actually exist in the world around us.

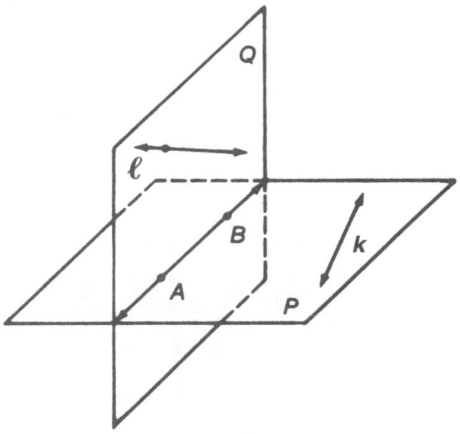

FIGURE 1.1

DEFINED TERMS

Table 1.2 lists some geometric terms and their definitions.

TABLE 1.2

Term	Definition	Illustration
1. Line segment	A *line segment* is a part of a line consisting of two points, called *end points*, and the set of all points between them.	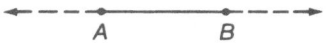 NOTATION: $\overline{AB}$

TABLE 1.2 *(continued)*

Term	Definition	Illustration
2. Ray	A *ray* is a part of a line consisting of a given point, called the *end point*, and the set of all points on one side of the end point.	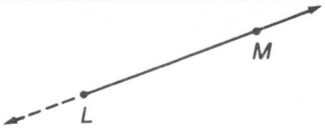 NOTATION: $\overrightarrow{LM}$ A ray is always named by using two points, the first of which must be the end point. The arrow drawn above always points to the right.
3. Opposite rays	*Opposite rays* are rays that have the same end point and that form a line.	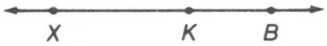 $\overrightarrow{KX}$ and $\overrightarrow{KB}$ are opposite rays.
4. Angle	An *angle* is the union of two rays having the same end point. The end point is called the *vertex* of the angle, and the rays are called the *sides* of the angle.	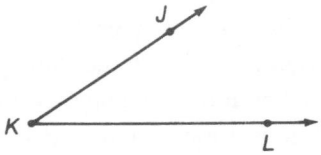 VERTEX: K SIDES: $\overrightarrow{KJ}$ and $\overrightarrow{KL}$

NAMING ANGLES

An angle may be named in one of three ways:

1. Using three letters, the center letter corresponding to the vertex of the angle and the other letters representing points on the sides of the angle. For example, in Figure 1.2, the name of the angle whose vertex is T can be angle RTB ($\measuredangle RTB$) or angle BTR ($\measuredangle BTR$).
2. Placing a number at the vertex and in the *interior* of the angle. The angle may then be referred to by the number. For example, in Figure 1.3, the name of the angle whose vertex is T can be $\measuredangle 1$ or $\measuredangle RTB$ or $\measuredangle BTR$.

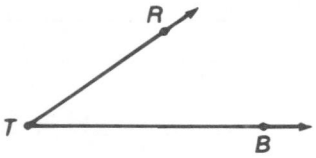

FIGURE 1.2

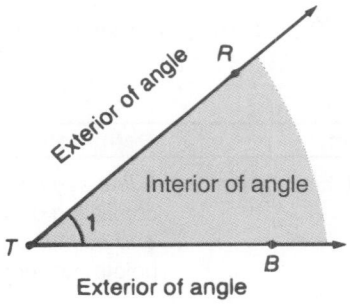

FIGURE 1.3

3. Using a single letter that corresponds to the vertex, provided that this does not cause any confusion. There is no question which angle on the diagram corresponds to angle A in Figure 1.4, but which angle on the diagram is angle D? Actually three angles are formed at vertex D:

- Angle ADB
- Angle CDB
- Angle ADC

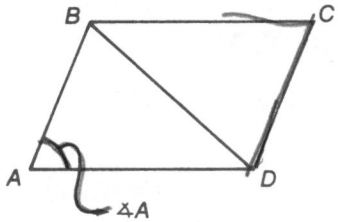

FIGURE 1.4

In order to uniquely identify the angle having D as its vertex, we must either name the angle using three letters or introduce a number into the diagram.

EXAMPLE 1.1

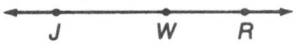

a. Name the accompanying line in three different ways.
b. Name three different segments.
c. Name four different rays.
d. Name a pair of opposite rays.

SOLUTION

a. $\overleftrightarrow{JW}$, $\overleftrightarrow{WR}$, and $\overleftrightarrow{JR}$
b. $\overline{JW}$, $\overline{WR}$ and $\overline{JR}$
c. $\overrightarrow{JR}$, $\overrightarrow{WR}$, $\overrightarrow{RJ}$ and $\overrightarrow{WJ}$
d. $\overrightarrow{WJ}$ and $\overrightarrow{WR}$

EXAMPLE 1.2

Use three letters to name each of the numbered angles in the accompanying diagram.

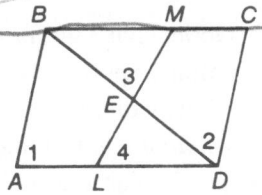

SOLUTION

$\measuredangle 1 = \measuredangle BAD$ or $\measuredangle DAB$ or $\measuredangle LAB$
$\measuredangle 2 = \measuredangle CDB$ or $\measuredangle BDC$ (NOTE: Letter E may be used instead of letter B.)
$\measuredangle 3 = \measuredangle BEM$ or $\measuredangle MEB$
$\measuredangle 4 = \measuredangle DLM$ or $\measuredangle MLD$ (NOTE: Letter E may be used instead of letter M.)

Definitions and Postulates

DEFINITIONS

The purpose of a definition is to make the meaning of a term clear. A good definition must:

- Clearly identify the word (or expression) that is being defined.
- State the distinguishing characteristics of the term being defined, using only words that are commonly understood or that have been previously defined.
- Be expressed in a grammatically correct sentence.

As an example, consider the term *collinear*. In Figure 1.5, points *A, B,* and *C* are collinear. In Figure 1.6, points *R, S,* and *T* are *not* collinear.

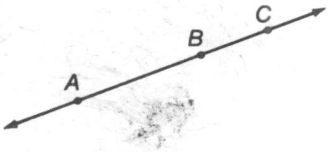

FIGURE 1.5

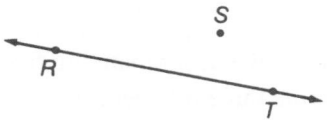

FIGURE 1.6

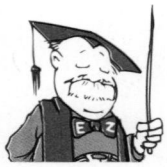

DEFINITIONS OF COLLINEAR AND NONCOLLINEAR POINTS

- *Collinear points* are points that lie on the same line.
- *Noncollinear points* are points that do not lie on the same line.

Notice that a definition begins by identifying the term being defined. The definition of collinear points uses only geometric terms (*points* and *line*) that have been previously discussed. Contrast this definition with the following one:

An *apothem* is a line segment drawn from the center of a regular polygon perpendicular to a side of the polygon.

Is this a good definition? No, it is not clear what an apothem is since several terms, including *regular polygon* and *perpendicular*, which have not been explained, are used in the definition.

Much of geometry involves building on previously discussed ideas. For example, we can use our current knowledge of geometric terms to arrive at a definition of a triangle. How would you draw a triangle? If you start with three noncollinear points (Figure 1.7a) and connect them with line segments (Figure 1.7b), a *triangle* is formed.

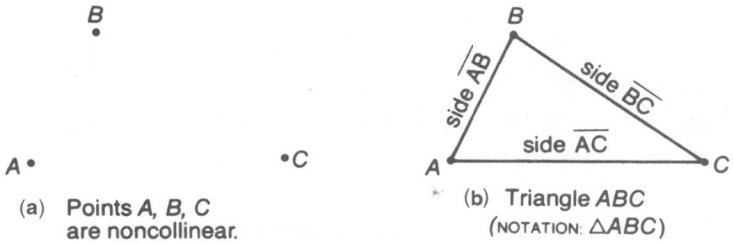

(a) Points *A, B, C* are noncollinear.

(b) Triangle *ABC* (NOTATION: $\triangle ABC$)

FIGURE 1.7

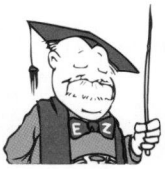

DEFINITION OF TRIANGLE

A *triangle* is a figure formed by connecting three noncollinear points with three different line segments each of which has two of these points as end points.

Notice that the definition uses the term *noncollinear*, which has been defined. Is it necessary to include that the three noncollinear points are connected by line segments? Yes; observe (in Figure 1.8) that it is possible to join three noncollinear points without using line segments.

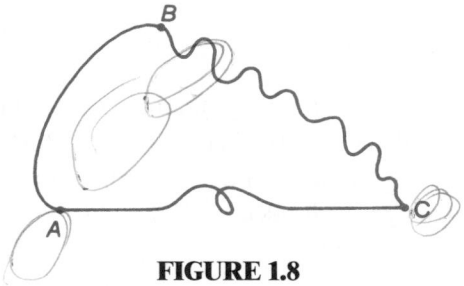

FIGURE 1.8

A good definition must be reversible as shown in the following table.

Definition	Reverse of the Definition
Collinear points are points that lie on the same line.	Points that lie on the same line are collinear points.
A *right angle* is an angle whose measure is 90 degrees.	An angle whose measure is 90 degrees is a right angle.
A *line segment* is a set of points.	A set of points is a line segment.

The first two definitions are reversible since the reverse of the definition is a true statement. The reverse of the third "definition" is false since the points may be scattered as in Figure 1.9.

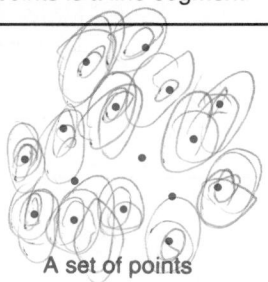

A set of points

FIGURE 1.9

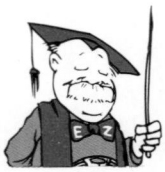

The Reversibility Test

The reverse of a definition must be true. If the reverse of a statement which is being offered as a definition is false, then the statement is not a good definition.

The reverse of a definition will prove useful in our later work when attempting to establish geometric properties of lines, segments, angles, and figures. For example, a *midpoint* of a segment may be defined as a point that divides a segment into two segments of equal length. In Figure 1.10, how can we prove that point M is the midpoint of $\overline{AB}$? We must apply the *reverse* of the definition of a midpoint: a point that divides a segment into two segments of equal length is the midpoint of the segment. In other words, we must first show that $AM = MB$. Once this is accomplished, we are entitled to conclude that point M is the midpoint of $\overline{AB}$.

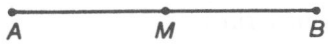

FIGURE 1.10

As another illustration, we may define an *even* integer as an integer that leaves a remainder of 0 when divided by 2. How can we prove that a particular integer is an even number? Simple—we use the reverse of the definition to show that, when the integer is divided by 2, the remainder is 0. If this is true, then the integer must be an even number.

INITIAL POSTULATES

In building a geometric system, not everything can be proved since there must be some basic assumptions, called *postulates* (or *axioms*), that are needed as a beginning. Here are our first two postulates.

POSTULATE 1.1

Two points *determine* a line.

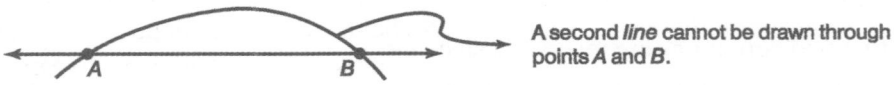

POSTULATE 1.2

Three noncollinear points *determine* a plane.

Postulate 1.1 implies that through two points exactly one line may be drawn while Postulate 1.2 asserts that a plane is defined when a third point not on this line is given.

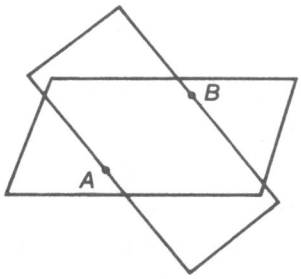

More than one plane may contain *two* points.

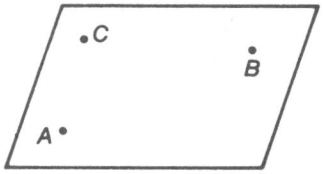

Exactly one plane contains *three* noncollinear points.

Inductive Versus Deductive Reasoning

Consider the result of accumulating consecutive odd integers beginning with 1.

String of Odd Integers	Sum
1 + 3	4
1 + 3 + 5	9
1 + 3 + 5 + 7	16
1 + 3 + 5 + 7 + 9	25

Do you notice a pattern? It appears that the sum of consecutive odd integers, beginning with 1, will always be a *perfect square*. (A perfect square is a number that can be expressed as the product of two identical numbers.) If, on the basis of this evidence, we now conclude that this relationship will always be true, regardless of how many terms are added, we have engaged in *inductive reasoning*. Inductive reasoning involves examining a few examples, observing a pattern, and then assuming that the pattern will never end. Inductive reasoning is *not* a valid method of proof, although it often suggests statements that can be proved by other methods.

Deductive reasoning may be considered to be the opposite of inductive reasoning. Rather than begin with a few specific instances as is common with inductive processes, deductive reasoning uses accepted *facts* (i.e., undefined terms, defined terms, postulates, and previously established theorems) to reason in a step-by-step fashion until a desired conclusion is reached.

EXAMPLE 1.3 Assume the following two postulates are true. (1) All last names that have seven letters with no vowels are the names of Martians. (2) All Martians are 3 feet tall. Prove that Mr. Xhzftlr is 3 feet tall.

SOLUTION

For illustrative purposes we will use the two-column proof format that will be explained more fully in subsequent chapters.

PROOF:

Statements	Reasons
1. The name is Mr. Xhzftlr.	1. Given.
2. Mr. Xhzftlr is a Martian.	2. All last names that have seven letters with no vowels are the names of Martians. (See Postulate 1.)
3. Mr. Xhzftlr is 3 feet tall.	3. All Martians are 3 feet tall. (See Postulate 2.)

Notice that each statement has a corresponding justification.

The IF . . . THEN . . . Sentence Structure

Consider the statement, "If I graduate from high school with an average greater than 90, then my parents will buy me a car." Will the student receive a car as a graduation present? The statement in the "If" clause identifies the *condition* that must be met in order for the student to get the car as a present, while the statement in the "then" clause gives the consequence.

Theorems in geometry are usually expressed as conditional statements in "If. . . then . . ." form.

- *After* a theorem is proved, the "then" clause represents the fact that you are allowed to apply whenever the condition in the "if" clause is true. A theorem given in a later chapter is "If a figure is a rectangle, then its diagonals have the same length." After this theorem has been proved, you can use the fact that its diagonals have the same length whenever that property of rectangles is needed.
- *Before* a proposed theorem is proved, the "if" clause contains what we know and the "then" clause identifies what we need to prove. A theorem that will be proved in a later chapter is

"If $\underbrace{\text{two sides of a triangle are equal in length}}_{\text{Given}}$, then $\underbrace{\text{the angles opposite them are equal}}_{\text{To Prove}}$."

Here is one possible format that can be used when organizing a proof:

GIVEN: $AB = CB$.
PROVE: Angle A = Angle C.

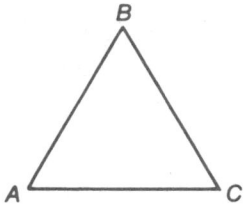

PROOF:

Statements	Reasons
1. $AB = CB$.	1. Given.
2.	2. ____
3.	3. ____
4.	4. ____

Undefined terms and previously established definitions, postulates, and theorems may be entered.

Once this proof is complete, the theorem is taken as fact and can then be used to prove other theorems.

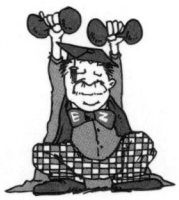

REVIEW EXERCISES FOR CHAPTER 1

1. For the accompanying diagram:

   (a) Name four rays each of which has point B as an end point.
   (b) Name line ℓ in three different ways.
   (c) Name line m in three different ways.
   (d) Name four angles that have the same vertex.
   (e) Name two pairs of opposite rays.

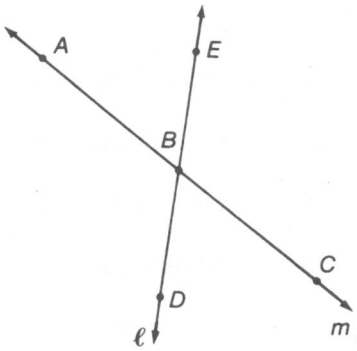

Use the following diagram for Exercises 2 and 3.

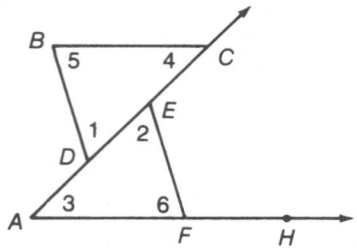

2. Name the vertex of each angle: (a) 1 (b) 3 (c) 5.

3. Use three letters to name each angle: (a) 2 (b) 4 (c) 6.

For Exercises 4 to 11, use the following diagram.

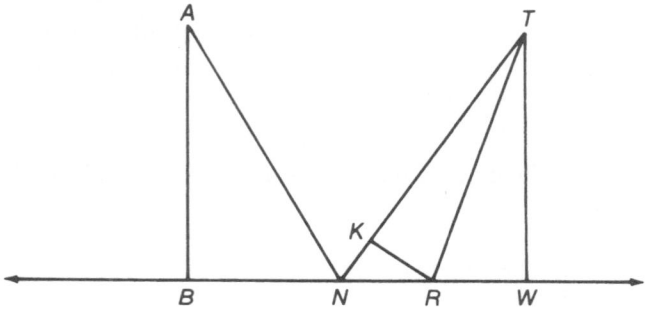

4. Name four collinear points.

5. If point N is the midpoint of $\overline{BW}$, name two segments that have the same length.

6. Name the different triangles that appear in the diagram.

7. Name each angle that has point R as its vertex.

8. Name an angle that is not an angle of a triangle.

9. Name two pairs of opposite rays.

10. Name a segment that is a side of two different triangles.

11. To prove R is the midpoint of $\overline{WN}$, which two segments must be demonstrated to have the same length?

12. Write the reverse of each of the following definitions:

    (a) An acute angle is an angle whose measure is less than 90.
    (b) An equilateral triangle is a triangle having three sides equal in length.
    (c) A bisector of an angle is the ray (or segment) which divides the angle into two congruent angles.

13. Identify each of the following as an example of inductive or deductive reasoning.

    (a) The sum of 1 and 3 is an even number; the sum of 3 and 5 is an even number; the sum of 5 and 7 is an even number; the sum of 7 and 31 is an even number; the sum of 19 and 29 is an even number. *Conclusion:* The sum of any two odd numbers is an even number.
    (b) All students in Mr. Euclid's geometry class are 15 years old. John is a member of Mr. Euclid's geometry class. *Conclusion:* John is 15 years old.
    (c) It has rained on Monday, Tuesday, Wednesday, Thursday, and Friday. *Conclusion:* It will rain on Saturday.
    (d) The sum of the measures of a pair of complementary angles is 90. Angle A and angle B are complementary. The measure of angle A is 50. *Conclusion:* The measure of angle B is 40.

14. A *median* of a triangle is a segment drawn from a vertex of the triangle to the midpoint of the opposite side of the triangle. Draw several large *right* (90-degree) triangles. See the diagram. For each triangle, locate the midpoint of the hypotenuse (the side opposite the 90-degree angle). Draw the median to the hypotenuse. Using a ruler, compare the lengths of the median and the hypotenuse in each triangle drawn. Use inductive reasoning to draw an appropriate conclusion. Note in the diagram that M is the midpoint of AB if $\overline{AM}$ and $\overline{BM}$ measure the same length.

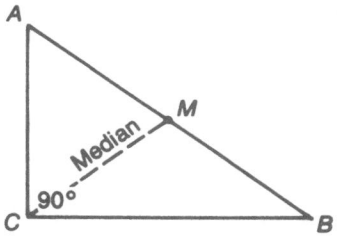

15. Draw several large triangles (not necessarily right triangles). In each triangle locate the midpoint of each side. Draw the three medians of each triangle. Use inductive reasoning to draw a conclusion related to where the medians intersect.

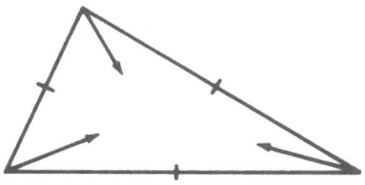

16. Use deductive reasoning to arrive at a conclusion based on the assumptions given.

    (a) ASSUMPTIONS
    1. All Martians have green eyes.
    2. Henry is a Martian.

    (b) ASSUMPTIONS
    1. The sum of the measures of the angles of a triangle is 180.
    2. In a particular triangle, the sum of the measures of two angles is 100.

17. A prime number is any whole number that is divisible only by itself and 1. For example, 7, 11, and 13 are prime numbers. Evaluate the formula $n^2 + n + 17$ using all integer values of n from 0 to 9, inclusive. Do you notice a pattern?

    (a) Using inductive reasoning, draw a conclusion.
    (b) Is your conclusion true for all values of n? Test $n = 16$.

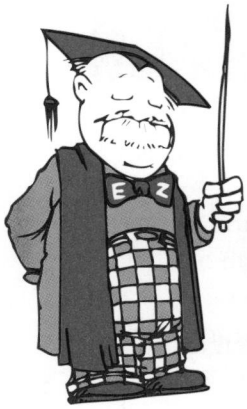

2

Measure and Congruence

WHAT YOU WILL LEARN

This chapter focuses on measuring angles and segments. In this chapter you will learn:

- definitions and terms related to segments and angles that have the same measures;
- the properties of equality that are useful in working with the lengths of line segments and the degree measures of angles;
- the way to draw conclusions using the properties of equality and congruence;
- the format of a formal two-column geometric proof.

SECTIONS IN THIS CHAPTER

- Measuring Segments and Angles
- Betweenness of Points and Rays
- Congruence
- Midpoint and Bisector
- Diagrams and Drawing Conclusions
- Properties of Equality and Congruence
- Additional Properties of Equality
- The Two-Column Proof Format

Measuring Segments and Angles

We often describe the size of something by comparing it to another thing we are already familiar with. "She is as thin as a rail" creates the image of an underweight person, but it is not very exact. In geometry we must be precise. How could we determine the exact weight of a person? We might use a measurement instrument that is specifically designed for this purpose—the scale. To determine the length of a segment or the measure of an angle, we must also use special measurement instruments—the *ruler* for measuring the length of a segment and the *protractor* for measuring an angle.

The units of measurement that we choose to express the length of a segment are not important, although they should be convenient. It would not be wise, for example, to try to measure and express the length of a postage stamp in terms of kilometers or miles. As illustrated in Figure 2.1, a segment is measured by lining up the end points of the segment with convenient markings of a ruler. In this example, the length or measure of line segment AB is 2 inches. We abbreviate this by writing $m\overline{AB} = 2$, read as "The measure of line segment AB is 2." Alternatively, we could write $AB = 2$, read as "The distance between points A and B is 2." It is customary to use the expressions $m\overline{AB}$ and AB (no bar over the letters A and B) interchangeably and to interpret each as the length of line segment $\overline{AB}$. *Caution:* It is incorrect to write $\overline{AB} = 2$, which implies that the infinite set of points that make up segment $\overline{AB}$ is equal to 2.

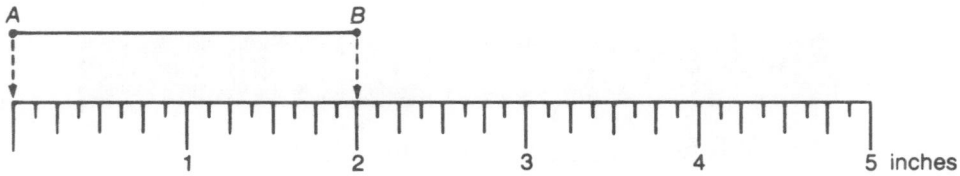

FIGURE 2.1

USING A PROTRACTOR

To measure an angle we use a protractor (see Figure 2.2), where the customary unit of measure is the degree. In our example, the measure of angle ABC is 60 degrees. We abbreviate this by writing $m\measuredangle ABC = 60$, read as "The measure of angle ABC is 60." It is customary to omit the degree symbol (°). Thus, we *never* write $m\measuredangle ABC = 60°$ or $\measuredangle ABC = 60$ (omitting the "m").

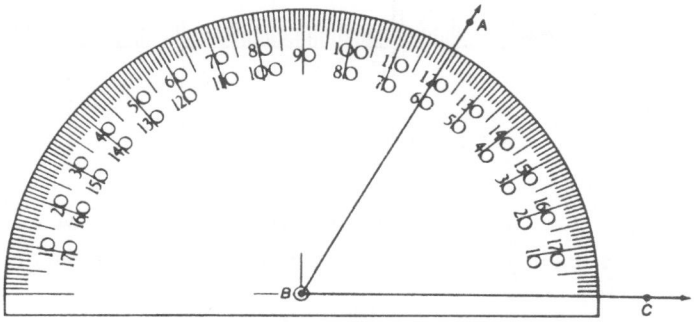

FIGURE 2.2

In this course, however, we will assume that the measure of an angle corresponds to some number on the protractor that is greater than 0 and less than *or* equal to 180.

EXAMPLE 2.1 Find the measures of these angles:

a. $\measuredangle APZ$
b. $\measuredangle FPZ$
c. $\measuredangle WPB$
d. $\measuredangle ZPB$
e. $\measuredangle SPZ$

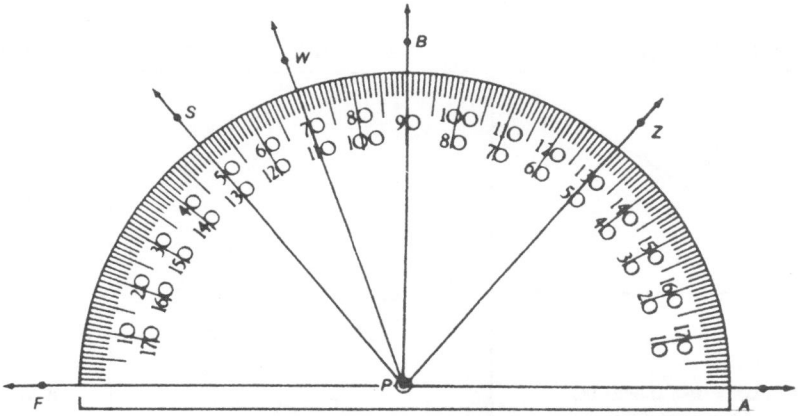

SOLUTION

a. $m\measuredangle APZ$ = 50 (read lower scale when the measure of the angle is less than 90)

b. $m\measuredangle FPZ$ = 130 (read upper scale when the measure of the angle is greater than 90)

c. $m\measuredangle WPB$ = $110 - 90 = 20$

d. $m\measuredangle ZPB$ = $90 - 50 = 40$

e. $m\measuredangle SPZ$ = $130 - 50 = 80$

EXAMPLE 2.2 Find m$\overline{RS}$.

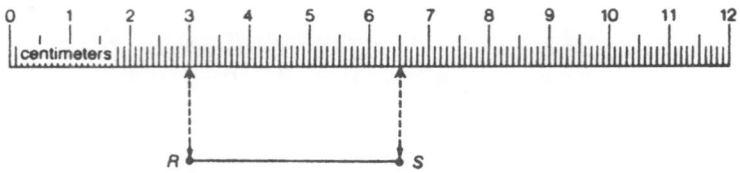

SOLUTION

m$\overline{RS}$ = 6.5 − 3 = 3.5 cm. Since length is a positive quantity, we must always subtract the smaller reading on the ruler (called a *coordinate*) from the larger coordinate.

CLASSIFYING ANGLES

Angles may be classified by comparing their measures to the number 90. An angle whose measure is less than 90 (but greater than 0) is called an *acute angle*. An L-shaped angle is called a *right angle* and its measure is exactly equal to 90. An angle whose measure is greater than 90 (but less than 180) is called an *obtuse angle*. See Figure 2.3.

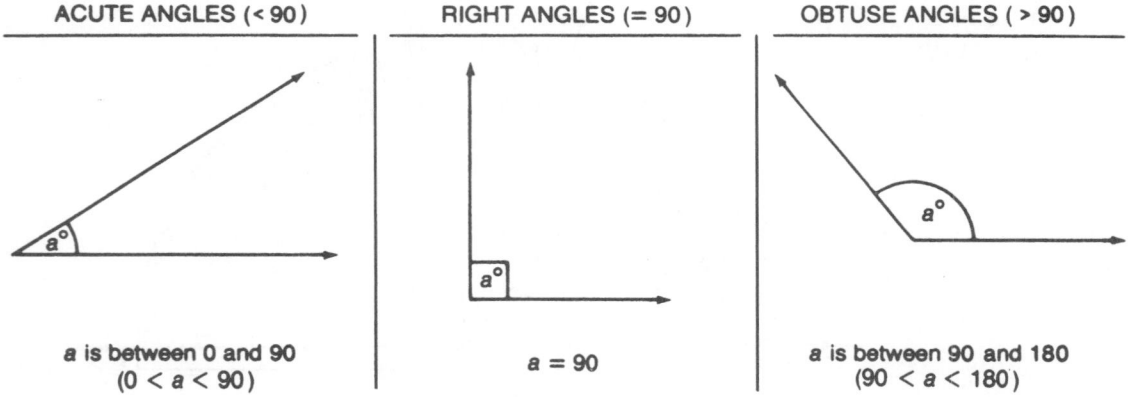

FIGURE 2.3

Betweenness of Points and Rays

Paul is standing on a line for theater tickets; he is standing *between* his friends Allan and Barbara. We represent this situation geometrically in Figure 2.4. We would like to be able to define this notion formally. The phrasing of the definition should eliminate the possibility that Paul may be standing "off" the ticket line or both Allan and Barbara are in front of Paul, or behind him, on the ticket line. (See Figure 2.5.)

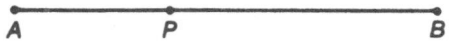

FIGURE 2.4

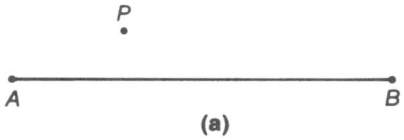

(a)

(Paul is not on line.)

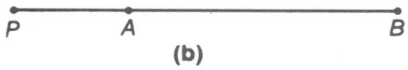

(b)

(Paul is behind both Allan and Barbara.)

FIGURE 2.5

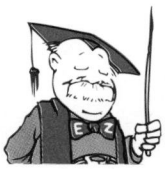

DEFINITION OF BETWEENNESS

Point P is *between* points A and B if both of the following conditions are met:

1. Points A, P, and B are three different collinear points.
2. $AB = AP + PB$.

Condition 1 of the definition of betweenness eliminates Figure 2.5*a* as a possibility, while condition 2 eliminates the possibility of Figure 2.5*b*.

EXAMPLE 2.3 Point Q is between points W and H. If $WQ = 2$ and $QH = 7$, find WH.

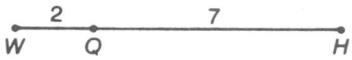

SOLUTION

$WH = 2 + 7 = 9$

EXAMPLE 2.4 $RT = 3$, $RS = 3$ and $ST = 1$, and points R, S, and T are collinear. Which of the points is between the other two?

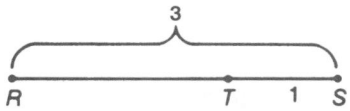

SOLUTION

Point T is between points R and S.

The analogous situation with angles occurs when a ray, say $\overrightarrow{OP}$, lies in the interior of an angle, say $\measuredangle AOB$, between its sides. The sum of the measures of the component angles of $\measuredangle AOB$ must equal the measure of the original angle. See Figure 2.6.

NOTE: If $m\measuredangle AOP = 40$ and $m\measuredangle POB = 10$ then $m\measuredangle AOB = 50$. This somewhat obvious relationship is given a special name: the Angle Addition Postulate.

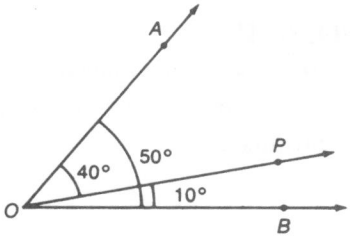

FIGURE 2.6

POSTULATE

ANGLE ADDITION POSTULATE

If ray $\overrightarrow{OP}$ in the interior of angle AOB, then

$$m\measuredangle AOB = m\measuredangle AOP + m\measuredangle POB$$

The Angle Addition Postulate may be expressed in two equivalent forms:

$$m\measuredangle AOP = m\measuredangle AOB - m\measuredangle POB$$

and

$$m\measuredangle POB = m\measuredangle AOB - m\measuredangle AOP$$

EXAMPLE 2.5

$\overrightarrow{BG}$ lies in the interior of $\measuredangle ABC$.
If $m\measuredangle ABG = 25$ and $m\measuredangle CBG = 35$, find $m\measuredangle ABC$.

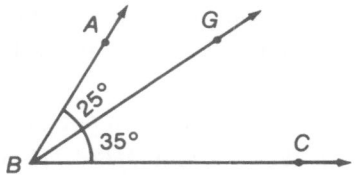

SOLUTION

$m\measuredangle ABC = 25 + 35 = 60$

EXAMPLE 2.6

$\overrightarrow{KM}$ lies in the interior of $\measuredangle JKL$.
If $m\measuredangle JKM = 20$ and $m\measuredangle LKJ = 50$, find $m\measuredangle MKL$.

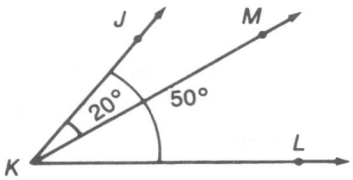

SOLUTION

$m\measuredangle MKL = 50 - 20 = 30$

EXAMPLE 2.7

In the accompanying figure the Angle Addition Postulate is contradicted; the measure of the largest angle is *not* equal to the sum of the measures of the two smaller angles. Explain.

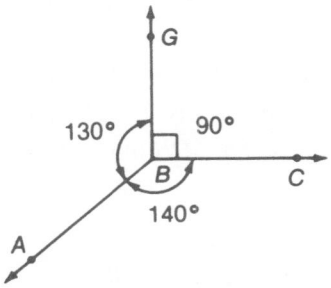

SOLUTION

$\overrightarrow{BG}$ is not in the interior of angle ABC, thus violating the assumption (hypothesis) of the Angle Addition Postulate.

Congruence

If the windshield of a car shatters or a computer disk drive is not working, we usually do not buy a new car or computer. Instead, we replace the broken parts. How do we know that the replacement parts will exactly fit where the old parts were removed? The new parts will fit because they have been designed to be interchangeable; they have been manufactured to have exactly the *same size and shape*. Figures that have the same size and shape are said to be *congruent*.

Figures may agree in one or more dimensions, yet *not* be congruent. Diagrams *ABCD* and *JKLM* (Figure 2.7) each have four sides that are identical in length, but the figures are not congruent since their corresponding angles are not identical in measure.

REMEMBER
Figures are congruent only if they agree in all their dimensions.

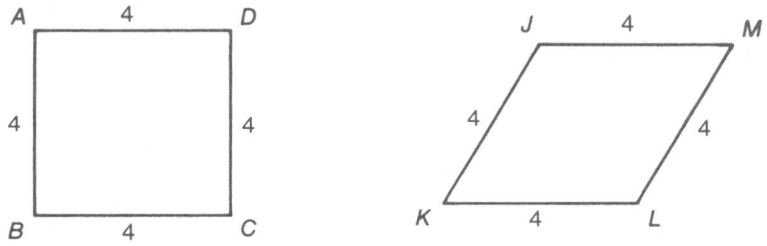

FIGURE 2.7

A line segment has a single dimension—its length. Two segments are congruent, therefore, if they have the same length. If line segments $\overline{AB}$ and $\overline{RS}$ have the same length, then they are congruent. We show that these segments are congruent by using the notation $\overline{AB} \cong \overline{RS}$. Similarly, if two angles have the same measure, then they are congruent. If angle X has the same measure as angle Y, we write $\measuredangle X \cong \measuredangle Y$. See Figure 2.8.

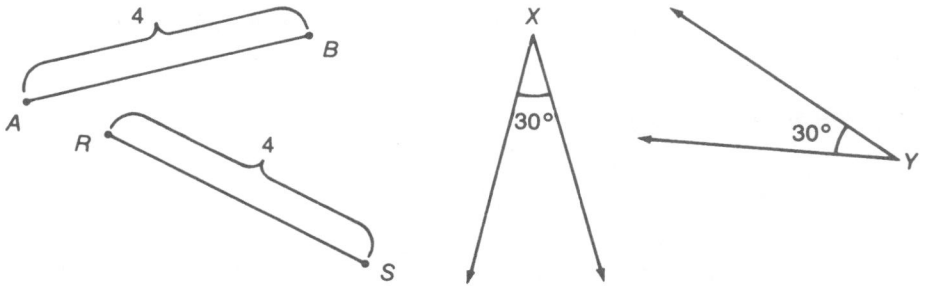

FIGURE 2.8

Congruence is one of the fundamental concepts of geometry. The problem of establishing that two triangles are congruent will be considered in Chapters 6 and 7.

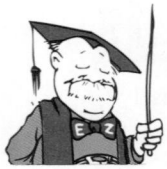

DEFINITION OF CONGRUENT SEGMENTS OR ANGLES

Segments (or angles) are *congruent* if they have the same measure.

The notation $\overline{AB} \cong \overline{RS}$ is read as "Line segment AB is congruent to line segment RS."

Midpoint and Bisector

Consider Figures 2.9 and 2.10. In Figure 2.9, $AM = MB = 3$. Since point M divides $\overline{AB}$ into two congruent segments ($\overline{AM} \cong \overline{MB}$), M is said to be the *midpoint* of $\overline{AB}$. Observe that the measure of each of the congruent segments is one-half the measure of the original segment, $\overline{AB}$. In Figure 2.10, $\overleftrightarrow{XY}$ intersects AB at point M, the midpoint of $\overline{AB}$. $\overleftrightarrow{XY}$ is said to *bisect* AB; a line, ray, or segment that bisects a segment is called a *bisector*.

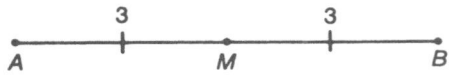

FIGURE 2.9

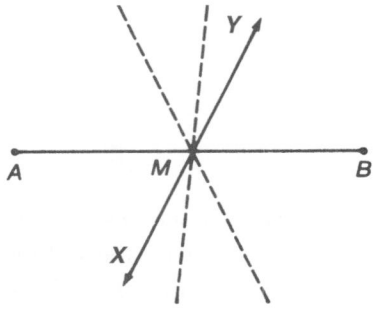

FIGURE 2.10

Since an infinite number of lines, rays, or segments may be drawn through the midpoint of a segment, a line segment possesses an infinite number of bisectors. The terms *midpoint* and *segment bisector* may be formally defined as follows.

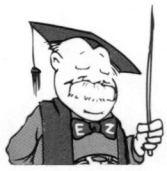

DEFINITION OF MIDPOINT

Point M is the *midpoint* of $\overline{AB}$ if

1. M is between A and B and 2. $AM = MB$.

If M is the midpoint of $\overline{AB}$, then M bisects $\overline{AB}$ and the following relationships involving the lengths of the segments thus formed are true:

- $AM = MB$
- $AM = \frac{1}{2}AB$ or $AB = 2AM$
- $MB = \frac{1}{2}AB$ or $AB = 2MB$

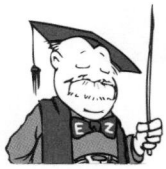

DEFINITION OF A SEGMENT BISECTOR

A *bisector* of a line segment AB is *any* line, ray, or segment that passes through the midpoint of $\overline{AB}$. Thus, a segment bisector divides a segment into two congruent segments.

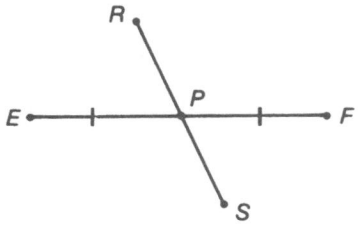

EXAMPLE 2.8

$\overline{RS}$ bisects $\overline{EF}$ at point P.

a. If $EF = 12$, find PF.

b. If $EP = 4$, find EF.

c. If $EP = 4x - 3$ and $PF = 2x + 15$, find EF.

SOLUTION

a. $PF = \frac{1}{2}EF = \frac{1}{2}(12) = 6$.

b. $EF = 2EP = 2(4) = 8$.

c. Since $EP = PF$,

$$4x - 3 = 2x + 15$$
$$4x = 2x + 18$$
$$2x = 18 \text{ and } x = 9$$
$$EP = 4x - 3 = 4(9) - 3 = 36 - 3 = 33$$
$$EF = 2(EP) = 2(33) = 66$$

Similarly, any ray that lies in the interior of an angle in such a way that it divides the original angle into two congruent angles is the *bisector* of the angle.

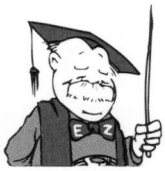

DEFINITION OF ANGLE BISECTOR

$\overrightarrow{BM}$ is the *bisector* of $\measuredangle ABC$ if M lies in the interior of $\measuredangle ABC$ and $\measuredangle ABM \cong \measuredangle CBM$. Thus, an angle bisector divides an angle into two congruent angles.

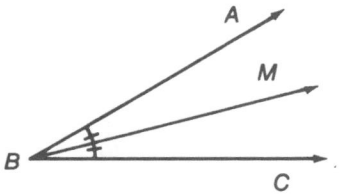

The measure of each of the angles formed by the bisector is one-half the measure of the original angles:

$$m\measuredangle ABM = m\measuredangle CBM \qquad m\measuredangle ABM = \tfrac{1}{2}m\measuredangle ABC \qquad m\measuredangle CBM = \tfrac{1}{2}m\measuredangle ABC$$

Diagrams and Drawing Conclusions

Which line segment is longer, $\overline{AB}$ or $\overline{CD}$?

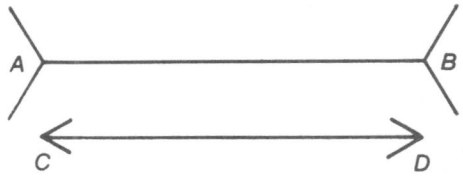

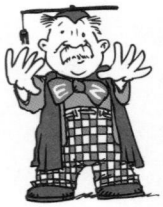

WARNING!

Actually both have the same length, although $\overline{AB}$ may give the illusion of being longer than $\overline{CD}$. When given a geometric diagram, we must exercise extreme caution in drawing conclusions based on the diagram—pictures can be deceiving! In general, we may assume only collinearity and betweenness of points. We may not make any assumptions regarding the measures of segments or angles unless they are given to us. In Figure 2.11*a*, although segments $\overline{AD}$ and $\overline{DC}$ appear to be equal in length, we may *not* conclude that $AD = DC$. The only assumption that we are entitled to make is that point D lies *between* points A and C. If, however, we are told that $AD = DC$, then we write this *given* information next to the figure. (See Figure 2.11*b*.)

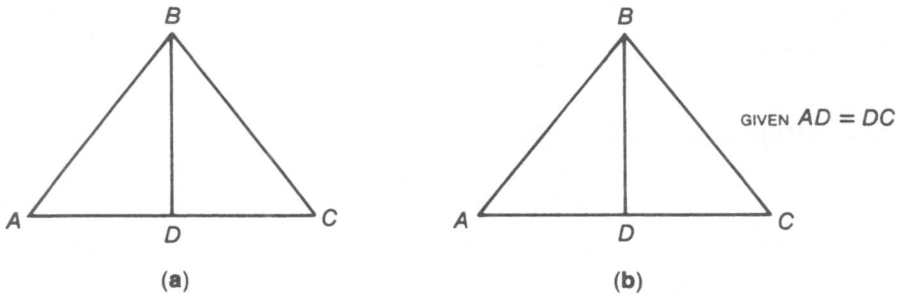

FIGURE 2.11

To indicate the equal segments on the diagram, draw a single vertical bar through each segment as in Figure 2.12. An angle may be assumed from the diagram to be a right angle only if the angle contains the "corner" marking (⊾).

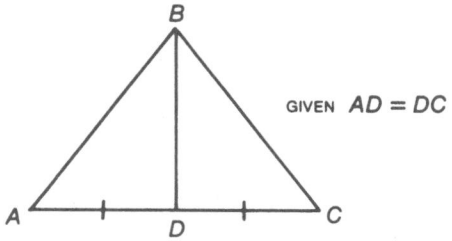

FIGURE 2.12

Properties of Equality and Congruence

John is taller than Kevin and Kevin is taller than Louis. How do the heights of John and Louis compare? We can analyze the situation with the aid of a simple diagram. (See Figure 2.13.) This leads us to conclude that John must be taller than Louis.

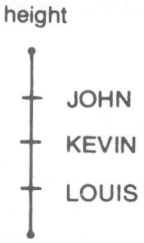

FIGURE 2.13

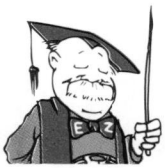

THIS IS THE KEY TO THE METHOD!

Using the mathematical symbol for "greater than," >, we can represent the height relationships by the following series of inequality statements:

if $J > K$
and $K > L$
then $J > L$

Without directly comparing John with Louis, we have used a *transitive property* to conclude that John's height is greater than Louis's height. The "greater than" relation is an example of a relation that possesses the transitive property. Is friendship a transitive relation? If Alice is Barbara's friend and Barbara is Carol's friend, does that mean that Alice and Carol must also be friends? Obviously, no. Some relations possess the transitive property, while others do not.

The equality (=) and congruence (≅) relations possess the transitive property. For example, if angle A is congruent to angle B and angle B is congruent to angle C, then angle A must be congruent to angle C. (See Figure 2.14.) Another way of looking at this interrelationship among angles A, B, and C is that angles A and C are each congruent to angle B and must, therefore, be congruent to each other.

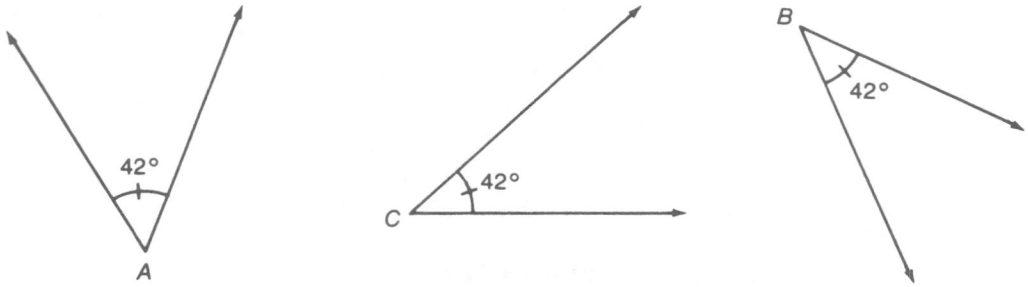

FIGURE 2.14

The equality and congruence relations also enjoy some additional properties. These are summarized in Table 2.1.

TABLE 2.1

Property	Equality Example	Congruence Example
Reflexive The identical expression may be written on either side of the $=$ or $\cong$ symbol. Any quantity is equal (congruent) to itself.	1. $9 = 9$. 2. $AB = AB$.	$\measuredangle ABC \cong \measuredangle ABC$.
Symmetric The positions of the expressions on either side of the $=$ or $\cong$ symbol may be reversed. Quantities may be "flip-flopped" on either side of an $=$ or $\cong$ sign.	1. If $4 = x$, then $x = 4$. 2. If $AB = CD$, then $CD = AB$	If $\measuredangle ABC \cong \measuredangle XYZ$, then $\measuredangle XYZ \cong \measuredangle ABC$.
Transitive If two quantities are equal (congruent) to the same quantity, then they are equal (congruent) to each other.	If $AB = CD$ and $CD = PQ$, then $AB = PQ$.	If $\measuredangle X \cong \measuredangle Y$ and $\measuredangle Y \cong \measuredangle Z$, then $\measuredangle X \cong \measuredangle Z$.

Another useful property of the equality relation is the *substitution* property. If $AB = 2 + 3$, then an equivalent number may be substituted in place of the numerical expression on the right side of the equation. We may substitute 5 for $2 + 3$, and write $AB = 5$. Here is a geometric illustration of this often-used property:

GIVEN: $m\measuredangle 1 + m\measuredangle 2 = 90$,
$m\measuredangle 2 = m\measuredangle 3$.
CONCLUSION: $m\measuredangle 1 + m\measuredangle 3 = 90$.
REASON: Substitution property. The $m\measuredangle 3$ replaces its equal ($m\measuredangle 2$) in the first equation stated in the Given.

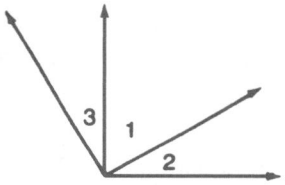

In each of the following examples, identify the property used to draw the conclusion as either the transitive or substitution property.

EXAMPLE 2.9

GIVEN: $\measuredangle 1 \cong \measuredangle 2$,
$\measuredangle 2 \cong \measuredangle 3$.
CONCLUSION: $\measuredangle 1 \cong \measuredangle 3$.
PROPERTY: ?

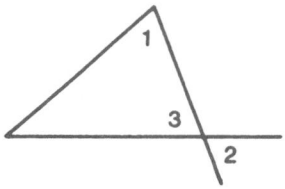

SOLUTION

Since both angles 1 and 3 are congruent to the same angle, angle 2, they must be congruent to each other. This is the *transitive property of congruence*. Since we may only substitute equals in equations, we do *not* have a substitution property of *congruence*.

EXAMPLE 2.10

GIVEN: $m\measuredangle 1 = m\measuredangle 4$,
$m\measuredangle 3 = m\measuredangle 5$,
$m\measuredangle 4 + m\measuredangle 2 + m\measuredangle 5 = 180$.
CONCLUSION: $m\measuredangle 1 + m\measuredangle 2 + m\measuredangle 3 = 180$.
PROPERTY: ?

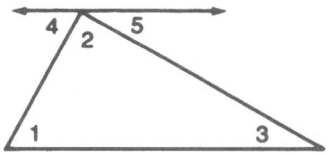

SOLUTION

This is the *substitution property*. In the last equation stated in the Given, the measures of angles 4 and 5 are replaced by their equals, the measures of angles 1 and 3, respectively.

EXAMPLE 2.11

GIVEN: $RS = SM$. (1)
$TW = SM$. (2)
CONCLUSION: $RS = TW$.
PROPERTY: ?

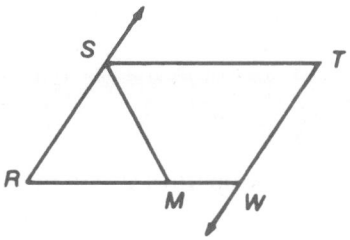

SOLUTION

Since RS and TW are both equal to the same quantity, SM, they must be equal to each other. This is the *transitive property*.

or

In Equation (1), SM may be replaced by its equal, TW. We are using the information in Equation (2) to make a substitution in Equation (1). Hence, the conclusion can be justified also by using the *substitution property*.

EXAMPLE 2.12

GIVEN: C is the midpoint of $\overline{AD}$,
$AC = CE$.
CONCLUSION: $CD = CE$.
PROPERTY: ?

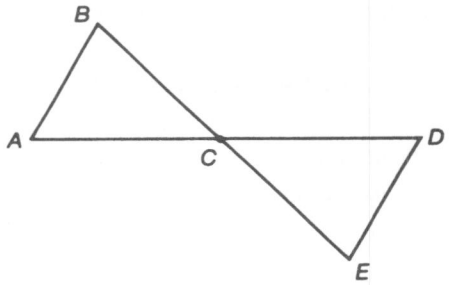

SOLUTION

$AC = CD$, since point C is the midpoint of $\overline{AD}$.
We now have the set of relationships:

$$AC = CD \quad (1)$$
$$AC = CE \quad (2)$$

Since CD and CE are both equal to the same quantity (AC) they must be equal to each other. Hence, $CD = CE$ by the *transitive property of equality*.

or

We may replace AC by CE in Equation (1), also reaching the desired conclusion.

Examples 2.11 and 2.12 illustrate that the transitive and substitution properties of equality, in certain situations, may be used interchangeably. In each of these examples, *two equations* state that two quantities are each equal to the same quantity, thus leading to either the substitution or transitive property of *equality*.

Additional Properties of Equality

Several properties of equality encountered in elementary algebra prove useful when working with measures of segments and angles. Table 2.2 reviews these properties in their algebraic context.

TABLE 2.2

Property	Algebraic Example	Formal Statement
Addition (+) The same (or =) quantities may be added to both sides of an equation.	Solve for x: $x - 3 = 12$ $x - 3 + 3 = 12 + 3$ same $x = 15$	If equals are added to equals, their sums are equal. or If $a = b$, then $a + c = b + c$.
Subtraction (–) The same (or =) quantities may be subtracted from both sides of an equation.	Solve for n: $n + 5 = 11$ $n + 5 - 5 = 11 - 5$ same $n = 6$	If equals are subtracted from equals, their differences are equal. or If $a = b$, then $a - c = b - c$.
Multiplication (×) The same quantity may be used to multiply both sides of an equation.	Solve for y: $\frac{y}{3} = 7$ same $3\left(\frac{y}{3}\right) = 3(7)$ $y = 21$	If equals are multiplied by equals, their products are equal. or If $a = b$, then $ac = bc$.

These equality properties may be summarized as follows: "Whatever you do to one side of an *equation*, do the same thing to the other side." The addition, subtraction, and multiplication properties may be applied also to geometric situations. The following examples illustrate how these properties of equality can be used to draw conclusions about the measures of segments and angles. In each of the first four examples, the numerical value in the conclusion is based on the measurements given in the accompanying diagram.

- USING THE ADDITION PROPERTY

a. GIVEN:

$$\begin{aligned} AB &= AC \\ +\ BD &= +\ CE \\ \hline \underbrace{AB + BD}_{\substack{\| \\ AD \\ (7}} &= \underbrace{AC + CE}_{\substack{\| \\ AE \\ 7)}} \end{aligned}$$

CONCLUSION:

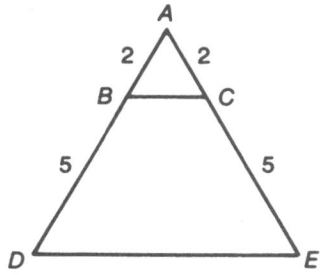

b. GIVEN:

$$\begin{aligned} m\measuredangle JXK &= m\measuredangle MXL \\ +\ m\measuredangle KXL &= +\ m\measuredangle KXL \\ \hline m\measuredangle JXL &= m\measuredangle KXM \\ (90 &= 90) \end{aligned}$$

CONCLUSION:

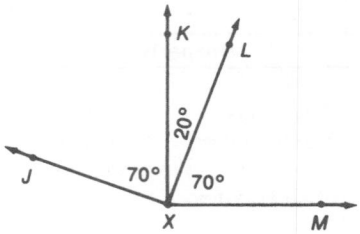

- USING THE SUBTRACTION PROPERTY

a. GIVEN:

$$\begin{aligned} m\measuredangle BAD &= m\measuredangle DCB \\ -\ m\measuredangle PAD &= -\ m\measuredangle BCQ \\ \hline m\measuredangle BAP &= m\measuredangle DCQ \\ (40 &= 40) \end{aligned}$$

CONCLUSION:

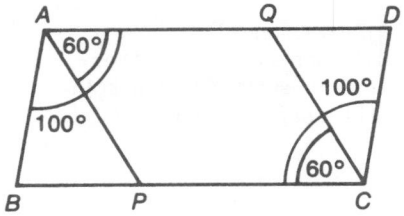

b. GIVEN:

$$\begin{aligned} VI &= NE \\ -\ EI &= -\ EI \\ \hline VI - EI &= NE - EI \\ \| \qquad & \qquad \| \\ VE &= NI \\ (3 &= 3) \end{aligned}$$

CONCLUSION:

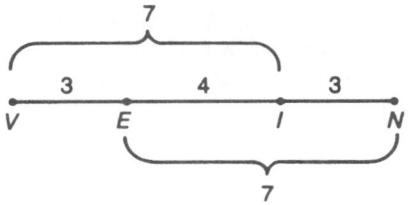

- USING THE MULTIPLICATION PROPERTY

GIVEN: $AB = CB$,
$AR = \frac{1}{2}AB$,
$CT = \frac{1}{2}CB$.

CONCLUSION: $AR = CT$. Why?

REASONING: Since we are multiplying equals ($AB = CB$) by the same number ($\frac{1}{2}$), their products must be equal:

$$\tfrac{1}{2}AB = \tfrac{1}{2}CB$$

By substitution,

$$AR = CT$$

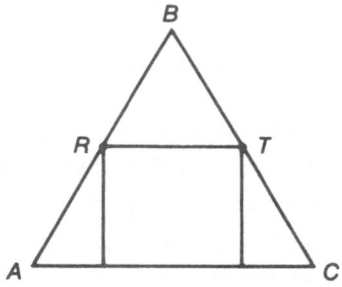

This chain of reasoning, in which the multiplying factor is $\frac{1}{2}$, is used so often that we give it a special name, "*halves of equals are equal*."

The Two-Column Proof Format

In the previous examples, stating the property involved and explaining the reasoning that justifies each conclusion forms a mathematical argument or "proof" in which the "conclusion" represents what you were required to prove.

THIS IS THE KEY TO THE METHOD!

A proof in geometry usually includes these four elements:

GIVEN: { The set of facts that you can use.

PROVE: { What you need to show.

PROOF: { Step-by-step reasoning that leads from what is "Given" to what you must "Prove."

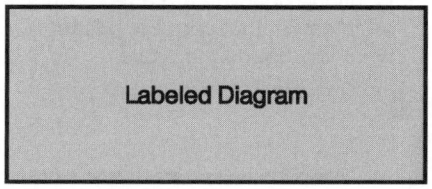

Greek mathematicians wrote proofs in paragraph form. Most beginning geometry students, however, find it helpful to organize and record their thinking using a table-like format, as shown in Table 2.3 and further illustrated in Example 2.13. The left column explains your reasoning as a set of numbered statements. The right column gives the corresponding reasons. This format is sometimes referred to as a "two-column proof."

EXAMPLE 2.13

GIVEN: $m\measuredangle RST = m\measuredangle WTS$,
$\overline{PS}$ bisects $\measuredangle RST$,
$\overline{PT}$ bisects $\measuredangle WTS$.

CONCLUSION: $m\measuredangle 1 = m\measuredangle 2$.

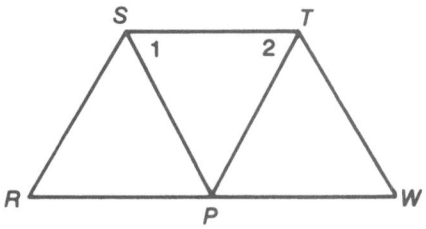

SOLUTION

PROOF:

Statements	Reasons
1. $m\measuredangle RST = m\measuredangle WTS$.	1. Given.
2. $\overline{PS}$ bisects $\measuredangle RST$, $\overline{PT}$ bisects $\measuredangle WTS$.	2. Given.
3. $m\measuredangle 1 = \frac{1}{2}\, m\measuredangle RST$. $m\measuredangle 2 = \frac{1}{2} m\measuredangle WTS$.	3. Definition of angle bisector.
4. $m\measuredangle 1 = m\measuredangle 2$.	4. Halves of equals are equal.

TABLE 2.3

Statements	Reasons
• Start with the Given or a fact that can be deduced from an accompanying diagram. Label the first statement number 1, and continue to number statements sequentially. • Develop a chain of mathematical reasoning, writing each deductive step as a separate statement with its justification in the "Reasons" column. • Continue until you are able to write a statement that corresponds to what you needed to prove.	• Using a matching number, write the reason that corresponds to each statement. • Restrict the "Reasons" to the following types of statements: - Given - Definition - Postulate - Theorem - Algebraic property

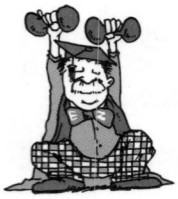

REVIEW EXERCISES FOR CHAPTER 2

1. In the accompanying diagram, classify each of the following angles as acute, right, obtuse, or straight:

   (a) $\measuredangle TOM$
   (b) $\measuredangle LOM$
   (c) $\measuredangle SOM$
   (d) $\measuredangle LOR$
   (e) $\measuredangle ROT$
   (f) $\measuredangle LOT$
   (g) $\measuredangle ROS$
   (h) $\measuredangle MOR$

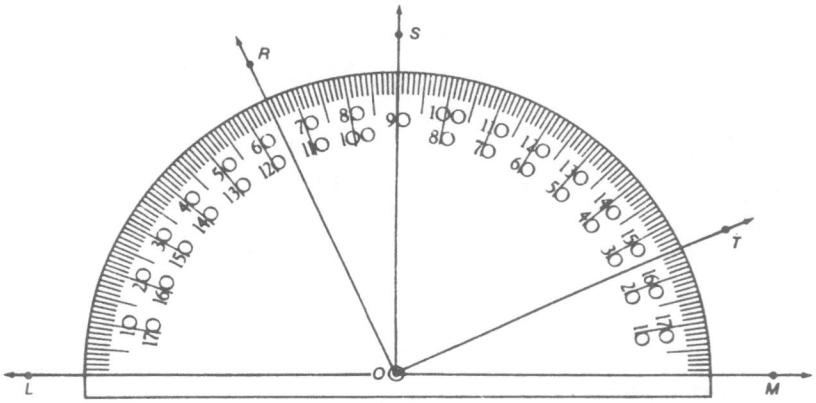

2. Point P is between points H and G. If $HP = 3$ and $PG = 5$, find HG.

3. If points M, I, and Z are collinear and $IZ = 8$. $MI = 11$, and $MZ = 3$, which point is between the other two?

4. $\overrightarrow{PL}$ lies in the interior of angle RPH; $m\measuredangle RPL = x - 5$ and $m\measuredangle LPH = 2x + 18$. If $m\measuredangle HPR = 58$, find the measure of the smallest angle formed.

5. XY bisects $\overline{RS}$ at point M. If $RM = 6$, find the length of RS.

6. $\overrightarrow{PQ}$ bisects $\measuredangle HPJ$. If $m\measuredangle HPJ = 84$, find $m\measuredangle QPJ$.

7. $\overrightarrow{BP}$ bisects $\measuredangle ABC$. If $m\measuredangle ABP = 4x + 5$ and $m\measuredangle CBP = 3x + 15$, classify angle ABC as acute, right, or obtuse.

8. If R is the midpoint of $\overline{XY}$, $XR = 3a + 1$, and $YR = 16 - 2a$, find the length of $\overline{XY}$.

9. In the accompanying diagram, pairs of angles and segments are indicated as congruent. Use the letters in the diagram to write the appropriate congruence relation.

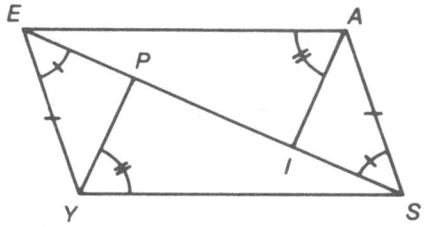

In Exercises 10 to 12, use the diagram and any information given to mark the diagram with the Given and to draw the appropriate conclusion.

10. GIVEN: $\overline{BF}$ bisects $\overline{AC}$.
CONCLUSION: ?

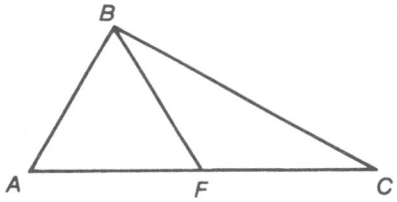

11. GIVEN: $\overline{PT}$ bisects $\measuredangle STO$.
CONCLUSION: ?

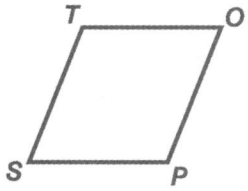

12. GIVEN: $\overline{AC}$ bisects $\overline{BD}$,
$\overline{BD}$ bisects $\measuredangle ADC$.
CONCLUSION: ?

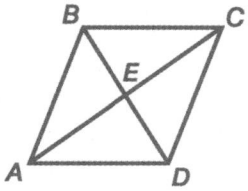

In Exercises 13 to 16, justify the conclusion drawn by identifying the property used to draw the conclusion as reflexive, transitive, symmetric, or substitution.

13. GIVEN: $\overline{LM} \cong \overline{GH}$,
$\overline{GH} \cong \overline{FV}$.
CONCLUSION: $\overline{LM} \cong \overline{FV}$.

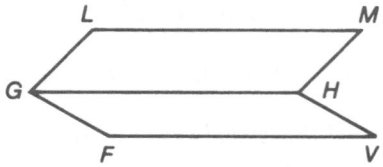

14. GIVEN: Quadrilateral $ABCD$.
CONCLUSION: $\overline{AC} \cong \overline{AC}$.

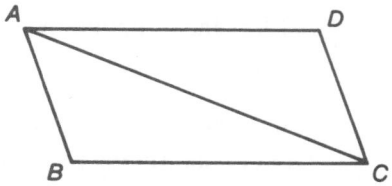

15. GIVEN: $\overline{TW}$ bisects $\measuredangle STV$,
$\measuredangle 1 \cong \measuredangle 3$.
CONCLUSION: $\measuredangle 2 \cong \measuredangle 3$.

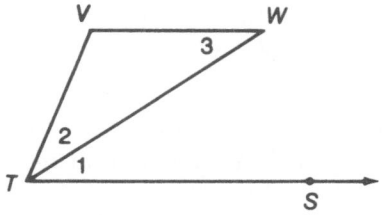

16. GIVEN: $m\measuredangle 1 + m\measuredangle 2 = 90$,
$m\measuredangle 1 = m\measuredangle 3$.
CONCLUSION: $m\measuredangle 3 + m\measuredangle 2 = 90$.

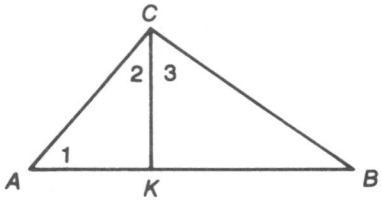

In Exercises 17 to 24, indicate on the diagrams the corresponding pairs of equal or congruent parts. Then state the property of equality that can be used to justify each conclusion.

17. GIVEN: $AC = BT$.
CONCLUSION: $AT = BC$.
PROPERTY: ?

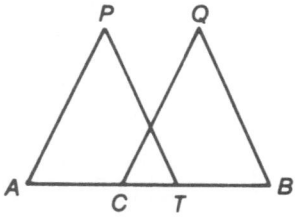

18. GIVEN: $m\measuredangle KPN = m\measuredangle LPM$.
CONCLUSION: $m\measuredangle KPM = m\measuredangle LPN$.
PROPERTY: ?

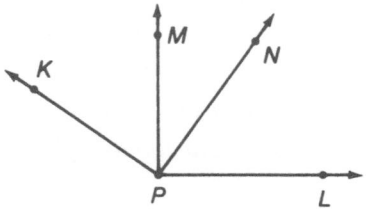

19. GIVEN: $AE = BE$,
$CE = DE$.
CONCLUSION: $AC = BD$.
PROPERTY: ?

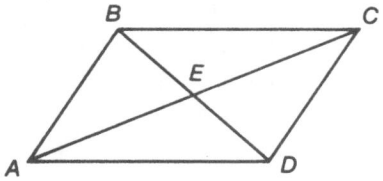

Figure not drawn to scale.

20. GIVEN: $m\measuredangle 1 = m\measuredangle 3$,
$m\measuredangle 2 = m\measuredangle 4$.
CONCLUSION: $m\measuredangle STA = m\measuredangle ARS$.
PROPERTY: ?

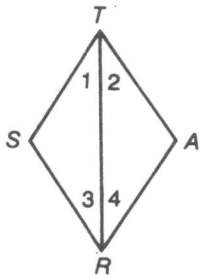

21. GIVEN: $m\measuredangle WXY = m\measuredangle ZYX$,
HX bisects $\measuredangle WXY$,
HY bisects $\measuredangle ZYX$.
CONCLUSION: $m\measuredangle 1 = m\measuredangle 2$.
PROPERTY: ?

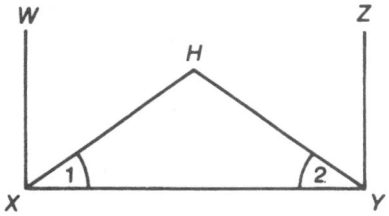

22. GIVEN: $JQ = LP$.
CONCLUSION: $JP = LQ$.
PROPERTY: ?

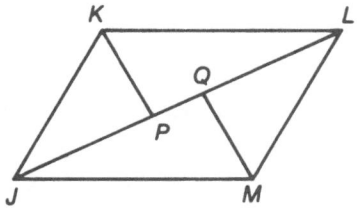

23. GIVEN: $m\measuredangle SBO = m\measuredangle TBH$.
CONCLUSION: $m\measuredangle SBH = m\measuredangle TBO$.
PROPERTY: ?

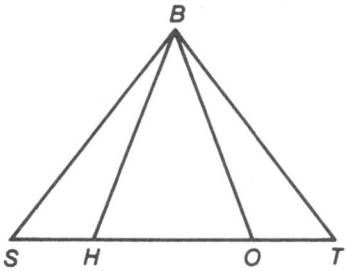

24. GIVEN: $AX = YS$,
$XB = RY$.
CONCLUSION: $AB = RS$.
PROPERTY: ?

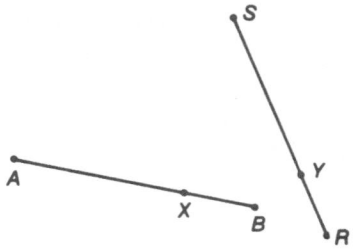

In Exercises 25 to 28, supply the missing reasons.

25. GIVEN: $\overline{PJ} \cong \overline{LR}$.
CONCLUSION: $\overline{PR} \cong \overline{LJ}$.

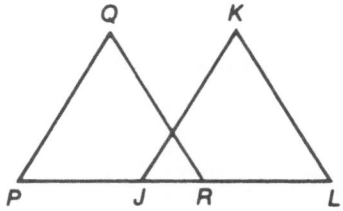

PROOF:

Statements	Reasons
1. $\overline{PJ} \cong \overline{LR}$.	1. Given.
2. $PJ = LR$.	2. If two segments are congruent, they are equal in length.
3. $JR = JR$.	3. ?
4. $PJ + JR = LR + JR$.	4. ?
5. $PR = LJ$.	5. Substitution property of equality.
6. $\overline{PR} \cong \overline{LJ}$.	6. If two segments are equal in measure, they are congruent.

26. GIVEN: $\measuredangle RLM \cong \measuredangle ALM$,
$\measuredangle 1 \cong \measuredangle 2$.
CONCLUSION: $\measuredangle 3 \cong \measuredangle 4$.

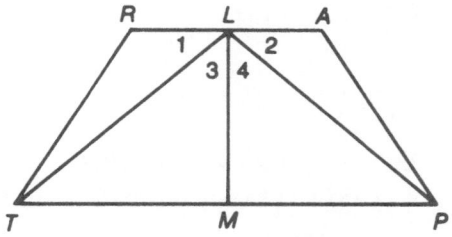

PROOF:

Statements	Reasons
1. $\measuredangle RLM \cong \measuredangle ALM$, $\measuredangle 1 \cong \measuredangle 2$.	1. ?
2. $m\measuredangle RLM = m\measuredangle ALM$, $m\measuredangle 1 = m\measuredangle 2$.	2. ?
3. $m\measuredangle RLM - m\measuredangle 1 = m\measuredangle ALM - m\measuredangle 2$.	3. ?
4. $m\measuredangle 3 = m\measuredangle 4$.	4. ?
5. $\measuredangle 3 \cong \measuredangle 4$.	5. ?

27. GIVEN: $m\measuredangle TOB = m\measuredangle WOM$, $TB \cong WM$.

PROVE: (a) $m\measuredangle TOM = m\measuredangle WOB$.
(b) $TM \cong WB$.

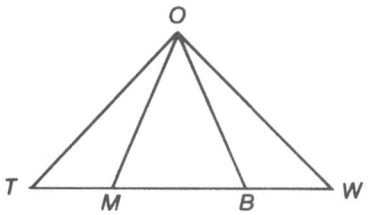

PROOF:

Statements	Reasons
PROOF OF PART A:	
1. $m\measuredangle TOB = m\measuredangle WOM$.	1. ?
2. $m\measuredangle MOB = m\measuredangle MOB$.	2. ?
*3. $m\measuredangle TOB - m\measuredangle MOB = m\measuredangle WOM - m\measuredangle MOB$.	3. ?
*4. $m\measuredangle TOM = m\measuredangle WOB$.	4. Angle addition postulate (and substitution property of equality).
PROOF OF PART B:	
5. $\overline{TB} \cong \overline{WM}$	5. ?
6. $TB = WM$.	6. ?
7. $MB = MB$.	7. ?
*8. $TB - MB = WM - MB$.	8. ?
*9. $TM = WB$.	9. Definition of betweenness (and substitution property of equality).
10. $\overline{TM} \cong \overline{WB}$.	10. ?

*In later work, these steps are sometimes consolidated.

28. GIVEN: $\overline{KB}$ bisects $\measuredangle SBF$,
$\overline{KB}$ bisects $\measuredangle SKF$,
$\measuredangle SKF \cong \measuredangle SBF$.
PROVE: $\measuredangle 1 \cong \measuredangle 2$.

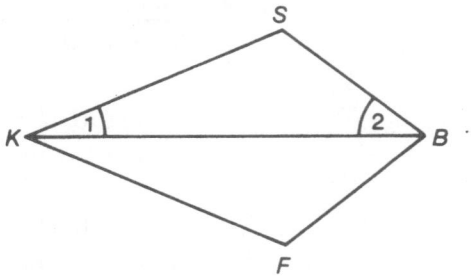

PROOF:

Statements	Reasons
1. $\measuredangle SKF \cong \measuredangle SBF$.	1. Given.
2. $m\measuredangle SKF = m\measuredangle SBF$.	2. ?
3. KB bisects $\measuredangle SBF$, KB bisects $\measuredangle SKF$.	3. ?
4. $m\measuredangle 1 = \frac{1}{2}m\measuredangle SKF$. $m\measuredangle 2 = \frac{1}{2}m\measuredangle SBF$.	4. ?
5. $m\measuredangle 1 = m\measuredangle 2$.	5. ?
6. $\measuredangle 1 \cong \measuredangle 2$.	6. ?

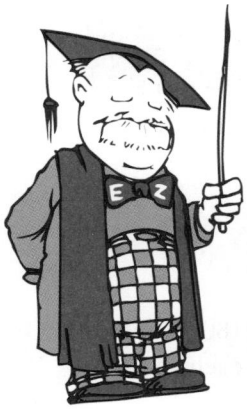

3

Angle Pairs and Perpendicular Lines

WHAT YOU WILL LEARN

This chapter looks at special pairs of angles. In this chapter you will learn:

- definitions and theorems involving complementary angles, supplementary angles, and vertical angles;
- definitions and theorems involving lines that intersect at 90° angles;
- alternative forms of a proof.

SECTIONS IN THIS CHAPTER

- Supplementary and Complementary Angle Pairs
- Adjacent and Vertical Angle Pairs
- Theorems Relating to Supplementary, Complementary, and Vertical Angles
- Definitions and Theorems Relating to Right Angles and Perpendiculars
- A Word About the Format of a Proof

Supplementary and Complementary Angle Pairs

Pairs of angles may be related in special ways. Supplementary and complementary angle pairs are of particular importance in the study of geometry.

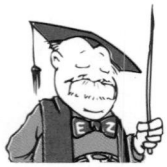

DEFINITIONS OF SUPPLEMENTARY AND COMPLEMENTARY ANGLES

- Two angles are *supplementary* if the sum of their measures is 180. If angle A is supplementary to angle B, then $m\angle A + m\angle B = 180$, and each angle is called the *supplement* of the other angle.
- Two angles are *complementary* if the sum of their measures is 90. If angle A is complementary to angle B, then $m\angle A + m\angle B = 90$, and each angle is called the *complement* of the other angle.

EXAMPLE 3.1 GIVEN:

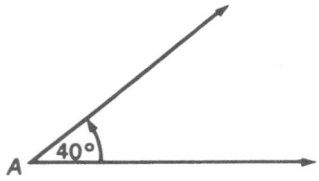

a. Determine the measure of the supplement of angle A.
b. Determine the measure of the complement of angle A.

SOLUTION

a. The supplement of angle A has measure 140.
b. The complement of angle A has measure 50.

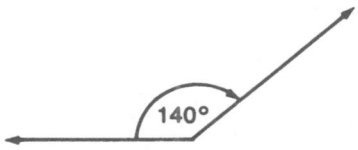

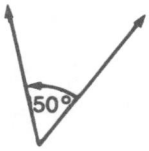

EXAMPLE 3.2 In triangle ABC, angle A is complementary to angle B. Find the measures of angles A and B.

SOLUTION

$$\begin{aligned} 2x + 3x &= 90 \\ 5x &= 90 \\ x &= 18 \\ m\angle A &= 2x = 2(18) = 36 \\ m\angle B &= 3x = 3(18) = 54 \end{aligned}$$

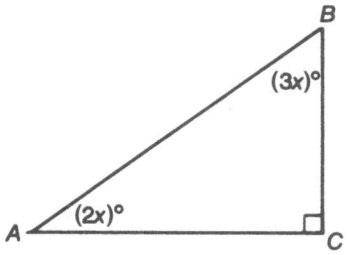

EXAMPLE 3.3 The measures of an angle and its supplement are in the ratio of 1:8 Find the measure of the angle.

SOLUTION

Method 1: Let x = measure of angle.
Then $180 - x$ = measure of supplement of angle.

$$\frac{x}{180 - x} = \frac{1}{8}$$
$$180 - x = 8x$$
$$9x = 180$$
$$x = 20$$

Method 2: Let x = measure of angle.
Then $8x$ = measure of supplement of angle.

$$x + 8x = 180$$
$$9x = 180$$
$$x = 20$$

Therefore, measure of angle = 20.

EXAMPLE 3.4 Determine the measure of an angle if it exceeds twice the measure of its complement by 30.

SOLUTION

Let x = measure of angle.
Then $90 - x$ = measure of complement of angle

$$\begin{aligned} x &= 2(90 - x) + 30 \\ &= 180 - 2x + 30 \\ &= 210 - 2x \\ 3x &= 210 \\ x &= 70 \end{aligned}$$

Therefore, measure of angle = 70.

Adjacent and Vertical Angle Pairs

Adjacent means "next to." But how close do two angles have to be in order to be considered adjacent? Figure 3.1 contrasts four pairs of angles; only the first pair is considered to be adjacent.

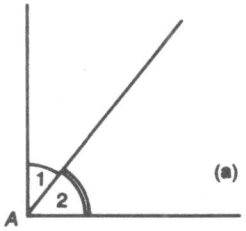

(a) Angles are adjacent.

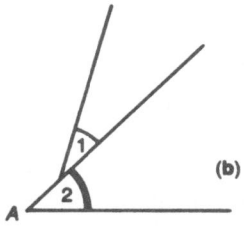

(b) Angles do not share the same vertex.

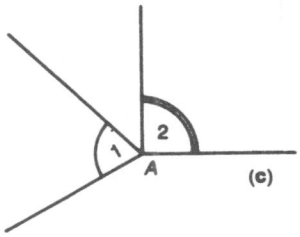

(c) Angles do not share a common side.

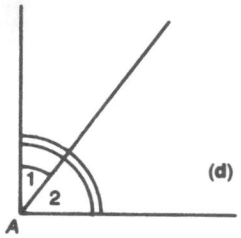

(d) Angles "overlap" — they have interior points in common.

FIGURE 3.1 Adjacent versus nonadjacent angle pairs.

Figure 3.1 suggests that we make the following definition:

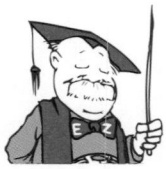

DEFINITION OF ADJACENT ANGLE PAIRS

Two angles are an *adjacent pair* if they:

- Have the same vertex.
- Share a common side.
- Have no interior points in common.

If two angles are adjacent, the two sides that are not shared are sometimes referred to as the *exterior* sides of the adjacent angles.

EXAMPLE 3.5 A beginning student of geometry wonders whether the following two assertions are true:

a. If a pair of angles are supplementary, then they must be adjacent.

b. If the exterior sides of a pair of adjacent angles form a straight line, then the angles are supplementary.

Comment on whether you think the statements are true or false. If you suspect that one or both are false, produce a diagram to help support your belief.

SOLUTION

a. A pair of supplementary angles, as the figures below illustrate, may be either adjacent or nonadjacent. The assertion is, therefore, false.

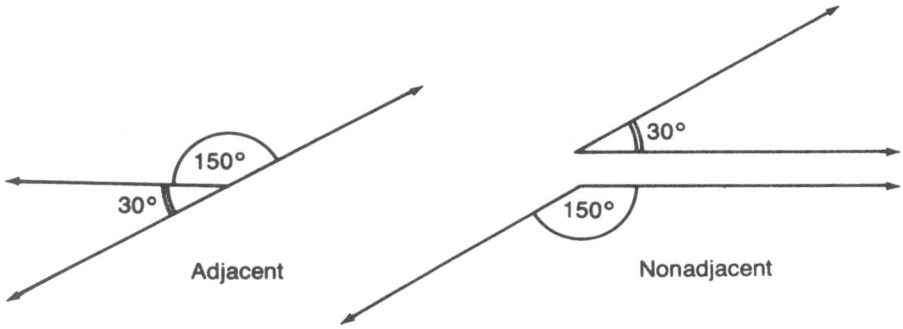

b. The assertion is true since, as illustrated below, a straight line is formed, implying that the sum of the measures of the adjacent angles is 180. Consequently, the angles are supplementary.

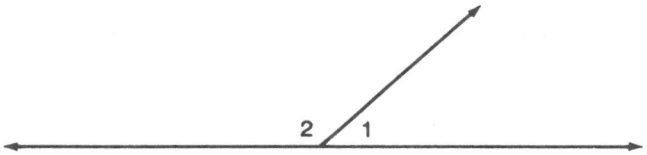

The statement presented in Example 3.5*b* may be formally stated as a theorem.

THEOREM 3.1

If the exterior sides of a pair of adjacent angles form a straight line, then the angles are supplementary.

If two lines intersect, four angles are formed, as shown. Angles 1 and 3 are called vertical angles; angles 1 and 4 are *not* vertical angles since they are adjacent. Notice that vertical angles are "opposite" one another. Angles 2 and 4 are also vertical angles.

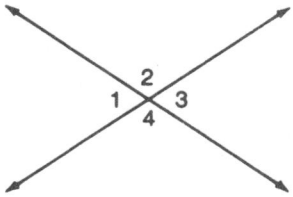

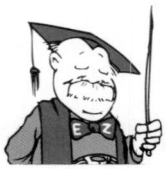

DEFINITION OF VERTICAL ANGLES

Vertical angles are pairs of nonadjacent (opposite) angles formed by two intersecting lines.

EXAMPLE 3.6 Name all pairs of vertical angles in the accompanying diagram.

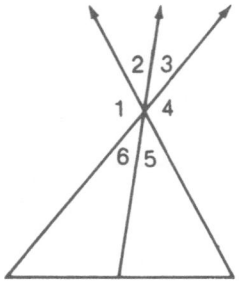

SOLUTION

Angle pairs 1 and 4, 2 and 5, 3 and 6.

Theorems Relating to Complementary, Supplementary, and Vertical Angles

In the diagrams below, angles A and B are each complementary to angle C.

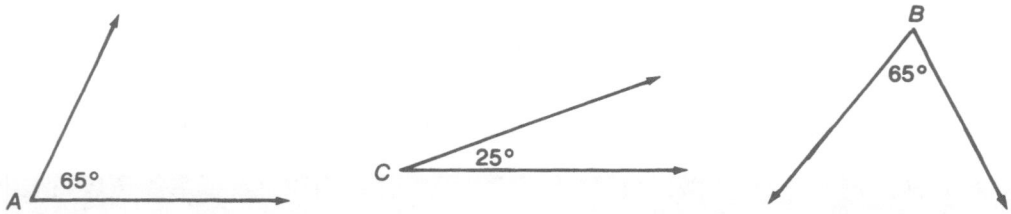

We may conclude that angle A must be equal in measure (or congruent to) angle B. By the same reasoning, if angles A and B were each *supplementary* to angle C, they would necessarily be congruent to each other.

Now, suppose angle A is complementary to angle C and angle B is complementary to angle D; furthermore, angles C and D are congruent:

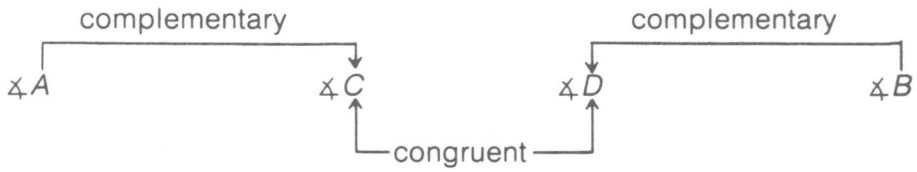

If $m\measuredangle C = 20$, then $m\measuredangle A = 70$. Since angles C and D are congruent, $m\measuredangle D = 20$ and $m\measuredangle B = 70$. Hence, angles A and B are congruent. If the original relationship specified that the angles were *supplementary* to a pair of congruent angles, the identical conclusion would result.

THEOREM 3.2

If two angles are complementary (or supplementary) to the same angle or to congruent angles, then they are congruent.

EXAMPLE 3.7 Present a formal two-column proof.

GIVEN: $\overline{LM}$ bisects $\measuredangle KMJ$,
$\measuredangle 1$ is complementary to $\measuredangle 2$,
$\measuredangle 4$ is complementary to $\measuredangle 3$.
PROVE: $\measuredangle 1 \cong \measuredangle 4$.

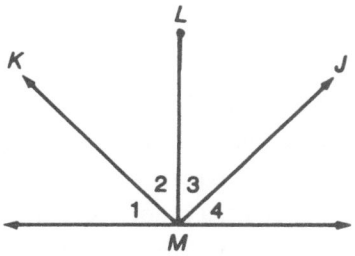

SOLUTION

PROOF:

Statements	Reasons
1. $\overline{LM}$ bisects $\measuredangle KMJ$.	1. Given.
2. $\measuredangle 2 \cong \measuredangle 3$.	2. A bisector divides an angle into two congruent angles.
3. $\measuredangle 1$ is complementary to $\measuredangle 2$. $\measuredangle 4$ is complementary to $\measuredangle 3$.	3. Given.
4. $\measuredangle 1 \cong \measuredangle 4$.	4. If two angles are complementary to congruent angles, then they are congruent.

EXAMPLE 3.8 Present a formal two-column proof.

GIVEN: Lines ℓ and m intersect at point P.
PROVE: $\measuredangle 1 \cong \measuredangle 3$.

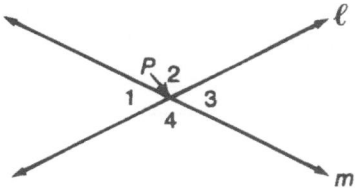

SOLUTION

PROOF:

Statements	Reasons
1. Lines ℓ and m intersect at point P.	1. Given.
2. $\measuredangle 1$ is supplementary to $\measuredangle 2$, $\measuredangle 3$ is supplementary to $\measuredangle 2$.	2. If the exterior sides of a pair of adjacent angles form a straight line, then the angles are supplementary.
3. $\measuredangle 1 \cong \measuredangle 3$.	3. If two angles are supplementary to the same angle, then they are congruent.

Notice, in the formal proofs presented in Examples 3.7 and 3.8, that factual statements are placed in their logical sequence and numbered in the "Statements" column. The reason used to support each statement receives a corresponding number and is written in the "Reasons" column. In addition to the Given, the "Reasons" column may include only previously stated definitions, postulates, and theorems. Keep in mind that, once a theorem is proved, it may be included in the repertoire of statements that may be used in the "Reasons" column of subsequent proofs.

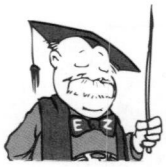

THIS IS THE KEY TO THE METHOD!

In approaching a proof, you need to plan thoughtfully and then organize the necessary steps in the "Statements" column. To maintain a logical train of thought, it is sometimes helpful to concentrate first on completing the *entire* "Statement" column; then, to complete the proof, the corresponding reasons may be entered.

In the phrasing of theorems written in support of statements in the "Reasons" column, we will sometimes use the expressions "congruent" and "equal in measure" interchangeably. This will simplify our work, avoiding the need to change from one expression to the other and then back again. For example, Theorem 3.2 may, if convenient, be used in the following form:

If two angles are complementary (or supplementary) to the same angle or to congruent angles, then they are *equal in measure*.

Returning to Example 3.8, we observe that angles 1 and 3 are vertical angles and are congruent. Using the same approach, we could establish that vertical angles 2 and 4 are congruent, leading to Theorem 3.3.

THEOREM 3.3

Vertical angles are congruent.

EXAMPLE 3.9

a. Find the value of x.

b. Find the measures of angles AEC, DEB, DEA, and BEC.

SOLUTION

a. By Theorem 3.3:

$$3x - 18 = 2x + 5$$
$$3x = 2x + 23$$
$$x = 23$$

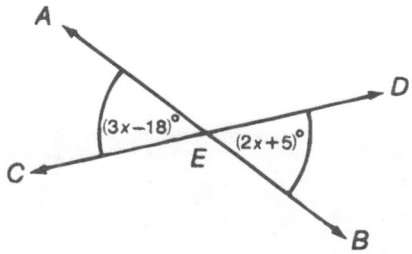

b. $m\measuredangle AEC = m\measuredangle DEB = 3x - 18$
$= 3(23) - 18$
$= 51$

Since angles AEC and DEA are supplementary,
$m\measuredangle DEA = 180 - 51 = 129$
$m\measuredangle DEA = m\measuredangle BEC = 129$

EXAMPLE 3.10

Present a formal two-column proof.

GIVEN: $\measuredangle 2 \cong \measuredangle 3$.
PROVE: $\measuredangle 1 \cong \measuredangle 3$.

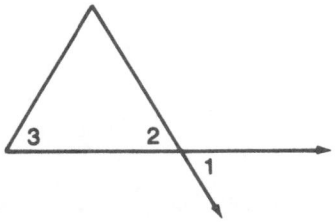

SOLUTION

PROOF:

Statements	Reasons
1. $\measuredangle 1 \cong \measuredangle 2$.	1. Vertical angles are congruent.
2. $\measuredangle 2 \cong \measuredangle 3$.	2. Given.
3. $\measuredangle 1 \cong \measuredangle 3$.	3. Transitive property of congruence.

SUMMARY

Two angles are congruent if they are:

- Vertical angles formed by two intersecting lines.
- Complements of the same or of congruent angles.
- Supplements of the same or of congruent angles.

Definitions and Theorems Relating to Right Angles and Perpendiculars

Recall that a *right angle* is an angle of measure 90. The following theorems are useful in proving other theorems:

THEOREM 3.4

All right angles are congruent.

INFORMAL PROOF: Let angles 1 and 2 be right angles. We must show that $\measuredangle 1 \cong \measuredangle 2$. Since angles 1 and 2 are right angles, $m\measuredangle 1 = 90$ and $m\measuredangle 2 = 90$. Therefore, $m\measuredangle 1 = m\measuredangle 2$, so $\measuredangle 1 \cong \measuredangle 2$.

THEOREM 3.5

If two angles are congruent and supplementary, then each is a right angle.

INFORMAL PROOF: Let angles 1 and 2 be congruent and supplementary. We must show that angles 1 and 2 are right angles.

- Since they are supplementary, $m\measuredangle 1 + m\measuredangle 2 = 180$
- By substitution, $m\measuredangle 1 + m\measuredangle 1 = 2\, m\measuredangle 1 = 180$
- Therefore, $m\measuredangle 1 = 90 = m\measuredangle 2$
- Hence, angles 1 and 2 are right angles.

Two lines, two segments, or a line and a segment that intersect to form a right angle are said to be *perpendicular*. If line ℓ is perpendicular to line m, we may write $\ell \perp m$, where the symbol $\perp$ is read as "is perpendicular to." See Figure 3.2.

A *perpendicular bisector* of a line segment is a line or segment that is perpendicular to the given segment at its midpoint. See Figure 3.3.

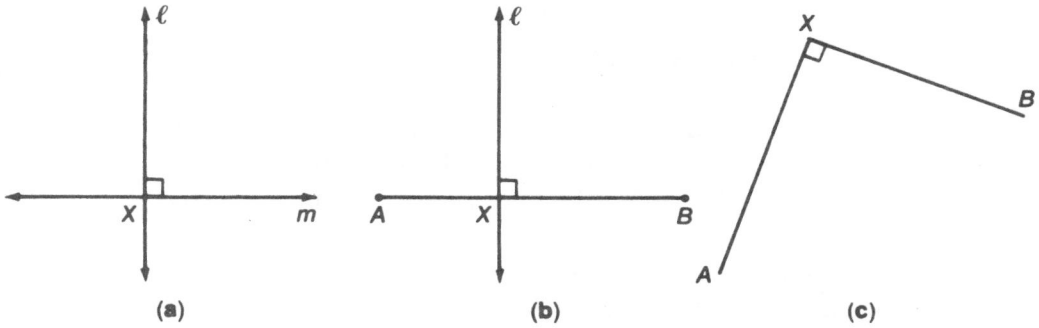

FIGURE 3.2 Perpendiculars.

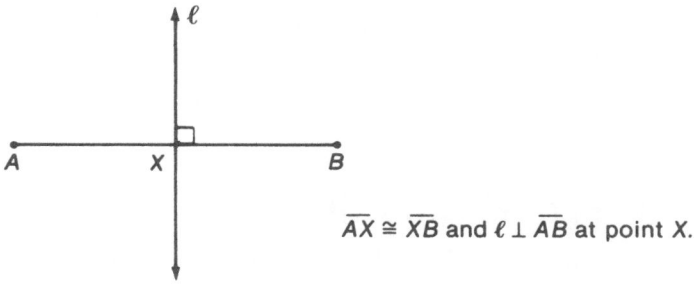

FIGURE 3.3 A perpendicular bisector.

DEFINITION OF PERPENDICULAR LINES

Perpendicular lines are lines that intersect to form right angles. If a line is perpendicular to a segment and intersects the segment at its midpoint, then the line is called the *perpendicular bisector* of the segment.

EXAMPLE 3.11

GIVEN: $\overline{AC} \perp \overline{BC}$, angles 1 and 2 are adjacent.

PROVE: $m\measuredangle 1 + m\measuredangle 2 = 90$.

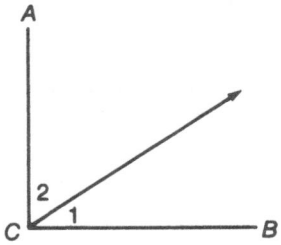

SOLUTION

PROOF:

Statements	Reasons
1. $\overline{AC} \perp \overline{BC}$	1. Given.
2. $\measuredangle ACB$ is a right angle.	2. Perpendicular segments intersect to form a right angle.
3. $m\measuredangle ACB = 90$.	3. A right angle has measure 90.
4. $m\measuredangle ACB = m\measuredangle 1 + m\measuredangle 2$.	4. Angle Addition Postulate.
5. $m\measuredangle 1 + m\measuredangle 2 = 90$.	5. Substitution.

Since the sum of the measures of angles 1 and 2 is 90, angles 1 and 2 are complementary. Example 3.11 proves the following theorem:

THEOREM 3.6

If the exterior sides of a pair of adjacent angles are perpendicular, then the angles are complementary.

EXISTENCE OF PERPENDICULARS

How many perpendiculars can be drawn to a given line? A line has an infinite number of perpendiculars; however, it can be proven that through a particular point *on* the line, there exists exactly one perpendicular to the given line. At a point *not* on a line, we postulate that there exists exactly one line (or segment) that passes through the point and is perpendicular to the line. These situations are represented in Figure 3.4*a*, *b*, and *c*, respectively. Observe that Figure 3.4*c* also illustrates that when a perpendicular intersects a line, *four* right angles are formed.

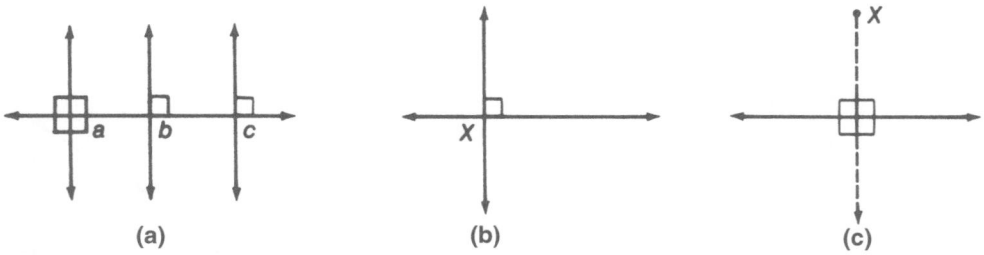

FIGURE 3.4 Existence of perpendiculars.

These results are summarized below.

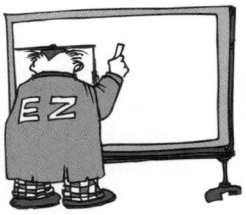

NOTE: FACTS ABOUT PERPENDICULARS

Theorem 3.7 Perpendicular lines intersect to form four right angles.

Theorem 3.8 Through a given point on a line, there exists exactly one perpendicular to the given line.

Postulate 3.1 Through a given point *not* on a line, there exists exactly one perpendicular to the given line.

DISTANCE

The term *distance* in geometry is always interpreted as the *shortest* path between two points. In Figure 3.5 the distance between points P and Q is the length of segment $\overline{PQ}$. A point that is exactly the same distance from two other points is said to be *equidistant* from the two points. The midpoint of a segment, for example, is equidistant from the end points of the segment. See Figure 3.6.

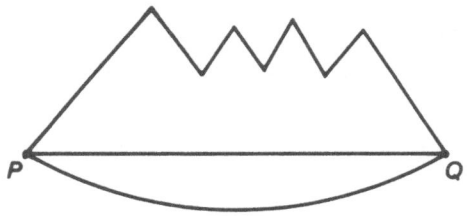

FIGURE 3.5 The shortest path between two points is represented by the segments joining the points, rather than any zigzag or circular route.

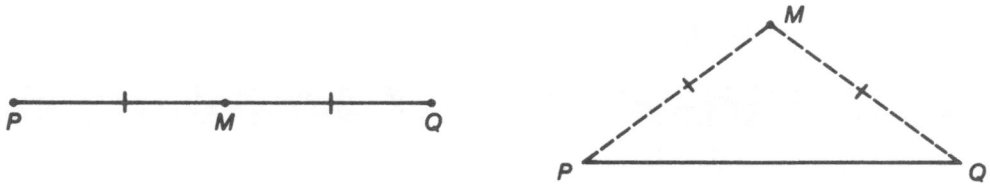

FIGURE 3.6 In each figure, point M is equidistant from point P and Q.

The shortest distance from a point not on a line to the line is measured by the length of the segment drawn from the point perpendicular to the line.

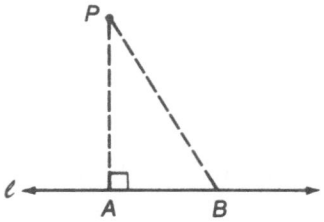

FIGURE 3.7 The distance from point P to line ℓ is the perpendicular segment $\overline{PA}$ rather than any other segment such as $\overline{PB}$.

DEFINITIONS OF DISTANCE

- The *distance between two points* is the length of the segment joining the points.
- The *distance between a line and a point not on the line* is the length of the perpendicular segment drawn from the point to the line.

METHODS FOR PROVING LINES ARE PERPENDICULAR

By using the definition of perpendicularity, we may conclude that two lines are perpendicular if they intersect to form a right angle. It will sometimes be convenient in subsequent work with perpendicular lines to use an alternative method. In Figure 3.8 it seems intuitively clear that, if we continue to rotate line ℓ in a clockwise fashion, eventually lines ℓ and m will be perpendicular, and this will be true when angles 1 and 2 are congruent.

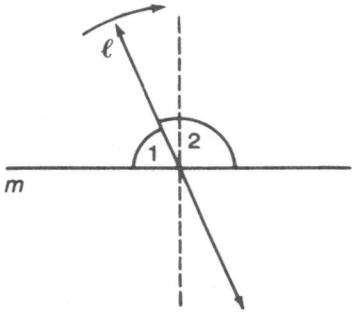

FIGURE 3.8

This leads to the following theorem:

THEOREM 3.9

If two lines intersect to form congruent adjacent angles, then the lines are perpendicular.

EXAMPLE 3.12 Prove Theorem 3.9.

SOLUTION

In approaching the proof of a theorem in which a diagram is not provided, we must:

1. Identify the Given from the information contained in the "if clause" of the statement of the theorem
2. Identify what is to be proved from the relationship proposed in the "then clause" of the theorem
3. Use the information obtained in Steps 1 and 2 to draw an appropriate diagram
4. Organize our thoughts by planning the sequence of statements that must be followed in logically progressing from the Given to the final conclusion
5. Write the formal proof

GIVEN: Lines ℓ and m intersect at k, $\measuredangle 1 \cong \measuredangle 2$.
PROVE: $\ell \perp m$.

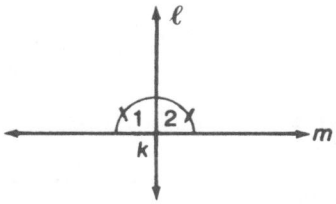

PROOF:

Statements	Reasons
1. Lines ℓ and m intersect at k so that $\measuredangle 1 \cong \measuredangle 2$.	1. Given.
2. $\measuredangle 1$ is supplementary to $\measuredangle 2$.	2. If the exterior sides of a pair of adjacent angles form a straight line, then the angles are supplementary.
3. $\measuredangle 1$ is a right angle.	3. If two angles are congruent and supplementary, then each is a right angle.
4. $\ell \perp m$	4. If two lines intersect to form a right angle, then the lines are perpendicular. (This follows from the reverse of the definition of perpendicularity.)

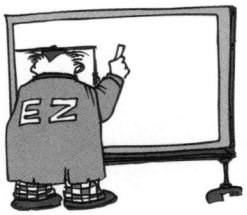

TO PROVE THAT THE TWO LINES ARE PERPENDICULAR:

- Show that the lines intersect to form a right angle.
- Show that the lines intersect to form a pair of congruent adjacent angles (Theorem 3.9).

A Word About the Format of a Proof

The two-column deductive geometric proof is a "formal" type of proof, since each statement is numbered and listed with a corresponding reason using a structured format. Sometimes it is convenient to use an "informal" proof in which the key steps of a proof are summarized in paragraph form. Informal proofs were given for Theorems 3.4 and 3.5.

Another type of proof that lends itself to a paragraph format is a *proof by counterexample*. This type of proof is used to *disprove* a statement by providing a *counterexample*—a single, specific instance that contradicts a proposed generalization. For example, Jessica claims that if two angles are supplementary, then each is a right angle. To prove this statement is false, all that is necessary is to provide a counterexample in which two angles are supplementary but are not right angles. If $m\angle A = 100$ and $m\angle B = 80$, then angles A and B are supplementary but are not right angles. Hence, Jessica's claim is incorrect.

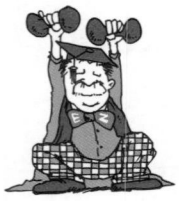

REVIEW EXERCISES FOR CHAPTER 3

1. Two angles are complementary. The measure of one angle exceeds two times the measure of the other angle by 21. What is the degree measure of the smaller angle?

   (1) 23 (2) 53 (3) 67 (4) 127

2. If the complement of $\angle A$ is greater than the supplement of $\angle B$, which statement must be true?

   (1) $m\angle A + m\angle B = 180$
   (2) $m\angle A + m\angle B = 90$
   (3) $m\angle B > m\angle A + 90$
   (4) $m\angle A > m\angle B$

3. In the accompanying figure, what is the value of y?

(1) 20 (2) 30 (3) 45 (4) 60

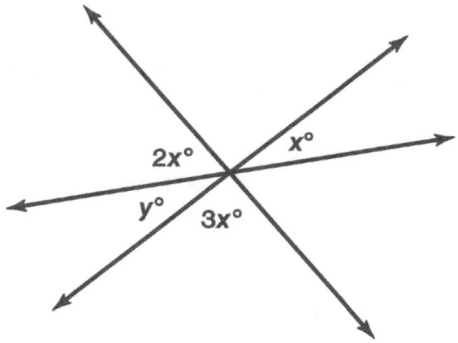

4. In the accompanying figure, line m is perpendicular to line p. What is the value of x?

(1) 15 (2) 20 (3) 24 (4) 30

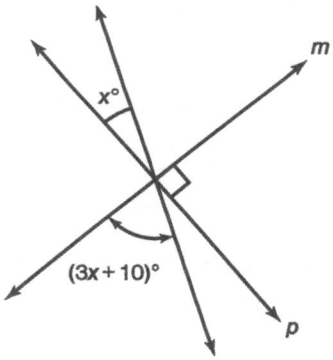

5. The measure of an angle is 5 times as great as the measure of its complement. Find the measure of the angle.

6. The measure of an angle exceeds 3 times its supplement by 4. Find the measure of the angle.

7. The measure of the supplement of an angle is 3 times as great as the measure of the angle's complement. Find the measure of the angle.

8. The difference between the measures of an angle and its complement is 14. Find the measure of the angle and its complement.

9. Find the measure of an angle if it is 12 less than twice the measure of its complement.

10. Find the value of x:

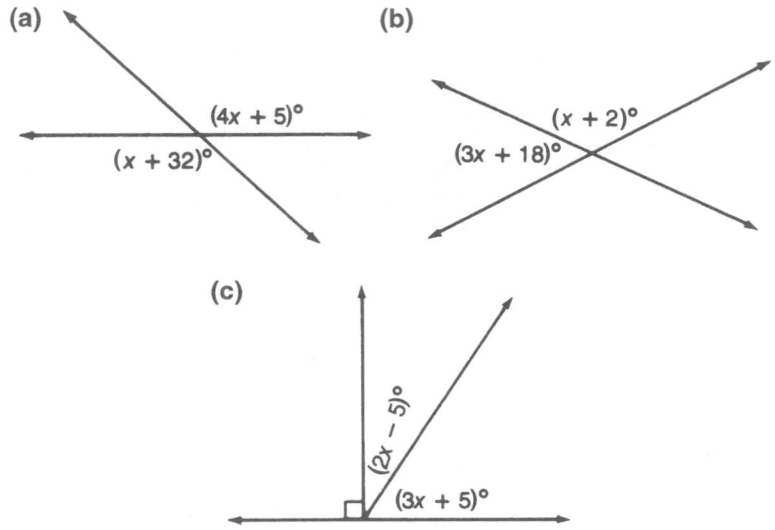

11. $\overleftrightarrow{XY}$ and $\overleftrightarrow{AB}$ intersect at point C. If $m\measuredangle XCB = 4x - 9$ and $m\measuredangle ACY = 3x + 29$, find $m\measuredangle XCB$.

In Exercises 12 and 13, state whether each statement is true or false. Prove a statement is false by providing a counterexample.

12. If an angle is congruent to its supplement, then it is a right angle.

13. If a point C is equidistant from points A and B, then point C is the midpoint of the segment that joins A and B.

14. GIVEN: $\measuredangle 1 \cong \measuredangle 4$.
PROVE: $\measuredangle 2 \cong \measuredangle 3$.

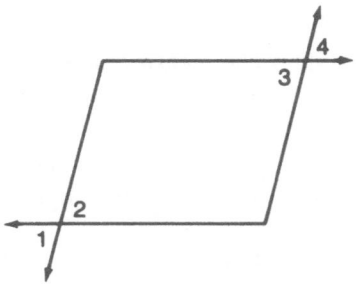

15. GIVEN: $\overline{BD}$ bisects $\measuredangle ABC$.

PROVE: $\measuredangle 1 \cong \measuredangle 2$.

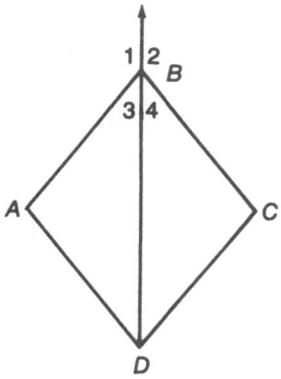

16. GIVEN: $\measuredangle 3$ is complementary to $\measuredangle 1$,
$\measuredangle 4$ is complementary to $\measuredangle 2$.

PROVE: $\measuredangle 3 \cong \measuredangle 4$.

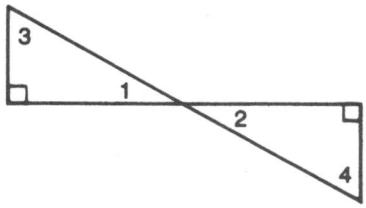

17. GIVEN: $\overline{AB} \perp \overline{BD}$, $\overline{CD} \perp \overline{BD}$,
$\measuredangle 2 \cong \measuredangle 4$.

PROVE: $\measuredangle 1 \cong \measuredangle 3$.

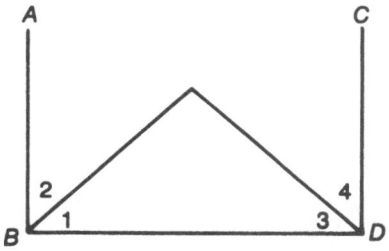

18. GIVEN: $\overline{KL} \perp \overline{JM}$

$\overline{KL}$ bisects $\measuredangle PLQ$.

PROVE: $\measuredangle 1 \cong \measuredangle 4$.

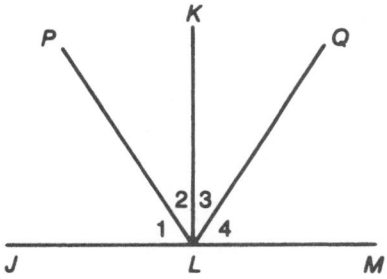

19. GIVEN: $\overline{NW} \perp \overline{WT}$
$\overline{WB} \perp \overline{NT}$,
$\measuredangle 4 \cong \measuredangle 6$.
PROVE: $\measuredangle 2 \cong \measuredangle 5$.

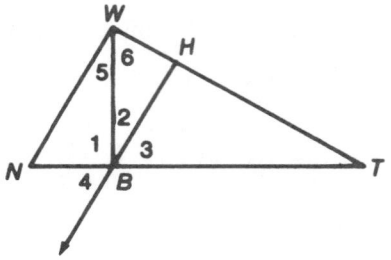

20. GIVEN: $\overline{MT}$ bisects $\measuredangle ETI$,
$\overline{KI} \perp \overline{TI}, \overline{KE} \perp \overline{TE}$,
$\measuredangle 3 \cong \measuredangle 1$, $\measuredangle 5 \cong \measuredangle 2$.
PROVE: $\measuredangle 4 \cong \measuredangle 6$.

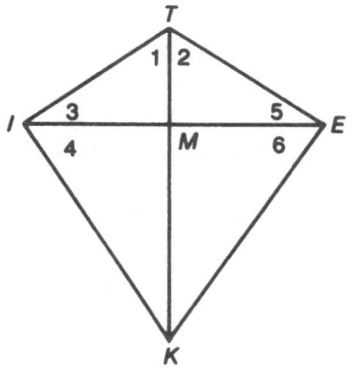

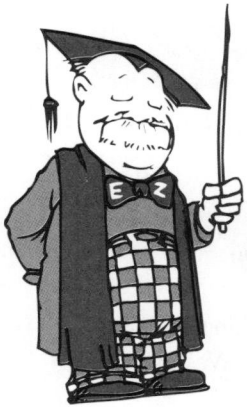

4

Parallel Lines

WHAT YOU WILL LEARN

Lines in a plane either intersect or are parallel. When two parallel lines are cut by a third line, called a *transversal*, eight angles are formed. In this chapter you will learn:

- that special relationships exist between pairs of angles formed when parallel lines are cut by a transversal;
- how you can prove two lines are parallel;
- that the controversial Parallel Postulate guarantees that, through a given point, a line can be drawn parallel to another line.

SECTIONS IN THIS CHAPTER

- Planes and Lines
- Properties of Parallel Lines
- Converses and Methods of Proving Lines Parallel
- The Parallel Postulate

Planes and Lines

Any flat surface such as a chalkboard or floor may be thought of as a *plane*. A plane has these properties:

1. It has length and width, but not depth or thickness.
2. Its length and width may be extended indefinitely in each direction.

Figure 4.1 illustrates that two lines may lie in same plane or in different planes. Lines, segments, rays, or points that lie in the same plane are said to be *coplanar*. We assume that lines are coplanar unless otherwise stated.

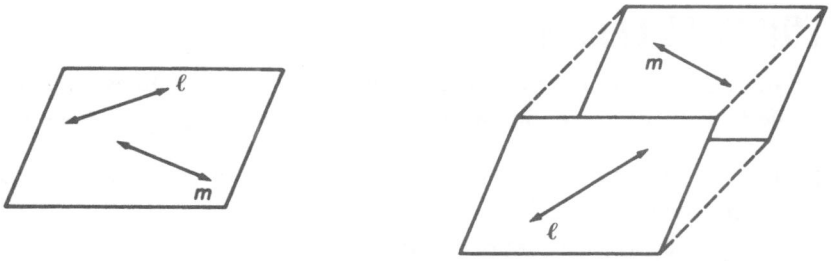

Coplanar Lines

Noncoplanar Lines (Called *Skew* Lines)

FIGURE 4.1

When a pair of lines are drawn, the plane is divided into distinct regions. The region bounded by both lines is referred to as the *interior* region; the remaining outside areas are *exterior* regions. See Figure 4.2.

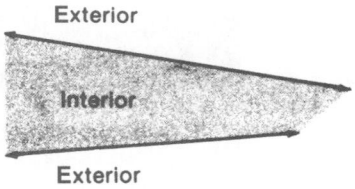

FIGURE 4.2

A line that intersects two or more lines in different points is called a *transversal*. In Figure 4.3 line t represents a transversal since it intersects lines ℓ and m at two distinct points. Notice that at each point of intersection four angles are formed. The angles that lie in the interior region—angles 1, 2, 5, and 6—are referred to as *interior angles*. Angles 3, 4, 7, and 8 lie in the exterior region and are called *exterior angles*.

Table 4.1 further classifies angle pairs according to their position relative to the transversal.

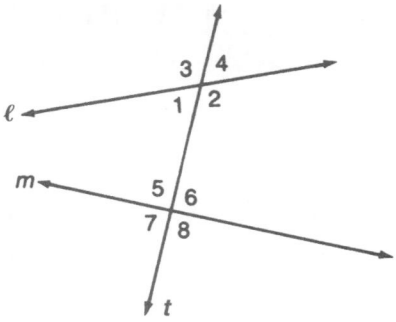

FIGURE 4.3

TABLE 4.1

Typing of Angle Pair	Distinguishing Features	Examples in Figure 4.3
Alternate interior angles	Angles are interior angles. Angles are on opposite sides of the transversal. Angles do not have the same vertex.	Angles 1 and 6; Angles 2 and 5.
Corresponding angles	One angle is an interior angle; the other angle is an exterior angle. Angles are on the same side of the transversal. Angles do not have the same vertex.	Angles 3 and 5; Angles 4 and 6; Angles 1 and 7; Angles 2 and 8.
Alternate exterior angles	Angles are exterior angles. Angles are on opposite sides of the transversal. Angles do not have the same vertex.	Angles 3 and 8; Angles 4 and 7.

EXAMPLE 4.1

Referring to the diagram, name all pairs of:

a. Alternate interior angles
b. Corresponding angles
c. Alternate exterior angles

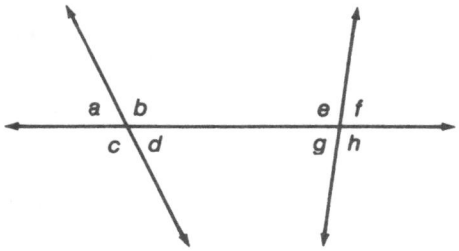

SOLUTION

a. Alternate interior angle pairs: b and g; d and e.
b. Corresponding angle pairs: a and e; b and f; c and g; d and h.
c. Alternate exterior angle pairs: a and h; c and f.

In analyzing diagrams, alternate interior angle pairs can be recognized by their Z shape; corresponding angles form an F shape. The Z and F shapes, however, may be turned so that the letter may appear reversed or upside down. See Figure 4.4.

Properties of Parallel Lines

Wires strung on telephone poles, bars on a prison cell, and lines drawn on a sheet of note paper are examples of *parallel* lines. We postulate that a pair of coplanar lines, if extended indefinitely, will eventually either intersect or never meet. Lines that never intersect are defined to be parallel.

In Figure 4.5, we may concisely express the fact that line ℓ *is parallel to* line m by using the notation $\ell \parallel m$; in Figure 4.6, the notation $\ell \nparallel m$ expresses the fact that line ℓ is *not* parallel to line m.

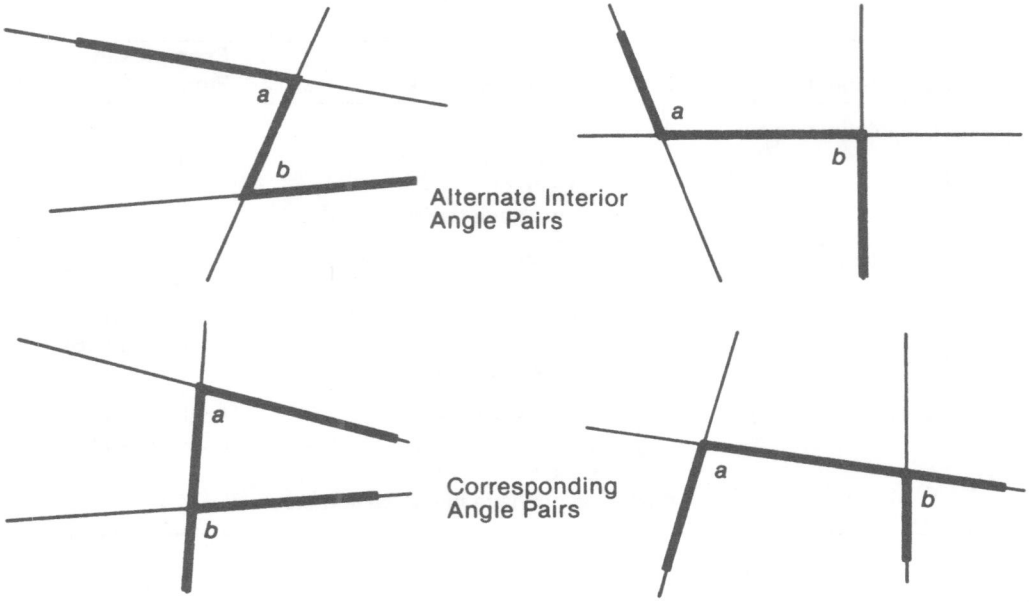

FIGURE 4.4

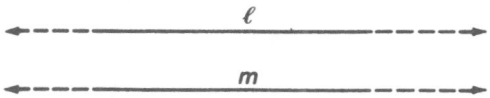

FIGURE 4.5

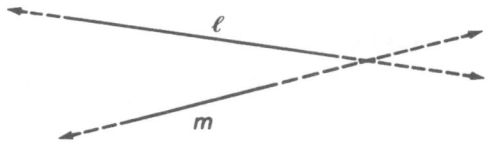

FIGURE 4.6

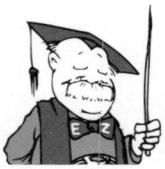

DEFINITION OF PARALLEL LINES

Parallel lines are lines in the same plane that never intersect.

NOTATION: $\ell \| m$ is read as "Line ℓ is parallel to line m."

It follows from the preceding definition that:

- Portions (segment or rays) of parallel lines are parallel.

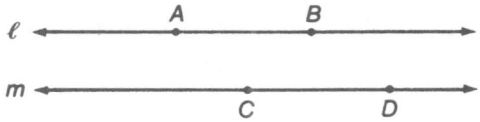

If line $\ell \parallel m$, then $\overline{AB} \parallel \overline{CD}$.

- Extensions of parallel segments or rays are parallel.

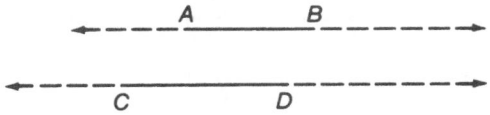

If $\overline{AB} \parallel \overline{CD}$, then $\overleftrightarrow{AB} \parallel \overleftrightarrow{CD}$.

Suppose, in Figure 4.7, that $\overline{AD} \parallel \overline{BC}$ and $\overline{AB} \parallel \overline{CD}$. To indicate this information in the diagram, arrowheads that point in the same direction are used. See Figure 4.8.

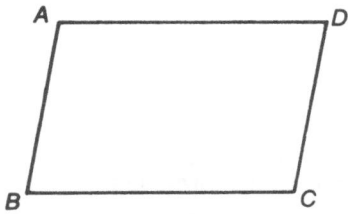

FIGURE 4.7

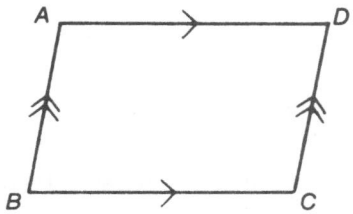

FIGURE 4.8

EXAMPLE 4.2 Mark the diagram with the information supplied

GIVEN: $\overline{SR} \parallel \overline{TW}$,
$\overline{RL} \parallel \overline{TM}$,
$\overline{ST} \cong \overline{RW}$.

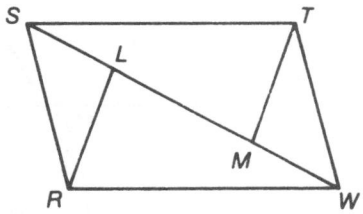

SOLUTION

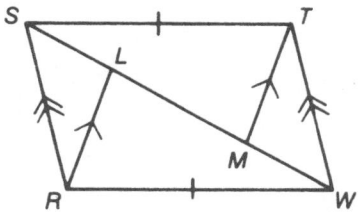

In Figure 4.9*a*, lines ℓ and m are not parallel and the alternate interior angles do *not* measure the same. On the other hand, in Figure 4.9*b*, where line ℓ has been drawn parallel to line m, all pairs of acute angles look congruent and all pairs of obtuse angles appear congruent. This suggests that there may be a relationship between certain pairs of angles formed by parallel lines. Since this observation cannot be proved with our existing knowledge, it is postulated.

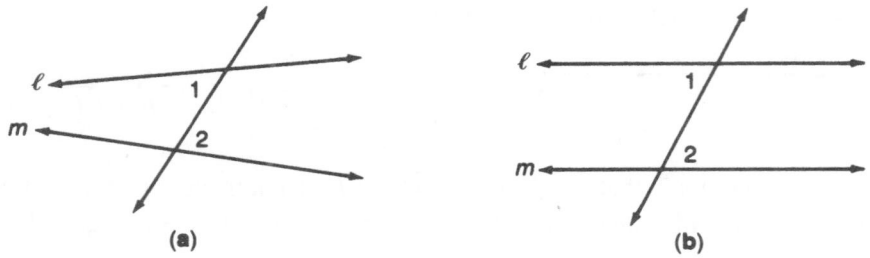

FIGURE 4.9

POSTULATE 4.1

If two lines are parallel, then their alternate interior angles are congruent.

EXAMPLE 4.3 Given that the indicated lines are parallel, determine the value of x.

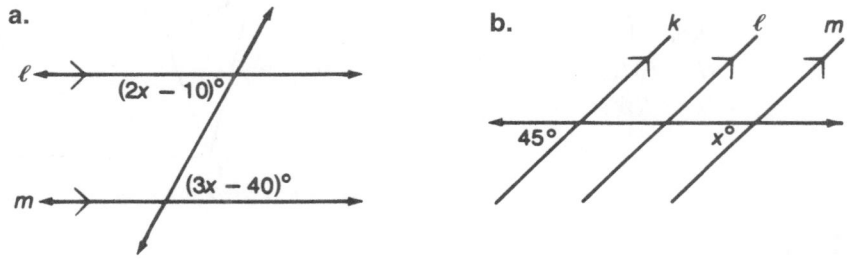

SOLUTION

a. From Postulate 4.1,

$$\begin{aligned} 3x - 40 &= 2x - 10 \\ 3x &= 2x + 30 \\ x &= 30 \end{aligned}$$

b. Using vertical angles and Postulate 4.1:

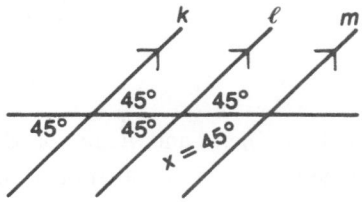

EXAMPLE 4.4

If lines ℓ and m are parallel, and $m\measuredangle a = 60$, find the measure of each of the numbered angles.

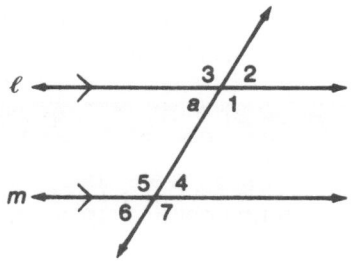

SOLUTION

Angles formed at the intersection of line ℓ and the transversal are either vertical or supplementary angle pairs.

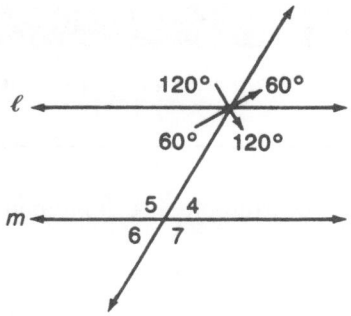

By Postulate 4.1, alternate interior angles are equal in measure:

$$m\measuredangle a = m\measuredangle 4 = 60$$
$$m\measuredangle 1 = m\measuredangle 5 = 120$$

And since vertical angles are equal in measure,

$$m\measuredangle 6 = m\measuredangle 4 = 60$$
$$m\measuredangle 7 = m\measuredangle 5 = 120$$

The completed diagram is as follows:

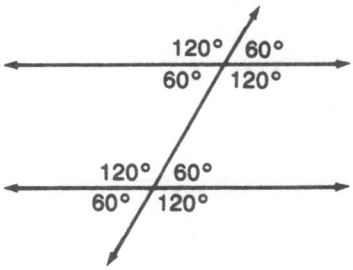

Look closely at the relationships between the special angle pairs derived in Example 4.4. The following theorems are suggested:

THEOREM 4.1

If two lines are cut by a transversal, then any pair of angles are either congruent or supplementary.

THEOREM 4.2

If two lines are parallel, then their corresponding angles are congruent.

For example, if $\ell \parallel m$, then $\measuredangle a \cong \measuredangle b$.

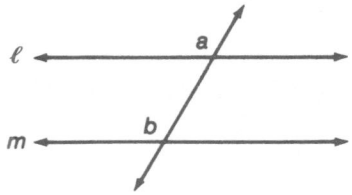

THEOREM 4.3

If two lines are parallel, then interior angles on the same side of the transversal are supplementary.

For example, if $\ell \parallel m$, then $\measuredangle e$ and $\measuredangle f$ are supplementary.

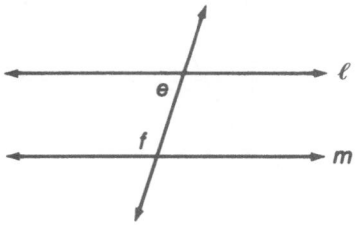

EXAMPLE 4.5 In the accompanying diagram, parallel lines HE and AD are cut by transversal, $\overleftrightarrow{BF}$ at points G and C, respectively. If $m\angle HGF = 5n$ and $m\angle BCD = 2n + 66$, find the value of n.

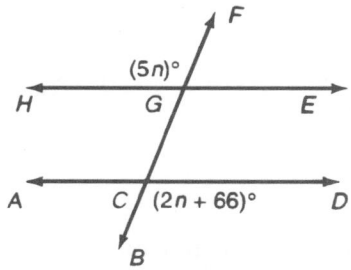

SOLUTION

Since vertical angles are equal in measure,

$$m\angle EGC = m\angle FGH = 5n.$$

Angles EGC and BCD are corresponding angles and are equal in measure since $\overleftrightarrow{HE}$ and $\overleftrightarrow{AD}$ are parallel. Therefore:

$$\begin{aligned} m\angle EGC &= m\angle BCD \\ 5n &= 2n + 66 \\ 5n - 2n &= 66 \\ \frac{\cancel{3}n}{\cancel{3}} &= \frac{66}{3} \\ n &= 22 \end{aligned}$$

EXAMPLE 4.6 Prove Theorem 4.2.

GIVEN: $\ell \parallel m$.

PROVE: $\measuredangle a \cong \measuredangle b$.

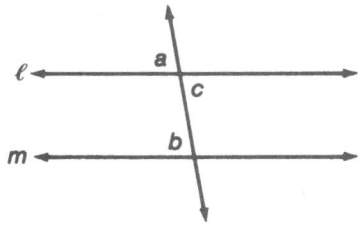

SOLUTION

PLAN: Introduce $\measuredangle c$ in the diagram so that angles b and c are alternate interior angles. Then use Postulate 4.1 and vertical angles to establish the conclusion.

PROOF:

Statements	Reasons
1. $\ell \parallel m$.	1. Given.
2. $\measuredangle a \cong \measuredangle c$.	2. Vertical angles are congruent.
3. $\measuredangle c \cong \measuredangle b$.	3. If two lines are parallel, then their alternate interior angles are congruent.
4. $\measuredangle a \cong \measuredangle b$.	4. Transitive property.

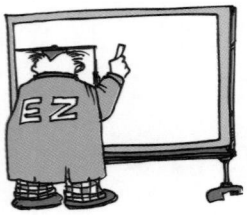

NOTE

We have introduced the notion of including a formal statement of a plan of attack. An indication of a "plan" and the corresponding annotation of the diagram is suggested before you begin to write the statements and reasons in the two-column proof.

EXAMPLE 4.7

GIVEN: $\overline{RS} \parallel \overline{TW}, \overline{TS} \parallel \overline{WX}$.

PROVE: $\measuredangle S \cong \measuredangle W$.

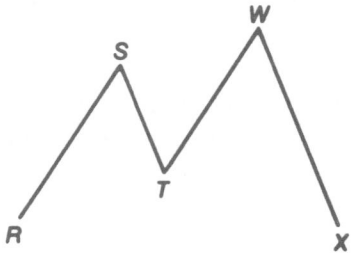

SOLUTION

PLAN: By alternate interior angles, $\measuredangle S \cong \measuredangle T$ and $\measuredangle T \cong \measuredangle W$. Therefore, $\measuredangle S \cong \measuredangle W$.

PROOF:

Statements	Reasons
1. $\overline{RS} \parallel \overline{TW}$.	1. Given.
2. $\measuredangle S \cong \measuredangle T$.	2. If two lines are parallel, then their alternate interior angles are congruent.
3. $\overline{TS} \parallel \overline{WX}$.	3. Given.
4. $\measuredangle T \cong \measuredangle W$.	4. Same as reason 2.
5. $\measuredangle S \cong \measuredangle W$.	5. Transitive property of congruence.

A few words of warning: Alternate interior and corresponding exterior angle pairs are formed whenever two or more lines are intersected by a transversal, regardless of whether the lines in the original set are parallel. These angle pairs, therefore, may or may *not* be congruent. We are entitled to conclude that alternate interior, corresponding and alternate exterior angle pairs are congruent only if it is first determined that the lines intersected by the transversal are parallel.

SUMMARY

If two lines are parallel, then any pair of angles formed are either congruent or supplementary:

- All acute angles are congruent, and all obtuse angles are congruent.
- Any pair of acute and obtuse angles are supplementary.

In particular:

- Alternate interior angles are congruent.

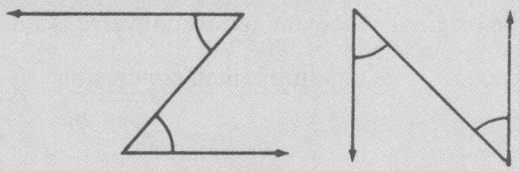

- Corresponding angles are congruent.

- Alternate exterior angles are congruent.

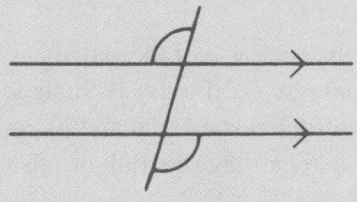

- Interior angles on the same side of the transversal are supplementary.

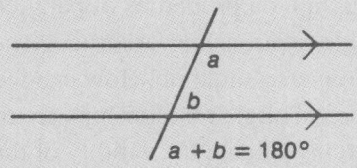

Converses and Methods of Proving Lines Parallel

Geometry depends on preciseness of language as well as on logical reasoning. Often the meaning of a sentence can be changed dramatically simply by switching the positions of words or phrases in the sentence. As an example, consider the statement:

If I am 12 years old, then I am not eligible to vote.

A new statement can be formed by interchanging the single-underlined words in the "if clause" with the double-underlined words in the "then clause:"

If I am not eligible to vote, then I am 12 years old.

The new statement formed is called the *converse* of the original statement. The relationship between the original statement and its converse is illustrated below.

Original Statement: If condition, then conclusion

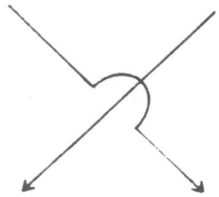

Converse: If _____, then _____

Our example illustrates that the converse of a true statement may not be assumed to be true. If I am not eligible to vote, I can be 13 years old, 14 years old, or perhaps not a citizen.

As another example, form the converse of the statement "If yesterday was Saturday, then today is Sunday." The converse is "If today is Sunday, then yesterday was Saturday." In this instance the original statement and its converse are true. The only generalization that can be made regarding the truth or falsity of the converse is that it may be either true or false.

Why then do we mention converses? Converses are important in mathematics since they provide *clues*—they suggest avenues to explore that may lead to the discovery of true relationships. In the discussion on properties of parallel lines we studied how to prove that certain types of angle relationships are true, *given* that lines are parallel. In this section we examine the converse situation. How can we prove that lines are parallel, *given* that certain types of angle relationships are true?

A logical starting point is to investigate the validity of the statement obtained by taking the converse of Postulate 4.1:

POSTULATE 4.1	CONVERSE
If two lines are parallel, then their alternate interior angles are congruent.	If a pair of alternate interior angles are congruent, then the lines are parallel.

In forming the converse, the phrase "a pair of" was inserted; if one pair of alternate interior angles are congruent, so must the other pair be congruent.

The truth of the converse is consistent with experience—if it is known that a pair of alternate interior angles are congruent, then the lines must be parallel. See Figure 4.10.

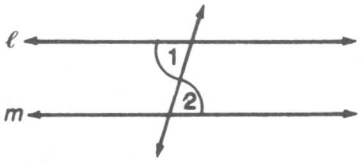

FIGURE 4.10

On the basis of our current knowledge, the converse cannot be proved. We therefore postulate it.

POSTULATE 4.2 CONGRUENT ALTERNATE INTERIOR ANGLES IMPLY PARALLEL LINES POSTULATE

If a pair of alternate interior angles are congruent, then the lines are parallel.

EXAMPLE 4.8

GIVEN: $\overline{BC} \parallel \overline{AD}$ $\measuredangle 2 \cong \measuredangle 3$.

PROVE: $\overline{AB} \parallel \overline{CD}$.

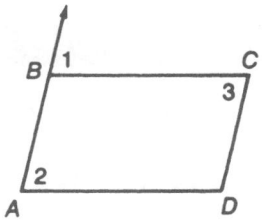

SOLUTION

PLAN: To show $\overline{AB} \parallel \overline{CD}$, first prove $\measuredangle 1 \cong \measuredangle 2$ and then use the Given to conclude that $\measuredangle 1 \cong \measuredangle 3$. By Postulate 4.2 the line segments must be parallel.

PROOF:

Statements	Reasons
1. $\overline{BC} \parallel \overline{AD}$.	1. Given.
2. $\measuredangle 1 \cong \measuredangle 2$.	2. If two lines are parallel, then their corresponding angles are congruent.
3. $\measuredangle 2 \cong \measuredangle 3$.	3. Given.
4. $\measuredangle 1 \cong \measuredangle 3$.	4. Transitive property of congruence.
5. $\overline{AB} \parallel \overline{CD}$.	5. If a pair of alternate interior angles are congruent, then the lines are parallel (Postulate 4.2).

Additional methods of proving lines parallel may be derived from the converses of Theorems 4.2 and 4.3. The following theorems can be proved using Postulate 4.2:

THEOREM 4.4

If a pair of corresponding angles are congruent, then the lines are parallel.

THEOREM 4.5

If a pair of alternate exterior angles are congruent, then the lines are parallel.

THEOREM 4.6

If a pair of interior angles on the same side of the transversal are supplementary, then the lines are parallel.

EXAMPLE 4.9

GIVEN: $\overleftrightarrow{AB} \perp t$, $\overleftrightarrow{CD} \perp t$.

PROVE: $\ell \parallel m$.

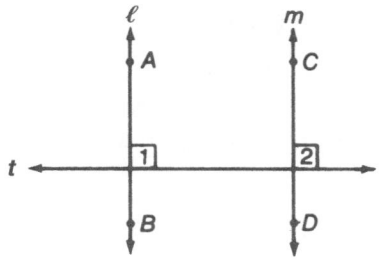

SOLUTION

PLAN: Perpendicular lines intersect to form right angles; all right angles are congruent. Therefore, the lines are parallel by either Theorem 4.4 or Postulate 4.2. (Theorem 4.6 may also be used.)

PROOF:

Statements	Reasons
1. $\overleftrightarrow{AB} \perp t$ and $\overleftrightarrow{CD} \perp t$.	1. Given.
2. Angles 1 and 2 are right angles.	2. Perpendicular lines intersect to form right angles.
3. $\measuredangle 1 \cong \measuredangle 2$.	3. All right angles are congruent.
4. $\ell \parallel m$.	4. If a pair of corresponding angles are congruent, the lines are parallel.

The Parallel Postulate

Euclid, a Greek mathematician who lived in approximately 300 B.C., is credited with collecting and organizing the postulates and theorems that are studied in beginning geometry courses. The Parallel Postulate represents one of the most controversial assumptions made by Euclid.

Euclid's Parallel Postulate

If two lines, when cut by another line, form two same side interior angles whose measures sum to less than 180°, then the two lines, if extended indefinitely, intersect. Figure 4.11 illustrates this situation where $a + b < 180°$ implies that lines ℓ and m eventually intersect.

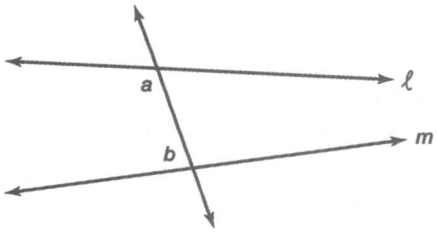

FIGURE 4.11

Over the years Euclid's Parallel Postulate has troubled mathematicians. The thought that lines may intersect at possibly infinite distances was an idea that some mathematicians were reluctant to accept without proof.

In 1795, the mathematician John Playfair devised an alternative formulation of Euclid's Parallel Postulate called *Playfair's Axiom*. Playfair's Axiom is more useful than Euclid's Parallel Postulate, as it answers the question, "How many lines can be drawn through a point not on a line and, at the same time, parallel to that line?"

POSTULATE 4.3 PLAYFAIR'S AXIOM

Postulate 4.3 [Playfair's Axiom]
Exactly one line can be drawn through a point not on a given line and parallel to the given line (see Figure 4.12).

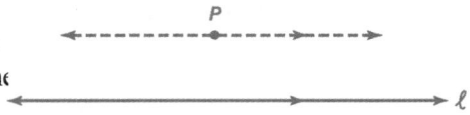

FIGURE 4.12

In most beginning geometry courses, Euclid's Parallel Postulate and Playfair's Axiom are used interchangeably. A development of geometry in which the Parallel Postulate or Playfair's Axiom do *not* hold is known as *non-Euclidean geometry*. Non-Euclidean geometries are studied in more advanced geometry courses.

EXAMPLE 4.10

GIVEN: $\ell \parallel m$.
FIND: $m\measuredangle ABC$.

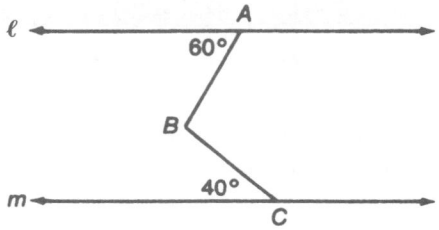

SOLUTION

Using the Parallel Postulate, through point B draw a line parallel to line ℓ (and, therefore, line m).

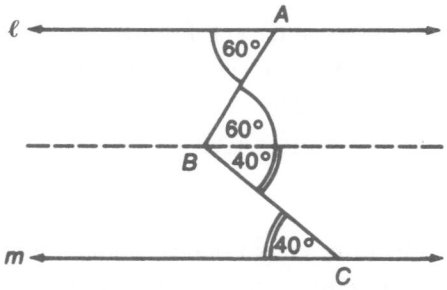

$\therefore$ m $\measuredangle ABC = 100$

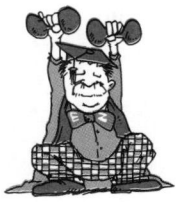

REVIEW EXERCISES FOR CHAPTER 4

1. In the accompanying diagram, line ℓ is parallel to line m, and line t is a transversal. Which must be a true statement?

   (1) $m\angle 1 + m\angle 4 = 180$
   (2) $m\angle 1 + m\angle 8 = 180$
   (3) $m\angle 3 + m\angle 6 = 180$
   (4) $m\angle 2 + m\angle 5 = 180$

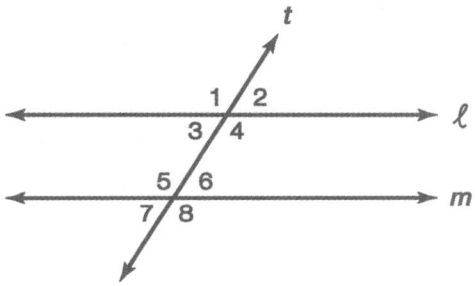

2. In the accompanying diagram, $\overleftrightarrow{AB} \parallel \overleftrightarrow{CD}$. From point E on $\overleftrightarrow{AB}$, transversals $\overleftrightarrow{EF}$ and $\overleftrightarrow{EG}$ are drawn, intersecting $\overleftrightarrow{CD}$ at H and I, respectively. If $m\angle CHF = 20$ and $m\angle DIG = 60$, what is $m\angle HEI$?

   (1) 60 (2) 80 (3) 100 (4) 120

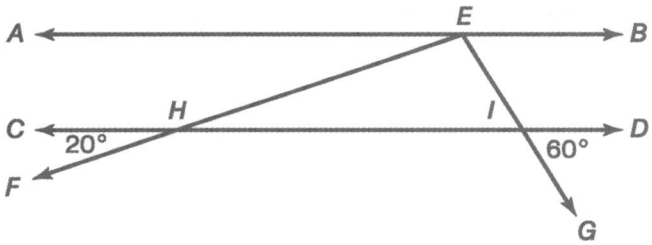

3. In the accompanying diagram, transversal $\overleftrightarrow{EF}$ intersects parallel lines $\overleftrightarrow{AB}$ and $\overleftrightarrow{CD}$ at G and H, respectively. If $m\angle EGB = 2x + 40$ and $m\angle FHC = 3x - 10$, what is the value of x?

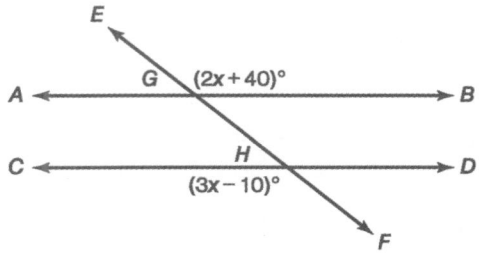

(1) 42 (2) 50 (3) 124 (4) 140

4. Find the values of x and y.

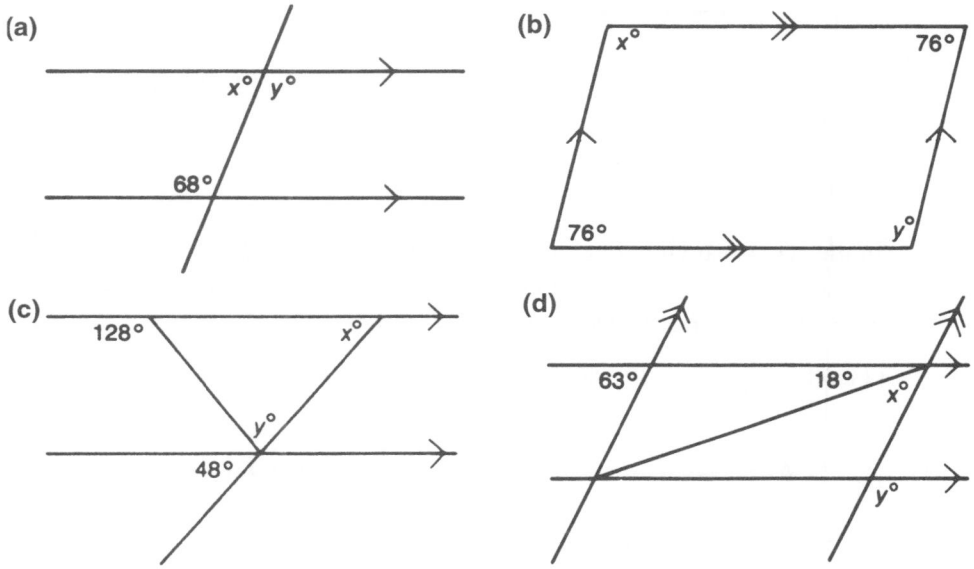

5. Given $\ell \parallel m$, find the value of x.

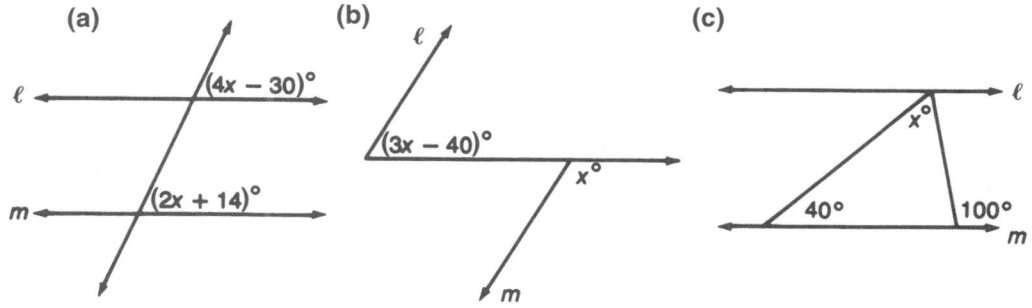

6. Two parallel lines are cut by a transversal. Find the measures of the angles if a pair of interior angles on the same side of the transversal:

   (a) Are represented by $(5x - 32)°$ and $(x + 8)°$.
   (b) Have measures such that one angle is 4 times the measure of the other.

7. Find the value of x.

(a)

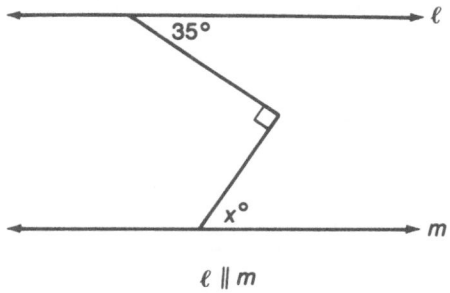

(b)

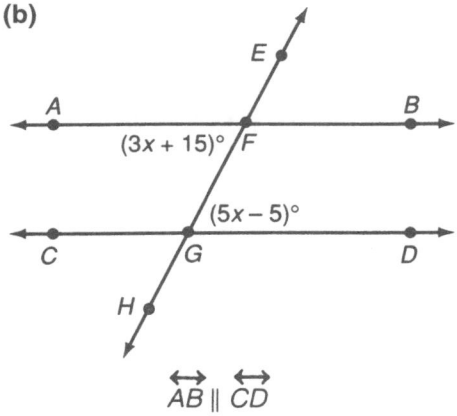

8. In the accompanying diagram, $\overleftrightarrow{AB} \parallel \overleftrightarrow{CD}$ and $\overrightarrow{FG}$ bisects $\angle EFD$. If $m\angle EFG = x$ and $m\angle FEG = 4x$, find x.

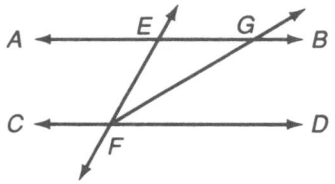

9. In the accompanying diagram, $\overleftrightarrow{AB} \parallel \overleftrightarrow{CD}$ and $\overline{EF}$ bisects $\measuredangle AFG$.

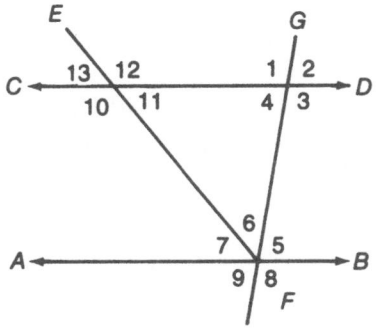

   (a) If $m\measuredangle 1 = 100$, find the measure of each of the numbered angles.
   (b) If $m\measuredangle 3 = 4x - 9$ and $m\measuredangle 5 = x + 19$, find the measure of each of the numbered angles.

10. Form the converse of each of the following statements. State whether the converse is true or false.

    (a) If I live in New York, then I live in the United States.
    (b) If two angles are congruent, then the angles are equal in measure.
    (c) If two angles are vertical angles, then they are congruent.
    (d) If two angles are complementary, then the sum of their measures is 90.
    (e) If two angles are adjacent, then they have the same vertex.
    (f) If two lines are perpendicular to the same line, then they are parallel.

11. GIVEN: $\overline{BA} \parallel \overline{CF}, \overline{BC} \parallel \overline{ED}$.
    PROVE: $\measuredangle 1 \cong \measuredangle 2$.

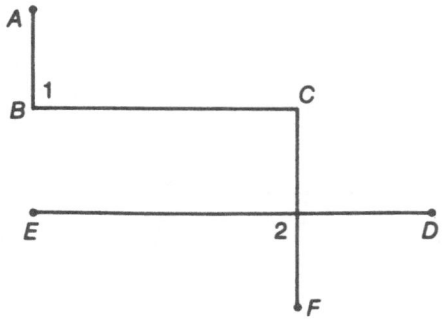

12. GIVEN: $\overline{LJ} \parallel \overline{WK} \parallel \overline{AP}, \overline{PL} \parallel \overline{AG}$.
    PROVE: $\measuredangle 1 \cong \measuredangle 2$.

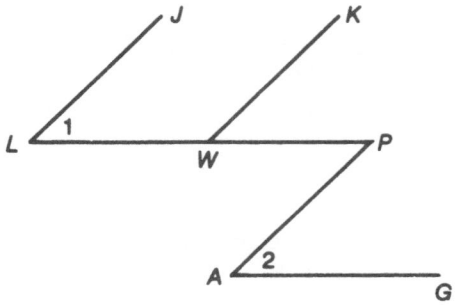

For Exercise 13, choose the correct numbered response to complete the sentence.

13. Two parallel lines are cut by a transversal. The bisectors of a pair of interior angles on the same side of the transversal intersect to form an angle that is

    (1) always acute
    (2) always right
    (3) always obtuse
    (4) either acute or obtuse, but never right

14. GIVEN: $\overline{QD} \parallel \overline{UA}$,
    $\overline{QU} \parallel \overline{DA}$.
    PROVE: $\measuredangle QUA \cong \measuredangle ADQ$.

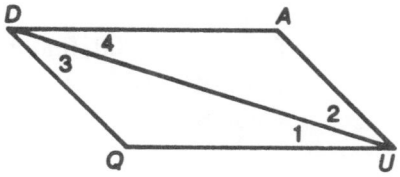

15. GIVEN: $\overline{AT} \parallel \overline{MH}$,
$\measuredangle M \cong \measuredangle H$.
PROVE: $\measuredangle A \cong \measuredangle T$.

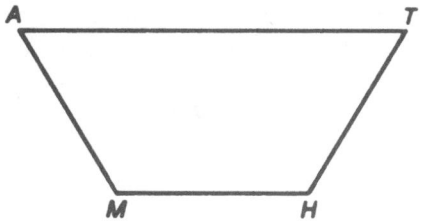

16. GIVEN: $\overline{IB} \parallel \overline{ET}$,
$\overline{IS}$ bisects $\measuredangle EIB$,
$\overline{EC}$ bisects $\measuredangle TEI$.
PROVE: $\measuredangle BIS \cong \measuredangle TEC$.

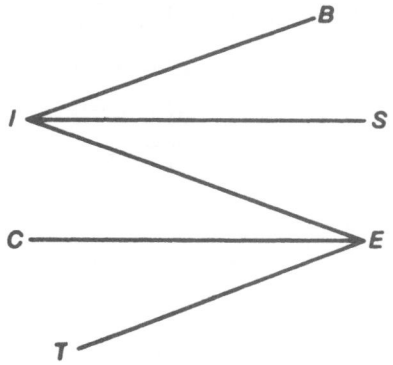

17. GIVEN: $\measuredangle B \cong \measuredangle D, \overline{BA} \parallel \overline{DC}$.
PROVE: $\overline{BC} \parallel \overline{DE}$.

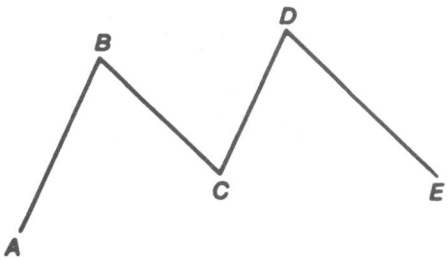

18. GIVEN: $k \parallel \ell$, $\measuredangle 5 \cong \measuredangle 8$.
PROVE: $j \parallel \ell$.

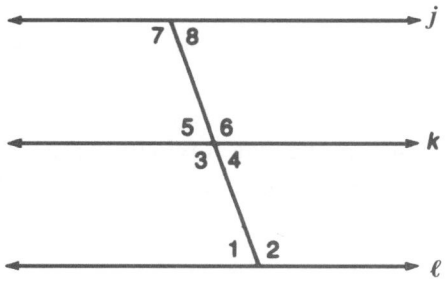

19. GIVEN: $\measuredangle K \cong \measuredangle P$,

$\measuredangle J + m\measuredangle P = 180$.

PROVE: $\overline{KL} \parallel \overline{JP}$.

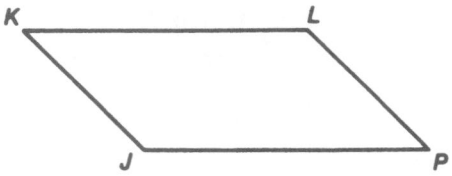

20. GIVEN: $\overline{AB} \perp \overline{BC}$,

$\measuredangle ACB$ is complementary

to $\measuredangle ABE$.

PROVE: $\overleftrightarrow{AC} \parallel \overleftrightarrow{EBD}$.

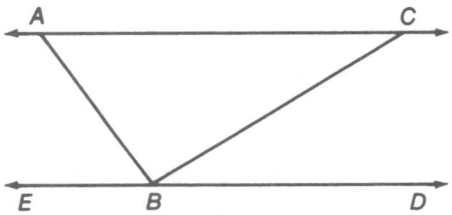

21. GIVEN: $\overline{AG} \parallel \overline{BC}$,

$KH \parallel BC$,

$\measuredangle 1 \cong \measuredangle 2$.

PROVE: $\overline{HK} \perp \overline{AB}$.

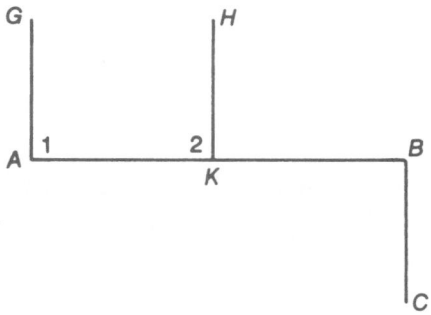

22. Prove that, if two lines are each parallel to a third line, then the lines are parallel to each other.

23. Prove that, if the rays that bisect a pair of alternate interior angles are parallel, then the lines are parallel.

24. Prove that, if two lines are parallel, then the rays that bisect a pair of corresponding angles are parallel.

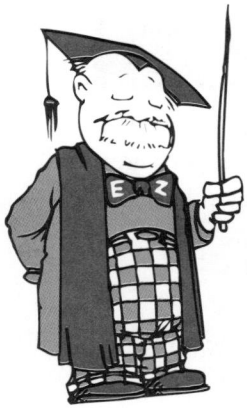

5

Angles of a Polygon

WHAT YOU WILL LEARN

A triangle is the simplest type of polygon. This chapter looks at angle-sum relationships in a triangle. These relationships are then generalized to other polygons, which may have any number of sides. In particular, you will learn:

- the way to name and classify polygons;
- the way to use the properties of parallel lines to prove that the measures of the three angles of a triangle must add up to 180;
- the relationship between the measures of the exterior and interior angles of a triangle;
- the way to find the sum of the measures of the interior and exterior angles of a polygon of any number of sides;
- the special angle relationships in equilateral and equiangular polygons.

SECTIONS IN THIS CHAPTER

- The Anatomy of a Polygon
- Angles of a Triangle
- Exterior Angles of a Triangle
- Angles of a Polygon

The Anatomy of a Polygon

Until now we have focused on points, segments, lines, rays, and angles. In this chapter we begin our study of geometric figures. A "closed" geometric figure whose sides are line segments is called a *polygon*, provided that each side intersects another side at its end point.

A polygon is named by choosing any vertex and then writing the letters corresponding to each vertex consecutively, proceeding in either a clockwise or counterclockwise fashion. In Figure 5.1, *ABCDE* and *BCDEA* name the same polygon.

PARTS OF A POLYGON

There are four important terms related to a polygon: *interior angle, side, vertex*, and *diagonal*. These terms are illustrated in Figures 5.1 and 5.2.

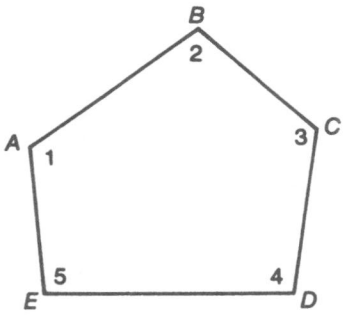

VERTICES: A, B, C, D, and E

SIDES: $\overline{AB}, \overline{BC}, \overline{CD}, \overline{DE}$ and $\overline{AD}$

INTERIOR ANGLES: 1, 2, 3, 4, and 5

FIGURE 5.1 Some parts of a polygon.

A segment that joins a pair of *nonadjacent* vertices of a polygon is called a *diagonal*. See Figure 5.3. Clearly, there exists a relationship between the number of sides of a polygon and the number of diagonals that can be drawn.

DIAGONALS: $\overline{AC}, \overline{BD}, \overline{EC}, \overline{EB}$ and $\overline{AD}$

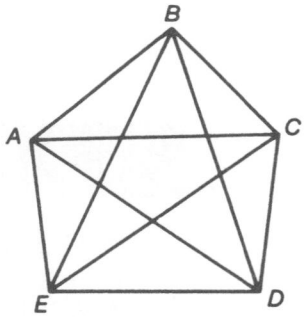

FIGURE 5.2

THEOREM 5.1

If a polygon has N sides, then $\frac{1}{2}N(N-3)$ diagonals can be drawn.

INFORMAL PROOF Consider a polygon having N sides. Draw segments from any vertex, say A, to each of the other $N - 1$ vertices, resulting in $N - 1$ segments. However, this number also includes the segments that coincide with sides $\overline{AB}$ and $\overline{AC}$. Hence, subtracting the two segments, $N - 3$ segments remain that correspond to diagonals. This analysis may be repeated with each of the N vertices, yielding a total of $N(N - 3)$ diagonals. But, each diagonal has been counted twice, since its end points are a vertex of the polygon. Therefore, the total number of distinct diagonals is $\frac{1}{2}N(N - 3)$.

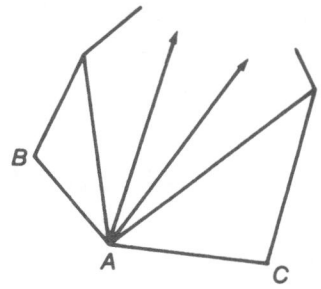

EXAMPLE 5.1 How many diagonals can be drawn in a polygon having seven sides?

SOLUTION

Number of diagonals $= \frac{1}{2}N(N - 3)$

For $N = 7$,

Number of diagonals $= \frac{1}{2}(7)(7 - 3) = \frac{1}{2}(7)(4) = 14$

In Figure 5.3(a), each diagonal has points only in the interior of the polygon. As a result $PQRS$ is a *convex* polygon. Each interior angle of a convex polygon has a measure less than 180.

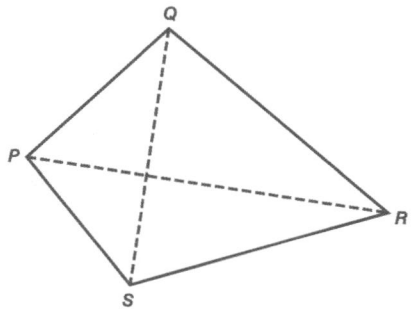

(a) Convex polygon $PQRS$

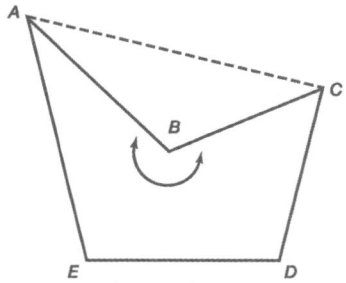

(b) Concave polygon $ABCDE$

FIGURE 5.3

A diagonal of a polygon may contain points that are not enclosed by the polygon as illustrated in Figure 5.3(b). Polygon *ABCDE* is a *concave* polygon. The interior angle at vertex *B* has a measure greater than 180. In this course only convex polygons are studied.

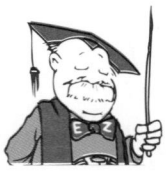

DEFINITIONS OF CONVEX AND CONCAVE POLYGONS

A polygon is *convex* if no diagonal contains points in the exterior of the polygon. A polygon is *concave* if at least one of its diagonals contains points *not* enclosed by the polygon.

CLASSIFYING POLYGONS

A polygon may be classified according to the number of sides it has. The most commonly referred to polygons are as follows:

Polygon Name	Number of Sides
Triangle	3
Quadrilateral	4
Pentagon	5
Hexagon	6
Octagon	8
Decagon	10
Duodecagon	12

A polygon may also be categorized according to whether all of its angles are equal in measure and/or all of its sides have the same length.

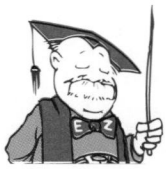

DEFINITIONS OF POLYGONS HAVING PARTS EQUAL IN MEASURE

- An *equiangular* polygon is a polygon in which each angle has the same measure.
- An *equilateral* polygon is a polygon in which each side has the same length.
- A *regular* polygon is a polygon that is both equiangular and equilateral.

EXAMPLE 5.2 Give an example of a regular quadrilateral.

SOLUTION

A square is both equiangular and equilateral. A square is, therefore, a regular quadrilateral.

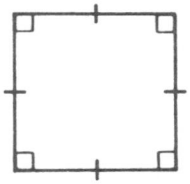

EXAMPLE 5.3 Give an example of a quadrilateral that is equiangular but *not* necessarily equilateral.

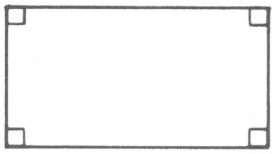

SOLUTION

A rectangle.

Angles of a Triangle

One of the most familiar geometric relationships is that the sum of the measures of the angles of a triangle is 180. See Figure 5.4. An experiment that informally proves this relationship can be performed by drawing any triangle, tearing off angles 1 and 3 (see Figure 5.5), and then aligning the edges (sides) of the angle about angle 2 so that a straight line is formed. Since a straight angle has measure 180, the sum of the measures of the angles of the triangle must be 180.

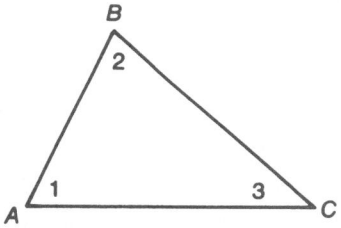

FIGURE 5.4 $m\measuredangle 1 + m\measuredangle 2 + m\measuredangle 3 = 180$.

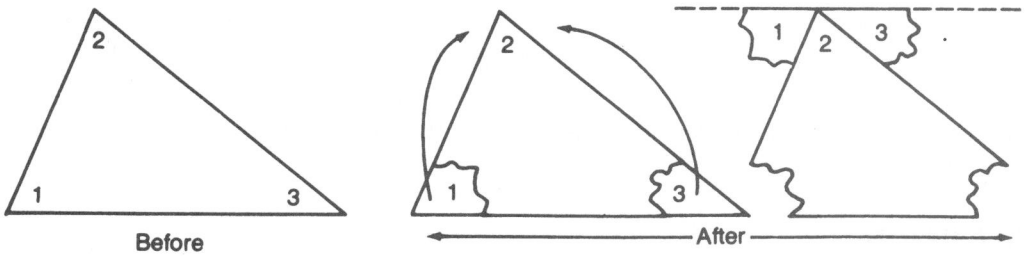

FIGURE 5.5 "Proving" that the sum of the measures of the angles of a triangle is 180. It is shown that angles 1, 2, and 3 can be aligned to form a straight angle.

THE TRIANGLE ANGLE-SUM THEOREM

THEOREM 5.2 SUM OF THE ANGLES OF A TRIANGLE

The sum of the measures of the angles of a triangle is 180.

Since in reality we cannot "tear up" a triangle, the proof of Theorem 5.2 introduces the concept of drawing an extra (auxiliary) line in the original diagram.

GIVEN: $\triangle ABC$ (read as "triangle ABC").

PROVE: $m\measuredangle 1 + m\measuredangle 2 + m\measuredangle 3 = 180$.

PLAN: Draw an auxiliary line ℓ through B and parallel to $\overline{AC}$. The sum of the measures of angles 4, 2, and 5 is 180. Use alternate interior angles and substitution to obtain the desired relationship.

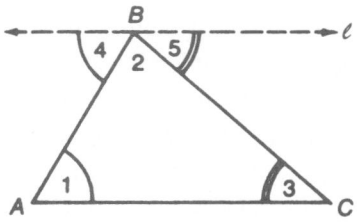

PROOF:

Statements	Reasons
1. Through point B, draw line ℓ parallel to $\overline{AC}$	1. Through a point not on a line exactly one line may be drawn parallel to the line.
2. $m\measuredangle 4 + m\measuredangle 2 + m\measuredangle 5 = 180$.	2. A straight angle has measure 180.
3. $m\measuredangle 1 = m\measuredangle 4$, $m\measuredangle 3 = m\measuredangle 5$.	3. If two lines are parallel, then their alternate interior angles are equal in measure
4. $m\measuredangle 1 + m\measuredangle 2 + m\measuredangle 3 = 180$.	4. Substitution.

EXAMPLE 5.4 The measures of the angles of a triangle are in the ratio 2:3:4. Find the measure of each angle.

SOLUTION

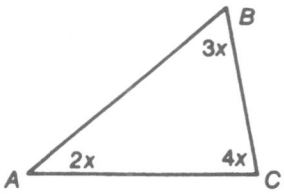

$$
\begin{aligned}
2x + 3x + 4x &= 180 \\
9x &= 180 \\
x &= 20 \\
m\measuredangle A &= 2x = 40 \\
m\measuredangle B &= 3x = 60 \\
m\measuredangle C &= 4x = 80
\end{aligned}
$$

EXAMPLE 5.5 Find the value of x.

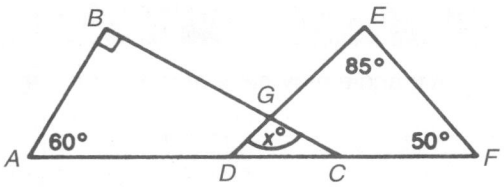

SOLUTION

$m\measuredangle BCA = 30,$
$m\measuredangle EDF = 45,$
$m\measuredangle DGC = 105,$
so $x = 105.$

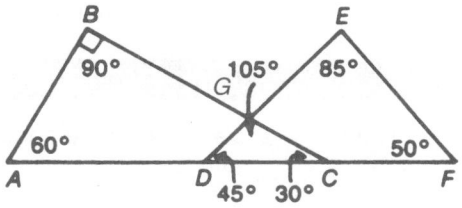

EXAMPLE 5.6 Given $\ell \parallel m$ and $\overline{CD} \perp \overline{AB}$, find the value of x.

SOLUTION

$m\measuredangle A = 35, m\measuredangle ADC = 90$
Hence $35 + 90 + x = 180$
$x = 180 - 125$
$= 55$

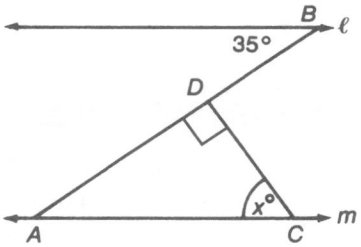

COROLLARIES OF THE TRIANGLE ANGLE-SUM THEOREM

A *corollary* is a theorem that is closely related to a previously proved theorem.

COROLLARY 5.2.1

The measure of each angle of an equiangular triangle is 60.

INFORMAL PROOF

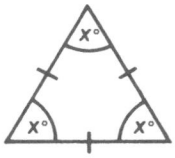

$$3x = 180$$
$$x = \frac{180}{3} = 60$$

COROLLARY 5.2.2

A triangle may have at most one right or one obtuse angle.

INFORMAL PROOF

Use an indirect method of proof—suppose a triangle had two such angles. Then the sum of their measures would be greater than or equal to 180. What about the measure of the third angle of the triangle?

COROLLARY 5.2.3

The acute angles of a right triangle are complementary.

INFORMAL PROOF

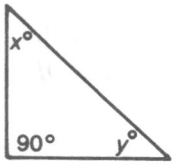

$$90 + x + y = 180$$
$$x + y = 90$$

COROLLARY 5.2.4

If two angles of a triangle are congruent to two angles of another triangle, then the remaining pair of angles are also congruent.

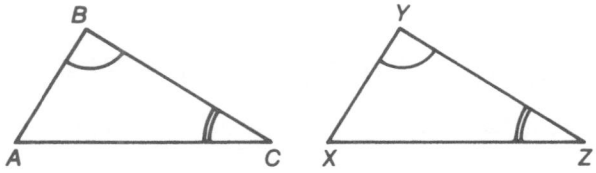

INFORMAL PROOF

$$\begin{array}{lllll} m\measuredangle A & & + m\measuredangle B & + m\measuredangle C & = 180 \\ -m\measuredangle X & & + m\measuredangle Y & + m\measuredangle Z & = 180 \\ \hline m\measuredangle A & - m\measuredangle X & + 0 & + 0 & = 0 \end{array}$$

Hence, $m\measuredangle A = m\measuredangle X$.

EXAMPLE 5.7 Use Corollary 5.2.4 to help construct a formal proof for the following problem:

GIVEN: $\overline{BD}$ bisects $\measuredangle ABC$,
$\overline{BD} \perp \overline{AC}$.

PROVE: $\measuredangle A \cong \measuredangle C$.

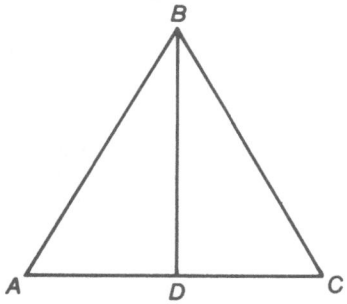

SOLUTION

PLAN:
1. Mark the diagram with the given information. Number the angles to make it easier to refer to them.
2. Since two angles of $\triangle ADB$ are congruent to two angles of $\triangle CDB$, the third pair of angles must be congruent. Hence $\measuredangle A \cong \measuredangle C$.

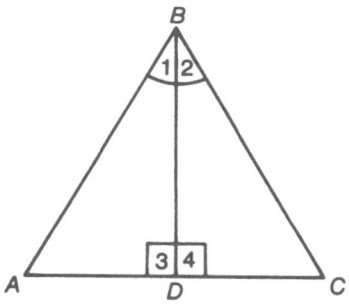

PROOF:

Statements	Reasons
1. $\overline{BD}$ bisects $\measuredangle ABC$.	1. Given.
2. $\measuredangle 1 \cong \measuredangle 2$.	2. A bisector divides an angle into two congruent angles.
3. $\overline{BD} \perp \overline{AC}$.	3. Given.
4. Angles 3 and 4 are right angles.	4. Perpendicular lines intersect to form right angles.
5. $\measuredangle A \cong \measuredangle C$.	5. If two angles of a triangle are congruent to two angles of another triangle, then the remaining pair of angles are congruent.

Exterior Angles of a Triangle

At each vertex of the triangle drawn in Figure 5.6, the sides have been extended, forming additional angles. Angles that are adjacent and supplementary to the interior angle at each vertex are called *exterior* angles of the triangle. Notice that at each vertex there are two such exterior angles:

Vertex	Interior Angle	Exterior Angles*
A	1	4 and 6
B	2	7 and 9
C	3	10 and 12

* Angles 5, 8, and 11 form a vertical pair with the interior angle and are therefore *not* exterior angles.

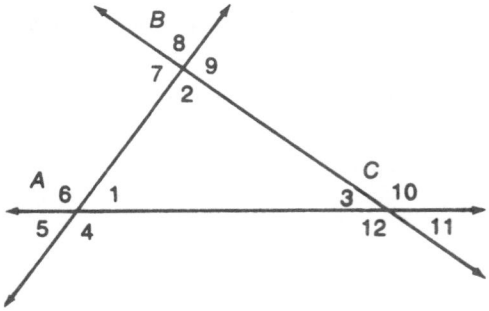

FIGURE 5.6 Exterior angles of a triangle.

Although it is possible to draw two exterior angles at each vertex, when we refer to the exterior angles of a polygon, we will normally mean only one at each vertex—that is, three in the case of a triangle. The fact that this may be done in several different ways is unimportant. Notice in Figure 5.7 that the exterior angle and the interior angle at a vertex form an adjacent pair of angles such that their exterior sides form a straight line.

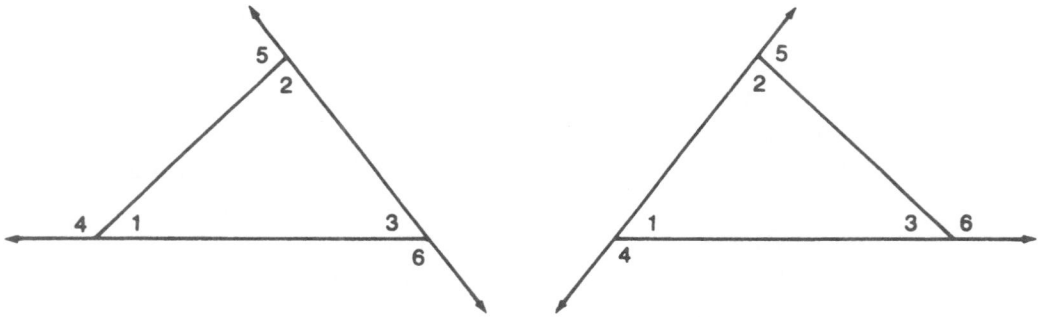

FIGURE 5.7

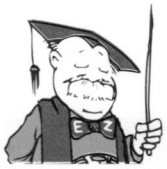

DEFINITION OF EXTERIOR ANGLE OF A TRIANGLE

An *exterior angle* of a triangle (or polygon) is an angle adjacent to an interior angle formed in such a way that their exterior sides form a straight line.

EXAMPLE 5.8 The measures of two angles of a triangle are 80 and 60. Find the sum of the measures of the exterior angles (one exterior angle at a vertex) of the triangle.

SOLUTION

First we note that the third angle of the triangle has measure 40 (180 – 140). Using the fact that, at each vertex, interior and exterior angles are supplementary, we deduce that the measures of the exterior angles are as shown in the diagram:

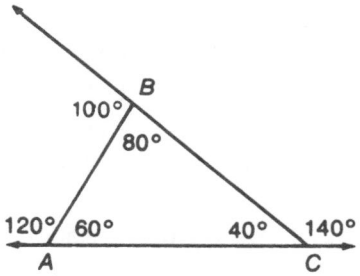

Then we sum: 120 + 100 + 140 = 360.

In the next section we will state and prove that the sum of the measures of the exterior angles of any polygon, regardless of the number of sides, is 360.

In Example 5.8 observe that the measure of each exterior angle is equal to the sum of the measures of the two nonadjacent (remote) interior angles.

THEOREM 5.3 EXTERIOR ANGLE OF A TRIANGLE THEOREM

The measure of an exterior angle of a triangle is equal to the sum of the measures of the two nonadjacent interior angles.

EXAMPLE 5.9 Find the value of x.

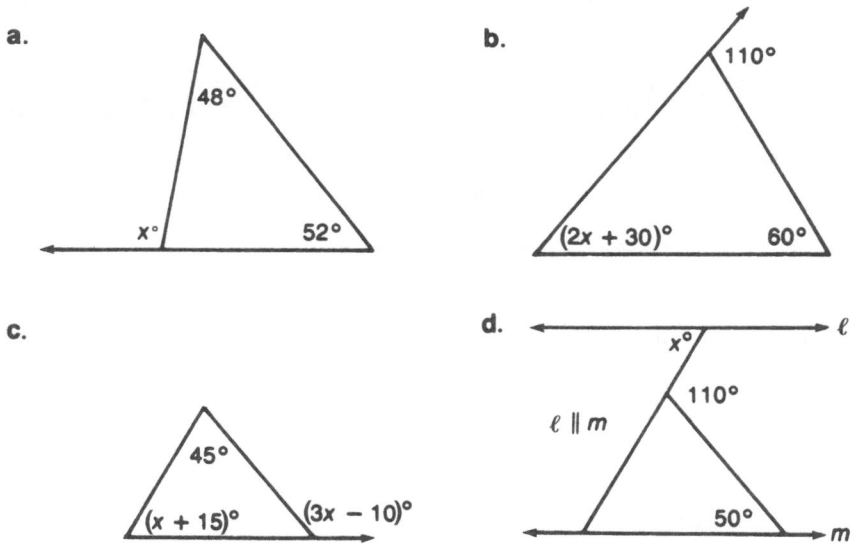

SOLUTION

a. $x = 48 + 52 = 100$

b. $110 = 2x + 30 + 60$
$110 = 2x + 90$
$20 = 2x$
$x = 10$

c. $3x - 10 = x + 15 + 45$
$= x + 60$
$3x = x + 70$
$2x = 70$
$x = 35$

d. $x + 50 = 110$
$x = 60$

We have already established that $m\measuredangle 1 = m\measuredangle b + m\measuredangle c$, as shown in the figure at the right. Since $m\measuredangle b$ and $m\measuredangle c$ represent positive numbers, if we delete one of these quantities from the equation, we no longer have an equality. The following inequalities result:

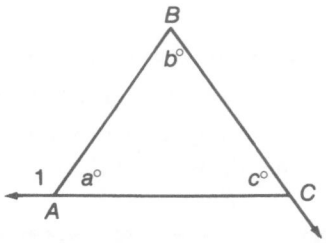

$m\measuredangle 1 = m\measuredangle b + m\measuredangle c$, so $m\measuredangle 1 > m\measuredangle b$
$m\measuredangle 1 = m\measuredangle b + m\measuredangle c$, so $m\measuredangle 1 > m\measuredangle c$

The symbol > is read as "is greater than" (the symbol < is read as "is less than"). We now state the corresponding theorem, which will prove useful in future work involving inequality relationships in triangles.

THEOREM 5.4 EXTERIOR ANGLE INEQUALITY THEOREM

The measure of an exterior angle of a triangle is greater than the measure of either nonadjacent interior angle.

EXAMPLE 5.10 Inside the box ▭ insert the correct symbol (either >, <, or =) so that the resulting statement is true. If it is not possible to draw a conclusion based on the given diagram, then write "cannot be determined."

a. $m\measuredangle 4$ ▭ $m\measuredangle 1$
b. $m\measuredangle 2$ ▭ $m\measuredangle 6$
c. $m\measuredangle 3$ ▭ $m\measuredangle 2$
d. $m\measuredangle 7$ ▭ $m\measuredangle 1$
e. $m\measuredangle 5$ ▭ $m\measuredangle 2$
f. $m\measuredangle 4$ ▭ $m\measuredangle 1 + m\measuredangle 2$

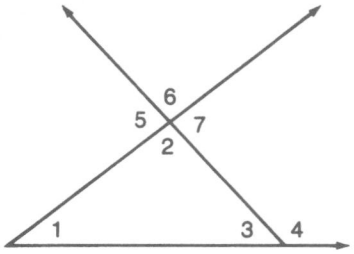

SOLUTION

a. >
b. =
c. Cannot be determined
d. >
e. Cannot be determined
f. =

SUMMARY: ANGLE RELATIONSHIPS IN A TRIANGLE

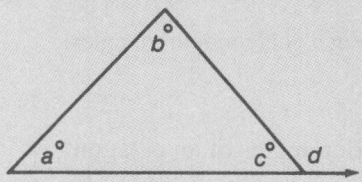

$a + b + c = 180$
$d = a + b$
$d > a$ and $d > b$

Angles of a Polygon

The sum of the measures of the angles of a polygon with three sides is 180. What is the sum of the measures of the angles of a polygon having four sides? Five sides? One hundred sides? To answer these questions, we need to derive a formula that gives the relationship between the sum of the angles of a polygon and the number of its sides. Our strategy is to break down any polygon into an equivalent set of triangles, as shown in Figure 5.8.

Figure 5.8 suggests that a polygon of n sides can be separated into $n - 2$ triangles. Since there are 180 degrees in the sum of the angle measures of each such triangle, it follows that the sum of the angles in the polygon can be found by multiplying $(n - 2)$ by 180.

THEOREM 5.5 SUM OF THE INTERIOR ANGLES OF A POLYGON THEOREM

The sum of the measures of the interior angles of a polygon having n sides is $180(n - 2)$.

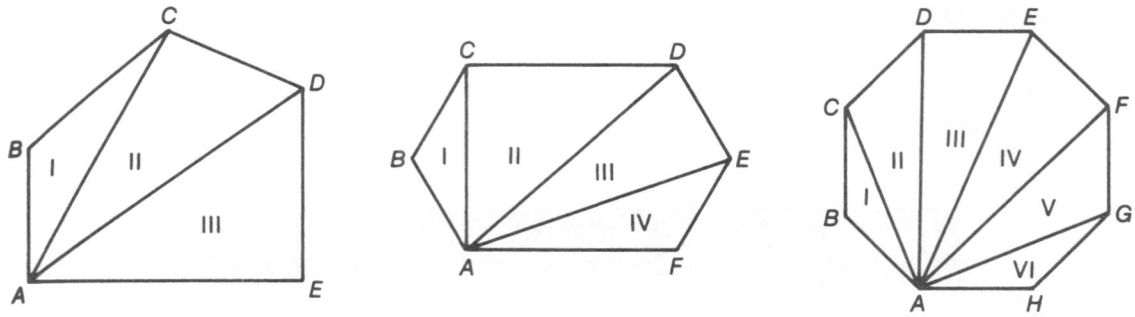

FIGURE 5.8 Separating a polygon into triangles.

EXAMPLE 5.11 Find the sum of the measures of the interior angles of an octagon.

SOLUTION

For $n = 8$, sum $= 180(n - 2) = 180(8 - 2) = 180(6) = 1{,}080$.

EXAMPLE 5.12 If the sum of the measures of the angles of a polygon is 900, determine the number of sides.

SOLUTION

For sum $= 900$,

$$\begin{aligned} 900 &= 180(n-2) \\ \frac{900}{180} &= n-2 \\ 5 &= n-2 \\ n &= 7 \text{ sides} \end{aligned}$$

We know that the sum of the measures of the *exterior* angles of a triangle is 360. Let's prove that this sum remains the same for any polygon.

THEOREM 5.6 SUM OF THE EXTERIOR ANGLES OF A POLYGON THEOREM

The sum of the measures of the exterior angles of any polygon (one exterior angle at a vertex) is 360.

INFORMAL PROOF An n-sided polygon has n vertices. At each vertex the sum of the interior and exterior angles is 180:

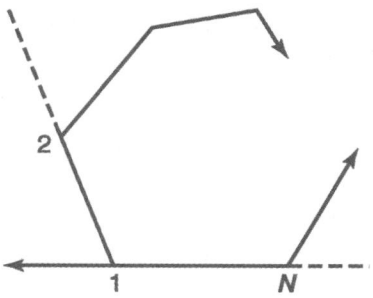

At vertex 1: Interior angle 1 + exterior angle 1 = 180
At vertex 2: Interior angle 2 + exterior angle 2 = 180

• • • •
• • • •
• • • •

At vertex N: Interior angle n + exterior angle n = 180

Adding:	Sum of interior angles	+	sum of exterior angles	=	$180n$
Substituting:	$180(n-2)$	+	sum of exterior angles	=	$180n$
Simplifying:	$180n - 360$	+	sum of exterior angles	=	$180n$
	-360	+	sum of exterior angles	=	0

Sum of exterior angles = 360

If a polygon is equiangular, then each interior angle will have the same measure, implying that each exterior angle has the same measure. Thus, in a *regular* polygon the measure of each exterior angle is equal to the sum of the measures of the exterior angles, 360, divided by the number of sides.

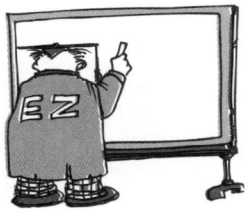

FOR REGULAR POLYGONS

- Exterior angle $= \dfrac{360}{n}$
- Interior angle = 180 – Exterior angle

EXAMPLE 5.13 Find the measure of each interior angle and each exterior angle of a regular decagon.

SOLUTION

Method 1: Sum $= 180(n-2)$
$= 180(10-2)$
$= 180(8)$
$= 1{,}440$

Since there are 10 interior angles, each of which is identical in measure,

Interior angle $= \dfrac{1{,}440}{10} = 144$

Since interior and exterior angles are supplementary,

Exterior angle = 180 – 144 = 36

Method 2: First determine the measure of an exterior angle:

Exterior angle $= \dfrac{360}{10} = 36$

Interior angle = 180 – 36 = 144

EXAMPLE 5.14 The measure of each interior angle of a regular polygon is 150. Find the number of sides.

SOLUTION

We use a method similar to the approach illustrated in Method 2 of Example 5.13. Since the measure of each interior angle is 150, the measure of an exterior angle is 180 – 150, or 30. Therefore,

$$30 = \frac{360}{n}, \quad \text{so } n = 12$$

SUMMARY OF GENERAL PRINCIPLES

- An n-sided polygon has n vertices and n interior angles. At each vertex an exterior angle may be drawn by extending one of the sides.
- If the polygon is regular, then the measures of the interior angles are equal and the measures of the exterior angles are equal.
- The sum of the measures of the *interior* angles of *any* polygon is given by this formula:

$$\text{Sum} = 180(n - 2).$$

The sum of the measures of the *exterior* angles of any polygon, regardless of the number of sides, is 360.

- To find the measures of the interior and exterior angles of a *regular* polygon, given the number of sides (or vice versa), we use the following relationships:

$$\text{Exterior angle} = \frac{360}{n}$$

and

$$\text{Interior angle} = 180 - \text{exterior angle}$$

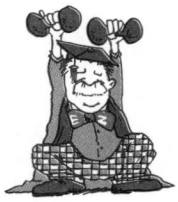

REVIEW EXERCISES FOR CHAPTER 5

1. What is the sum of the measures of the interior angles of a polygon with 13 sides?

   (1) 1800 (2) 1980 (3) 2340 (4) 2700

2. If the sum of the measures of a polygon with n sides is 2,160, then $n =$

   (1) 11 (2) 12 (3) 13 (4) 14

3. Which of the following *cannot* represent the measure of an exterior angle of a regular polygon?

   (1) 72 (2) 15 (3) 27 (4) 45

4. A stop sign in the shape of a regular octagon is resting on a brick wall, as shown in the accompanying diagram. What is the measure of angle x?

   (1) 45° (2) 60° (3) 120° (4) 135°

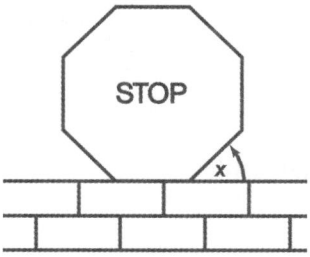

5. The accompanying figure represents a section of bathroom floor tiles shaped like regular hexagons. What is the measure of angle ABC?

   (1) 60° (2) 90° (3) 120° (4) 150°

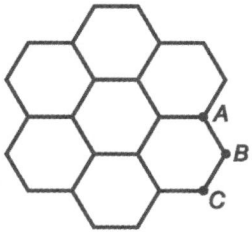

6. Find the value of x:

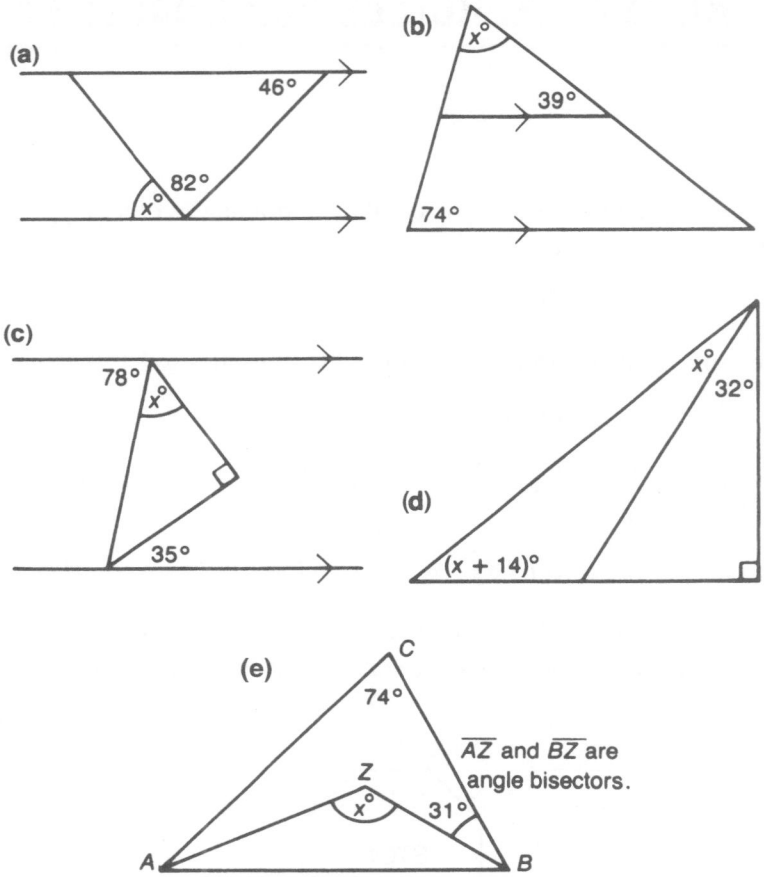

7. Find the measure of angle RWT.

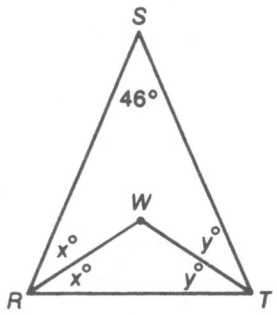

8. Find the sum of the measures of the interior angles of a polygon having:

(a) 4 sides (b) 6 sides (c) 9 sides (d) 13 sides

9. Find the number of sides of a polygon if the sum of the measures of the interior angles is:

   (a) 1,800 (b) 2,700 (c) 540 (d) 2,160

10. Find the measure of the remaining angle of each of the following figures, given the measures of the other interior angles.

    (a) Quadrilateral: 42, 75, and 118
    (b) Pentagon: 116, 138, 94, 88
    (c) Hexagon: 95, 154, 80, 145, 76

11. Find the measure of each interior angle of a regular polygon having:

    (a) 5 sides (b) 24 sides (c) 8 sides (d) 15 sides

12. Find the number of sides of a regular polygon if the measure of an interior angle is:

    (a) 162 (b) 144 (c) 140 (d) 168

13. Find the number of sides in a polygon if the sum of the measures of the interior angles is 4 times as great as the sum of the measures of the exterior angles.

14. Find the number of sides in a regular polygon if:
    (a) The measure of an interior angle is 3 times the measure of an exterior angle.
    (b) The measure of an interior angle equals the measure of an exterior angle.
    (c) The measure of an interior angle exceeds 6 times the measure of an exterior angle by 12.

15. GIVEN: $\overline{AB} \perp \overline{BD}$,
    $\overline{ED} \perp \overline{BD}$.
    PROVE: $\measuredangle A \cong \measuredangle E$.

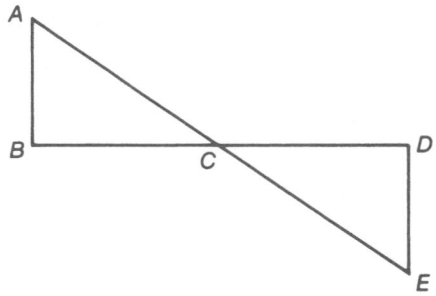

16. GIVEN: $\overline{AC} \perp \overline{CB}$,
    $\overline{DE} \perp \overline{AB}$.
    PROVE: $\measuredangle 1 \cong \measuredangle 2$.

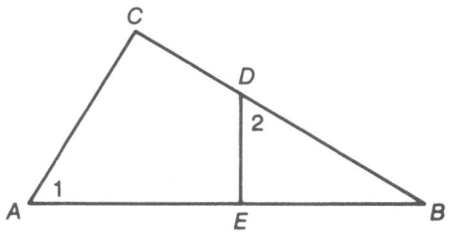

17. GIVEN: $\overline{JP} \perp \overline{PR}$,
$\overline{LK} \perp \overline{KM}$,
$\overline{KM} \parallel \overline{PR}$.
PROVE: $\overline{KL} \parallel \overline{PJ}$.

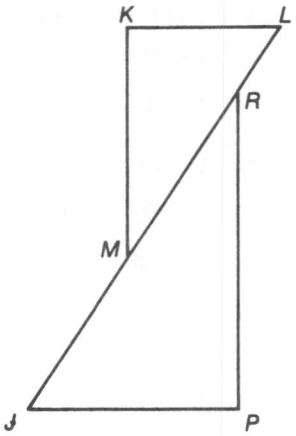

18. GIVEN: $\overline{AB} \perp \overline{BC}$,
$\overline{DC} \perp \overline{BC}$,
$\overline{DE} \perp \overline{AC}$.
PROVE: $\measuredangle 1 \cong \measuredangle 2$.

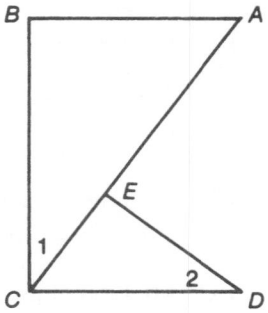

19. GIVEN: $\overleftrightarrow{AX} \parallel \overleftrightarrow{CY}$,
$\overline{BA}$ bisects $\measuredangle CAX$,
$\overline{BC}$ bisects $\measuredangle ACY$.
PROVE: $\measuredangle ABC$ is a right angle.

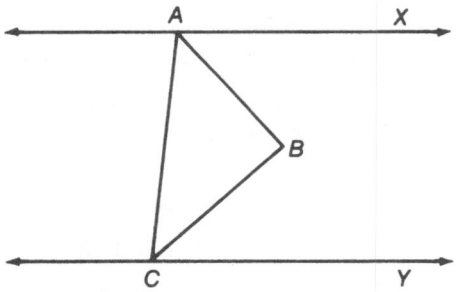

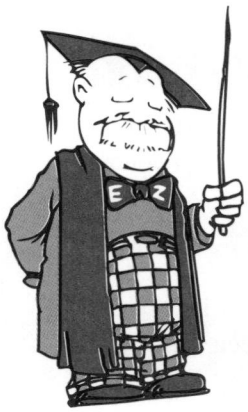

6

Proving Triangles Are Congruent

WHAT YOU WILL LEARN

Triangles that can be made to coincide by placing one on top of the other are said to be *congruent*. If the triangles cannot be "moved," how can we tell whether two triangles are congruent? Do we need to know that all of the angles and all of the sides of one triangle are congruent to the corresponding parts of a second triangle before we can conclude that the two triangles are congruent? In this chapter you will learn:

- different shortcut methods for proving triangles congruent that require only that a particular combination of angles and sides of one triangle be proved congruent to the corresponding combination of parts of the second triangle;
- the fact that two *right triangles* are congruent if the hypotenuse and leg of one right triangle are congruent to the corresponding parts of the second right triangle.

SECTIONS IN THIS CHAPTER

- Correspondences and Congruent Triangles
- Proving Triangles Congruent: SSS, SAS, and ASA Postulates
- Proving Overlapping Triangles Congruent
- Proving Triangles Congruent: AAS and Hy-Leg Methods
- When Two Triangles Are *NOT* Congruent

Correspondences and Congruent Triangles

On a triple blind date, Steve, Bob, and Charles accompanied Jane, Lisa, and Kris. To indicate which young man from the set of gentlemen is being paired with which young lady, the following notation may be used:

Steve ↔ Jane Bob ↔ Lisa Charles ↔ Kris

Such a pairing of the members of one group with the members of another group is called a *correspondence*. A correspondence may also be established between the vertices of two triangles, as shown in Figure 6.1.

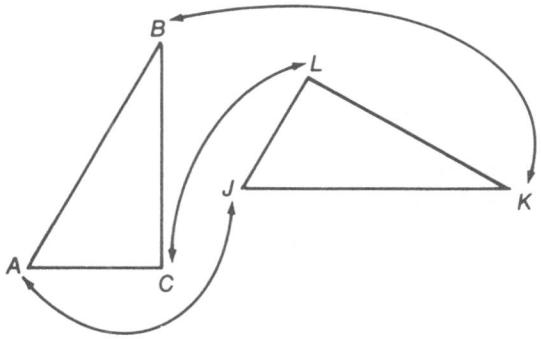

FIGURE 6.1 Correspondence: $A \leftrightarrow J, B \leftrightarrow K, C \leftrightarrow L$.

This correspondence can be expressed concisely by using the notation: $\triangle ABC \leftrightarrow \triangle JKL$. The order in which the vertices are written matters since it defines the pairing of vertices:

$$\triangle A\,B\,C \leftrightarrow \triangle J\,K\,L$$

The correspondence may be written in more than one way. For example, $\triangle CAB \leftrightarrow \triangle LJK$ defines the same correspondence as $\triangle ABC \leftrightarrow \triangle JKL$ since, in each case, the same pair of vertices are matched. On the other hand, $\triangle ABC \leftrightarrow \triangle JKL$ and $\triangle ACB \leftrightarrow \triangle LKJ$ define two different correspondences.

A specified correspondence also serves to define a set of corresponding angles and a set of corresponding sides. If $\triangle RST \leftrightarrow \triangle XYZ$, angle pairs R and X, S and Y, and T and Z are corresponding angles. Segments whose end points are corresponding vertices determine the corresponding sides:

$\triangle RST \leftrightarrow \triangle XYZ$ $\quad$ $\overline{RS} \leftrightarrow \overline{XY}$

$\triangle RST \leftrightarrow \triangle XYZ$ $\quad$ $\overline{ST} \leftrightarrow \overline{YZ}$

$\triangle RST \leftrightarrow \triangle XYZ$ $\quad$ $\overline{RT} \leftrightarrow \overline{XZ}$

REMEMBER
Corresponding sides lie opposite corresponding angles.

Corresponding sides are $\overline{RS}$ and $\overline{XY}$, $\overline{ST}$ and $\overline{YZ}$, and $\overline{RT}$ and $\overline{XZ}$. Drawing the actual triangles in Figure 6.2, we see that *corresponding sides lie opposite corresponding angles.*

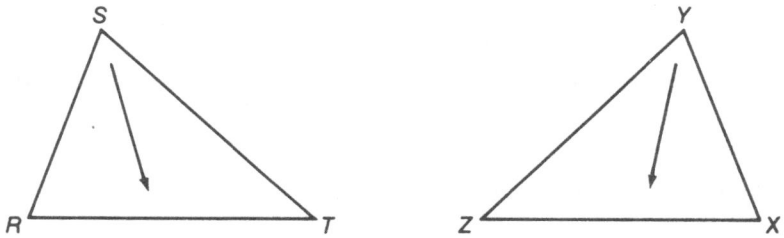

FIGURE 6.2 Corresponding sides lie opposite corresponding angles.

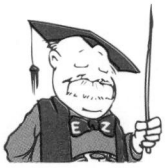

DEFINITIONS OF CORRESPONDING PARTS OF TRIANGLES

- Corresponding *angles* are pairs of angles whose vertices are paired in a given correspondence between two triangles.
- Corresponding *sides* are pairs of segments whose end points are vertices that are paired in a given correspondence between two triangles. Corresponding sides of triangles lie opposite corresponding angles.

EXAMPLE 6.1 Given the correspondence $\triangle BAW \leftrightarrow \triangle TFK$, name all pairs of corresponding angles and sides.

SOLUTION

$\measuredangle B \leftrightarrow \measuredangle T$ $\qquad$ $\overline{BA} \leftrightarrow \overline{TF}$

$\measuredangle A \leftrightarrow \measuredangle F$ $\qquad$ $\overline{AW} \leftrightarrow \overline{FK}$

$\measuredangle W \leftrightarrow \measuredangle K$ $\qquad$ $\overline{BW} \leftrightarrow \overline{TK}$

EXAMPLE 6.2 Rewrite the correspondence given in Example 6.1 so that the same pairing of elements is maintained.

SOLUTION

There is more than one way of writing the same correspondence including, $\triangle WAB \leftrightarrow \triangle KFT$.

CONGRUENT TRIANGLES

REMEMBER
If geometric figures have the same size and shape, they are congruent.

Congruence means "same size" and "same shape." Plane congruent figures can be made to coincide. Two segments, for example, can be made to coincide (assuming, of course, that we could move the segments) if they have the same length. Two angles can be made to coincide if they have the same shape, that is, if they have the same measure. In order for geometric *figures* to be congruent, they must have the same size *and* the same shape.

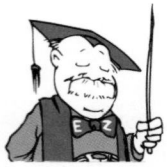

DEFINITION OF CONGRUENT TRIANGLES

Two triangles are *congruent* if their vertices can be paired so that two conditions are met:

1. All pairs of corresponding angles are congruent.
2. All pairs of corresponding sides are congruent.

The definition states that, in order for two triangles to be congruent, six pairs of parts must be congruent: three pairs of angles and three pairs of sides. A similar definition may be used for congruent polygons of any number of sides.

EXAMPLE 6.3 Are the following two triangles congruent? If they are, write the correspondence between the triangles that establishes the congruence.

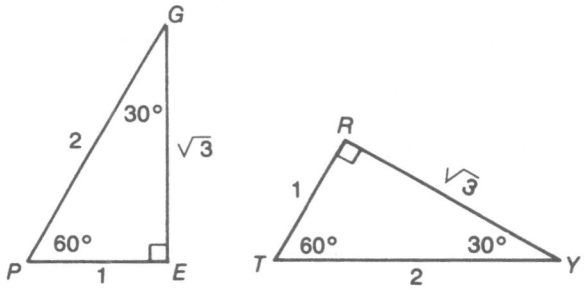

SOLUTION

Yes. $\triangle PEG \cong \triangle TRY$.

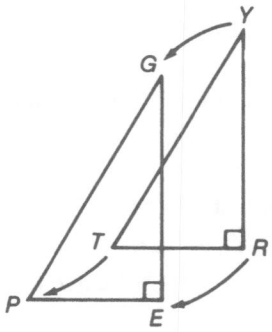

INCLUDED ANGLES AND SIDES

With respect to angles A and C in Figure 6.3, side $\overline{AC}$ is called an *included* side; or side $\overline{AC}$ is said to be *included* between angle A and angle C. With respect to angles B and C, $\overline{BC}$ is the included side. What is the included side between angles A and B?

The answer is $\overline{AB}$.

Since sides $\overline{AB}$ and $\overline{CB}$ intersect at B, angle B is called an *included angle*; or angle B is said to be *included* between side $\overline{AB}$ and side $\overline{CB}$. With respect to sides $\overline{BA}$ and $\overline{CA}$, angle A is the included angle. What angle is included between sides $\overline{BC}$ and AC? If you answer "angle C," you have the right idea.

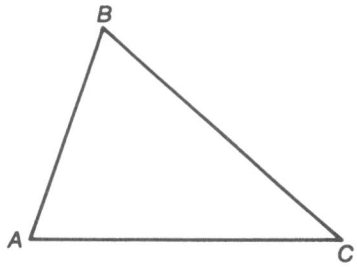

FIGURE 6.3

EXAMPLE 6.4 For the accompanying figure:

a. Use three letters to name the angle included between sides:
  (i) $\overline{RS}$ and $\overline{WS}$
  (ii) $\overline{SW}$ and $\overline{TW}$

b. Name the segment that is included between:
  (i) Angles WST and WTS
  (ii) Angles SWR and SRW

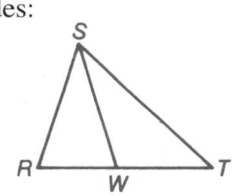

SOLUTION

a. **(i)** $\measuredangle RSW$ **(ii)** $\measuredangle TWS$ **b.** **(i)** $\overline{TS}$ **(ii)** $\overline{RW}$

Proving Triangles Congruent: SSS, SAS, and ASA Postulates

Much of our work in geometry will be devoted to trying to prove that two triangles are congruent, given certain facts about the triangles. Fortunately, there are several shortcut methods for proving triangles congruent. Rather than proving triangles congruent by demonstrating that they agree in *six* pairs of parts, it is possible to conclude that a pair of triangles are congruent if they agree in *three* pairs of parts, *provided that these parts are a particular set of three pairs of congruent angles and sides.*

TRIANGLES THAT AGREE IN THREE SIDES

Using a ruler, draw two triangles so that the three sides of the first triangle have the same length as the corresponding sides of the second triangle. For example, see Figure 6.4.

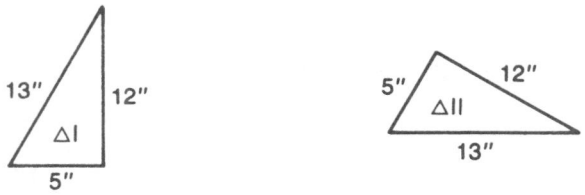

FIGURE 6.4

After the triangles are drawn, use a pair of scissors to cut out △II. You should be able to demonstrate that the two triangles can be made to coincide. In other words, if three sides of one triangle are congruent to three sides of another triangle, the corresponding angles of the triangles are "forced" to be congruent; therefore, the triangles are congruent. This is postulated as follows:

SIDE-SIDE-SIDE (SSS) POSTULATE

If the vertices of two triangles can be paired so that three sides of one triangle are congruent to the corresponding sides of the second triangle, then the two triangles are congruent.

EXAMPLE 6.5

GIVEN: $\overline{AB} \cong \overline{BC}$,
M is the midpoint of $\overline{AC}$.

PROVE: $\triangle AMB \cong \triangle CMB$.

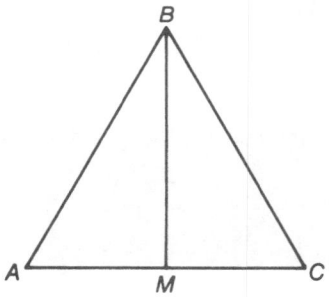

SOLUTION

PLAN:
1. Mark the diagram with the Given. Note that an "x" is used to indicate that side $\overline{BM}$ is congruent to itself
2. After the three pairs of congruent sides are identified, write the formal proof.

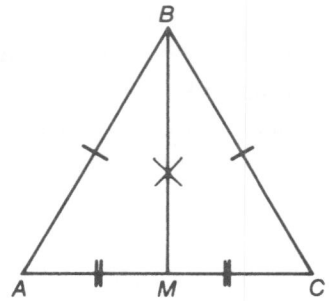

PROOF:

Statements	Reasons
1. $\overline{AB} \cong \overline{BC}$. (Side)	1. Given.
2. M is the midpoint of $\overline{AC}$.	2. Given.
3. $\overline{AM} \cong \overline{CM}$. (Side)	3. A midpoint divides a segment into two congruent segments.
4. $\overline{BM} \cong \overline{BM}$. (Side)	4. Reflexive property of congruence.
5. $\triangle AMB \cong \triangle CMB$.	5. SSS Postulate.

THE SIDE-ANGLE-SIDE (SAS) AND THE ANGLE-SIDE-ANGLE (ASA) POSTULATES

In addition to the SSS Postulate, several other shortcut methods can be used to prove triangles congruent. The Side-Angle-Side Postulate is illustrated in Figure 6.5. If it is known that two sides and the included angle of one triangle are congruent to the corresponding parts of another triangle, then we may conclude that the triangles are congruent. Because the congruent angle must be sandwiched in between the congruent pairs of sides, this method is commonly referred to as the SAS Postulate.

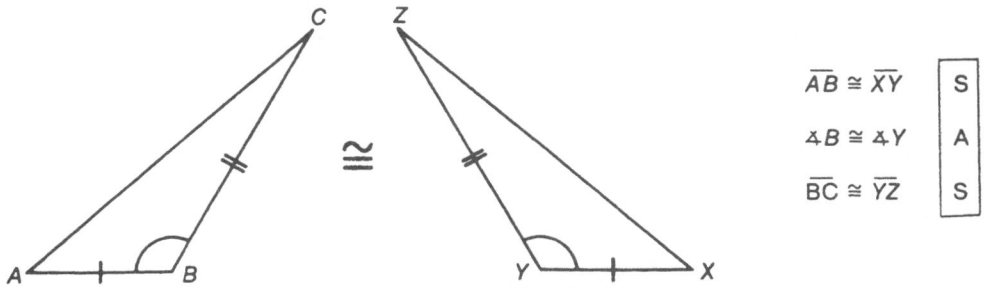

FIGURE 6.5

SIDE-ANGLE-SIDE (SAS) POSTULATE

If the vertices of two triangles can be paired so that two sides and the included angle of one triangle are congruent to the corresponding parts of the second triangle, then the two triangles are congruent.

Similarly, if two angles and the included side (ASA) of one triangle are congruent to the corresponding parts of another triangle, then the triangles are congruent. See Figure 6.6.

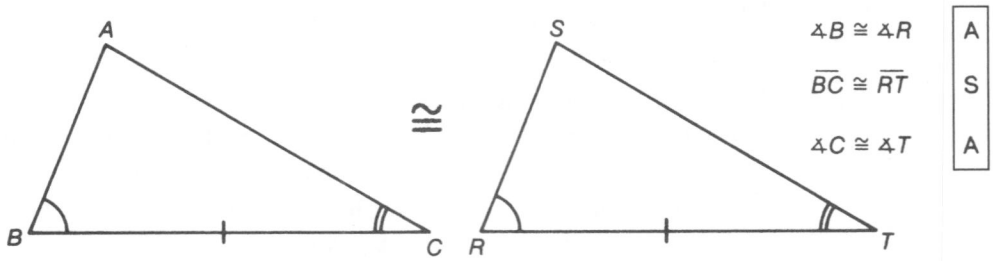

FIGURE 6.6

ANGLE-SIDE-ANGLE (ASA) POSTULATE

If the vertices of two triangles can be paired so that two angles and the included side of one triangle are congruent to the corresponding parts of the second triangle, then the two triangles are congruent.

EXAMPLE 6.6

GIVEN: C is the midpoint of $\overline{BE}$
$\measuredangle B \cong \measuredangle E$.

PROVE: $\triangle ABC \cong \triangle DEC$.

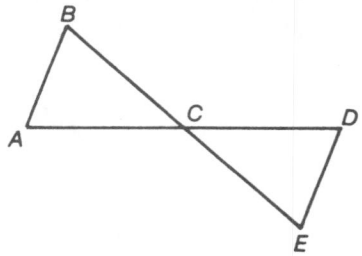

SOLUTION

PLAN:

1. Mark the diagram with the Given *and* any additional parts, such as the vertical angle pair, that can be deduced to be congruent based on the diagram.
2. Examine the diagram marked in Step 1 to decide which method of proving triangles congruent to use. In this case, use ASA.
3. Write the formal proof.

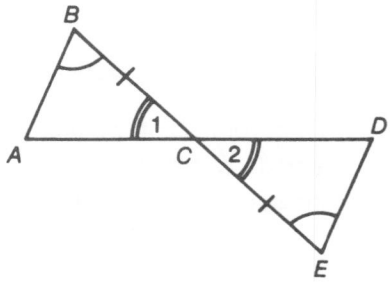

PROOF:

Statements	Reasons
1. $\measuredangle B \cong \measuredangle E$. (Angle)	1. Given.
2. C is the midpoint of $\overline{BE}$.	2. Given.
3. $\overline{BC} \cong \overline{EC}$. (Side)	3. A midpoint divides a segment into two congruent segments.
4. $\measuredangle 1 \cong \measuredangle 2$. (Angle)	4. Vertical angles are congruent.
5. $\triangle ABC \cong \triangle DEC$.	5. ASA Postulate.

EXAMPLE 6.7

GIVEN: $\overline{AB} \parallel \overline{CD}$, $\overline{AB} \cong \overline{CD}$

PROVE: $\triangle DAB \cong \triangle BCD$.

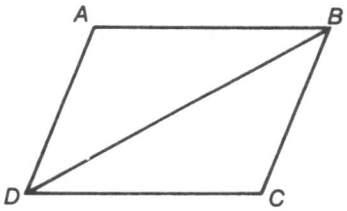

SOLUTION

PLAN:
1. Mark the diagram.
2. Decide on the method to use. In this example, SAS is appropriate.
3. Write the formal proof.

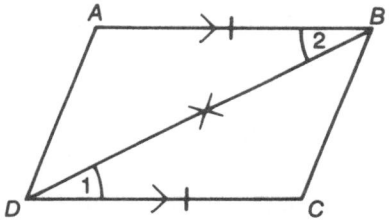

PROOF:

Statements	Reasons
1. $\overline{AB} \cong \overline{CD}$. (Side)	1. Given.
2. $\overline{AB} \parallel \overline{CD}$.	2. Given.
3. $\measuredangle 1 \cong \measuredangle 2$. (Angle)	3. If two lines are parallel, then their alternate interior angles are congruent.
4. $\overline{BD} \cong \overline{BD}$. (Side)	4. Reflexive property of congruence.
5. $\triangle DAB \cong \triangle BCD$.	5. SAS Postulate.

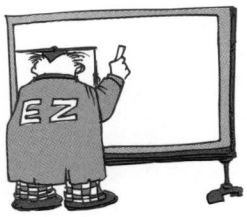

TO PROVE TWO TRIANGLES CONGRUENT

1. Mark the diagram with the Given.
2. Mark any additional parts that are congruent such as vertical angles, or sides (or angles) shared by both triangles.
3. If helpful, label angles with numbers. This will make the job of writing the proof easier.
4. Decide which method of proving triangles congruent to use.

5. Write the formal proof. Next to each statement in which a required side or angle is established as being congruent, write "(Side)" or "(Angle)." When you look back at your proof, this notation will help you verify that you have satisfied the necessary conditions of the congruence postulate being used.

Proving Overlapping Triangles Congruent

Sometimes proofs appear to be more complicated than they actually are simply because the triangles are drawn so that one overlaps the other. For example, consider the following problem:

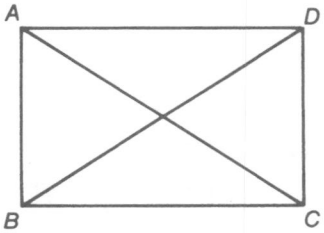

GIVEN: $\overline{AB} \cong \overline{DC}$,
$\overline{AB} \perp \overline{BC}$ $\overline{DC} \perp \overline{BC}$.

PROVE: $\triangle ABC \cong \triangle DCB$.

Often the analysis of the problem can be made easier by using pencils of different colors to outline the triangles to be proved congruent. Alternatively, it is sometimes helpful to redraw the diagram, "sliding" the triangles apart and then proceeding to mark the diagram:

PLAN: Show $\triangle ABC \cong \triangle DCB$ by using SAS. Notice that $\overline{BC}$ is a side of both triangles. Overlapping triangles often share the same side or angle.

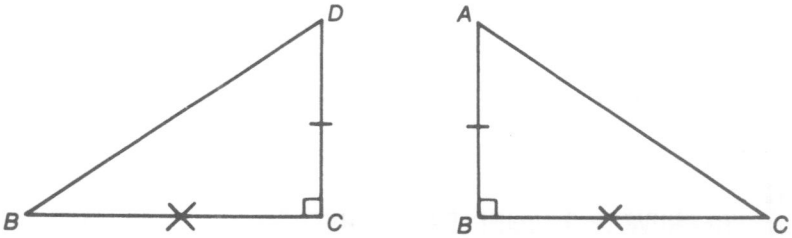

EXAMPLE 6.8

GIVEN: $\overline{RA} \perp \overline{PQ}$, $\overline{QB} \perp \overline{PR}$, $\overline{PA} \cong \overline{PB}$.

PROVE: $\triangle PAR \cong \triangle PBQ$.

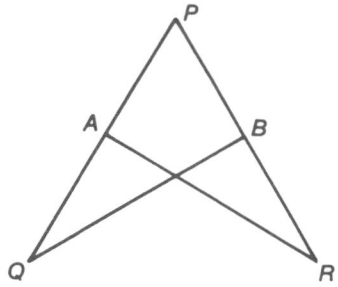

SOLUTION

PLAN: 1. "Separate" the triangles, and mark corresponding congruent parts:

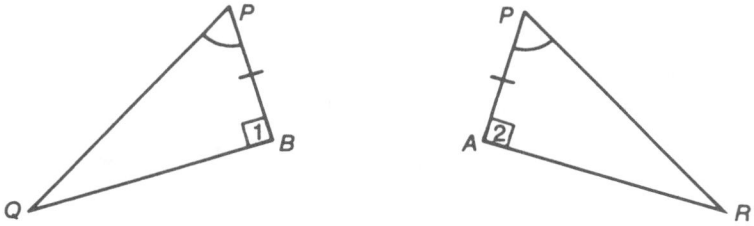

2. Note that angle P is common to both triangles.
3. Prove triangles congruent by ASA.

PROOF:

Statements	Reasons
1. $\measuredangle P \cong \measuredangle P$. (Angle)	1. Reflexive property of congruence.
2. $\overline{PA} \cong \overline{PB}$. (Side)	2. Given
3. $\overline{RA} \perp \overline{PQ}$ and $\overline{QB} \perp \overline{PR}$.	3. Given.
4. Angles 1 and 2 are right angles.	4. Perpendicular lines intersect to form right angles.
5. $\measuredangle 1 \cong \measuredangle 2$. (Angle)	5. All right angles are congruent.
6. $\triangle PAR \cong \triangle PBQ$.	6. ASA Postulate.

Sometimes the addition or subtraction property must be applied to obtain a pair of congruent corresponding parts that will be needed in order to establish the congruence of a pair of triangles. Recall that we may only add or subtract the *measures* of segments or angles. To reduce the number of steps in a proof, we will freely convert from congruence to measure (in order to perform the arithmetic operation) and then convert back to congruence. This procedure is illustrated in Steps 3 and 6 of the following example.

EXAMPLE 6.9

GIVEN: $\overline{AB} \cong \overline{DE}$, $\overline{AD} \cong \overline{FC}$, $\overline{AB} \parallel \overline{DE}$.

PROVE: $\triangle ABC \cong \triangle DEF$.

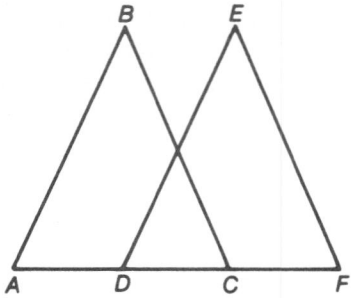

SOLUTION

PLAN: Mark the diagram with the Given.
NOTE: We may use SAS, *but* we must first establish within the proof that $\overline{AC} \cong \overline{DF}$ by adding the measure of $\overline{DC}$ to AD and CF.

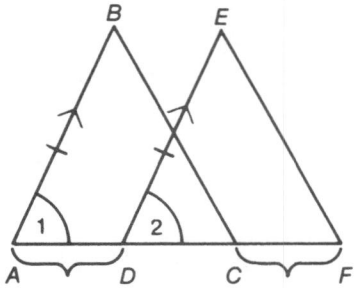

PROOF:

Statements	Reasons
1. $\overline{AB} \cong \overline{DE}$ (Side)	1. Given.
2. $\measuredangle 1 \cong \measuredangle 2$. (Angle)	2. If two lines are parallel, then their corresponding angles are congruent.
3. $AD = FC$.	3. Given.
4. $DC = DC$.	4. Reflexive property of equality.
5. $AD + DC = FC + DC$.	5. Addition property of equality.
6. $\overline{AC} \cong \overline{FD}$. (Side)	6. Substitution.
7. $\Delta ABC \cong \Delta DEF$.	7. SAS Postulate.

EXAMPLE 6.10

GIVEN: $\measuredangle ABE \cong \measuredangle CBD$, $\measuredangle BDE \cong \measuredangle BED$, $\overline{BD} \cong \overline{BE}$.

PROVE: $\triangle DAB \cong \triangle ECB$.

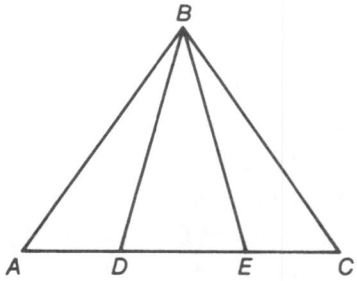

SOLUTION

PLAN: Mark the diagram. NOTE: We may use ASA, but we must establish within the proof that $\measuredangle 3 \cong \measuredangle 4$ by subtracting the measure of angle *DBE* from the measures of angles *ABE* and *CBD*.

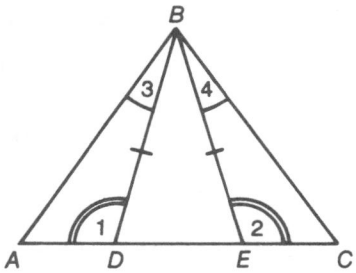

PROOF:

Statements	Reasons
1. $\measuredangle BDE \cong \measuredangle BED$.	1. Given.
2. $\measuredangle 1 \cong \measuredangle 2$. (Angle)	2. Supplements of congruent angles are congruent.
3. $\overline{BD} \cong \overline{BE}$. (Side)	3. Given.
4. $m\measuredangle ABE = m\measuredangle CBD$.	4. Given.
5. $m\measuredangle DBE = m\measuredangle DBE$.	5. Reflexive property of equality.
6. $m\measuredangle ABE - m\measuredangle DBE = m\measuredangle CBD - m\measuredangle DBE$.	6. Subtraction property of equality.
7. $\measuredangle 3 \cong \measuredangle 4$. (Angle)	7. Substitution.
8. $\triangle DAB \cong \triangle ECB$.	8. ASA Postulate.

Proving Triangles Congruent: AAS and Hy-Leg Methods

Consider the following pair of triangles with the corresponding pairs of congruent parts already marked:

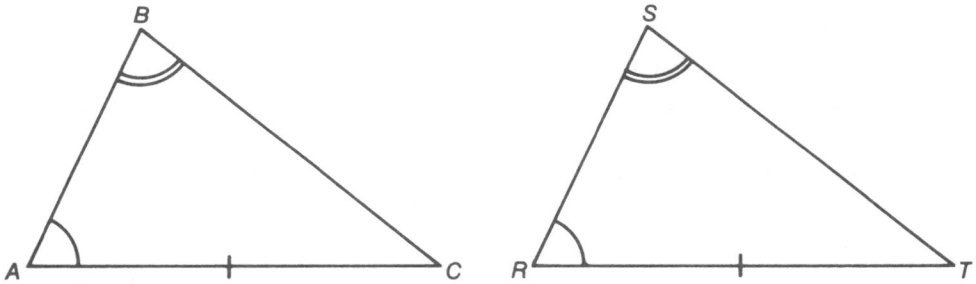

Are the triangles congruent? The triangles agree in two angles and the side *opposite* one of them (AAS), so SSS, SAS, or ASA cannot be directly applied in establishing the congruence of the triangles. However, by Corollary 5.2.4, the third pair of angles of the triangles must be congruent. It follows that the triangles are congruent by application of the ASA Postulate. Thus, angle-angle-side $\cong$ angle-angle-side means that the two triangles are congruent.

ANGLE-ANGLE-SIDE (AAS) THEOREM

If the vertices of two triangles can be paired so that two angles and the side opposite one of them in one triangle are congruent to the corresponding parts of the second triangle, then the two triangles are congruent.

EXAMPLE 6.11

GIVEN: $\overline{AB} \perp \overline{BD}$, $\overline{AC} \perp \overline{CD}$, $\measuredangle 1 \cong \measuredangle 2$.

PROVE: $\triangle ABD \cong \triangle ACD$.

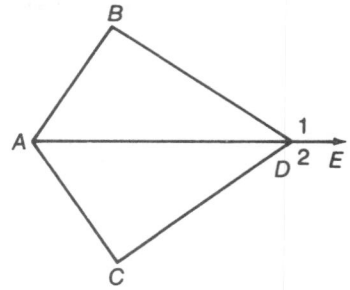

SOLUTION

OUTLINE OF PROOF: Angles 3 and 4 are congruent since they are supplements of congruent angles. Angles 5 and 6 are congruent since all right angles are congruent. The triangles are, therefore, congruent by the AAS Theorem.

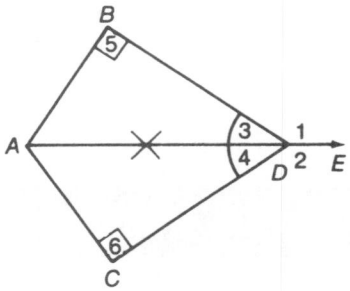

Two definitions related to a right triangle are necessary before we consider the next (and final) method for proving triangles congruent.

In a right triangle, the side opposite the right angle is called the *hypotenuse* and each of the two remaining sides is called a leg. Notice in Figure 6.7 that the legs are perpendicular to each other.

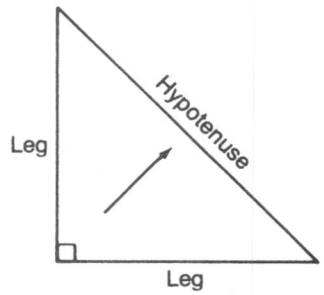

FIGURE 6.7

Next, consider two right triangles in which the hypotenuses and one of the legs (Figure 6.8) have the same lengths. This information is sufficient to conclude that the two *right* triangles are congruent. This way of proving triangles congruent is called the Hypotenuse-Leg (Hy-Leg) method.

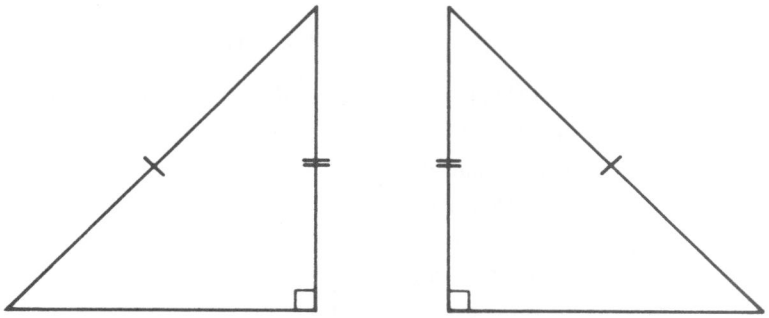

FIGURE 6.8

HYPOTENUSE-LEG (HY-LEG) POSTULATE

If the vertices of two *right* triangles can be paired so that the hypotenuse and leg of one triangle are congruent to the corresponding parts of the second *right* triangle, then the two *right* triangles are congruent.

This method is applicable only to pairs of *right* triangles. Before the Hy-Leg Postulate can be applied, you need to know that the triangles involved are right triangles.

EXAMPLE 6.12

GIVEN: $\overline{AB} \perp \overline{BC}$, $\overline{AD} \perp \overline{DC}$, $\overline{AB} \cong \overline{AD}$.

PROVE: $\triangle ABC \cong \triangle ADC$.

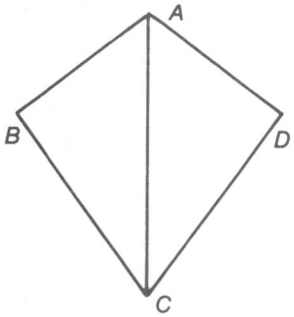

SOLUTION

PLAN: 1. Mark the diagram.
2. Use the Hy-Leg Postulate.

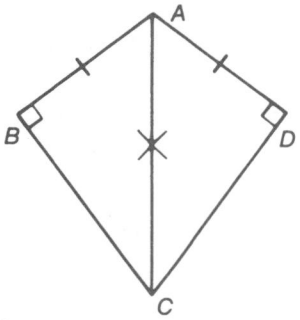

PROOF:

Statements	Reasons
1. $\overline{AC} \cong \overline{AC}$. (Hy)	1. Reflexive property of congruence.
2. $\overline{AB} \cong \overline{AD}$. (Leg)	2. Given.
NOTE: Before we can use the Hy-Leg Postulate we must establish that triangles *ABC* and *ADC* are right triangles.	
3. $\overline{AB} \perp \overline{BC}$ $\overline{AD} \perp \overline{DC}$	3. Given.
4. Angles *B* and *D* are right angles.	4. Perpendicular lines intersect to form right angles.
5. Triangles *ABC* and *ADC* are right triangles.	5. A triangle that contains a right angle is a right triangle.
6. $\triangle ABC \cong \triangle ADC$.	6. Hy-Leg Postulate.

SUMMARY

- You may conclude that two triangles are congruent if it can be shown that:

1. Three sides of one triangle are congruent to the corresponding parts of the other triangle.

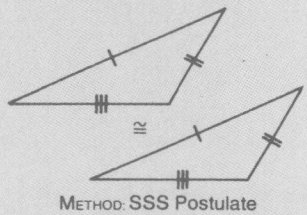

METHOD: SSS Postulate

2. Two sides and the included angle of one triangle are congruent to the corresponding parts of the other triangle.

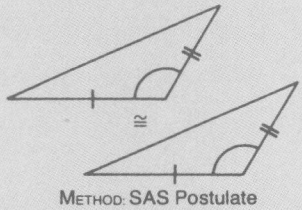

METHOD: SAS Postulate

SUMMARY (*Continued*)

3. Two angles and the included side of one triangle are congruent to the corresponding parts of the other triangle.

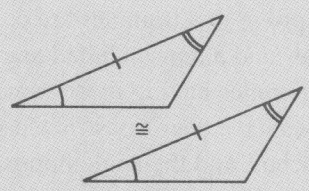

METHOD: ASA Postulate

4. Two angles and the side opposite one of them are congruent to the corresponding parts of the other triangle.

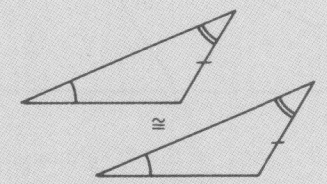

METHOD: AAS Theorem

- You may conclude that two right triangles are congruent if the hypotenuse and either leg of one triangle are congruent to the corresponding parts of the other right triangle.

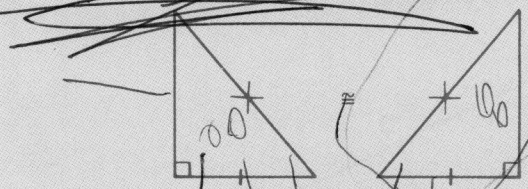

METHOD: Hy-Leg Postulate

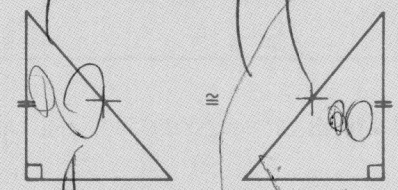

METHOD: Hy-Leg Postulate

When Two Triangles Are *NOT* Congruent

When *ASA*, *SAS*, *SSS*, *AAS*, or *HL* measurements are given, exactly one triangle can be constructed using those measurements. For that reason, when two triangles agree in those measurements, the two triangles must be congruent.

- Do *not* use $SSA \cong SSA$ to prove two triangles are congruent.
  If the measures of two sides and a non-included angle (*SSA*) are given, there may exist two non-congruent triangles having those same measurements. In Figure 6.9, two non-congruent triangles have been constructed given that the measures of two sides are 3 inches and 5 inches, and the angle opposite the 3-inch side measures 30°.

REMEMBER
Do NOT use SSA ≅ SSA or AAA ≅ AAA to prove two triangles are congruent.

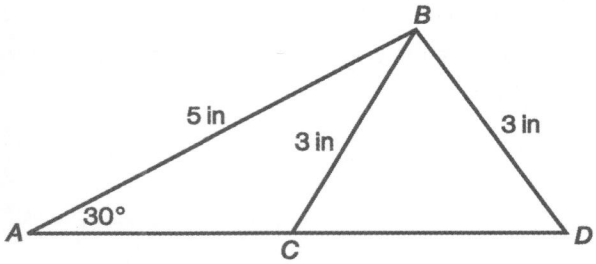

FIGURE 6.9 $\triangle ABC \ncong \triangle ABD$ although $SSA \cong SSA$.

- Do *not* use $AAA \cong AAA$ to prove two triangles are congruent.

Given the measures of the three angles of a triangle, it is possible to construct two non-congruent triangles with those measurements. In Figure 6.10, $\overline{AB} \parallel \overline{DE}$. Because $\angle A \cong \angle 1$, $\angle B \cong \angle 2$, and $\angle C \cong \angle C$ triangles *ABC* and *DEC* agree in all of their angle measurements yet are *not* congruent.

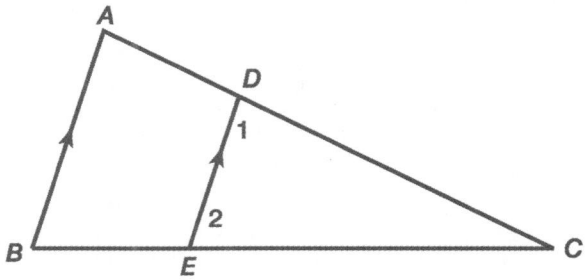

FIGURE 6.10 $\triangle ABC \ncong \triangle DEC$ although $AAA \cong AAA$.

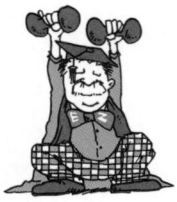

REVIEW EXERCISES FOR CHAPTER 6

1. In the accompanying diagram, $\overline{HK}$ bisects $\overline{IL}$ and $\angle H \cong \angle K$. What is the most direct approach that could be used to prove $\triangle HIJ \cong \triangle KLJ$?

   (1) $HL \cong HL$ (2) $SAS \cong SAS$ (3) $AAS \cong AAS$ (4) $ASA \cong ASA$

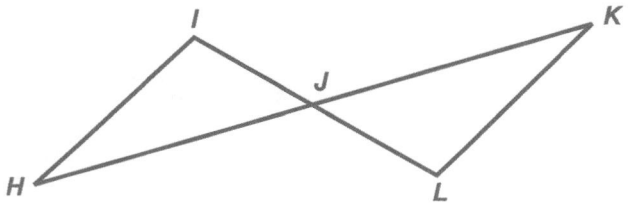

2. In the accompanying diagram, $\overline{CA} \perp \overline{AB}$, $\overline{ED} \perp \overline{DF}$, $\overline{ED} \parallel \overline{AB}$, $\overline{CE} \cong \overline{BF}$, $\overline{AB} \cong \overline{ED}$, and $m\angle CAB = m\angle FDE = 90$. Which statement would *not* be used to prove $\triangle ABC \cong \triangle DEF$?

   (1) $SSS \cong SSS$ (2) $SAS \cong SAS$ (3) $AAS \cong AAS$ (4) $HL \cong HL$

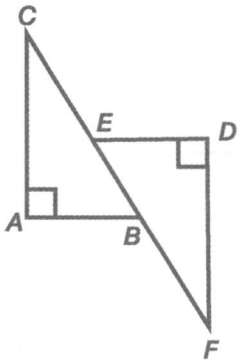

3. GIVEN: $\overline{BM} \perp \overline{AC}$, M is the midpoint of $\overline{AC}$.
   PROVE: $\triangle ABM \cong \triangle CBM$.

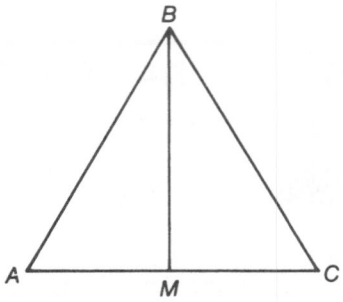

4. GIVEN: $\overline{RT}$ bisects angles STW and SRW.
   PROVE: $\triangle RST \cong \triangle RWT$.

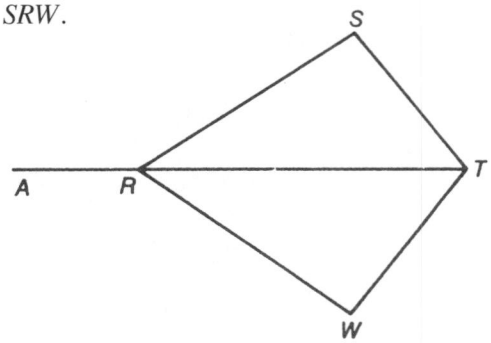

5. GIVEN: $\overline{EF} \parallel \overline{AB}$, $\overline{ED} \parallel \overline{BC}$,
   $\overline{AD} \cong \overline{FC}$.
   PROVE: $\triangle ABC \cong \triangle FED$.

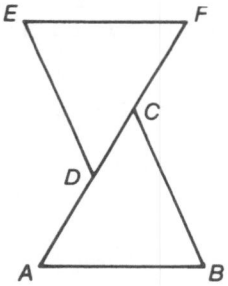

Use the accompanying diagram to solve Exercises 6 and 7.

6. GIVEN: $\measuredangle R \cong \measuredangle T$,
   $\overline{SR} \cong \overline{ST}$.
   PROVE: $\triangle SRH \cong \triangle STE$.

7. GIVEN: $\overline{TE} \perp \overline{RS}$, $\overline{RH} \perp \overline{ST}$,
   $\overline{EW} \cong \overline{HW}$.
   PROVE: $\triangle EWR \cong \triangle HWT$.

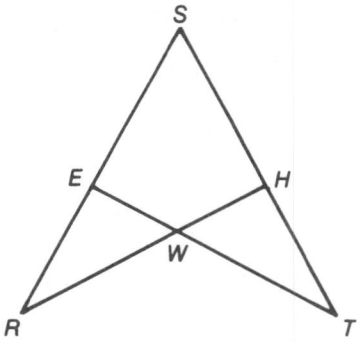

8. GIVEN: $\overline{QU} \parallel \overline{DA}, \overline{QU} \cong \overline{DA}$.
PROVE: $\triangle UXQ \cong \triangle DXA$.

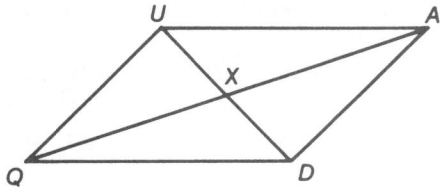

Use the accompanying diagram to solve Exercises 9 and 10.

9. GIVEN: $\overline{JK} \perp \overline{KT}, \overline{ET} \perp \overline{KT}$,
$\overline{KV} \cong \overline{TL}, \overline{JL} \cong \overline{EV}$.
PROVE: $\triangle JKL \cong \triangle ETV$.

10. GIVEN: $\overline{JK} \perp \overline{KT}, \overline{ET} \perp \overline{KT}$,
$\measuredangle 1 \cong \measuredangle 2, \overline{KL} \cong \overline{TV}$.
PROVE: $\triangle \overline{JKL} \cong \triangle \overline{ETV}$.

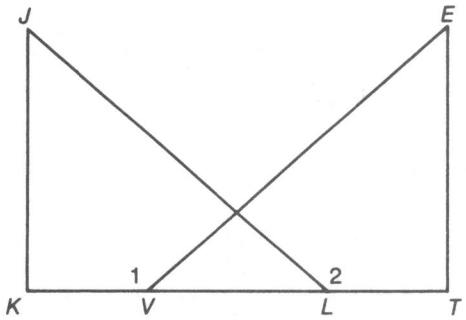

11. GIVEN: $\overline{AR}$ bisects $\measuredangle FRI$,
$\measuredangle 1 \cong \measuredangle 2$,
$\measuredangle RFI \cong \measuredangle RIF$.
PROVE: $\triangle AFR \cong \triangle AIR$.

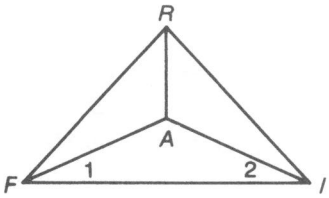

Use the accompanying diagram to solve Exercises 12 and 13.

12. GIVEN: S is the midpoint of $\overline{RT}$,
$\overline{SW} \cong \overline{SP}$ $\measuredangle RSW \cong \measuredangle TSP$.
PROVE: $\triangle TSW \cong \triangle SRP$.

13. GIVEN: $\overline{TW} \cong \overline{SP}, \overline{RP} \parallel \overline{SW}$,
$\overline{SP} \parallel \overline{TW}$.
PROVE: $\triangle TSW \cong \triangle SRP$.

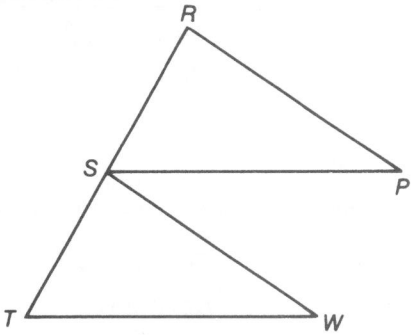

Use the accompanying diagram to solve Exercises 14 and 15.

14. GIVEN: $\overline{AB} \cong \overline{DE}$ M is the midpoint of $\overline{BE}$,
$\overline{AM} \cong \overline{KM}, \overline{DM} \cong \overline{KM}$.
PROVE: $\triangle ABM \cong \triangle DEM$.

15. GIVEN: $\overline{KM}$ is the $\perp$ bisector of $\overline{BE}$,
$\overline{KM}$ bisects $\measuredangle AMD$,
$\overline{AB} \parallel \overline{KM} \parallel \overline{DE}$.
PROVE: $\triangle ABM \cong \triangle DEM$.

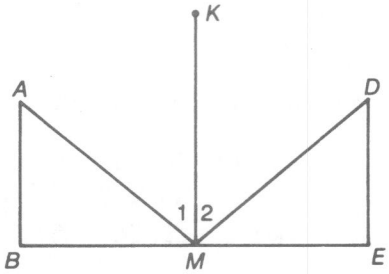

Use the accompanying diagram to solve Exercises 16 and 17.

16. GIVEN: $\overline{BD} \perp \overline{AC}, \overline{AF} \perp \overline{BC}$,
$\overline{FC} \cong \overline{DC}$.
PROVE: $\triangle AFC \cong \triangle BDC$.

17. GIVEN: $\overline{AD} \cong \overline{BF}$,
$\measuredangle BAD \cong \measuredangle ABF$.
PROVE: $\triangle BAD \cong \triangle ABF$.

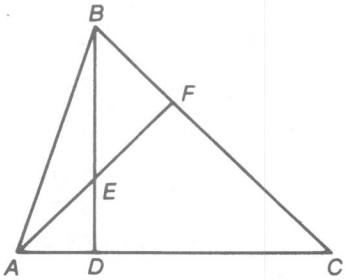

18. GIVEN: $\overline{BF} \perp \overline{AC}, \overline{DE} \perp \overline{AC}$,
$\overline{AB} \cong \overline{DC}, \overline{AE} \cong \overline{CF}$.
PROVE: $\triangle AFB \cong \triangle CED$.

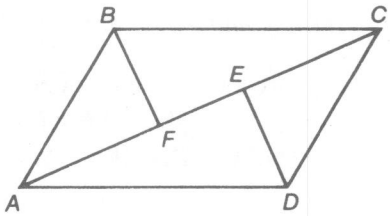

Use the accompanying diagram to solve Exercises 19 and 20.

19. GIVEN: $\overline{JK} \perp \overline{KG}, \overline{KL} \perp \overline{JG}$,
$\overline{KL} \cong \overline{JR}$.
PROVE: $\triangle KLG \cong \triangle JRG$.

20. GIVEN: $\overline{KG} \cong \overline{JG}, \overline{JR} \perp \overline{KG}, \overline{KL} \perp \overline{JG}$
R is the midpoint of $\overline{KG}$,
L is the midpoint of $\overline{JG}$.
PROVE: $\triangle KOR \cong \triangle JOL$.

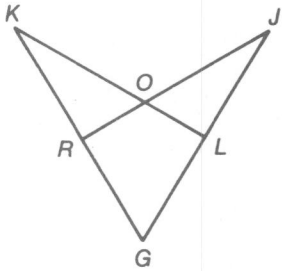

21. GIVEN: $\overline{AB} \perp \overline{BC}, \overline{DC} \perp \overline{BC}$
$\overline{DB}$ bisects $\measuredangle ABC$,
$\overline{AC}$ bisects $\measuredangle DCB$,
$\overline{EB} \cong \overline{EC}$
PROVE: $\triangle BEA \cong \triangle CED$.

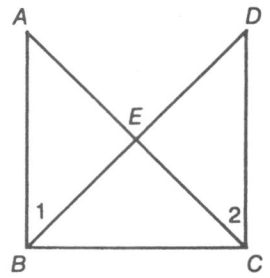

22. Prove that, if two triangles are congruent to the same triangle, then they are congruent to each other.

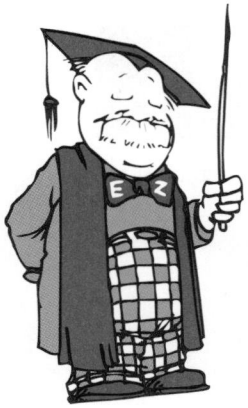

7

Applying Congruent Triangles

WHAT YOU WILL LEARN

You can prove that two segments or two angles are congruent if you can first show that these pairs of segments or angles are contained in congruent triangles. In this chapter you will learn:

- the way to prove pairs of segments or angles congruent using congruent triangles;
- the difference between drawing segments determined, overdetermined, and underdetermined;
- that every triangle has three altitudes and three medians;
- that the base angles of an isosceles triangle are congruent, and, conversely, the sides opposite congruent angles of a triangle are congruent;
- the way to use one pair of congruent triangles to prove a second pair of triangles congruent.

SECTIONS IN THIS CHAPTER

- Using Congruent Triangles to Prove Segments and Angles Congruent
- Using Congruent Triangles to Prove Special Properties of Lines
- Classifying Triangles and Special Segments
- The Isosceles Triangle
- Double Congruence Proofs

Using Congruent Triangles to Prove Segments and Angles Congruent

Based on the information marked in Figure 7.1, what conclusion can be drawn about how the measures of angles R and H compare?

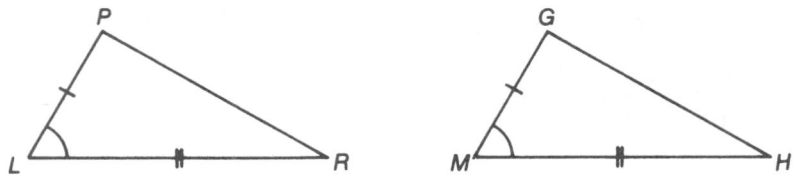

FIGURE 7.1

At first glance, there may seem to be insufficient information to draw any conclusion about these angles. However, by the SAS Postulate, $\triangle PLR \cong \triangle GMH$. We need to recall the *definition* of congruent triangles: If two triangles have all *six* pairs of corresponding parts congruent, then the triangles are congruent. In this case, since we know that the two triangles are congruent, we may use the reverse of the definition of congruent triangles to conclude that *all* pairs of corresponding sides and *all* pairs of corresponding angles must be congruent. Since angles R and H are corresponding angles, they must be congruent.

This reasoning represents an extremely useful method for proving segments or angles congruent, provided that they are parts of triangles that can be proved congruent. *After* a pair of triangles are proved congruent, then *any* pair of corresponding parts may be stated to be congruent, based on the principle that *c*orresponding *p*arts of *c*ongruent *t*riangles are *c*ongruent, abbreviated as CPCTC.

EXAMPLE 7.1

GIVEN: $\overline{AB} \cong \overline{AD}$,
$\overline{BC} \cong \overline{DC}$.

PROVE: $\measuredangle B \cong \measuredangle D$.

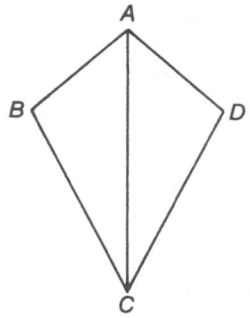

SOLUTION

PLAN: Angles B and D are corresponding parts of triangles ABC and ADC, respectively. After proving these triangles congruent, we may then conclude that the desired pair of angles are congruent. Marking the diagram suggests that the triangles can be proved congruent by the SSS Postulate.

PROOF:

Statements	Reasons
1. $\overline{AB} \cong \overline{AD}$ and $\overline{BC} \cong \overline{DC}$	1. Given
2. $\overline{AC} \cong \overline{AC}$	2. Reflexive property of congruence.
3. $\triangle ABC \cong \triangle ADC$.	3. SSS Postulate.
4. $\measuredangle B \cong \measuredangle D$.	4. CPCTC.

EXAMPLE 7.2

GIVEN: $\overline{AB} \cong \overline{DC}$,
$\overline{AB} \perp \overline{BC}, \overline{DC} \perp \overline{BC}$.

PROVE: $\overline{AC} \cong \overline{DB}$

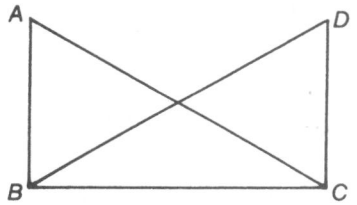

SOLUTION

PLAN: $\overline{AC}$ and $\overline{DB}$ are sides of triangles ABC and DCB, respectively. After proving these triangles congruent, we may then conclude that the desired pair of segments are congruent. Marking the diagram suggests that the SAS method be applied.

PROOF:

Statements	Reasons
1. $\overline{AB} \cong \overline{DC}$. (Side)	1. Given.
2. $\overline{AB} \perp \overline{BC}$ and $\overline{DC} \perp \overline{BC}$.	2. Given.
3. Angles ABC and DCB are right angles.	3. Perpendicular lines intersect to form right angles.
4. $\measuredangle ABC \cong \measuredangle DCB$. (Angle)	4. All right angles are congruent.
5. $\overline{BC} \cong \overline{BC}$. (Side)	5. Reflexive property of congruence.
6. $\triangle ABC \cong \triangle DCB$.	6. SAS Postulate.
7. $\overline{AC} \cong \overline{DB}$	7. CPCTC.

As the diagrams and problems become more complicated, we apply the following four-step procedure.

TO PROVE SEGMENTS AND/OR ANGLES CONGRUENT USING CONGRUENT TRIANGLES

1. IDENTIFY. Identify the pair of triangles that contain the parts that need to be proved congruent.
2. PLAN. Plan how to prove the selected pair of triangles congruent by first marking the diagram with the Given. Then mark any additional pairs of parts that may be congruent as a result of vertical angles, perpendicular lines, parallel lines, supplements (complements) of the same (or congruent) angles, or a common angle or side.
3. SELECT. Select the method of congruence to be used.
4. WRITE. Write the proof.

The next example illustrates this procedure.

EXAMPLE 7.3

GIVEN: $\overline{MP} \cong \overline{ST}$,
$\overline{MP} \parallel \overline{ST}$,
$\overline{PL} \cong \overline{RT}$.

PROVE: $\overline{RS} \parallel \overline{LM}$.

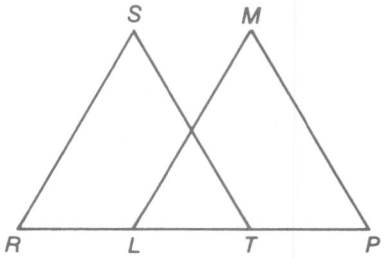

SOLUTION

To prove a pair of line segments parallel, we must usually prove that an appropriate pair of angles are congruent.

1. Looking at the diagram, we see that, if $\overline{RS}$ is to be parallel to $\overline{LM}$, we must prove that $\measuredangle SRT \cong \measuredangle MLP$. This implies that the triangles that contain these angles must be proved congruent. Therefore, we must prove $\triangle RST \cong \triangle LMP$.
2. Mark the diagram with the Given.
3. Use the SAS Postulate.
4. Write the proof.

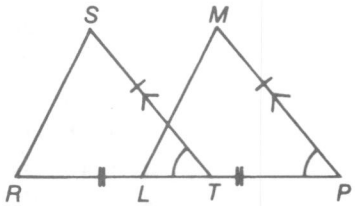

PROOF:

Statements	Reasons
1. $\overline{MP} \cong ST$. (Side)	1. Given.
2. $\overline{MP} \parallel \overline{ST}$	2. Given.
3. $\measuredangle MPL \cong \measuredangle STR$. (Angle)	3. If two lines are parallel, then their corresponding angles are congruent.
4. $\overline{PL} \cong \overline{RT}$. (Side)	4. Given.
5. $\triangle RST \cong \triangle LMP$.	5. SAS Postulate.
6. $\measuredangle SRT \cong \measuredangle MLP$.	6. CPCTC.
7. $\overline{RS} \parallel \overline{LM}$.	7. If two lines have their corresponding angles congruent, then they are parallel.

Notice that our approach is based on developing a plan that works backward, beginning with the Prove. Once the solution path is clear, we write the proof, working forward.

Using Congruent Triangles to Prove Special Properties of Lines

In the previous section we illustrated how congruent triangles can be used to prove a pair of lines are parallel. Similarly, by first showing that an appropriate pair of segments or angles are congruent, a line may be demonstrated to be an angle or segment bisector or a pair of lines may be proved to be perpendicular.

To prove that a line bisects an angle or segment, we must show that the line divides the angle or segment into two congruent parts. A pair of lines may be proved to be perpendicular by showing either of the following:

- The lines intersect to form right angles.
- The lines intersect to form a pair of congruent adjacent angles.

For example, from the accompanying diagram, we may conclude that:

- $\overline{BX}$ bisects $\overline{AC}$ by first proving that $\overline{AX} \cong \overline{XC}$
- $\overline{BX}$ bisects $\measuredangle ABC$ by first proving that $\measuredangle 1 \cong \measuredangle 2$.
- $\overline{BX} \perp \overline{AC}$ by first proving that $\measuredangle 3 \cong \measuredangle 4$ (that is, by showing a pair of adjacent angles are congruent).

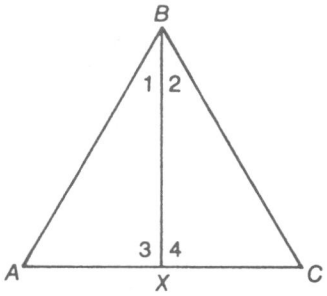

EXAMPLE 7.4

GIVEN: $\measuredangle 1 \cong \measuredangle 2$,
$\overline{RM} \cong \overline{TM}$.

PROVE: $\overline{SM}$ bisects $\measuredangle RST$.

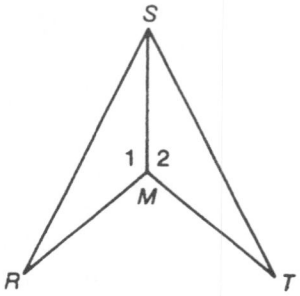

SOLUTION

PLAN: Our goal is to prove $\measuredangle RSM \cong \measuredangle TSM$ by proving $\triangle RSM \cong \triangle TSM$. Marking the diagram suggests that the SAS Postulate be applied.

PROOF:

Statements	Reasons
1. $\overline{RM} \cong \overline{TM}$. (Side)	1. Given.
2. $\measuredangle 1 \cong \measuredangle 2$. (Angle)	2. Given.
3. $\overline{SM} \cong \overline{SM}$. (Side)	3. Reflexive property of congruence.
4. $\triangle RSM \cong \triangle TSM$.	4. SAS Postulate.
5. $\measuredangle RSM \cong \measuredangle TSM$.	5. CPCTC.
6. $\overline{SM}$ bisects $\measuredangle RST$.	6. A segment that divides an angle into two congruent angles is an angle bisector. (NOTE. This is the *reverse* of the definition of angle bisector.)

EXAMPLE 7.5

GIVEN: $\overline{LH} \cong \overline{LN}$ $\overline{LB}$ bisects $\overline{HN}$.

PROVE: $\overline{LB} \perp \overline{HN}$.

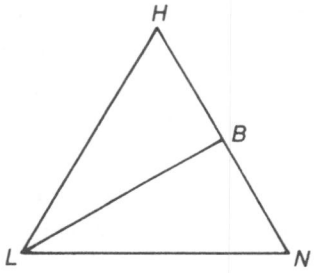

SOLUTION

PLAN: Show $\measuredangle LBH \cong \measuredangle LBN$ by showing $\triangle LBH \cong \triangle LBN$ using SSS.

PROOF:

Statements	Reasons
1. $\overline{LH} \cong \overline{LN}$. (Side)	1. Given.
2. $\overline{LB}$ bisects $\overline{HN}$.	2. Given.
3. $\overline{BH} \cong \overline{BN}$. (Side)	3. A bisector divides a segment into two congruent segments.
4. $\overline{LB} \cong \overline{LB}$. (Side)	4. Reflexive property of congruence.
5. $\triangle LBH \cong \triangle LBN$.	5. SSS Postulate.
6. $\measuredangle LBH \cong \measuredangle LBN$.	6. CPCTC.
7. $\overline{LB} \perp \overline{HN}$.	7. If two lines intersect to form congruent adjacent angles, then the lines are perpendicular.

SUMMARY

- To prove a line bisects a segment (or an angle) show that it divides the segment (or angle) into two congruent segments (or angles).
- To prove a line is perpendicular to another line show that the lines meet to form right angles or, equivalently, that a pair of adjacent angles are congruent.

Classifying Triangles and Special Segments

In addition to classifying a triangle as acute, right, or obtuse by the measures of its angles, we may classify a triangle according to the number of its sides that are congruent. See Figure 7.2.

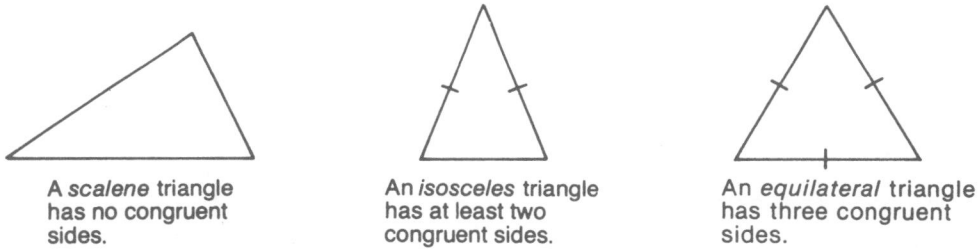

FIGURE 7.2

Notice that an equilateral triangle is also isosceles. Some further definitions regarding the parts of an isosceles triangle are given in Figure 7.3.

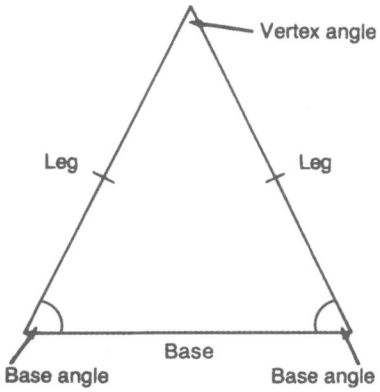

FIGURE 7.3 Legs are the congruent sides; vertex angle is the angle included between the legs; base is the side opposite the vertex angle; base angles are the angles that include the base and lie opposite the legs.

DRAWING AUXILIARY LINES (SEGMENTS)

To be able to complete a proof, it may be necessary to draw another line. This "extra" line is sometimes referred to as an *auxiliary line*. For example, an auxiliary line is needed to prove that the sum of the measures of the angles of a triangle is 180. Recall that this proof requires that a line be drawn that satisfies two conditions: it passes through a vertex of the given triangle, and at the same time it is parallel to the side opposite this vertex.

Sometimes it may be necessary to draw other types of auxiliary lines. Given one or more conditions, is it always possible to draw an auxiliary line that satisfies these conditions? If so, how many different lines can be drawn that satisfy the given set of conditions?

A particular set of geometric conditions may make the existence of a desired auxiliary line (or segment) *determined, underdetermined*, or *overdetermined*. An auxiliary line is said to be:

- *Determined* if exactly one line can be drawn that satisfies the given set of conditions. For example, an auxiliary line drawn so that it bisects a given angle is determined since every angle has exactly one bisector. Figure 7.4 illustrates that, through a point not on a line segment, a line can be drawn parallel to the segment, perpendicular to the segment, or to the midpoint of the segment.
- *Underdetermined* if too few conditions are given, so that more than one line can be drawn to satisfy the conditions. For example, in Figure 7.5 more than one line can be drawn through vertex B of triangle ABC so that the line intersects side $\overline{AC}$.

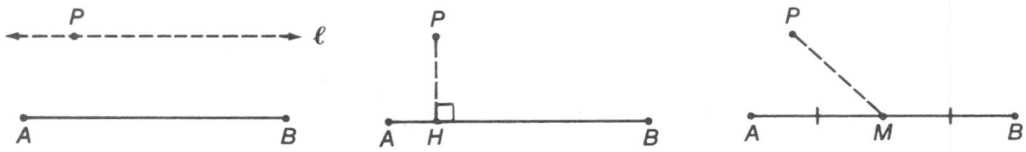

FIGURE 7.4 Each of these lines (or segments) is *determined* since, based on the given conditions, *exactly one* such line (segment) can be drawn.

- *Overdetermined* if too many conditions are given, so that it is not necessarily possible to draw one line that simultaneously meets all of these conditions. For example, in Figure 7.6 we cannot be sure that the line we draw through vertex B of triangle ABC will be a perpendicular bisector of side $\overline{AC}$.

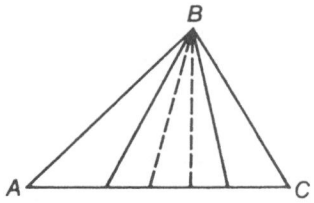

FIGURE 7.5 The line is *underdetermined* since more than one such line may be drawn.

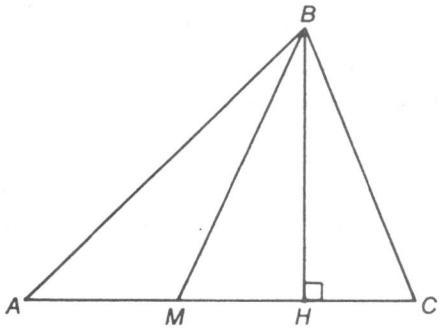

FIGURE 7.6 The line is *overdetermined* since we cannot be assured that the perpendicular dropped from point B will also intersect $\overline{AC}$ at its midpoint.

EXAMPLE 7.6 For the accompanying figure, classify each line you draw as underdetermined, determined, or overdetermined.

a. Draw $\overline{BP}$.
b. Through point B draw a line that bisects $\overline{AC}$.
c. Through point B draw a line that intersects $\overline{AC}$.
d. Draw a line through point A and parallel to $\overline{BC}$.
e. Draw $\overline{BP}$ so that $\overline{BP}$ bisects angle ABC.
f. Through point B draw a line that is perpendicular to $\overline{AC}$.

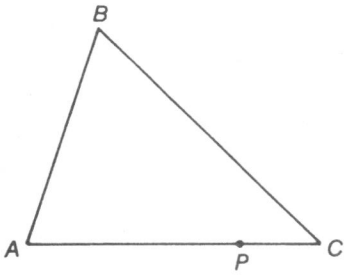

SOLUTION

(a) Determined	**(b)** Determined	**(c)** Underdetermined
(d) Determined	**(e)** Overdetermined	**(f)** Determined

ALTITUDES AND MEDIANS

In every triangle a median and an altitude can be drawn from any vertex to the side opposite that vertex.

- A *median* of a triangle is a segment drawn from a vertex of the triangle to the midpoint of the opposite side.

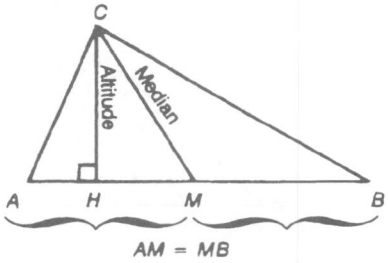

- An *altitude* of a triangle is a segment drawn from a vertex of the triangle perpendicular to the opposite side or, as in the accompanying diagram, to the opposite side extended.

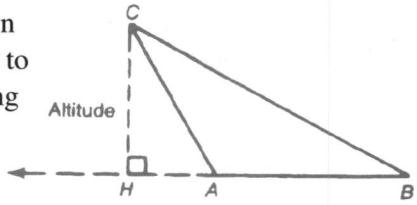

EXAMPLE 7.7 Prove that the median drawn to the base of an isosceles triangle bisects the vertex angle.

REMEMBER
Altitudes and medians of triangles are determined segments. In any triangle, three medians and three altitudes can be drawn.

SOLUTION

When confronted with a verbal statement of a problem requiring a formal proof, our first concern is to draw a suitable diagram. The following general approach is suggested:

1. Rewrite, if necessary, the problem statement in "if . . . then" form. For example,

   PROVE: If a median is drawn to the base of an isosceles triangle, then the vertex angle is bisected.

2. Identify the hypothesis (the Given), which is contained in the "If clause."
3. Identify the conclusion (the Prove), which is contained in the "then clause."
4. Draw and label an appropriate diagram:

   GIVEN: $\overline{AB} \cong \overline{AC}$,
   $\overline{AM}$ is a median to side $\overline{BC}$.
   PROVE: $\overline{AM}$ bisects $\measuredangle BAC$.

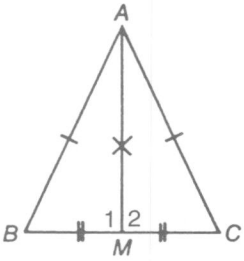

5. Proceed as usual by arriving at a plan before writing the formal two-column proof.

   PLAN: $\overline{BM} \cong \overline{CM}$ since a median bisects the segment to which it is drawn. $\triangle AMB \cong \triangle AMC$ by the SSS Postulate. Angles 1 and 2 are congruent by CPCTC. $\overline{AM}$ bisects angle BAC (a segment that divides an angle into two congruent angles is an angle bisector). The formal two-column proof is left for you.

The Isosceles Triangle

THE BASE ANGLES THEOREM

After drawing several isosceles triangles (Figure 7.7), you may suspect that there is a relationship between the measures of the angles that lie opposite the congruent sides

THEOREM 7.1 BASE ANGLES THEOREM

If two sides of a triangle are congruent, then the angles opposite those sides are congruent.

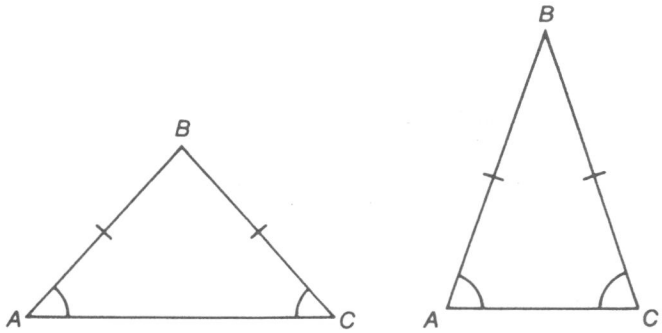

FIGURE 7.7

The proof is easy after an auxiliary line is drawn.

GIVEN: $\overline{AB} \cong \overline{CB}$.

PROVE: $\measuredangle A \cong \measuredangle C$.

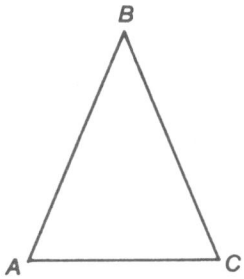

PLAN: From past experience we look to prove angle A congruent to angle C by proving that these angles are corresponding angles of congruent triangles. To form two triangles that can be proved congruent, we may draw the bisector of angle ABC, which will intersect the base at some point, say R. The resulting triangles may be proved congruent by the SAS Postulate. (NOTE: It is also possible to prove this theorem by drawing an altitude or median to side $\overline{AC}$.)

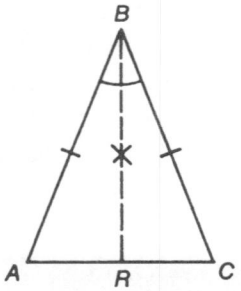

PROOF:

Statements	Reasons
1. $\overline{AB} \cong \overline{CB}$. (Side)	1. Given.
2. Draw the bisector of angle ABC, naming the point at which it intersects $\overline{AC}$, point R.	2. An angle has exactly one bisector.
3. $\measuredangle ABR \cong \measuredangle CBR$. (Angle)	3. An angle bisector divides an angle into two congruent angles.
4. $\overline{BR} \cong \overline{BR}$. (Side)	4. Reflexive property of congruence.
5. $\triangle ABR \cong \triangle CBR$.	5. SAS Postulate.
6. $\measuredangle A \cong \measuredangle C$.	6. CPCTC.

EXAMPLE 7.8 The measure of the vertex angle of an isosceles triangle is three times as great as the measure of a base angle. Find the measure of a base angle of the triangle.

SOLUTION

$$3x + x + x = 180$$
$$5x = 180$$
$$x = 36$$

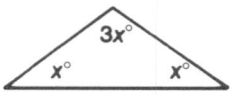

EXAMPLE 7.9

GIVEN: $\overline{SR} \cong \overline{ST}$,
$\overline{MP} \perp \overline{RS}$, $\overline{MQ} \perp \overline{ST}$,
M is the midpoint of $\overline{RT}$.

PROVE: $\overline{MP} \cong \overline{MQ}$.

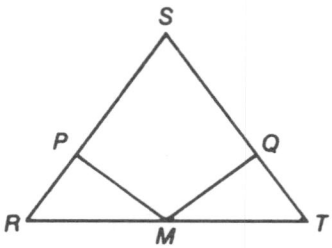

SOLUTION

PLAN: By application of Theorem 7.1, $\measuredangle R \cong \measuredangle T$. Marking the diagram suggests that triangles MPR and MQT may be proved congruent by using the AAS Theorem.

PROOF:

Statements	Reasons
1. $\overline{SR} \cong \overline{ST}$.	1. Given.
2. $\measuredangle R \cong \measuredangle T$. (Angle)	2. If two sides of a triangle are congruent, then the angles opposite those sides are congruent.
3. $\overline{MP} \perp \overline{RS}, \overline{MQ} \perp \overline{ST}$.	3. Given.
4. Angles MPR and MQT are right angles.	4. Perpendicular lines intersect to form right angles.
5. $\measuredangle MPR \cong \measuredangle MQT$. (Angle)	5. All right angles are congruent.
6. M is the midpoint of $\overline{RT}$.	6. Given.
7. $\overline{RM} \cong \overline{TM}$. (Side)	7. A midpoint divides a segment into two congruent segments.
8. $\triangle MPR \cong \triangle MQT$.	8. AAS Theorem.
9. $\overline{MP} \cong \overline{MQ}$.	9. CPCTC.

PROVING A TRIANGLE IS ISOSCELES

The converse of the Base Angles Theorem is also a useful theorem.

THEOREM 7.2 CONVERSE OF THE BASE ANGLES THEOREM

If two angles of a triangle are congruent, then the sides opposite are congruent.

OUTLINE OF PROOF

GIVEN: $\measuredangle A \cong \measuredangle C$.

PROVE: $\overline{AB} \cong \overline{CB}$.

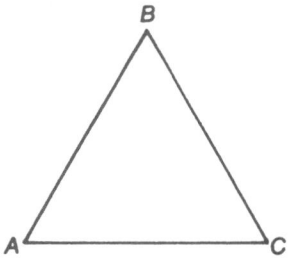

PLAN: Draw the bisector of angle B, intersecting side $\overline{AC}$ at R. The resulting pair of triangles may be proven congruent by the AAS Theorem. It follows that $\overline{AB} \cong \overline{CB}$ by CPCTC. (Other auxiliary lines, such as a median or altitude, may also be drawn.)

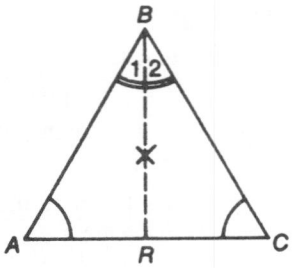

Theorem 7.2 is particularly useful in proving that a triangle is isosceles. To prove that a triangle is isosceles, show either of the following:

- A pair of sides are congruent.
- A pair of angles are congruent (since by Theorem 7.2 the sides opposite must be congruent).

EXAMPLE 7.10

GIVEN: $\overline{WR} \parallel \overline{ST}$,
$\overline{WR}$ bisects $\measuredangle SRE$.

PROVE: $\triangle SRT$ is isosceles.

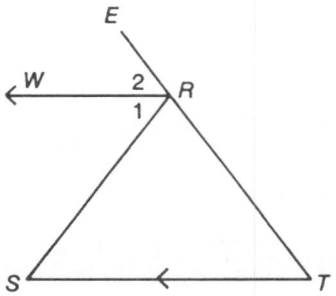

SOLUTION

PLAN: Show the base angles are congruent to each other by showing that each is congruent to one of the pairs of angles formed by the angle bisector. Since the base angles are congruent to congruent angles, they are congruent to each other and the triangle is isosceles.

PROOF:

Statements	Reasons
1. $\overline{WR}$ bisects $\measuredangle SRE$.	1. Given.
2. $\measuredangle 1 \cong \measuredangle 2$.	2. A bisector divides an angle into two congruent angles.
3. $\overline{WR} \parallel \overline{ST}$.	3. Given.
4. $\measuredangle 1 \cong \measuredangle S$.	4. If two lines are parallel, then their alternate interior angles are congruent.
5. $\measuredangle 2 \cong \measuredangle T$.	5. If two lines are parallel, then their corresponding angles are congruent.
6. $\measuredangle S \cong \measuredangle T$.	6. Transitive property of congruence.
7. $\triangle SRT$ is isosceles.	7. A triangle that has a pair of congruent angles is isosceles.

EXAMPLE 7.11 Prove that, if two altitudes of a triangle are congruent, then the triangle is isosceles.

SOLUTION

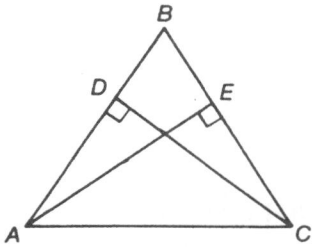

GIVEN: $\overline{CD}$ is the altitude to $\overline{AB}$,
$\overline{AE}$ is the altitude to $\overline{BC}$,
$\overline{CD} \cong \overline{AE}$.

PROVE: $\triangle ABC$ is isosceles.

PLAN: Our goal is to show $\measuredangle BAC = \measuredangle BCA$ by proving $\triangle ADC \cong \triangle CEA$. Marking the diagram suggests the Hy-Leg Postulate be used:

$$\overline{AC} \cong \overline{AC} \text{ (Hy)} \quad \text{and} \quad \overline{CD} \cong \overline{AE} \text{ (Leg)}$$

PROOF:

Statements	Reasons
1. $\overline{CD}$ is the altitude to $\overline{AB}$, $\overline{AE}$ is the altitude to $\overline{BC}$.	1. Given.
2. Triangles ADC and CEA are right triangles.	2. A triangle that contains a right angle is a right triangle. (NOTE: This step consolidates several obvious steps.)
3. $\overline{CD} \cong \overline{AE}$. (Leg)	3. Given.
4. $\overline{AC} \cong \overline{AC}$. (Hy)	4. Reflexive property of congruence.
5. $\triangle ADC \cong \triangle CEA$.	5. Hy-Leg Postulate.
6. $\measuredangle BAC \cong \measuredangle BCA$.	6. CPCTC.
7. $\triangle ABC$ is isosceles.	7. A triangle that has a pair of congruent angles is isosceles.

Double Congruence Proofs

In some problems it may appear that not enough information is provided in the Given to prove a pair of triangles congruent. Upon closer examination, however, it may be possible to prove another pair of triangles congruent in order to obtain congruent corresponding parts, which can then be used to prove the original pair of triangles congruent. These problems tend to be difficult and will require a certain amount of trial-and-error work on your part.

EXAMPLE 7.12

GIVEN: $\overline{AB} \cong \overline{CB}$
E is the midpoint of $\overline{AC}$.
PROVE: $\triangle AED \cong \triangle CED$.

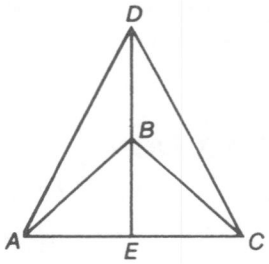

SOLUTION

PLAN:
1. Based on the Given, $\triangle AED$ could be proved congruent to $\triangle CED$ if it was known that $\measuredangle AED$ was congruent to $\measuredangle CED$.
2. Triangles AEB and CEB contain these angles as parts and can be proved congruent by the SSS Postulate.
3. By CPCTC, $\measuredangle AED \cong \measuredangle CED$.
4. $\triangle AED \cong \triangle CED$ by SAS.

PROOF:

Statements	Reasons
Part I. To Prove $\triangle AEB \cong \triangle CEB$:	
1. $\overline{AB} \cong \overline{CB}$. (Side)	1. Given.
2. E is the midpoint of $\overline{AC}$.	2. Given.
3. $\overline{AE} \cong \overline{CE}$. (Side)	3. A midpoint divides a segment into two congruent segments.
4. $\overline{BE} \cong \overline{BE}$.	4. Reflexive property of congruence.
5. $\triangle AEB \cong \triangle CEB$.	5. SSS Postulate.
Part II. To Prove $\triangle AED \cong \triangle CED$:	
6. $\measuredangle AED \cong \measuredangle CED$. (Angle)	6. CPCTC.
7. $\overline{DE} \cong \overline{DE}$. (Side)	7. Reflexive property of congruence.
8. $\triangle AED \cong \triangle CED$.	8. SAS Postulate.

EXAMPLE 7.13

GIVEN: $\overline{BC} \cong \overline{AD}$,
$\overline{BC} \parallel \overline{AD}$,
$\overline{AR} \cong \overline{CS}$.
PROVE: $\overline{BR} \cong \overline{DS}$.

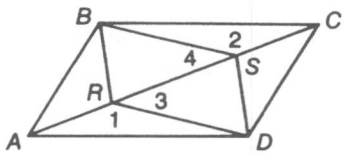

SOLUTION

PLAN: 1. The desired pair of segments can be proven congruent if it can be proven that $\triangle BRS \cong \triangle DSR$. (*NOTE:* Although the desired pair of segments are also contained in triangles ARB and CSD, efforts to prove these triangles congruent would prove fruitless.)

2. By first proving $\triangle ARD \cong \triangle CSB$ we may obtain the congruent parts necessary to prove the desired pair of triangles congruent. Triangles ARD and CSB are congruent by SAS:

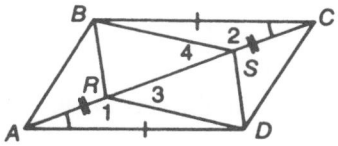

3. By CPCTC, $\overline{RD} \cong \overline{SB}$ and $\measuredangle 1 \cong \measuredangle 2$. Since supplements of congruent angles are congruent, angles 3 and 4 are congruent.
4. $\overline{RS} \cong \overline{RS}$ so $\triangle BRS \cong \triangle DSR$ by SAS, and $\overline{BR} \cong \overline{DS}$ by CPCTC.

The formal proof is left for you.

EXAMPLE 7.14

GIVEN: $\overline{AB} \cong \overline{AC}$, $\overline{BD} \cong \overline{CE}$, $\overline{BF}$ and $\overline{CG}$ are drawn perpendicular to $\overline{AD}$ and $\overline{AE}$, respectively.

PROVE: $\overline{DF} \cong \overline{EG}$.

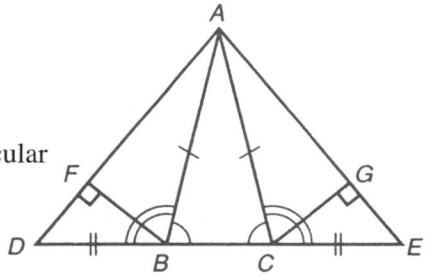

SOLUTION

First prove $\triangle ABD \cong \triangle ACE$ in order to obtain an additional pair of congruent parts needed to prove that $\triangle DFB \cong \triangle EGC$. Here is a detailed plan:

PLAN: 1. $\triangle ABD \cong \triangle ACE$ by SAS since $\overline{AB} \cong \overline{AC}$, $\angle ABD \cong \angle ACE$ (supplements of the congruent base angles are congruent), and $\overline{BD} \cong \overline{CE}$. Therefore, $\angle D \cong \angle E$.

2. $\triangle DFB \cong \triangle EGC$ by AAS since $\angle D \cong \angle E$, $\angle DFB \cong \angle EGC$ (right angles are congruent), and $\overline{BD} \cong \overline{CE}$. Therefore, $\overline{DF} \cong \overline{EG}$ by CPCTC.

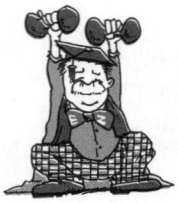

REVIEW EXERCISES FOR CHAPTER 7

1. In the accompanying diagram of $\triangle ABC$, $\overline{AB} \cong \overline{AC}$, $BD = \frac{1}{3}BA$, and $CE = \frac{1}{3}CA$. Triangle EBC can be proved congruent to triangle DCB by

   (1) $SAS \cong SAS$ (2) $ASA \cong ASA$ (3) $SSS \cong SSS$ (4) $HL \cong HL$

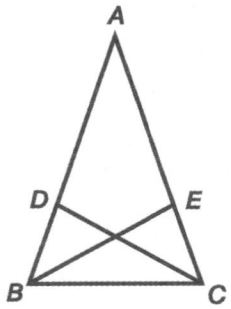

2. In the accompanying diagram of $\triangle BCD$, $\triangle ABC$ is an equilateral triangle and $AD = AB$. What is the value of x, in degrees?

   (1) 10 (2) 15 (3) 30 (4) 60

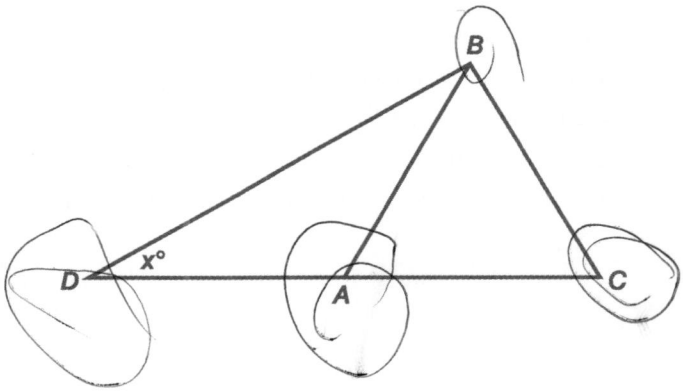

3. In the accompanying diagram, $\overline{BDE}$, $\overleftrightarrow{AB} \parallel \overleftrightarrow{CD}$, and $\overleftrightarrow{DC}$ bisects $\angle ADE$. Triangle ABD must be

(1) scalene (3) equiangular
(2) isosceles (4) right

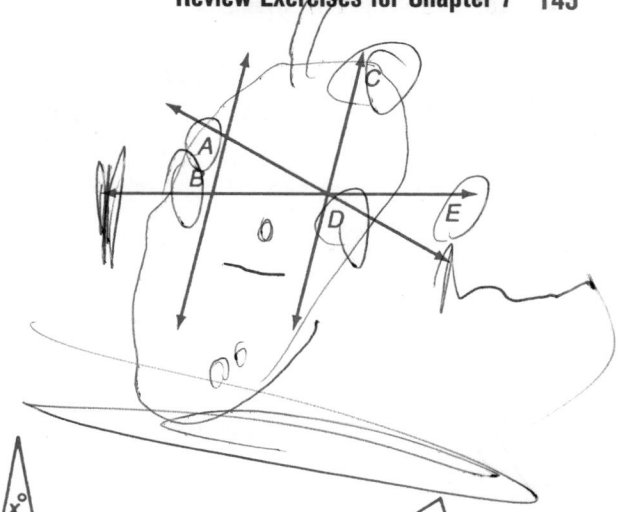

4. Find the value of x:

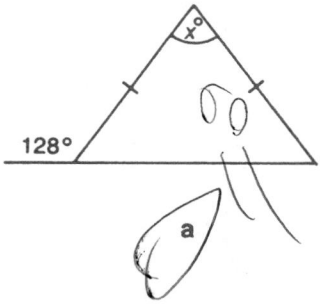

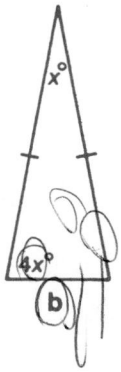

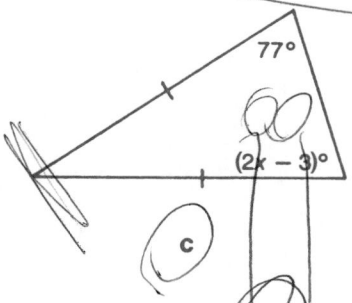

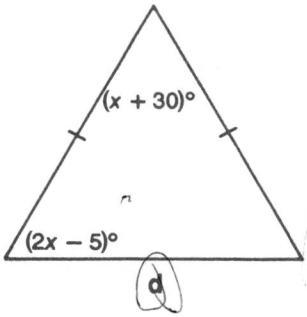

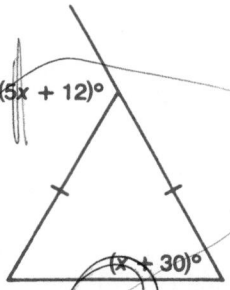

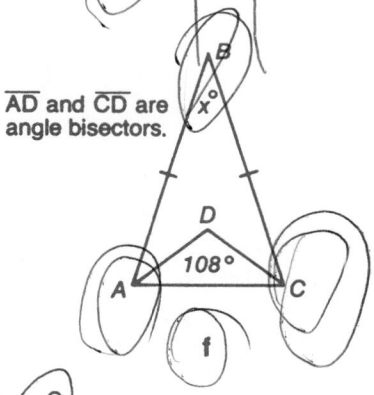

5. GIVEN: $\overline{FH}$ bisects $\measuredangle GHJ$, $\overline{GH} \cong \overline{JH}$.
PROVE: $\measuredangle 1 \cong \measuredangle 2$.

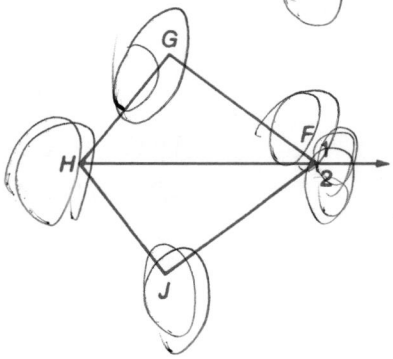

6. GIVEN: $\overline{AC} \cong \overline{BD}$,
$\overline{AB} \perp \overline{BC}, \overline{DC} \perp \overline{BC}$.
PROVE: $\measuredangle 1 \cong \measuredangle 2$.

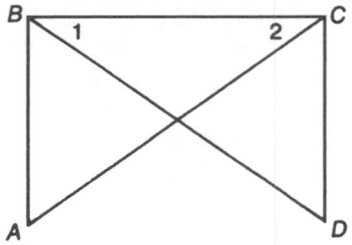

Use the accompanying diagram to solve Exercises 7 and 8.

7. GIVEN: $\measuredangle J \cong \measuredangle K$,
$\overline{PJ} \cong \overline{PK}$.
PROVE: $\overline{JY} \cong \overline{KX}$.

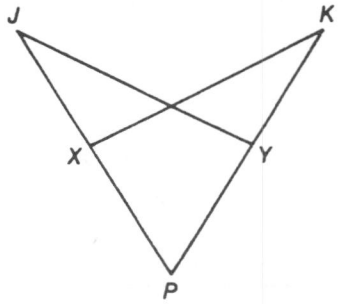

8. GIVEN: $\overline{PX} \cong \overline{PY}$,
$\overline{XJ} \cong \overline{YK}$.
PROVE: $\overline{KX} \cong \overline{JY}$.

9. GIVEN: $\overline{UT} \parallel \overline{DW}, \overline{UT} \cong \overline{DW}$,
$\overline{QW} \cong \overline{AT}$.
PROVE: $\overline{UQ} \parallel \overline{AD}$.

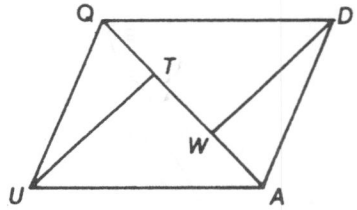

10. GIVEN: $\measuredangle S \cong \measuredangle H$,
$\overline{SR} \perp \overline{RW}, \overline{HW} \perp \overline{RW}$,
$\overline{ST} \cong \overline{HT}$.
PROVE: T is the midpoint of $\overline{RW}$.

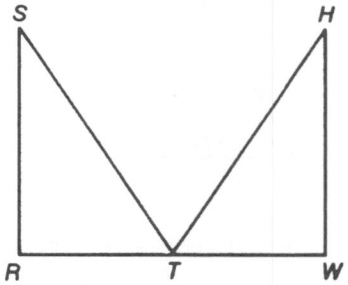

11. GIVEN: $\overline{AB} \cong \overline{CB}, \overline{AD} \cong \overline{CD}$.
PROVE: $\overline{DB}$ bisects $\measuredangle ADC$.

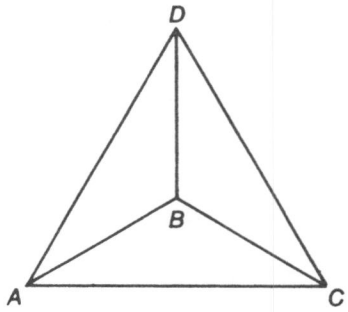

12. GIVEN: $\overline{BA} \perp \overline{MA}, \overline{CD} \perp \overline{MD}$,
M is the midpoint of $\overline{BC}$.
PROVE: $\overline{BC}$ bisects $\overline{AD}$.

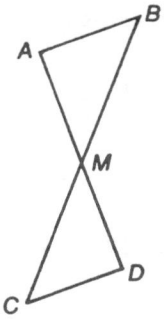

13. GIVEN: $\measuredangle 1 \cong \measuredangle 2$,
$\overline{HK}$ bisects $\measuredangle RHN$,
$\overline{HR} \cong \overline{HN}$.
PROVE: $\overline{HK} \perp \overline{RN}$.

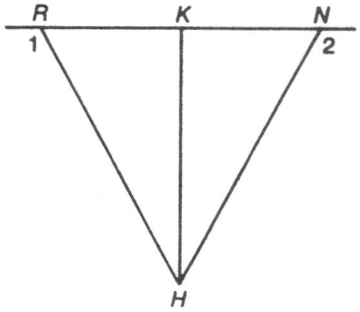

14. GIVEN: $\measuredangle 1$ is supplementary to $\measuredangle 2$,
$\overline{RC} \cong \overline{AT}$,
$\overline{SR} \cong \overline{AB}$,
$\overline{ST} \cong \overline{BC}$.
PROVE: 1. $\overline{ST} \perp \overline{TR}$.
2. $\overline{BC} \perp \overline{AC}$.

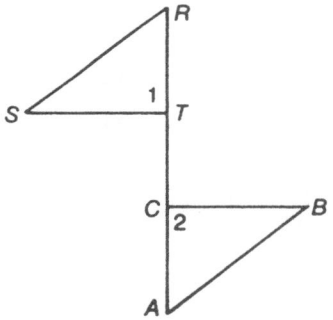

15. GIVEN: $\overline{RS} \perp \overline{SL}, \overline{RT} \perp \overline{LT}$,
$\overline{RS} \cong \overline{RT}$.
PROVE: 1. $\triangle RLS \cong \triangle RLT$.
2. $\overline{WL}$ bisects $\measuredangle SWT$.

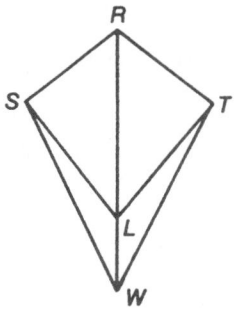

Use the accompanying diagram to solve Exercises 16 and 17.

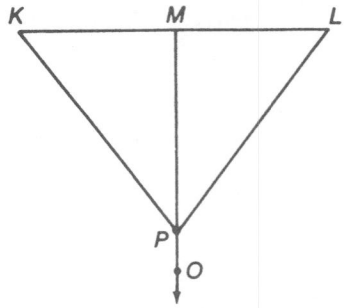

16. GIVEN: $\overline{PM}$ is the altitude to $\overline{KL}$,
$\overline{PK} \cong \overline{PL}$.
PROVE: M is the midpoint of $\overline{KL}$.

17. GIVEN: $\overline{PM}$ is the median to $\overline{KL}$,
$\overline{KP} \cong \overline{LP}$.
PROVE: $\overline{PM} \perp \overline{KL}$.

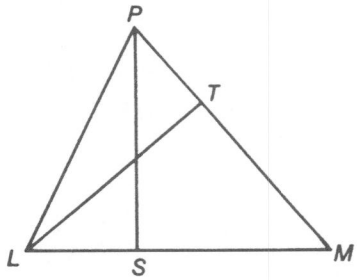

18. GIVEN: $\overline{PS}$ and $\overline{LT}$ are altitudes to sides $\overline{LM}$ and $\overline{PM}$, respectively.
$\measuredangle LPT \cong \measuredangle SLP$.
PROVE: $\overline{PS} \cong \overline{LT}$.

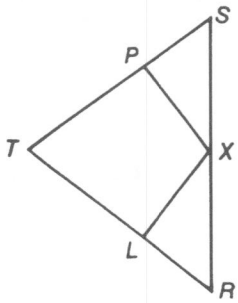

19. GIVEN: $\overline{XL} \cong \overline{XP}$,
$\overline{XL} \perp \overline{TR}, \overline{XP} \perp \overline{TS}$,
X is the midpoint of $\overline{RS}$.
PROVE: $\triangle RTS$ is isosceles.

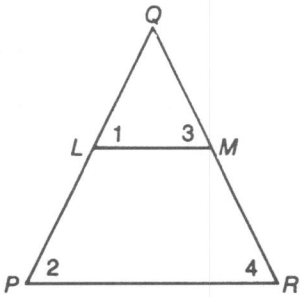

20. GIVEN: $\overline{QL} \cong \overline{QM}$,
$\overline{LM} \parallel \overline{PR}$.
PROVE: $\triangle PQR$ is isosceles.

21. GIVEN: $\overline{LW} \perp \overline{TW}, \overline{FX} \perp \overline{XP}$,
$\overline{TF} \cong \overline{PL}, \overline{WL} \cong \overline{XF}$.
PROVE: $\triangle FML$ is isosceles.

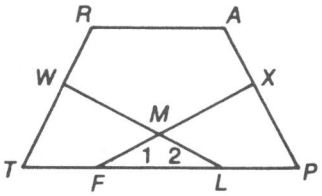

22. GIVEN: $\overline{SE} \cong \overline{SW}$,
$\measuredangle 1 \cong \measuredangle 2$,
$\overline{EL} \cong \overline{WB}$.
PROVE: $\overline{KL} \cong \overline{AB}$.

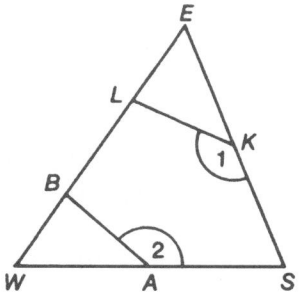

23. GIVEN: $\overline{OV} \cong \overline{LV}$,
$\overline{KO} \cong \overline{ZL}$.
PROVE: $\triangle KVZ$ is isosceles.

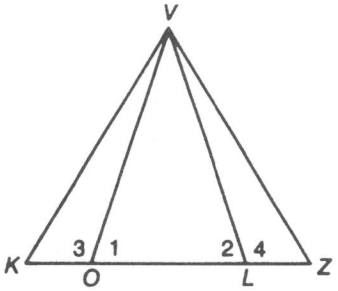

24. GIVEN: $\measuredangle JHL \cong \measuredangle JLH$,
$\overline{BH} \perp \overline{HJ}, \overline{KL} \perp \overline{LJ}$.
PROVE: $\overline{JB} \cong \overline{JK}$.

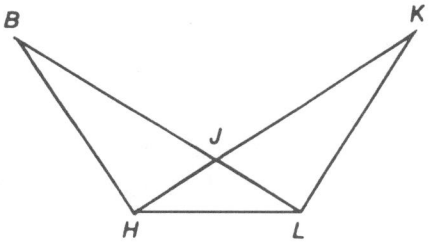

Use the accompanying diagram to solve Exercises 25 and 26.

25. GIVEN: $\overline{AB} \cong \overline{AD}$,
$\overline{EA}$ bisects $\measuredangle DAB$.
PROVE: $\triangle BCE \cong \triangle DCE$.

26. GIVEN: $\overline{BE} \cong \overline{DE}$,
$\overline{BC} \cong \overline{DC}$.
PROVE: $\overline{CA}$ bisects $\measuredangle DAB$.

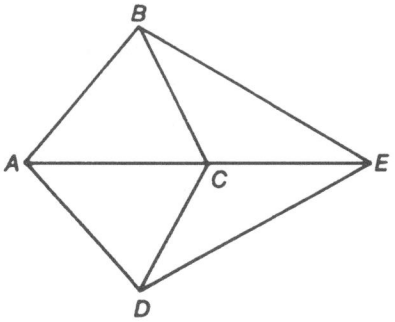

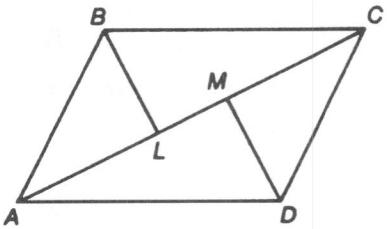

27. GIVEN: $\overline{AB} \parallel \overline{CD}$,
$\overline{AB} \cong \overline{CD}$,
$\overline{AL} \cong \overline{CM}$.
PROVE: $\measuredangle CBL = \measuredangle ADM$.

Use the accompanying diagram to solve Exercises 28 and 29.

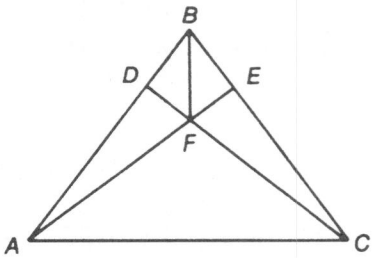

28. GIVEN: $\measuredangle FAC \cong \measuredangle FCA$,
$\overline{FD} \perp \overline{AB}, \overline{FE} \perp \overline{BC}$.
PROVE: $\overline{BF}$ bisects $\measuredangle DBE$.

29. GIVEN: $\overline{BD} \cong \overline{BE}$,
$\overline{FD} \cong \overline{FE}$.
PROVE: $\triangle AFC$ is isosceles.

30. Prove that, if two triangles are congruent, then the altitudes drawn to a pair of corresponding sides are congruent.

31. Prove that an equilateral triangle is equiangular.

32. Prove that the altitudes drawn to the legs of an isosceles triangle are congruent.

33. Prove that the medians drawn to the legs of an isosceles triangle are congruent.

34. Prove that any point on the perpendicular bisector of a line segment is equidistant from the endpoints of the segment.

35. Prove that, if two points are each equidistant from the endpoints of a line segment, then the two points determine the perpendicular bisector of the line segment.

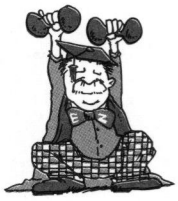

CUMULATIVE REVIEW EXERCISES: CHAPTERS 1–7

1. In an isosceles triangle, the measure of the vertex angle is 8 times the measure of one of the base angles. Find the number of degrees in the measure of a base angle of the triangle.

2. In the accompanying diagram, $\overleftrightarrow{AB}$, $\overleftrightarrow{CD}$, and $\overline{FH}$ intersect at E, and $\overleftrightarrow{AB}$ bisects $\measuredangle FEC$. If $m\measuredangle DEB = 25$, what is $m\measuredangle DEF$?

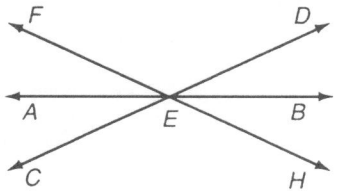

3. The measures of two complementary angles are represented by $2x$ and $3x - 10$. What is the value of x?

4. In the accompanying diagram, $\overleftrightarrow{AKB} \parallel \overleftrightarrow{CD}$, $\overline{AE} \perp \overline{CK}$, $m\measuredangle KCD = 2x$, and $m\measuredangle KAE = 3x$. What is $m\measuredangle CKB$?

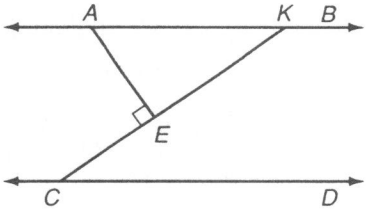

5. What is the measure of an exterior angle of a regular polygon having nine sides?

6. $\overleftrightarrow{AB} \parallel \overleftrightarrow{CD}$ and each line is intersected by $\overleftrightarrow{MN}$ at G and H, respectively. If $m\measuredangle BGH = 2x + 50$ and $m\measuredangle CHG = 5x - 70$, find x.

7. Dylan says that all isosceles triangles are acute triangles. Joan wants to prove that Dylan is not correct. Describe how Joan could prove that Dylan is *not* correct.

8. The perimeter of an isosceles triangle is 71 centimeters. The length of one of the sides is 22 centimeters. What are all the possible lengths of the other two sides?

Use the accompanying diagram to solve Exercises 9 and 10.

9. GIVEN: $\measuredangle 1 \cong \measuredangle 2$,
$\overline{AB} \cong \overline{BC}$,
F is the midpoint of $\overline{AB}$,
G is the midpoint of $\overline{BC}$.
PROVE: $\overline{FD} \cong \overline{GE}$.

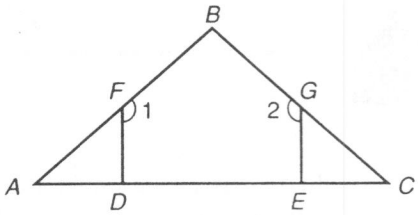

10. GIVEN: $\measuredangle 1 \cong \measuredangle 2$,
$\overline{FD} \perp \overline{AC}, \overline{GE} \perp \overline{AC}$,
$\overline{AE} \cong \overline{CD}$.
PROVE: $\triangle ABC$ is isosceles.

11. GIVEN: $\overline{RS}$ intersects $\overleftrightarrow{ARB}$ and $\overleftrightarrow{CTS}$,
$\overleftrightarrow{ARB} \parallel \overleftrightarrow{CTS}$,
$\overline{RT}$ bisects $\angle BRS$,
M is the midpoint of $\overline{RT}$,
$\overline{SM}$ is drawn.
PROVE: **a** $\overline{RS} \cong \overline{ST}$.
b $\overline{SM}$ bisects $\angle RST$.

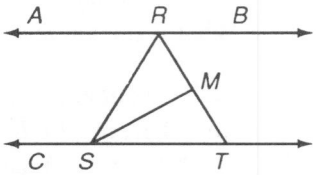

12. GIVEN: $\triangle ABC$, $\overline{CM}$ is the median to $\overline{AB}$, $\overline{CM}$ is extended to P so that $\overline{CM} \cong \overline{MP}$, and $\overline{AP}$ is drawn.
PROVE: $AP \parallel CB$.

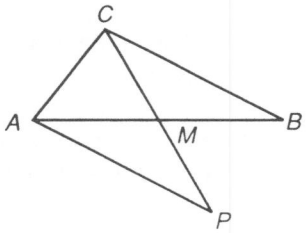

13. GIVEN: $\overline{ADFB}, \overline{AGC}, \overline{BEC}, \overline{AD} \cong \overline{BF}$,
$\measuredangle x \cong \measuredangle y$, and $\measuredangle A \cong \measuredangle B$.
PROVE: a $\triangle AFG \cong \triangle BDE$
b $\overline{GC} \cong \overline{EC}$.

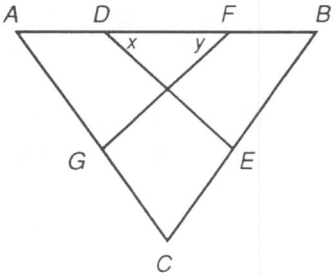

14. GIVEN: $\overline{KPQRL}, \overline{MQN}, \overline{KM}, \overline{NL}, \overline{MP}, \overline{NR}$, $\overline{KL}$ and $\overline{MN}$ bisect each other at Q, $\measuredangle 1 \cong \measuredangle 2$.

PROVE: $\overline{PM} \cong \overline{NR}$.

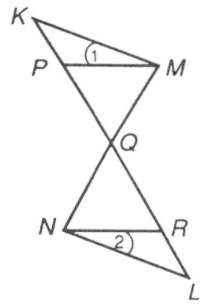

15. GIVEN: $\triangle ABC, \overline{CEA} \cong \overline{CDB}, \overline{AD}$ and $\overline{BE}$ intersect at P, and $\measuredangle PAB \cong \measuredangle PBA$.

PROVE: $\overline{PE} \cong \overline{PD}$.

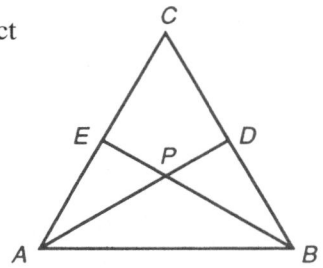

16. GIVEN: $\triangle ABC$,
$\overline{AC} \cong \overline{BC}, \overline{CE} \cong \overline{CD}$,
in $\triangle BCD, \overline{DF}$ is a median to $\overline{BC}$,
in $\triangle ACE, \overline{EG}$ is a median to $\overline{AC}$.

PROVE: $\overline{EG} \cong \overline{DF}$.

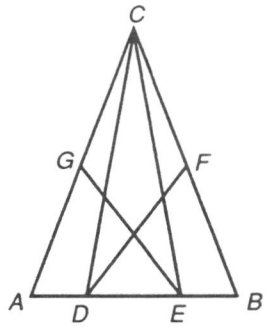

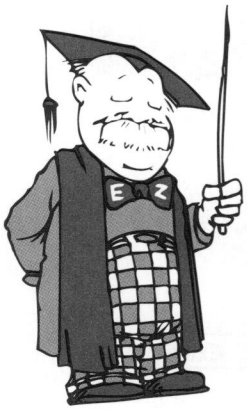

8

Geometric Inequalities

WHAT YOU WILL LEARN

If angles or sides are not congruent, it is often helpful to compare the measures of these angles or sides using inequality statements. Geometric inequalities are often proved by an indirect method. In this chapter you will learn:

- the inequality relationship between the measure of an exterior angle of a triangle and the measure of either of the two remote interior angles;
- the relationship between unequal angle measures and unequal side lengths in a triangle;
- the way the length of any side of a triangle compares to the sum of the lengths of the other two sides;
- the application of the indirect method of proof.

SECTIONS IN THIS CHAPTER

- Some Basic Properties of Inequalities
- Inequality Relationships in a Triangle
- The Indirect Method of Proof

Some Basic Properties of Inequalities

In everyday life we are constantly making comparisons. Which of two people earns more money? Which of two athletes can run faster? Which of two students has the higher grade average? Comparisons also play an important role in mathematics. In comparing two quantities, say a and b, there are exactly three possibilities, which are summarized in Table 8.1.

TABLE 8.1

Condition	Notation	Example
a is less than b.	$a < b$	$3 < 5$
a is equal to b.	$a = b$	$4 = 4$
a is greater than b.	$a > b$	$7 > 2$

The *direction* or *sense* of an inequality refers to whether the inequality symbol is pointing to the right (>) or to the left (<). For example:

- $a > b$ and $c > d$ are two inequality expressions that have the same direction or *sense* since they *both* indicate that one quantity is greater than the other.
- By changing $a > b$ to $a < b$ (or vice versa), we have reversed the direction or *sense* of the inequality.

We shall assume that a number of properties related to inequalities are true. See Table 8.2.

TABLE 8.2

Property	Formal Statement	Example
Addition	If $a < b$, then $a + c < b + c$.	$3 < 5$ $+\ 4 = 4$ $7 < 9$
	If $a < b$ and $c < d$, then $a + c < b + d$.	$3 < 5$ $+\ 6 < 10$ $9 < 15$
Subtraction	If $a < b$, then $a - c < b - c$.	$8 < 13$ $-\ 6 = -6$ $2 < 7$
Multiplication	If $a < b$ and $c > 0$, then $ac < bc$.	$5 < 8$ $2(5) < 2(8)$ or $10 < 16$
Transitive	If $a < b$ and $b < c$, then $a < c$.	$4 < 7$ and $7 < 10$ then $7 < 10$.
Substitution	If $a + b < c$, and $b = d$, then $a + d < c$.	$x + y < 9$ $y = 5$ $x + 5 < 9$

Keep in mind that:

- Although in Table 8.2 each property is expressed in terms of the less than relation (<), the properties clearly hold for the greater than (and equal) relation.
- As a corollary to the multiplication property, "halves of unequals are unequal." For example, if $a < b$, then $a/2 < b/2$.
- $a < b$ and $b > a$ are equivalent expressions.

Here are two geometric inequalities that should be obvious:

1. If point X is *between* points A and B, then $AX < AB$ and $XB < AB$.

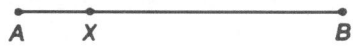

2. If ray $\overrightarrow{BX}$ is *between* rays $\overrightarrow{BA}$ and $\overrightarrow{BC}$ and $\overrightarrow{BX}$ lies in the interior of $\measuredangle ABC$, then $m\measuredangle ABX < m\measuredangle ABC$ and $m\measuredangle CBX < m\measuredangle ABC$.

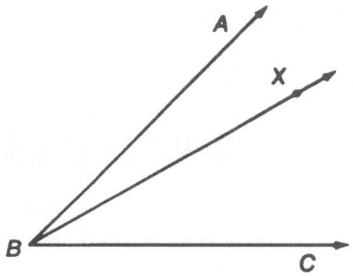

Inequality Relationships in a Triangle

In Figure 8.1, the shortest trip from city A to city B is the direct route represented by segment AB. This direct route must be less than any indirect route, such as that from city A to city C and then from city C to city B. In symbols:

$$AB < AC + CB$$

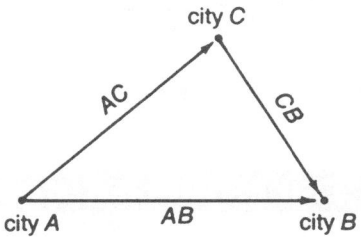

FIGURE 8.1

TRIANGLE INEQUALITY POSTULATE

The length of each side of a triangle must be less than the sum of the lengths of the other two sides.

This postulate provides us with a convenient method for determining whether a set of three numbers can represent the lengths of the sides of a triangle. All we need do is test that *each* number is less than the sum of the other two.

EXAMPLE 8.1 Which of the following sets of numbers *cannot* represent the sides of a triangle?
(1) 9, 40, 41 (2) 7, 7, 3 (3) 4, 5, 1 (4) 6, 6, 6

SOLUTION

Choice (3): $4 < 5 + 1$ and $1 < 4 + 5$, but 5 is *not* less than $4 + 1$.

COMPARING ANGLES OF A TRIANGLE

The Base Angles Theorem (Theorem 7.1) tells us that, if two sides of a triangle are congruent, then the angles opposite these sides are congruent (see Figure 8.2*a*). What conclusion can be drawn if the two sides of the triangle are *not* congruent? In Figure 8.2*b*, clearly the angles cannot be congruent. Furthermore, it appears that the greater angle lies opposite the longer side.

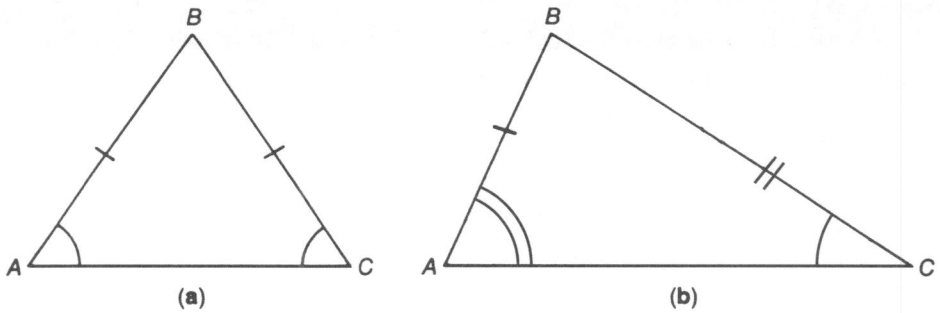

FIGURE 8.2

THEOREM 8.1 $\ncong$ SIDES IMPLIES $\ncong$ OPPOSITE ANGLES

If two sides of a triangle are not congruent, then the angles opposite these sides are not congruent, and the greater angle is opposite the longer side.

OUTLINE OF PROOF

GIVEN: $BC > BA$.

PROVE: $m\measuredangle BAC > m\measuredangle C$.

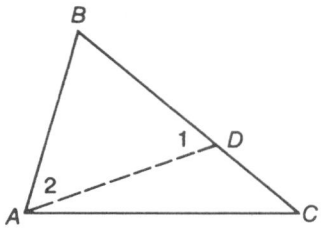

PLAN Since $BC > BA$, there exists a point D on $\overline{BC}$ such that $AB = DB$.

- Drawn $\overline{AD}$. (See diagram.)
- $m\measuredangle 1 > m\measuredangle C$ (Exterior Angle Inequality Theorem, Theorem 5.4.)
- $m\measuredangle 1 = m\measuredangle 2$ (Base Angles Theorem, Theorem 7.1.)
- $m\measuredangle 2 > m\measuredangle C$ (Substitution.)
- $m\measuredangle BAC > m\measuredangle 2$ (Definition of betweenness of rays; see diagram.)
- $m\measuredangle BAC > m\measuredangle C$ (Transitive property.)

The converse of Theorem 8.1 is also true.

THEOREM 8.2 ≠ ANGLES IMPLIES ≠ OPPOSITE SIDES

If two angles of a triangle are not congruent, then the sides opposite these angles are not congruent, and the longer side is opposite the greater angle.

EXAMPLE 8.2 In $\triangle ABC$, $AB = 3$, $BC = 5$, and $AC = 7$. What is the largest angle of the triangle? The smallest angle of the triangle?

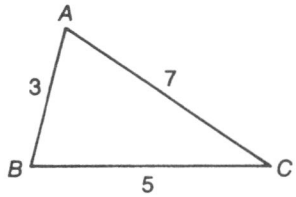

SOLUTION

$\measuredangle B$ is the largest angle.

$\measuredangle C$ is the smallest angle.

EXAMPLE 8.3 The measure of the vertex angle S of isosceles triangle RST is 80. What is the longest side of the triangle?

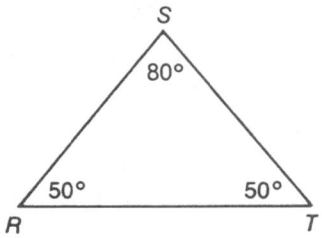

SOLUTION

$\overline{RT}$ is the longest side.

EXAMPLE 8.4

GIVEN: $\overline{BD}$ bisects $\measuredangle ABC$.
PROVE: $AB > AD$.

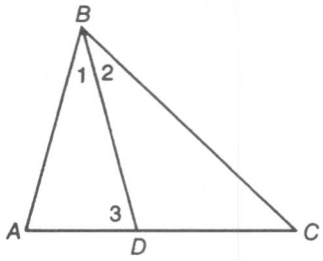

SOLUTION

PLAN: By Theorem 8.2, in order to prove $AB > AD$ we must first establish that $m\measuredangle 3$ (the angle opposite $\overline{AB}$) is greater than $m\measuredangle 1$ (the angle opposite $\overline{AD}$).

PROOF:

Statements	Reasons
1. $m\measuredangle 3 > m\measuredangle 2$.	1. The measure of an exterior angle of a triangle is greater than the measure of either nonadjacent interior angle.
2. $\overline{BD}$ bisects $\measuredangle ABC$.	2. Given.
3. $m\measuredangle 1 = m\measuredangle 2$.	3. A bisector divides an angle into two angles having the same measure.
4. $m\measuredangle 3 > m\measuredangle 1$.	4. Substitution property of inequalities.
5. $AB > AD$.	5. If two angles of a triangle are not equal in measure, then the sides opposite are not equal and the longer side is opposite the greater angle.

The Indirect Method of Proof

Sometimes it is too difficult or even impossible to prove a statement directly using deductive reasoning. In such situations it may be helpful to see what happens if the statement were *not* true. If you then discover that it cannot be the case that the statement is not true, then you may conclude that the statement is true as this is the only other possibility. This is the underlying principle of the indirect method of proof.

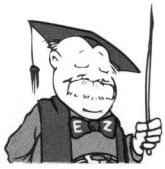

TO PROVE A STATEMENT INDIRECTLY

- Assume that the statement in the Prove is *not* true. This is equivalent to assuming that the opposite or negation of the statement in the Prove *is* true.
- Show that this assumption contradicts a known fact and, as a result, is false. The contradicted fact may be an earlier theorem or postulate, or may be part of the Given.
- Conclude that what you needed to prove, the opposite of the false assumption, is true.

An indirect proof is usually needed when the statement you need to prove involves the word "not."

EXAMPLE 8.5 Use an indirect method of proof to prove that a triangle cannot have more than one obtuse angle.

SOLUTION

- Assume a triangle can have more than one obtuse angle.
- If the triangle has two obtuse angles, then the sum of these two obtuse angles is greater than 180, which contradicts the fact that the sum of the three angles of a triangle is 180.
- Because of this contradiction, the original assumption is false. Since the assumption represents the opposite of what you are trying to prove and is false, the statement you were asked to prove must be true. That is, a triangle cannot have more than one obtuse angle.

EXAMPLE 8.6

GIVEN: $\measuredangle 1$ is not congruent to $\measuredangle 2$.
PROVE: Line ℓ is not parallel to line m.

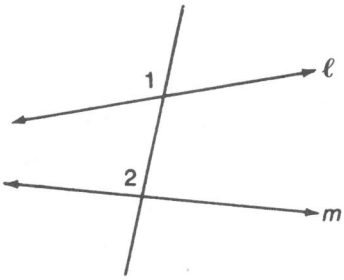

SOLUTION

Use an indirect method of proof:

1. Assume the opposite or negation of what you need to prove is true; that is, line ℓ *is* parallel to line m.
2. If the lines are parallel, then corresponding angles are congruent, so $\measuredangle 1 \cong \measuredangle 2$. But this contradicts the Given.

3. Since its negation is false, the statement "line ℓ is not parallel to line m" must be true.

EXAMPLE 8.7

GIVEN: $\overline{AB} \cong \overline{DB}$.
PROVE: $\overline{AB}$ is not $\cong$ to $\overline{BC}$.

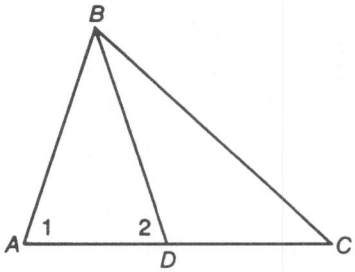

SOLUTION

INDIRECT PROOF Assume the negation of the desired conclusion is true; that is, assume $\overline{AB}$ *is* $\cong$ to $\overline{BC}$. If $\overline{AB} \cong \overline{BC}$, then $\measuredangle 1 \cong \measuredangle C$. From the Given, $\measuredangle 1 \cong \measuredangle 2$. Hence, $\measuredangle 2 \cong \measuredangle C$. But this is impossible since the measure of an exterior angle of a triangle must be greater than the measure of either nonadjacent interior angle. Hence, the assumption is false and the only other possibility, that $\overline{AB}$ is not $\cong$ to $\overline{BC}$, must be true.

EXAMPLE 8.8

GIVEN: $\overline{TW} \perp \overline{RS}$, $\measuredangle 1$ is not $\cong$ to $\measuredangle 2$.
PROVE: $\overline{TW}$ is not the median to side $\overline{RS}$.

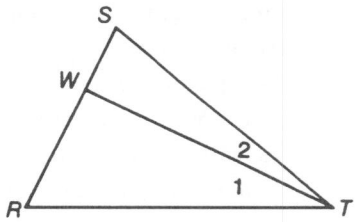

SOLUTION

INDIRECT PROOF: Assume the negation of the desired conclusion is true: assume $\overline{TW}$ is the median to side $\overline{RS}$. Then $\overline{WR} \cong \overline{WS}$ and $\triangle TWS$ is congruent to $\triangle TWR$ by SAS. By CPCTC, $\measuredangle 1 \cong \measuredangle 2$. But this contradicts the Given. Hence, the only remaining possibility, that $\overline{TW}$ is *not* the median to side $\overline{RS}$, must be true.

Proofs that rely on the indirect method may also be organized in our familiar two-column format. The two-column format of the proof for Example 8.8 is as follows:

PROOF:

Statements	Reasons
1. $\measuredangle 1$ is *not* $\cong$ to $\measuredangle 2$.	1. Given.
2. Either $\overline{TW}$ is *not* the median to side $\overline{RS}$ or $\overline{TW}$ *is* the median to side $\overline{RS}$. *Assume* $\overline{TW}$ *is* the median to side $\overline{RS}$.	2. A statement is either true or false.
3. $\overline{WR} \cong \overline{WS}$. (Side)	3. A median divides a side into two congruent segments.
4. $\overline{TW} \perp \overline{RS}$.	4. Given.
5. Angles TWS and TWR are right angles.	5. Perpendicular lines meet to form right angles.
6. $\measuredangle TWS \cong \measuredangle TWR$. (Angle)	6. All right angles are congruent.
7. $\overline{TW} \cong \overline{TW}$. (Side)	7. Reflexive property of congruence.
8. $\triangle TWS \cong \triangle TWR$.	8. SAS Postulate.
9. $\measuredangle 1 \cong \measuredangle 2$.	9. CPCTC.
10. $\overline{TW}$ is *not* the median to side $\overline{RS}$.	10. Statement 9 contradicts statement 1. The assumption made in statement 2 must therefore be false, so its opposite is true.

Another interesting application of the indirect method of proof provides an alternative means for developing the properties of parallel lines. Example 8.9 provides a proof for a statement that was originally postulated.

EXAMPLE 8.9 If two lines are parallel, then their alternate interior angles are congruent.

GIVEN: $\ell \parallel m$.

PROVE: $\measuredangle 1 \cong \measuredangle 2$.

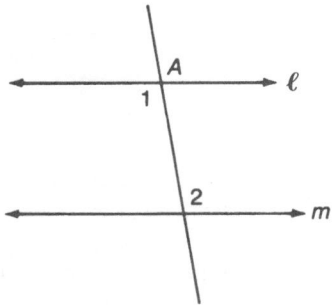

SOLUTION

Using an indirect method of proof, we assume that angle 1 is *not* congruent to angle 2. It is therefore possible to construct at point A a line k such that angle 3 is congruent to angle 2:

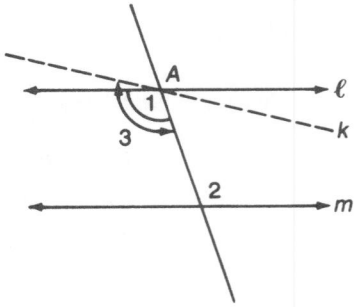

Since $\measuredangle 3 \cong \measuredangle 2$, line k is parallel to line m (if alternate interior angles are congruent, the lines are parallel). Through point A, line ℓ has been drawn parallel to line m (given) and line k has been drawn parallel to line m. This contradicts the Parallel Postulate, which says that exactly one such line can be drawn. Hence our assumption that angles 1 and 2 are not congruent leads to a contradiction, which implies that angles 1 and 2 must be congruent.

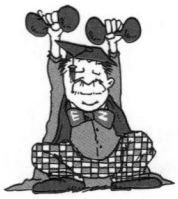

REVIEW EXERCISES FOR CHAPTER 8

1. If the lengths of two sides of a triangle are 4 and 10, what could be the length of the third side?

   (1) 6 (2) 8 (3) 14 (4) 16

2. On the banks of a river, surveyors marked locations A, B, and C, as shown in the accompanying diagram, where $m\angle ACB = 70$ and $m\angle ABC = 65$.

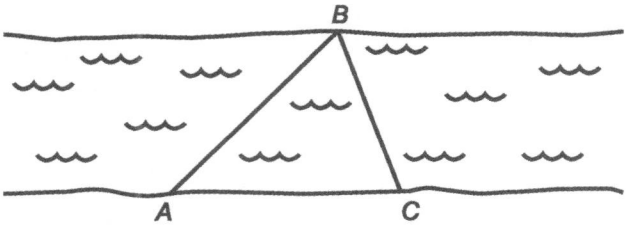

   Which expression shows the relationship between the lengths of the sides of this triangle?

   (1) $AB < BC < AC$
   (2) $AC < BC < AB$
   (3) $BC < AC < AB$
   (4) $AC < AB < BC$

3. If 3, 8, and x represent the lengths of the sides of a triangle, how many *integer* values for x are possible?

   (1) 7 (2) 6 (3) 5 (4) 4

4. In $\triangle ABC$, $m\angle A = 55$ and $m\angle B = 60$. Which statement about $\triangle ABC$ is true?

   (1) All the sides have different lengths, and $\overline{AC}$ is the longest side.
   (2) All the sides have different lengths, and $\overline{AB}$ is the longest side.
   (3) Sides $\overline{AB}$ and $\overline{AC}$ have the same length and are longer than $\overline{BC}$.
   (4) Sides $\overline{AB}$ and $\overline{BC}$ have the same length and are longer than side $\overline{AC}$.

5. If the integer lengths of the three sides of a triangle are 4, x, and 9, what is the smallest possible perimeter of the triangle?

   (1) 18 (2) 19 (3) 20 (4) 21

6. In $\triangle ABC$, $m\measuredangle A = 50$ and $m\measuredangle B = 60$. Which is the longest side of the triangle?

7. In $\triangle ABC$, $m\measuredangle A = 30$ and the measure of the exterior angle at B is 120. Which is the longest side of the triangle?

8. In right triangle ABC, altitude $\overline{CD}$ is drawn to hypotenuse $\overline{AB}$. Which is the longest side of $\triangle CDB$?

9. In $\triangle ABC$, $m\measuredangle B = 120$, $m\measuredangle A = 55$, and D is the point on $\overline{AC}$ such that $\overline{BD}$ bisects $\measuredangle ABC$. Which is the longest side of $\triangle ABD$?

10. An exterior angle formed at vertex angle J of isosceles triangle JKL by extending leg $\overline{LJ}$ has a degree measure of 115. Which is the longest side of the triangle?

11. In $\triangle ABC$, $BC > AB$ and $AC < AB$. Which is the longest side of the triangle?

12. In $\triangle RST$ $m\measuredangle R < m\measuredangle T$ and $m\measuredangle S > m\measuredangle T$. Which is the largest angle of the triangle?

13. GIVEN: $\overline{AB} \cong \overline{BD}$,
$m\measuredangle 5 > m\measuredangle 6$.

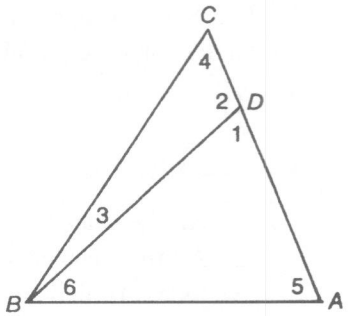

State whether each of the following inequality relationships is true or false.

(a) $m\measuredangle 1 > m\measuredangle 3$.
(b) $m\measuredangle 5 > m\measuredangle 1$.
(c) $m\measuredangle 3 < m\measuredangle ABC$.
(d) $m\measuredangle 2 > m\measuredangle 1$.
(e) $BC > BA$.
(f) $AB < AD$.

14. In $\triangle ABC$, $AB > AC$ and $BC > AC$. Name the smallest angle of $\triangle ABC$.

15. In $\triangle RST$, $ST > RT$ and $RT > RS$.

(a) If one of the angles of the triangle is obtuse, which angle must it be?
(b) If the measure of one of the angles of the triangle is 60, which angle must it be?

16. Determine whether each of the following sets of numbers can represent the lengths of the sides of a triangle.

(a) 8, 17, 15
(b) $\frac{1}{2}, \frac{1}{3}, \frac{1}{6}$
(c) 1, 1, 3
(d) 6, 6, 7

17. GIVEN: $\overline{AB} \cong \overline{CB}$.
PROVE: $AB > BD$.

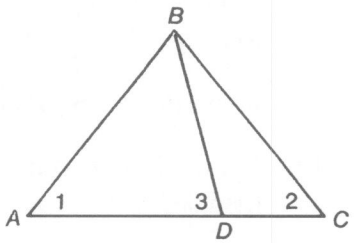

18. In the accompanying diagram, $\triangle ABC$ is *not* isosceles. Prove that if altitude $\overline{BD}$ were drawn, it would *not* bisect $\overline{AC}$.

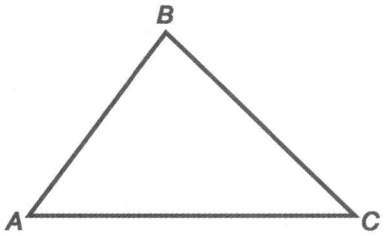

19. GIVEN: $m\measuredangle 1 = m\measuredangle 2$.
PROVE: $AD > ED$.

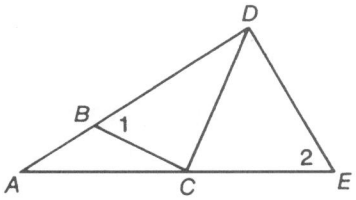

Use the accompanying diagram to solve Exercises 20 to 22.

20. GIVEN: Triangles AEC and ABC.
PROVE: $m\measuredangle 4 > m\measuredangle AEC$.

21. GIVEN: $AC > BC$.
PROVE: $AD > BD$.

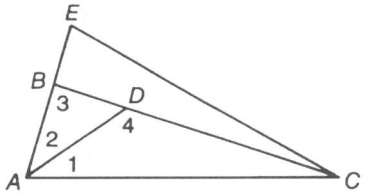

22. GIVEN: $AD > BD$,
AD bisects $\measuredangle BAC$.
PROVE: $AC > DC$.

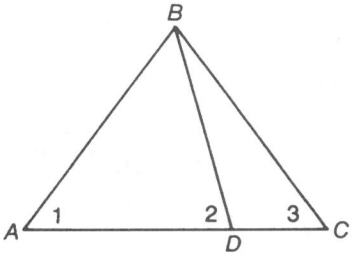

23. GIVEN: $\measuredangle 1 \cong \measuredangle 3$.
PROVE: $m\measuredangle 1 > m\measuredangle 2$.

24. GIVEN: $\measuredangle 1 \cong \measuredangle 2$.
PROVE: $\overline{AB} \ncong \overline{BC}$.

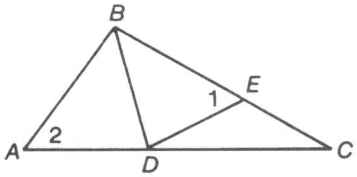

25. GIVEN: $RS = TS$.
PROVE: $RW \neq WL$.

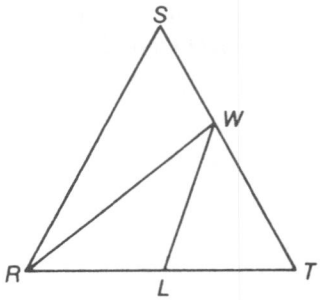

26. GIVEN: $\triangle ABC$ is *not* isosceles,
$\measuredangle ADB \cong \measuredangle CDB$.
PROVE: $\triangle ADC$ is *not* isosceles.

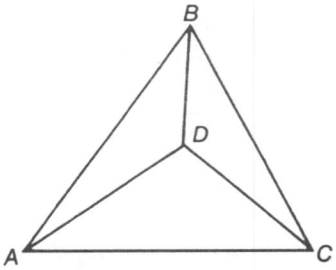

27. GIVEN: $AC > AB$,
$\overline{DE} \cong \overline{CE}$.
PROVE: $\overline{AB}$ is *not* parallel to $\overline{DE}$.

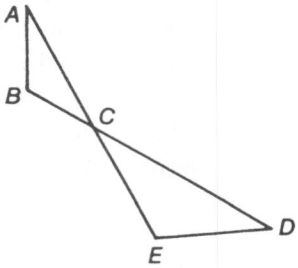

28. GIVEN: $\triangle ABC$ is scalene,
$\overline{BD}$ bisects $\measuredangle ABC$.
PROVE: $\overline{BD}$ is *not* $\perp \overline{AC}$.

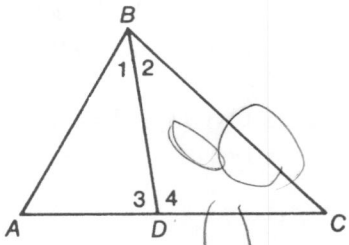

29. GIVEN: Quadrilateral $BCDE$, $\overline{BE}$ bisects $\measuredangle ABD$,
$\measuredangle BDC \cong \measuredangle C$, and $m\measuredangle E > m\measuredangle ABE$.
PROVE: $BC > ED$.

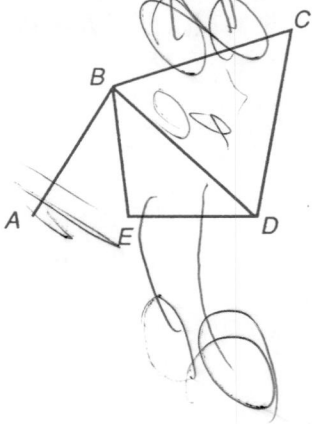

30. GIVEN: $\overline{AC} \cong \overline{BC}$, $\overline{AD} \cong \overline{BD}$, $\overline{AEC}$, and $\overline{BDE}$.

PROVE: **a** $\measuredangle CAD \cong \measuredangle CBD$.

b $AD > DE$.

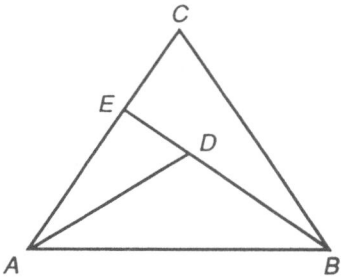

31. Prove that the length of the line segment drawn from any vertex of an equilateral triangle to a point on the opposite side is less than the length of any side of the triangle.

32. Prove that, if the vertex angle of an isosceles triangle is obtuse, then the base is longer than either leg.

33. Prove that the shortest distance from a point to a line is the length of the perpendicular segment from the point to the line.

34. Prove that an altitude of an acute scalene triangle *cannot* bisect the angle from whose vertex it is drawn.

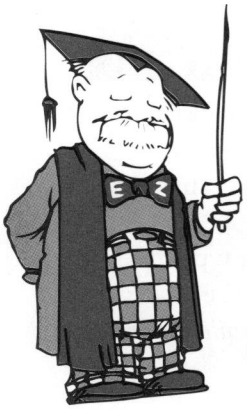

9

Special Quadrilaterals

WHAT YOU WILL LEARN

Quadrilaterals with one or two pairs of parallel sides have some interesting properties. In this chapter you will learn:

- the names and properties of special quadrilaterals;
- different ways of proving a quadrilateral is a parallelogram;
- different ways of proving a parallelogram is a rectangle, rhombus, or square;
- the relationship between a side of a triangle and the segment joining the midpoints of the two other sides;
- different ways of proving a trapezoid is isosceles.

SECTIONS IN THIS CHAPTER

- Classifying Quadrilaterals
- Properties of a Parallelogram
- Properties of Special Parallelograms
- Proving a Quadrilateral Is a Parallelogram
- Applications of Parallelograms
- Properties of a Trapezoid

Classifying Quadrilaterals

If you consider the relationships between the sides and the angle pairs of quadrilaterals to be genetic traits, then a family tree of quadrilaterals may be developed with a special name given to each type of quadrilateral that displays a special trait. This is illustrated in Figure 9.1.

The family tree of quadrilaterals shows that a quadrilateral has two major types of descendants. One, called a *trapezoid*, has exactly one pair of parallel sides. The other major type of quadrilateral has two pairs of parallel sides and is called a *parallelogram*.

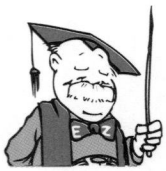

DEFINITION OF A PARALLELOGRAM

A *parallelogram* is a quadrilateral having two pairs of parallel sides.

NOTATION: ▱*ABCD* is read as "parallelogram *ABCD*." The letters *A*, *B*, *C*, and *D* represent consecutive vertices of the quadrilateral, and the symbol that precedes these letters is a miniature parallelogram.

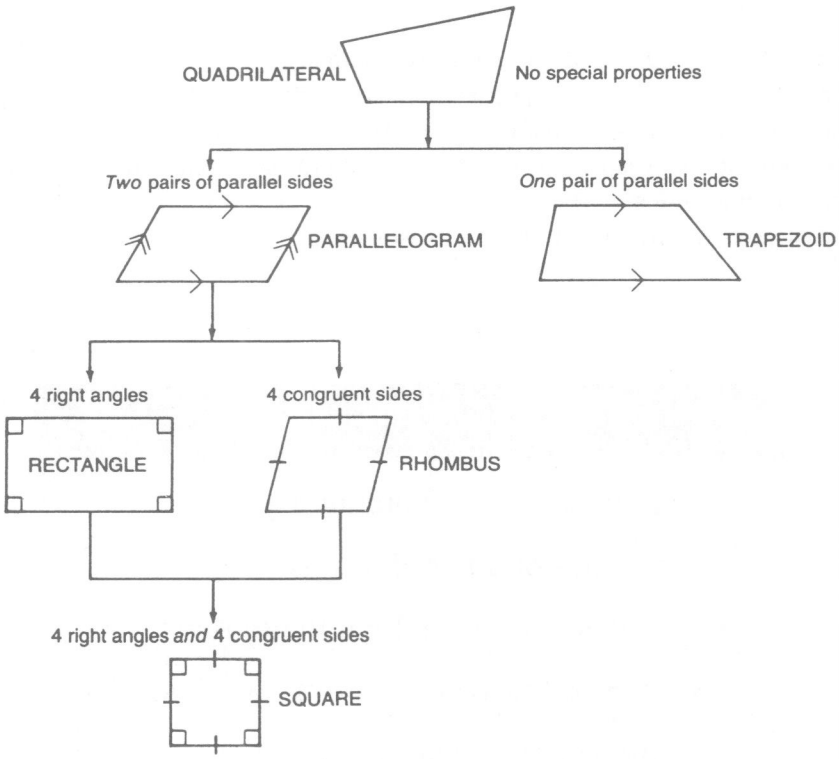

FIGURE 9.1

Of the special quadrilaterals illustrated in Figure 9.1, only the trapezoid is *not* a parallelogram. A rectangle, a rhombus, and a square are special types of parallelograms. The rest of this chapter is devoted to developing the properties of each of these figures.

Properties of a Parallelogram

ANGLES OF A PARALLELOGRAM

Consider parallelogram *ABCD* (Figure 9.2) in which the measure of angle *A* is 70. What are the measures of angles *B*, *C*, and *D*?

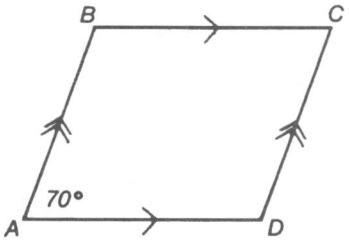

FIGURE 9.2

In parallelogram *ABCD*, $\overline{AB}$ may be considered to be a transversal, intersecting parallel line segments $\overline{AD}$ and $\overline{BC}$. Since interior angles on the same side of a transversal intersecting parallel lines are supplementary, $m\measuredangle B = 110$ $(180 - 70 = 110)$. Similarly, angles *B* and *C* are supplementary, so $m\measuredangle C = 70$, and angles *A* and *D* are supplementary, so $m\measuredangle D = 110$.

Notice also that, since angles *A* and *C* are each supplementary to angle *B* (and angle *D*), they are congruent (recall that "If two angles are supplementary to the same angle, then they are congruent"). For the same reason, opposite angles *B* and *D* are congruent. The following two theorems summarize these results.

ANGLES OF A PARALLELOGRAM THEOREMS

THEOREM 9.1 Consecutive pairs of angles of a parallelogram are supplementary.

THEOREM 9.2 Opposite angles of a parallelogram are congruent.

EXAMPLE 9.1 In parallelogram *ABCD* the measure of angle *B* is twice the measure of angle *A*. Find the measure of each angle of the parallelogram.

SOLUTION

Since angles A and B are consecutive angles of a parallelogram, by Theorem 9.1 they are supplementary. Hence, we may write

$$\begin{aligned} x + 2x &= 180 \\ 3x &= 180 \\ x &= 60 \\ m\measuredangle A = x &= 60 \\ m\measuredangle B = 2x &= 120 \\ m\measuredangle C = m\measuredangle A &= 60 \\ m\measuredangle D = m\measuredangle B &= 120 \end{aligned}$$

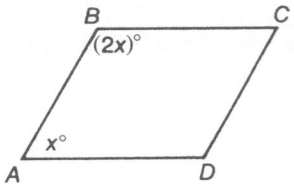

DIAGONALS AND SIDES OF A PARALLELOGRAM

If you draw either diagonal of a parallelogram, two triangles are formed. Is there a relationship between the triangles formed by a diagonal? Consider diagonal $\overline{BD}$, drawn in parallelogram $ABCD$, in Figure 9.3*a*. Notice that $\overline{BC}$, $\overline{BD}$, and $\overline{DA}$ form a Z. Since $\overline{BC}$ is parallel to $\overline{AD}$ $\measuredangle 1 \cong \measuredangle 2$. Similarly, $\overline{AB}$, $\overline{BD}$, and $\overline{DC}$ form a Z. Since, $\overline{AB} \parallel \overline{DC}$, $\measuredangle 3 \cong \measuredangle 4$. In addition, $\overline{BD} \cong \overline{BD}$, from which it follows that $\triangle BAD \cong \triangle DCB$ by ASA. Using the same approach it can be easily shown that in Figure 9.3*b* $\overline{AC}$ divides parallelogram $ABCD$ into two triangles such that $\triangle ABC \cong \triangle CDA$. These observations are summarized in Theorem 9.3.

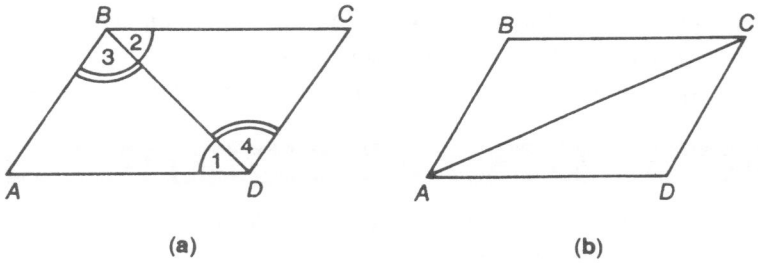

FIGURE 9.3

THEOREM 9.3

In a parallelogram either diagonal separates the parallelogram into two congruent triangles.

Theorem 9.3 allows us to draw some conclusions regarding the parts of the parallelogram. Applying the CPCTC principle, we may conclude that $\overline{AD} \cong \overline{BC}$ and $\overline{AB} \cong \overline{DC}$. This result is stated formally in Theorem 9.4.

THEOREM 9.4

Opposite sides of a parallelogram are congruent.

EXAMPLE 9.2

GIVEN: $\square ABCD$, diagonals $\overline{AC}$ and $\overline{BD}$ intersect at point E.

PROVE: **(a)** $\overline{AE} \cong \overline{EC}$

(b) $\overline{BE} \cong \overline{ED}$

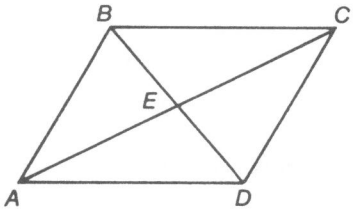

SOLUTION

PLAN: Prove $\triangle BEC \cong \triangle DEA$ by ASA.

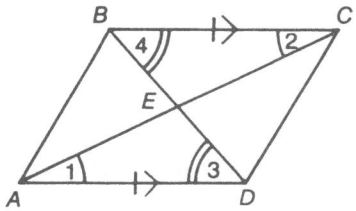

PROOF:

Statements	Reasons
1. $\square ABCD$.	1. Given.
2. $\overline{BC} \parallel \overline{AD}$.	2. Opposite sides of a parallelogram are parallel.
3. $\measuredangle 1 \cong \measuredangle 2$. (Angle)	3. If two lines are parallel, then their alternate interior angles are congruent.
4. $\overline{AD} \cong \overline{BC}$. (Side)	4. Opposite sides of a parallelogram are congruent.
5. $\measuredangle 3 \cong \measuredangle 4$. (Angle)	5. Same as reason 3.
6. $\triangle BEC \cong \triangle DEA$.	6. ASA Postulate.
7. $\overline{AE} \cong \overline{EC}$ and $\overline{BE} \cong \overline{ED}$.	7. CPCTC.

Example 9.2 establishes Theorem 9.5.

THEOREM 9.5

The diagonals of a parallelogram bisect each other.

SUMMARY OF PROPERTIES OF A PARALLELOGRAM

DEFINITION

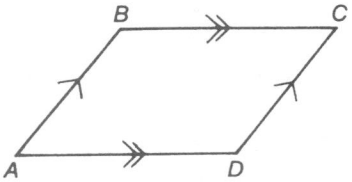

$\overline{AB} \parallel \overline{CD}$
and
$\overline{AD} \parallel \overline{BC}$

THEOREM 9.1

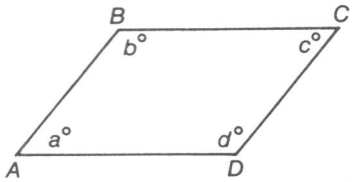

$a + b = 180$
$b + c = 180$
$c + d = 180$
$a + d = 180$

THEOREM 9.2

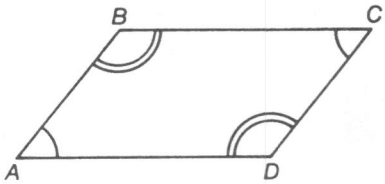

$\measuredangle A \cong \measuredangle C$
and
$\measuredangle B \cong \measuredangle D$

THEOREM 9.3

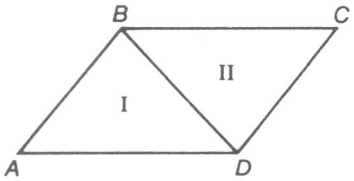

$\triangle\text{I} \cong \triangle\text{II}$

THEOREM 9.4

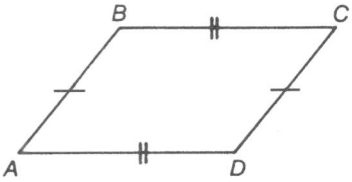

$\overline{AB} \cong \overline{CD}$
and
$\overline{AD} \cong \overline{BC}$

THEOREM 9.5

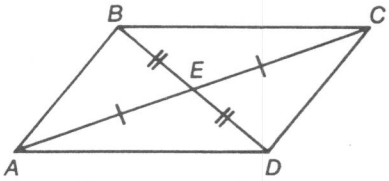

$\overline{AE} \cong \overline{EC}$
and
$\overline{BE} \cong \overline{ED}$

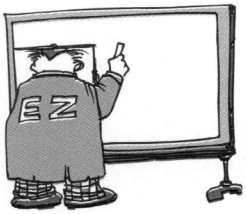

IN A PARALLELOGRAM

1. Opposite sides are parallel. (Definition of parallelogram)
2. Consecutive angles are supplementary. (Theorem 9.1)
3. Opposite angles are congruent. (Theorem 9.2)
4. Opposite sides are congruent. (Theorem 9.4)
5. Diagonals bisect each other. (Theorem 9.5)

Properties of Special Parallelograms

An equiangular parallelogram is called a *rectangle*. An equilateral parallelogram is a *rhombus*. A *square* is a parallelogram that is both equiangular and equilateral. It will be convenient to use the following definitions in our work with these figures.

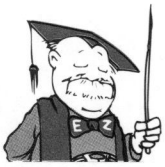

DEFINITIONS OF SPECIAL PARALLELOGRAMS

- A *rectangle* is a parallelogram having four right angles.
- A *rhombus* is a parallelogram having four congruent sides.
- A *square* is a rectangle having four congruent sides.

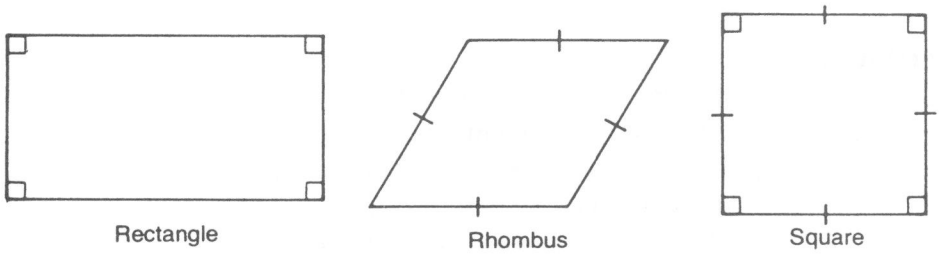

EXAMPLE 9.3

GIVEN: Rectangle $ABCD$.

PROVE: $\overline{AC} \cong \overline{DB}$.

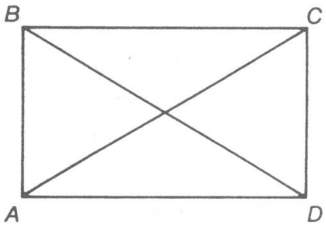

SOLUTION

OUTLINE OF PROOF: Prove $\triangle BAD \cong \triangle CDA$ by SAS.

- $\overline{AB} \cong \overline{CD}$ since opposite sides of a rectangle are congruent.
- Angles BAD and CDA are congruent since they are right angles. $\overline{AD} \cong \overline{AD}$.
- Since the triangles are congruent, $\overline{AC} \cong \overline{DB}$ by CPCTC.

Example 9.3 establishes Theorem 9.6.

THEOREM 9.6

The diagonals of a rectangle are congruent.

EXAMPLE 9.4

GIVEN: Rhombus $ABCD$.

PROVE: **a** $\measuredangle 1 \cong \measuredangle 2$.

b $\measuredangle 3 \cong \measuredangle 4$.

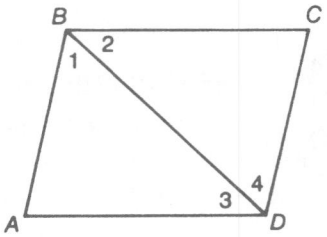

SOLUTION

OUTLINE OF PROOF: Prove $\triangle BAD \cong \triangle BCD$ by SAS.

- $\overline{AB} \cong \overline{CB}$ since a rhombus is equilateral.
- Angles A and C are congruent since opposite angles of a rhombus are congruent.
- $\overline{AD} \cong \overline{CD}$ since a rhombus is equilateral.
- The desired angle pairs are congruent by the CPCTC principle.

Notice that diagonal $\overline{BD}$ bisects angles B and D. Using a similar approach, we can show that diagonal $\overline{AC}$ bisects angles A and C. Thus, the *diagonals of a rhombus bisect the four angles of the rhombus*.

EXAMPLE 9.5 Prove the diagonals of a rhombus are perpendicular to each other.

SOLUTION

GIVEN: Rhombus $ABCD$,
diagonals $\overline{AC}$ and $\overline{BD}$ intersect at E.

PROVE: $\overline{AC} \perp \overline{BD}$.

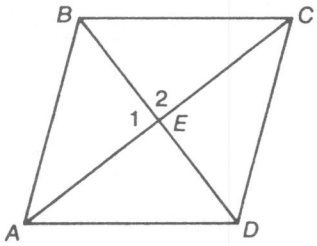

OUTLINE OF PROOF: Prove $\triangle AEB \cong \triangle CEB$ (other triangle pairs may be selected) by SAS.

- $\overline{AB} \cong \overline{CB}$ since a rhombus is equilateral.
- Angles ABE and CBE are congruent by Theorem 9.7 since a diagonal of a rhombus bisects the angle formed at each vertex.
- $\overline{BE} \cong \overline{BE}$. By CPCTC, angles 1 and 2 are congruent.
- $\overline{AC} \perp \overline{BD}$ since the lines intersect to form a congruent pair of adjacent angles.

DIAGONALS OF A RHOMBUS THEOREMS

THEOREM 9.7 The diagonals of a rhombus bisect the four angles of the rhombus (see Example 9.4)

THEOREM 9.8 The diagonals of a rhombus are perpendicular to each other (see Example 9.5)

Since a rhombus includes all the properties of a parallelogram, the diagonals of a rhombus bisect each other. Each diagonal of a rhombus is the *perpendicular bisector of the other diagonal*.

EXAMPLE 9.6

GIVEN: *ABCD* in the accompanying figure is a rhombus and m $\measuredangle$ 1 = 40. Find the measure of each of the following angles:

(a) $\measuredangle 2$ **(b)** $\measuredangle 3$ **(c)** $\measuredangle ADC$

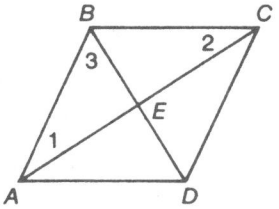

SOLUTION

a. Triangle *ABC* is isosceles since $\overline{AB} \cong \overline{BC}$. Hence, the base angles of the triangle must be congruent.

$$m\measuredangle 1 = m\measuredangle 2 = 40$$

b. In triangle *AEB*, angle *AEB* is a right angle since the diagonals of a rhombus are perpendicular to each other. Since the sum of the measures of the angles of a triangle is 180, the measure of angle 3 must be 50.

c. Since the diagonals of a rhombus bisect the angles of the rhombus, if $m\measuredangle 3 = 50$, then $m\measuredangle ABC = 100$. Since opposite angles of a rhombus are equal in measure, $m\measuredangle ADC$ must also equal 100.

REMEMBER
A rhombus is not necessarily a rectangle, and a rectangle is not necessarily a rhombus.

Keep in mind that a rhombus is *not* necessarily a rectangle since it may or may not contain four right angles. A rectangle is *not* necessarily a rhombus since it may or may not contain four congruent sides. A square combines the properties of a rectangle with the properties of a rhombus. The diagonals of a square are congruent, bisect its opposite angles, and intersect at right angles.

EXAMPLE 9.7

GIVEN: Square $ABCD$, $\overline{AE} \cong \overline{DF}$.

PROVE: $\overline{AF} \cong \overline{BE}$.

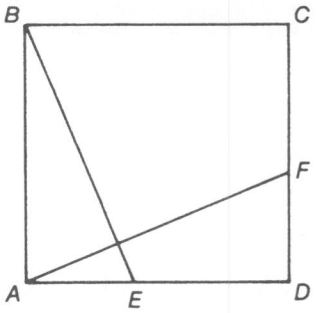

SOLUTION

PLAN: Prove $\triangle BAE \cong \triangle ADF$ by SAS.

PROOF:

Statements	Reasons
1. $ABCD$ is a square.	1. Given.
2. $\overline{BA} \cong \overline{DA}$. (Side)	2. A square is equilateral.
3. Angles A and D are right angles.	3. A square contains four right angles.
4. $\measuredangle A \cong \measuredangle D$. (Angle)	4. All right angles are congruent.
5. $\overline{AE} \cong \overline{DF}$. (Side)	5. Given.
6. $\triangle BAE \cong \triangle ADF$.	6. SAS Postulate.
7. $\overline{AF} \cong \overline{BE}$.	7. CPCTC.

SUMMARY OF PROPERTIES OF A RECTANGLE, RHOMBUS, AND SQUARE

Property	Rectangle	Rhombus	Square
1. All the properties of a parallelogram?	Yes	Yes	Yes
2. Equiangular (4 right angles)?	Yes	No	Yes
3. Equilateral (4 congruent sides)?	No	Yes	Yes
4. Diagonals congruent?	Yes	No	Yes
5. Diagonals bisect opposite angles?	No	Yes	Yes
6. Diagonals perpendicular?	No	Yes	Yes

Proving a Quadrilateral Is a Parallelogram

The preceding sections developed the properties of quadrilaterals that were known to be parallelograms. We now consider the other side of the coin. How can we prove that a quadrilateral is a parallelogram? What is the minimum information required to justify the conclusion that a quadrilateral is a parallelogram?

Using the reverse of the definition of a parallelogram, we know that, if *both pairs of sides of a quadrilateral are parallel, then the quadrilateral is a parallelogram*. We may use this fact to establish alternative methods for proving that a quadrilateral is a parallelogram. For example, we may try drawing a quadrilateral in which the same pair of sides are both congruent and parallel. Imposing this condition forces the remaining pair of sides to be parallel.

THEOREM 9.9

If a quadrilateral has one pair of sides that are both parallel and congruent, then the quadrilateral is a parallelogram.

GIVEN: Quadrilateral $ABCD$, $\overline{AD} \parallel \overline{BC}, \overline{AD} \cong \overline{BC}$.

PROVE: Quadrilateral $ABCD$ is a parallelogram.

PLAN: Show $\overline{AB} \parallel \overline{CD}$.

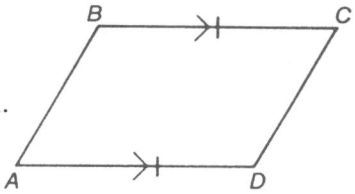

OUTLINE OF PROOF: Draw diagonal $\overline{AC}$. Prove $\triangle ABC \cong \triangle CDA$ by SAS.

- $\overline{BC} \cong \overline{AD}$ (Side)
- $\measuredangle 1 \cong \measuredangle 2$ (Angle)
- $\overline{AC} \cong \overline{AC}$ (Side)

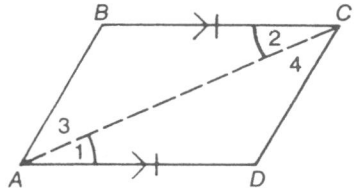

Angles 3 and 4 are congruent by CPCTC. This implies that $\overline{AB} \parallel \overline{CD}$ (since alternate interior angles are congruent, the lines are parallel). Since both pairs of sides of quadrilateral $ABCD$ are parallel, $ABCD$ is a parallelogram.

Are there any additional methods for proving a quadrilateral is a parallelogram? The converses of the theorems that state the properties of quadrilaterals that are parallelograms may offer some clues. Recall that these theorems take the general form

If a quadrilateral is a parallelogram, then a *certain property is true*.

The converse of this statement takes the form

If a quadrilateral has a *certain property*, then it is a parallelogram.

Keeping in mind that the converse of a theorem is *not* necessarily true, we need to investigate whether the converse holds for each special property of parallelograms. Table 9.1 summarizes the results of these investigations.

TABLE 9.1

If a Quadrilateral has . . .	Then is it a Parallelogram (?)
• Congruent opposite sides 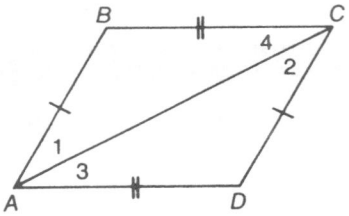	Yes, since $\triangle ABC \cong \triangle CDA$ by SSS. By CPCTC, angles 1 and 2 are congruent, which implies that $\overline{AB} \parallel \overline{CD}$. Similarly, angles 3 and 4 are congruent making $\overline{AD} \parallel \overline{BC}$. Since both pairs of sides are parallel, *ABCD* is a parallelogram.
• Congruent opposite angles 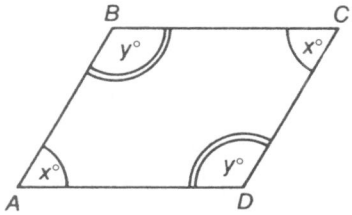	Yes. Since the sum of the interior angles of a quadrilateral is 360, $x + y + x + y = 360$ $2x + 2y = 360$ $2(x + y) = 360$ $x + y = 180$ If interior angles on the same side of a transversal are supplementary, the lines are parallel. It follows that $\overline{AB} \parallel \overline{DC}$ and $\overline{AD} \parallel \overline{BC}$. *ABCD* is therefore a parallelogram.
• Diagonals that bisect each other 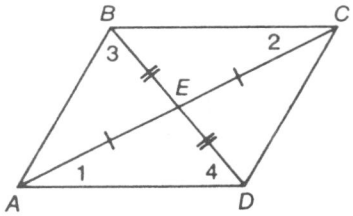	Yes. $\triangle AED \cong \triangle BEC$ by SAS. By CPCTC, $\overline{AD} \cong \overline{BC}$ and $\measuredangle 1 \cong \measuredangle 2$. It follows that $\overline{AD} \parallel \overline{BC}$. Since $\overline{AD}$ is both congruent and parallel to $\overline{BC}$, *ABCD* is a parallelogram (see Theorem 9.9).

Theorem 9.10 gives three additional methods for proving that a quadrilateral is a parallelogram.

THEOREM 9.10

A quadrilateral is a parallelogram if any one of the following is true:

- Opposite sides are congruent.
- Opposite angles are congruent.
- Diagonals bisect each other.

EXAMPLE 9.8 Draw a diagram to help prove or disprove that a quadrilateral is a parallelogram if:
(a) One pair of opposite sides are congruent.
(b) One pair of sides are parallel.
(c) Two pairs of sides are congruent.

SOLUTION

a. Not necessarily a parallelogram:

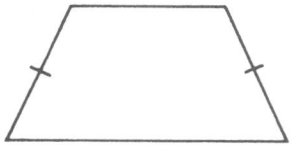

b. Not necessarily a parallelogram:

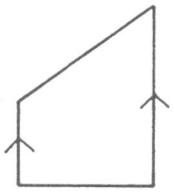

c. Not necessarily a parallelogram:
NOTE: This type of figure is referred to as a *kite*.

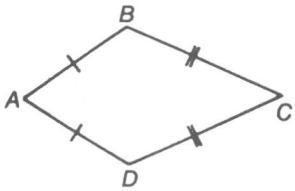

EXAMPLE 9.9

GIVEN: $\overline{BE} \perp \overline{AC}, \overline{DF} \perp \overline{AC}$,
$\overline{BE} \cong \overline{DF}$,
$\measuredangle EBC \cong \measuredangle FDA$.

PROVE: $ABCD$ is a parallelogram.

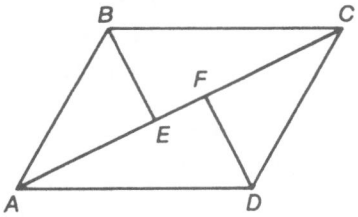

SOLUTION

PLAN: Prove $\overline{BC}$ is parallel and congruent to $\overline{AD}$ by first proving that $\triangle BEC \cong \triangle DFA$ by ASA.

PROOF:

Statements	Reasons
1. $\measuredangle EBC \cong \measuredangle FDA$. (Angle)	1. Given.
2. $\overline{BE} \cong \overline{DF}$. (Side)	2. Given.
3. $\overline{BE} \perp \overline{AC}$ and $\overline{DF} \perp \overline{AC}$.	3. Given.

4. $\measuredangle BEC \cong \measuredangle DFA$. (Angle)	4. Perpendicular lines intersect to form right angles. All right angles are congruent. (*NOTE*: We have consolidated steps.)
5. $\triangle BEC \cong \triangle DFA$.	5. ASA Postulate.
6. $\overline{AD} \cong \overline{BC}$ and $\measuredangle BCE \cong \measuredangle DAF$.	6. CPCTC.
7. $\overline{AD} \parallel \overline{BC}$.	7. If alternate interior angles are congruent, then the lines are parallel.
8. Quadrilateral $ABCD$ is a parallelogram.	8. If a quadrilateral has a pair of sides that are both parallel and congruent, then the quadrilateral is a parallelogram (Theorem 9.9).

EXAMPLE 9.10

GIVEN: $\square ABCD$,
$\overline{AE}$ bisects $\measuredangle BAD$,
$\overline{CF}$ bisects $\measuredangle BCD$.

PROVE: Quadrilateral $AECF$ is a parallelogram.

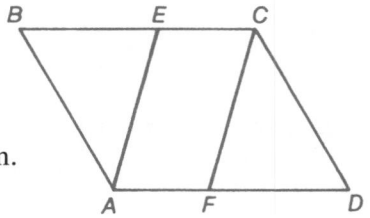

SOLUTION

PLAN: Prove $\triangle ABE \cong \triangle CDF$ by ASA. $AE = CF$ and $BE = DF$ by CPCTE. $EC = AF$ by subtraction. Hence, $AECF$ is a parallelogram since opposite sides have the same length.

PROOF:

Statements	Reasons
1. $ABCD$ is a parallelogram.	1. Given.
2. $m\measuredangle BAD = m\measuredangle DCB$.	2. Opposite angles of a parallelogram are equal in measure.
3. $\overline{AE}$ bisects $\measuredangle BAD$. $\overline{CF}$ bisects $\measuredangle BCD$.	3. Given.
4. $\measuredangle BAE \cong \measuredangle DCF$. (Angle)	4. Halves of equals are equal (and therefore congruent).
5. $\overline{AB} \cong \overline{DC}$. (Side)	5. Opposite sides of a parallelogram are congruent.
6. $\measuredangle B \cong \measuredangle D$. (Angle)	6. Same as reason 2.
7. $\triangle ABE \cong \triangle CDF$.	7. ASA Postulate.
8. $AE = FC$.	8. Corresponding sides of congruent triangles are equal in length.

To show $EC = AF$ *use subtraction*:	
9. $BE = DF$.	9. Same as reason 8.
10. $BC = AD$.	10. Opposite sides of a parallelogram are equal in length.
11. $EC = AF$.	11. Subtraction property of equality.
12. $AECF$ is a parallelogram.	12. If the opposite sides of a quadrilateral are equal in length (that is, congruent), then the quadrilateral is a parallelogram.

In the proof of Example 9.10, statements and corresponding reasons are expressed in terms of equality of measures rather than congruence. This is necessary since arithmetic operations (taking halves of equals and subtraction) were performed on these quantities.

SUMMARY

TO PROVE A QUADRILATERAL IS A PARALLELOGRAM

Show that any one of the following is true:

- Opposite sides are parallel.
- Opposite sides are congruent.
- Opposite angles are congruent.
- Diagonals bisect each other.
- A pair of sides are both parallel and congruent.

Sometimes we need to prove that a quadrilateral (or parallelogram) is a rectangle, rhombus, or square:

1. To prove a quadrilateral is a *rectangle*, show that it is a *parallelogram* having one of the following properties:
   - It contains a right angle.
   - The diagonals are congruent.
2. To prove a quadrilateral is a *rhombus*, show that it is a *parallelogram* having one of the following properties:
   - It contains a pair of congruent adjacent sides.
   - The diagonals intersect at right angles.
   - The diagonals bisect the vertex angles.

3. To prove that a quadrilateral is a *square*, show that it is either of the following:
   - A *rectangle* with an adjacent pair of congruent sides.
   - A *rhombus* with a right angle.

Applications of Parallelograms

In Figure 9.4*a*, *ABCD* is a rectangle. If $BD = 10$, what is the length of $\overline{AM}$? The diagonals of a rectangle are congruent, making $AC = 10$. Since the diagonals bisect each other, $AM = 5$ (and $MC = 5$). We can redraw the diagram as Figure 9.4*b* by deleting $\overline{BC}$, $\overline{MC}$, and $\overline{CD}$. Since M is the midpoint of $\overline{BD}$, $\overline{AM}$ is the median to hypotenuse $\overline{BD}$ of right triangle *BAD*. This suggests Theorem 9.11.

THEOREM 9.11

The length of the median drawn to the hypotenuse of a right triangle is one-half of the length of the hypotenuse.

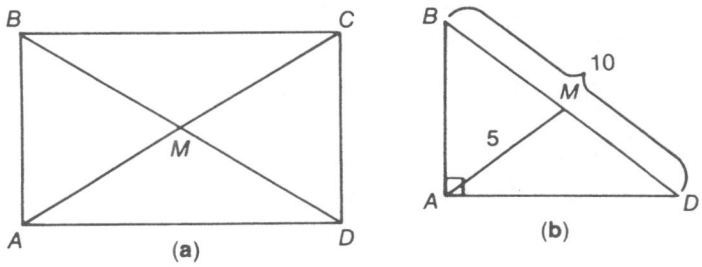

FIGURE 9.4

The theorem that we will now establish states that if the midpoints of *any* two sides of a triangle are connected by a line segment, then this segment must be parallel to the remaining side of the triangle. Furthermore, its length must be exactly one-half the length of the remaining side of the triangle. This theorem may be stated formally as follows.

THEOREM 9.12 MIDPOINTS OF A TRIANGLE THEOREM

The line segment joining the midpoints of two sides of a triangle is parallel to the third side and is one-half its length.

GIVEN: D and E are midpoints of sides $\overline{AB}$ and $\overline{CB}$, respectively.

PROVE: **(a)** $\overline{DE} \parallel \overline{AC}$.

(b) $DE = \frac{1}{2} AC$.

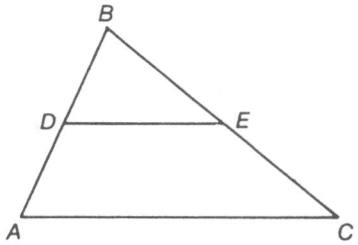

OUTLINE OF PROOF: Extend $\overline{DE}$ so that $\overline{DE} \cong \overline{EF}$, and then draw $\overline{CF}$.

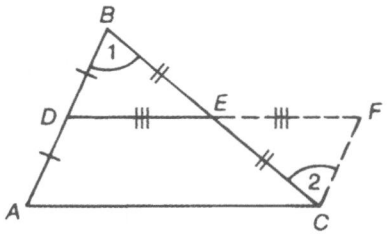

Prove $ADFC$ is a parallelogram by showing that $\overline{CF}$ is parallel and congruent to $\overline{AD}$ as follows:

- $\triangle DBE \cong \triangle FCE$ (by SAS).
- By CPCTC, angles 1 and 2 are congruent, implying $\overline{FC} \parallel \overline{AD}$. Also, $\overline{CF} \cong \overline{DB} \cong \overline{AD}$.
- Since $ADFC$ is a parallelogram, $\overline{DE} \parallel \overline{AC}$. Also,

$$DE = \frac{1}{2} DF = \frac{1}{2} AC.$$

EXAMPLE 9.11 In triangle RST, A is the midpoint of $\overline{RS}$ and B is the midpoint of $\overline{RT}$.

a. If $ST = 18$, find AB.

b. If $AB = 7$, find ST.

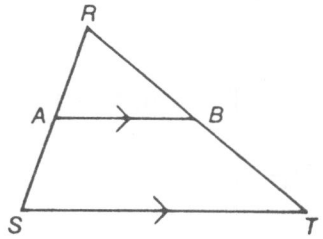

SOLUTION

a. $AB = \frac{1}{2} ST = \frac{1}{2}(18) = 9.$

b. $ST = 2(AB) = 2(7) = 14.$

EXAMPLE 9.12

GIVEN: Points Q, R, and S are midpoints.

PROVE: $PQRS$ is a parallelogram.

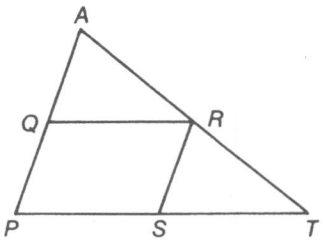

SOLUTION

PLAN: Show that $\overline{QR}$ is parallel and congruent to $\overline{PS}$.

PROOF:

Statements	Reasons
1. Points Q, R, and S are midpoints.	1. Given.
2. $\overline{QR} \parallel \overline{PT}$.	2. The line segment joining the midpoints of two sides of a triangle is parallel to the third side and is one-half its length.
3. $\overline{QR} \parallel \overline{PS}$.	3. Segments of parallel lines are parallel.
4. $QR = \frac{1}{2} PT$.	4. Same as reason 2.
5. $PS = \frac{1}{2} PT$.	5. A midpoint divides a segment into two congruent segments.
6. $\overline{QR} \cong \overline{PS}$.	6. Transitive property.
7. Quadrilateral $PQRS$ is a parallelogram.	7. If a quadrilateral has a pair of sides that are both parallel and congruent, then the quadrilateral is a parallelogram.

Properties of a Trapezoid

Unlike a parallelogram, a trapezoid has exactly one pair of parallel sides. In Figure 9.5, quadrilaterals *ABCD* and *RSTW* are trapezoids. Quadrilateral *JKLM* is *not* a trapezoid since it does not have one pair of parallel sides.

DEFINITION OF A TRAPEZOID

A *trapezoid* is a quadrilateral that has exactly *one* pair of parallel sides. The parallel sides are called the *bases* of the trapezoid. The nonparallel sides are referred to as the *legs* of the trapezoid.

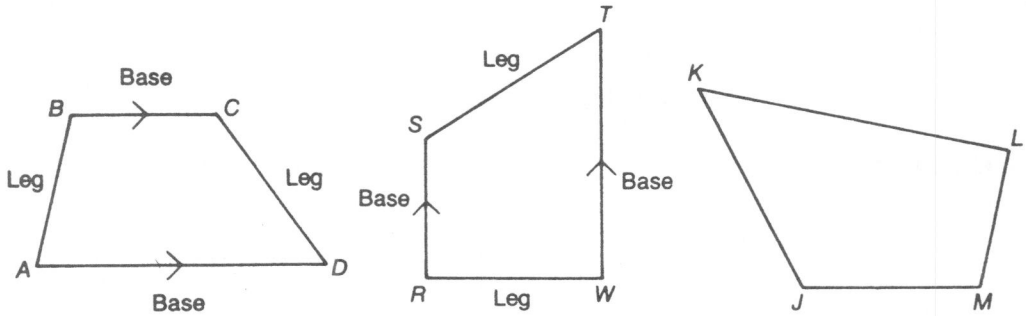

FIGURE 9.5

ALTITUDE AND MEDIAN OF A TRAPEZOID

Two additional terms that are used in connection with trapezoids are *altitude* and *median*. An *altitude* of a trapezoid (or for that matter, a parallelogram) is a segment drawn from any point on one of the parallel sides (base) perpendicular to the opposite side (other base). An infinite number of altitudes may be drawn in a trapezoid. See Figure 9.6.

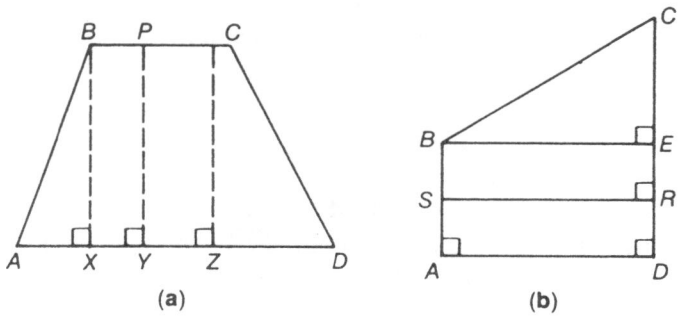

FIGURE 9.6 *(a)* $\overline{BX}$, $\overline{PY}$, and $\overline{CZ}$ are examples of *altitudes*.
(b) $\overline{AD}$, $\overline{SR}$, and $\overline{BE}$ are examples of *altitudes*.

A *median* of a trapezoid is the segment that joins the midpoints of the nonparallel sides (legs). A trapezoid has exactly one median. See Figure 9.7.

As Figure 9.7 seems to indicate, the median of a trapezoid appears to be parallel to the bases. In addition, there is a relationship between the length of the median and the lengths of the bases of the trapezoid. Our strategy in developing the properties of the median of a trapezoid is to apply Theorem 9.12. To accomplish this, we must draw an auxiliary segment so that the median of the trapezoid becomes the segment that joins the midpoints of a newly formed triangle.

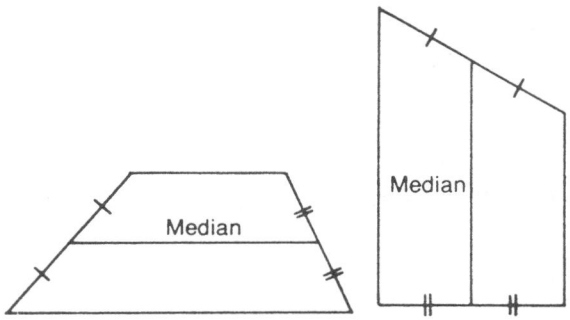

FIGURE 9.7

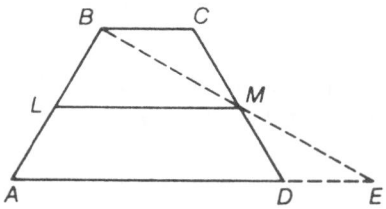

FIGURE 9.8

In trapezoid *ABCD* of Figure 9.8, median $\overline{LM}$ has been drawn. By drawing line segment $\overline{BM}$ and extending it to meet the extension of line segment $\overline{AD}$, we form triangle *ABE*. Point *L* is the midpoint of side $\overline{AB}$ of triangle *ABE*. If we can show that point M is the midpoint of side $\overline{BE}$ of triangle ABE, then we can apply Theorem 9.12, which states that the line segment joining the midpoints of two sides of a triangle is parallel to the third side of the triangle. This would establish that median $\overline{LM}$ is parallel to $\overline{BC}$ and $\overline{AD}$.

To show that M is the midpoint of side $\overline{BE}$, we consider triangles BMC and EMD. They are congruent by ASA:

$$\begin{aligned} \measuredangle BMC &\cong \measuredangle EMD \quad \text{(Angle)} \\ \overline{CM} &\cong \overline{DM} \quad \text{(Side)} \end{aligned}$$

and since $\overline{BC} \parallel \overline{AE}$,

$$\measuredangle BCM \cong \measuredangle EDM \quad \text{(Angle)}$$

By CPCTC, $\overline{BM} \cong \overline{EM}$. Hence, Theorem 9.12 applies.

Theorem 9.12 also informs us that the length of $\overline{LM}$ (the segment joining the midpoints of two sides of triangle ABE) must be one-half the length of the third side. Thus,

$$LM = \frac{1}{2}AE = \frac{1}{2}(AD + DE)$$

Since we have established that $\triangle BMC \cong \triangle EMD$, $\overline{DE} \cong \overline{BC}$. Applying the substitution principle, we have

$$LM = \frac{1}{2}(AD + BC) = \frac{1}{2}\text{(sum of lengths of bases)}$$

The result of our investigation is stated formally in Theorem 9.13.

THEOREM 9.13 MEDIAN OF A TRAPEZOID THEOREM

The median of a trapezoid is parallel to the bases and has a length equal to one-half the sum of the lengths of the bases.

EXAMPLE 9.13

In trapezoid $ABCD$, $\overline{BC} \parallel \overline{AD}$ and $\overline{RS}$ is the median.

a. If $AD = 13$ and $BC = 7$, find RS.

b. If $BC = 6$ and $RS = 11$, find AD.

SOLUTION

a. $RS = \frac{1}{2}(13 + 7) = \frac{1}{2}(20) = 10$

b. If median = $\frac{1}{2}$ (sum of lengths of bases), then

$$\begin{aligned} \text{Sum of lengths of bases} &= 2 \times \text{median} \\ AD + 6 &= 2 \times 11 \\ AD + 6 &= 22 \\ AD &= 22 - 6 \\ &= 16 \end{aligned}$$

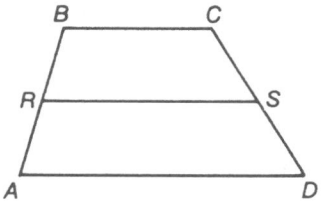

EXAMPLE 9.14 The length of the lower base of a trapezoid is 3 times the length of the upper base. If the median has a 24-inch length, find the lengths of the bases.

SOLUTION

$$\begin{aligned} \text{Let } a &= \text{length of upper base.} \\ \text{Then } 3a &= \text{length of lower base.} \\ \text{Since sum of lengths of bases} &= 2 \times \text{median,} \\ a + 3a &= 2 \times 24 \\ 4a &= 48 \\ a &= \frac{48}{4} \\ &= 12 \\ 3a &= 36 \end{aligned}$$

SUMMARY

SPECIAL MEDIAN RELATIONSHIPS

- The length of the median drawn to the hypotenuse of a right triangle is one-half the length of the hypotenuse.
- The length of the median of a trapezoid may be found by taking the average of the lengths of the two bases:

$$\text{Median} = \frac{1}{2}(\text{upper-base length} + \text{lower-base length})$$

- The median of a trapezoid is parallel to both bases.

THE ISOSCELES TRAPEZOID

If the legs of a trapezoid are congruent, then the trapezoid is called an *isosceles trapezoid*. See Figure 9.9. An *isosceles trapezoid* features some special properties not found in all trapezoids.

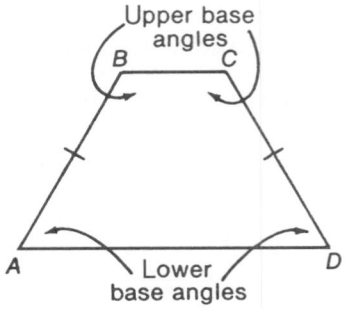

FIGURE 9.9

THEOREM 9.14

The lower (and upper) base angles of an isosceles trapezoid are congruent.

OUTLINE OF PROOF

GIVEN: Isosceles trapezoid $ABCD$.

PROVE: $\measuredangle A \cong \measuredangle D$.

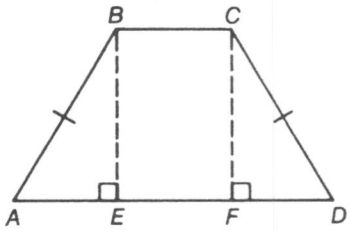

Draw altitudes $\overline{BE}$ and $\overline{CF}$, thereby forming triangles ABE and DCF. Prove these triangles congruent by applying the Hy-Leg method:

$$\overline{AB} \cong \overline{DC} \text{ (Hypotenuse)}$$

Since parallel lines are everywhere equidistant,

$$\overline{BE} \cong \overline{CF} \text{ (Leg)}$$

By CPCTC, $\measuredangle A \cong \measuredangle D$.

Since supplements of congruent angles are congruent, the upper base angles are congruent. We now turn our attention to the diagonals of an isosceles trapezoid.

THEOREM 9.15

The diagonals of an isosceles trapezoid are congruent.

OUTLINE OF PROOF

GIVEN: Isosceles trapezoid $ABCD$, diagonals $\overline{AC}$ and $\overline{DB}$.

PROVE: $\overline{AC} \cong \overline{DB}$.

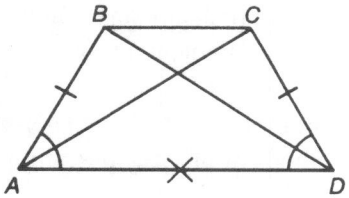

Diagonals $\overline{AC}$ and $\overline{DB}$ form a pair of overlapping triangles, $\triangle ACD$ and $\triangle DBA$. Prove these triangles congruent by using SAS. The included angles, angles BAD and CDA, are congruent as a result of Theorem 9.14. By CPCTC, $\overline{AC} \cong \overline{PB}$.

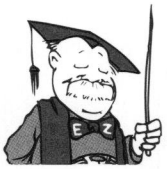

PROPERTIES OF AN ISOSCELES TRAPEZOID

1. It has all the properties of a trapezoid.
2. The legs are congruent. (Definition of isosceles trapezoid)
3. The base angles are congruent. (Theorem 9.14)
4. The diagonals are congruent. (Theorem 9.15)

EXAMPLE 9.15 Prove the opposite angles of an isosceles trapezoid are supplementary.

SOLUTION

GIVEN: Isosceles trapezoid $ABCD$.

PROVE: **a.** Angles A and C are supplementary.
b. Angles D and B are supplementary.

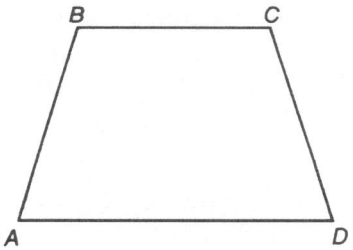

PROOF:

Statements	Reasons
1. $ABCD$ is an isosceles trapezoid.	1. Given.
2. $\overline{AD} \parallel \overline{BC}$.	2. The bases of a trapezoid are parallel.
3. $m\measuredangle D + m\measuredangle C = 180$.	3. If two lines are parallel, then interior angles on the same side of the transversal are supplementary.
4. $m\measuredangle A = m\measuredangle D$.	4. Base angles of an isosceles trapezoid are equal in measure.
5. $m\measuredangle A + m\measuredangle C = 180$.	5. Substitution.
6. Angles A and C are supplementary.	6. If the sum of the measures of two angles is 180, then the angles are supplementary.

In a similar fashion, it can be shown easily that angles D and B are supplementary.

For the sake of completeness we state without proof Theorem 9.16, which offers methods of proving that a trapezoid is an isosceles trapezoid.

THEOREM 9.16 WAYS OF PROVING A TRAPEZOID IS ISOSCELES

A trapezoid is an isosceles trapezoid if any one of the following is true:

1. The legs are congruent.
2. The base angles are congruent.
3. The diagonals are congruent.

The congruent-legs part of Theorem 9.16 follows directly from the reverse of the definition of an isosceles trapezoid. The congruent-base-angles part can be proved by dropping altitudes to the longer base, as shown in the diagram. If the base angles of this trapezoid are congruent, then $\triangle BEA \cong \triangle CFD$ by AAS since $\measuredangle A \cong \measuredangle D$, right angles BEA and CFD are congruent, and $\overline{BE} \cong \overline{CF}$ (parallel lines are everywhere equidistant). Therefore, $\overline{AB} \cong \overline{CD}$ by CPCTC and $ABCD$ is an isosceles trapezoid. The congruent-diagonals part of the theorem can be established by first proving $\triangle BED \cong \triangle CFA$ in order to obtain $\measuredangle 1 \cong \measuredangle 2$, which can then be used to help prove $\triangle ABD \cong \triangle DCA$. By CPCTC, $\overline{AB} \cong \overline{DC}$. You are asked to supply the details of this proof in Exercise 36 at the end of this chapter.

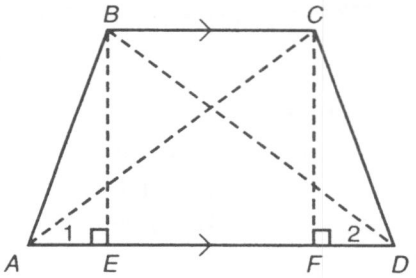

SUMMARY OF PROPERTIES OF DIAGONALS OF SPECIAL QUADRILATERALS

Special Quadrilateral	Diagonals Are Always		Diagonals Always Bisect	
	Congruent	Perpendicular	Each Other	Vertex Angles
Parallelogram	No	No	Yes	No
Rectangle	Yes	No	Yes	No
Rhombus	No	Yes	Yes	Yes
Square	Yes	Yes	Yes	Yes
Trapezoid	No	No	No	No
Isosceles Trapezoid	Yes	No	No	No

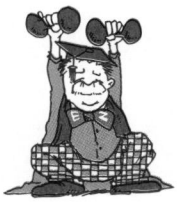

REVIEW EXERCISES FOR CHAPTER 9

1. In a certain quadrilateral, two opposite sides are parallel, and the other two opposite sides are not congruent. This quadrilateral could be a

   (1) rhombus (2) parallelogram (3) rectangle (4) trapezoid

2. In the accompanying diagram of parallelogram $ABCD$, $\overline{DE} \cong \overline{BF}$.

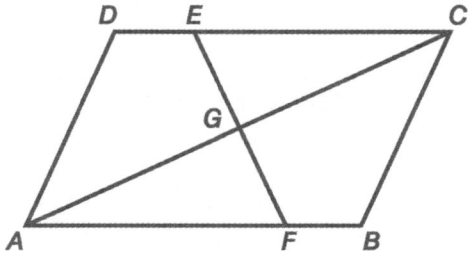

   Triangle EGC can be proved congruent to triangle FGA by

   (1) $HL \cong HL$ (2) $AAA \cong AAA$ (3) $AAS \cong AAS$ (4) $SSA \cong SSA$

3. Which quadrilateral must have diagonals that are congruent and perpendicular?

   (1) rhombus (2) square (3) trapezoid (4) parallelogram

4. Which statement about a diagonal of a parallelogram is *always* true?

   (1) It bisects the other diagonal of the parallelogram.
   (2) It bisects an angle of the parallelogram.
   (3) It is congruent to the other diagonal of the parallelogram.
   (4) It is perpendicular to the other diagonal of the parallelogram.

5. In rhombus $RSTW$, diagonal $\overline{RT}$ is drawn. If m $\measuredangle RST = 108$, find m $\measuredangle SRT$.

6. In parallelogram $MATH$ the measure of angle T exceeds the measure of angle H by 30. Find the measure of each angle of the parallelogram.

7. In parallelogram $TRIG$, m$\measuredangle R = 2x + 19$ and m$\measuredangle G = 4x - 17$. Find the measure of each angle of the parallelogram.

8. The length of the median drawn to the hypotenuse of a right triangle is represented by the expression $3x - 7$, while the length of the hypotenuse is represented by $5x - 4$. Find the length of the median.

9. In triangle RST, E is the midpoint of $\overline{RS}$ and F is the midpoint of $\overline{ST}$. If $EF = 5y - 1$ and $RT = 7y + 10$, find the lengths of $\overline{EF}$ and $\overline{RT}$.

10. In trapezoid $BYTE$, $\overline{BE} \parallel \overline{YT}$ and median $\overline{LM}$ is drawn.

    (a) $LM = 35$. If the length of $\overline{BE}$ exceeds the length of $\overline{YT}$ by 13, find the lengths of the bases.
    (b) If $YT = x + 9$, $LM = x + 15$, and $BE = 2x - 5$, find the lengths of $\overline{YT}$, $\overline{LM}$, and $\overline{BE}$.

11. In parallelogram $RSTW$ diagonals $\overline{RT}$ and $\overline{SW}$ intersect at point A. If $SA = x - 13$ and $AW = 2x - 37$, find SW.

12. The lengths of the sides of a triangle are 9, 40, and 41. Find the perimeter of the triangle formed by joining the midpoints of the sides.

13. GIVEN: $\square ABCD$, $\overline{AE} \cong \overline{CF}$.
    PROVE: $\measuredangle ABE \cong \measuredangle CDF$.

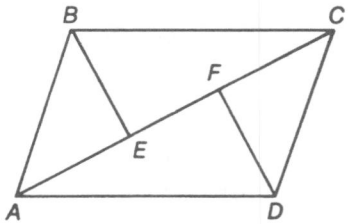

14. GIVEN: $\square ABCD$,
$\overline{EF} \cong \overline{HG}$.
PROVE: $\overline{AF} \cong \overline{CG}$.

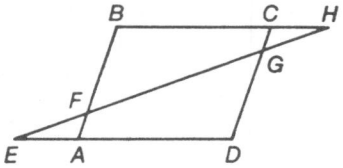

15. GIVEN: $\square ABCD$,
B is the midpoint of $\overline{AE}$.
PROVE: $\overline{EF} \cong \overline{FD}$.

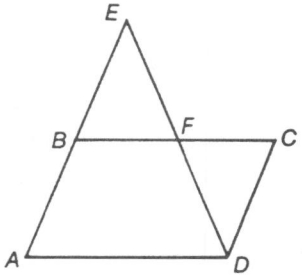

16. GIVEN: Rectangle $ABCD$,
M is the midpoint of $\overline{BC}$.
PROVE: $\triangle AMD$ is isosceles.

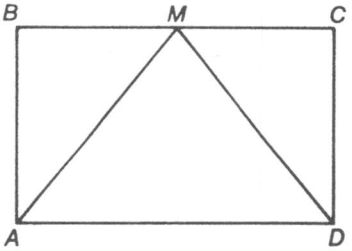

17. GIVEN: Rhombus $ABCD$.
PROVE: $\triangle ASC$ is isosceles.

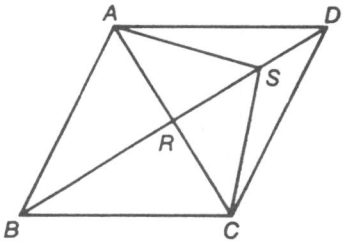

18. GIVEN: Rectangle $ABCD$,
$\overline{BE} \cong \overline{CE}$.
PROVE: $\overline{AF} \cong \overline{DG}$.

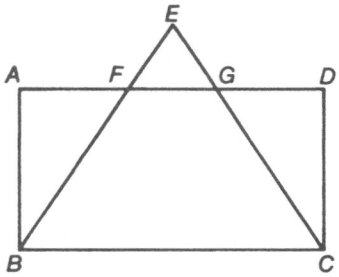

19. GIVEN: $ABCD$ is a parallelogram, $AD > DC$.
PROVE: $m\measuredangle BAC > m\measuredangle DAC$.

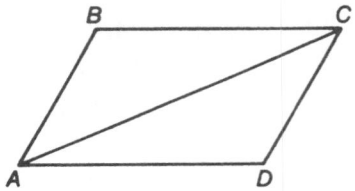

20. GIVEN: $ABCD$ is a parallelogram, $\overline{BR}$ bisects $\measuredangle ABC$, $\overline{DS}$ bisects $\measuredangle CDA$.
PROVE: $BRDS$ is a parallelogram.

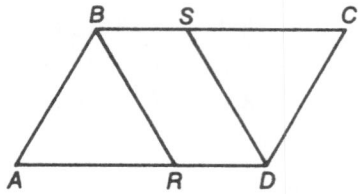

Use the accompanying diagram to solve Exercises 21 and 22.

21. GIVEN: $\square BMDL$, $\overline{AL} \cong \overline{CM}$.
PROVE: $ABCD$ is a parallelogram.

22. GIVEN: $\square ABCD$, $\measuredangle ABL \cong \measuredangle CDM$.
PROVE: $BLDM$ is a parallelogram.

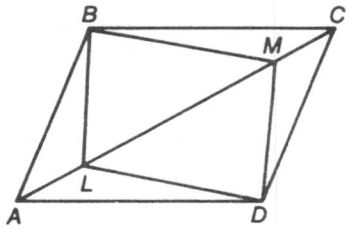

23. GIVEN: $\square ABCD$ is a rhombus, $\overline{BL} \cong \overline{CM}$, $\overline{AL} \cong \overline{BM}$.
PROVE: $ABCD$ is a square.

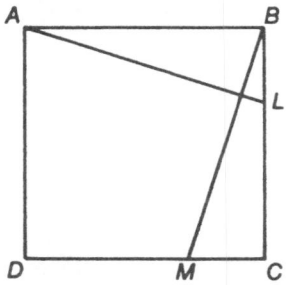

24. GIVEN: $\square ABCD$, $m\measuredangle 2 > m\measuredangle 1$.
PROVE: $\square ABCD$ is *not* a rectangle.

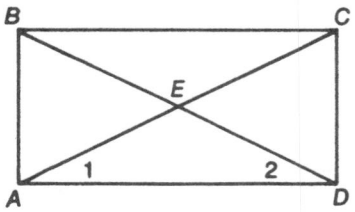

25. GIVEN: $\overline{DE} \cong \overline{DF}$,
D, E, and F are the midpoints of $\overline{AC}$, $\overline{AB}$, and $\overline{BC}$, respectively.
PROVE: $\triangle ABC$ is isosceles.

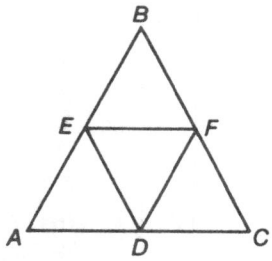

26. GIVEN: $\square RSTW$;
in $\triangle WST$, B and C are midpoints.
PROVE: $WACT$ is a parallelogram.

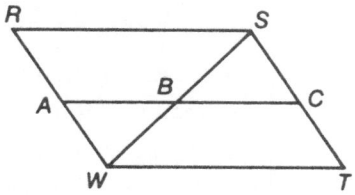

27. GIVEN: Isosceles trapezoid $RSTW$.
PROVE: $\triangle RPW$ is isosceles.

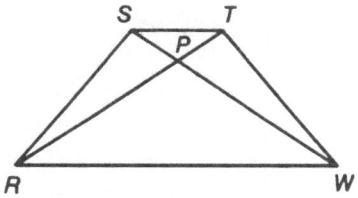

28. GIVEN: Trapezoid $ABCD$,
$\overline{EF} \cong \overline{EG}$,
$\overline{AF} \cong \overline{DG}$,
$\overline{BG} \cong \overline{CF}$.
PROVE: Trapezoid $ABCD$ is isosceles.

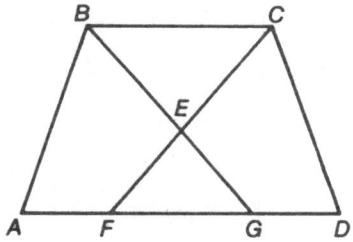

29. GIVEN: Trapezoid $ABCD$ with median $\overline{LM}$,
P is the midpoint of $\overline{AD}$,
$\overline{LP} \cong \overline{MP}$.
PROVE: Trapezoid $ABCD$ is isosceles.

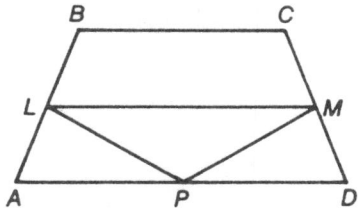

30. GIVEN: Isosceles trapezoid $ABCD$,
$\measuredangle BAK \cong \measuredangle BKA$.
PROVE: $BKDC$ is a parallelogram.

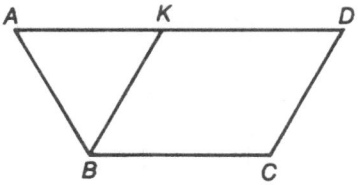

31. GIVEN: D is the midpoint of $\overline{AB}$,
F is the midpoint of $\overline{AC}$,
$\overline{BE} \cong \overline{CG}$, $\measuredangle B \cong \measuredangle C$,
$\measuredangle 1$ is supplementary to $\measuredangle 2$,
$\overline{FD} \cong \overline{FG}$.

PROVE: **(a)** $\triangle DEB \cong \triangle FGC$.
(b) Quadrilateral $DEGF$ is a square.

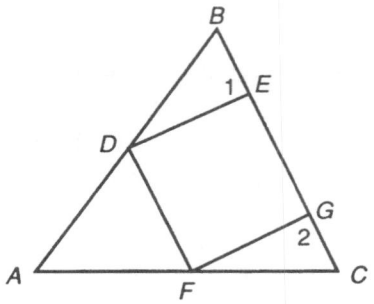

32. Prove that in a rhombus the longer diagonal lies opposite the larger angle of the rhombus. (*Hint:* Given rhombus $ABCD$ with diagonals $\overline{AC}$ and $\overline{DB}$ intersecting at point E, assume $m\measuredangle CDA > m\measuredangle BAD$. Prove $AC > BD$. Work with $\triangle AED$ and first establish that $AE > DE$.)

33. Prove parallel lines are everywhere equidistant. (*Hint:* Given any pair of parallel lines, select two distinct points on one of the lines and from each point draw a segment that is perpendicular to the other parallel line. Show that the lengths of these perpendicular segments are equal.)

34. Prove that the quadrilateral formed by joining consecutively the midpoints of the sides of a parallelogram is a parallelogram.

35. Prove that, if the midpoints of the sides of a rectangle are joined consecutively, the resulting quadrilateral is a rhombus.

36. Prove that, if the diagonals of a trapezoid are congruent, the trapezoid is isosceles.

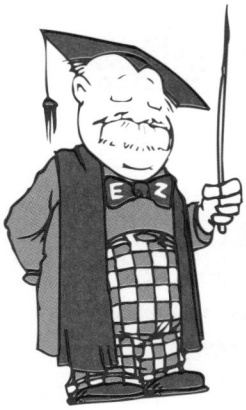

10

Ratio, Proportion, and Similarity

WHAT YOU WILL LEARN

Congruent figures have the same shape and exactly the same dimensions. We now consider *similar* figures, which are figures with the same shape but not necessarily the same dimensions. In this chapter you will learn:

- the ways that proportions arise in geometric situations;
- facts about figures that have the same shape but whose corresponding dimensions are in proportion rather than equal;
- the way to prove two triangles are similar;
- how to prove proportions and products of line segments equal;
- the special properties of perimeters, altitudes, and medians of similar triangles.

SECTIONS IN THIS CHAPTER

- Ratio and Proportion
- Proportions in a Triangle
- Defining Similar Polygons
- Proving Triangles Similar
- Proving Lengths of Sides of Similar Triangles in Proportion
- Proving Products of Segment Lengths Equal

Ratio and Proportion

RATIOS

Allan is 30 years old and Bob is 10 years old. How do their ages compare? Obviously, Allan is 20 years older than Bob. It is sometimes desirable, however, to compare two numbers by determining how many times larger (or smaller) one number is compared to a second number. This can be accomplished by dividing the first number by the second number:

$$\frac{\text{Allan's age}}{\text{Bob's age}} = \frac{30}{10} = \frac{3}{1} \quad \text{or} \quad 3$$

Allan is 3 times as old as Bob. The result of dividing two numbers is called a *ratio*.

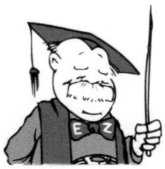

DEFINITION OF RATIO

The *ratio* of two numbers a and b ($b \neq 0$) is the quotient of the numbers. The numbers a and b are referred to as the *terms* of the ratio.

The ratio of two numbers, a and b, may be written in a variety of ways. For example,

$$\frac{a}{b} \qquad a + b \qquad a \text{ to } b \qquad a : b$$

We will normally express a ratio by using either the first or the last of these forms.

In writing the ratio of two numbers, it is usually helpful to express the ratio (fraction) in *simplest* form. For example, the ratio of 50 to 100 is expressed as follows:

$$\frac{50}{100} = \frac{1}{2} \quad \text{or} \quad 1 : 2 \text{ (read as "1 is to 2")}$$

In forming the ratio of two numbers, each number may be expressed in different units of measurement. For example, if a person travels 120 miles in a car in 3 hours, then the ratio of the distance traveled to the time traveled is

$$\frac{120 \text{ miles}}{3 \text{ hours}} = 40 \frac{\text{miles}}{\text{hour}}$$

The value 40 miles/hour is the average *rate* of speed during the trip.

If, however, a ratio is formed in order to determine how many times larger or smaller one value is than another, both quantities must be expressed in the same unit of measurement. For example, if the length of $\overline{AB}$ is 2 feet and the length of $\overline{XY}$ is 16 inches, then, to determine how many times larger $\overline{AB}$ is compared to $\overline{XY}$, we must convert one of the units of measurement into the other. In this example it is convenient to express feet in terms of inches. Since 2 feet is equivalent to 24 inches, we may write

$$\frac{AB}{XY} = \frac{24}{16} = \frac{3}{2}$$

The ratio of the length of $\overline{AB}$ to the length of $\overline{XY}$ is 3:2. Since the decimal representation of $\frac{3}{2}$ is 1.5, we may say that the length of $\overline{AB}$ is 1.5 times the length of $\overline{XY}$.

EXAMPLE 10.1 If the measure of angle A is 60 and angle B is a right angle, find the ratio of the measure of angle A to the measure of angle B.

SOLUTION

$$\frac{m\measuredangle A}{m\measuredangle B} = \frac{60}{90} = \frac{2}{3} \quad \text{or} \quad 2:3$$

EXAMPLE 10.2 Find each of the following ratios using the figures provided:
a $AB:XY$ **b** $BC:YZ$ **c** $XZ:AC$

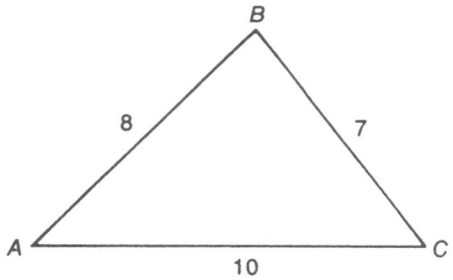

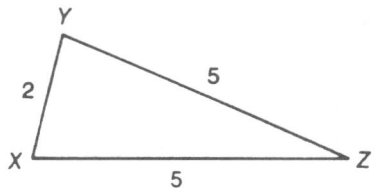

SOLUTION

a $\frac{8}{2} = 4:1$ **b** $\frac{7}{5} = 7:5$ **c** $\frac{5}{10} = 1:2$

EXAMPLE 10.3 The measures of a pair of consecutive angles of a parallelogram are in the ratio of 1:8. Find the measure of the smaller of these angles.

SOLUTION

Let x = measure of smaller angle.
Then $8x$ = measure of larger angle.

Since consecutive angles of a parallelogram are supplementary,

$$\begin{aligned} x + 8x &= 180 \\ 9x &= 180 \\ \frac{9x}{9} &= \frac{180}{9} \\ x &= 20 \end{aligned}$$

The measure of the smaller angle is 20.

EXAMPLE 10.4 In an isosceles triangle, the ratio of the measure of the vertex angle to the measure of a base angle is 2:5. Find the measure of each angle of the triangle.

SOLUTION

Let $2x$ = measure of vertex angle.
Then $5x$ = measure of one base angle,
and $5x$ = measure of the other base angle.

Since the sum of the measures of the angles of a triangle is 180,

$$\begin{aligned} 2x + 5x + 5x &= 180 \\ 12x &= 180 \\ x &= \frac{180}{12} = 15 \\ 2x &= 2(15) = 30 \\ 5x &= 5(15) = 75 \end{aligned}$$

The measures of the three angles of the triangle are 30, 75, and 75.

PROPORTIONS

The ratio $\frac{24}{16}$ may be simplified and written as $\frac{3}{2}$:

$$\frac{24}{16} = \frac{3}{2}$$

An equation that states that two ratios are equal is called a *proportion*. The preceding proportion may also be written in the form 24:16 = 3:2.

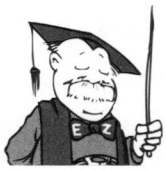

DEFINITION OF PROPORTION

A *proportion* is an equation that states that two ratios are equal:

$$\frac{a}{b} = \frac{c}{d} \quad \text{or} \quad a:b = c:d \quad (\text{provided } b \neq 0 \text{ and } d \neq 0)$$

Each term of a proportion is given a special name according to its position in the proportion.

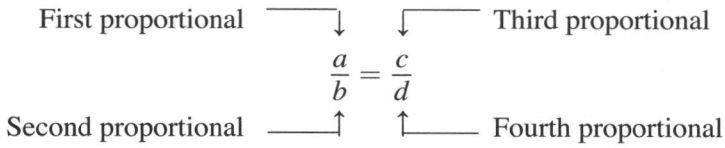

The pair of terms that form the first and fourth proportionals are referred to as the *extremes* of a proportion; the second and third proportionals of a proportion are called the *means* of a proportion.

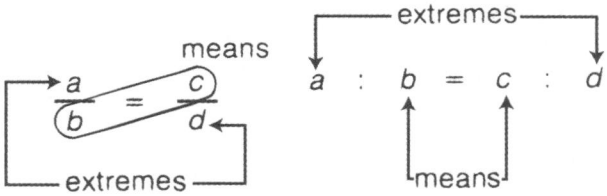

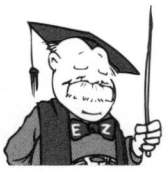

DEFINITION OF THE TERMS OF A PROPORTION

In the proportion $\frac{a}{b} = \frac{c}{d}$, a is called the *first* proportional, b is called the *second* proportional, c is called the *third* proportional, and d is called the *fourth* proportional. The pair of terms a and d are referred to as the *extremes* of the proportion; the pair of terms b and c are referred to as the *means* of the proportion.

Returning to the proportion $\frac{24}{16} = \frac{3}{2}$, we see that 24 and 2 (the outermost terms) are the extremes. The two innermost terms, 16 and 3, are the means. Notice that the *cross-products* of a proportion are equal:

$$\frac{24}{16} = \frac{3}{2}$$

$$16 \times 3 = 24 \times 2$$

$$48 = 48$$

THEOREM 10.1 EQUAL CROSS-PRODUCTS THEOREM

In a proportion the product of the means is equal to the product of the extremes. (Forming this product is sometimes referred to as cross-multiplying.)

EXAMPLE 10.5 Find the first proportional if the remaining terms of a proportion are 3, 14, and 21.

SOLUTION

Let x = first proportional. Then

$$\frac{x}{3} = \frac{14}{21}$$

By Theorem 10.1,

$$21x = 42$$
$$\frac{21x}{21} = \frac{42}{21}$$
$$x = 2$$

The first proportional is 2.

REMEMBER
To solve a proportion, cross-multiply.

Sometimes we may be able to work with smaller numbers by simplifying an arithmetic ratio in the original proportion before cross-multiplying: If in Example 10.5 we write $\frac{14}{21}$ as $\frac{2}{3}$, then:

$$\frac{x}{3} = \frac{14}{21}$$
$$\frac{x}{3} = \frac{2}{3}$$
$$3x = 6$$
$$x = 2$$

EXAMPLE 10.6 The first term of a proportion is 2 and the second and third terms are both 8. Find the fourth proportional.

SOLUTION

Let x = fourth proportional. Then,

$$\frac{2}{8} = \frac{8}{x}$$

Simplify *before* applying Theorem 10.1:

$$\frac{1}{4} = \frac{8}{x}$$

Cross-multiply:

$$x = 32$$

The fourth proportional is 32.

In Example 10.6, the means were equal since the second and third terms of the proportion were both equal to 8. Whenever the means of a proportion are identical, then the value that appears in the means is referred to as the *mean proportional* (or *geometric mean*) between the first and fourth terms of the proportion. In Example 10.6, 8 is said to be the mean proportional between 2 and 32.

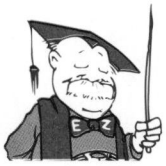

DEFINITION OF MEAN PROPORTIONAL

If the second and third terms of a proportion are the same, then either term is referred to as the *mean proportional* or *geometric mean* between the first and fourth terms of the proportion:

$$\frac{a}{m} = \frac{m}{d} \quad \text{mean proportional} \qquad \text{or} \qquad a : m = m : d$$

mean proportional

EXAMPLE 10.7 Find the mean proportional between each pair of extremes.

a. 3 and 27 **b.** 5 and 7

SOLUTION

a. Let m = mean proportional between 3 and 27. Then

$$\frac{3}{m} = \frac{m}{27}$$
$$m^2 = 3(27)$$
$$= 81$$
$$m = \sqrt{81} = 9$$

b. Let m = mean proportional between 5 and 7.

$$\frac{5}{m} = \frac{m}{7}$$
$$m^2 = 35$$
$$m = \sqrt{35}$$

It is sometimes useful to be able to determine whether a pair of ratios are in proportion. Two ratios are in proportion if *the product of the means of the resulting proportion is equal to the product of the extremes*. Are the ratios $\frac{2}{5}$ and $\frac{12}{30}$ in proportion? We write a tentative proportion and then determine whether the cross-products are equal:

$$\frac{2}{5} \stackrel{?}{=} \frac{12}{30}$$
$$5 \times 12 \stackrel{?}{=} 30 \times 2$$
$$60 = 60$$

Therefore, $\frac{2}{5}$ and $\frac{12}{30}$ are in proportion.

Now we will repeat this procedure, this time to investigate whether the ratios $\frac{8}{12}$ and $\frac{6}{8}$ are in proportion:

$$\frac{8}{12} \stackrel{?}{=} \frac{6}{8}$$
$$12 \times 6 \stackrel{?}{=} 8 \times 8$$
$$72 \neq 64$$

This result implies that $\frac{8}{12}$ and $\frac{6}{8}$ are *not* in proportion.

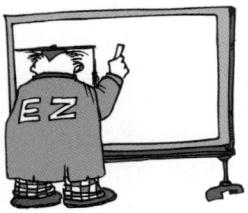

NOTE

Some algebraic properties of proportions are worth remembering.

- **PROPERTY 1** If the numerators and denominators of a proportion are switched, then an equivalent proportion results.

  If $\frac{a}{b} = \frac{c}{d}$, then $\frac{b}{a} = \frac{d}{c}$

  (provided a, b, c, and d are nonzero numbers).

- **PROPERTY 2** If either pair of opposite terms of a proportion are interchanged, then an equivalent proportion results.

  a If $\frac{a}{b} = \frac{c}{d}$, then $\frac{d}{b} = \frac{c}{a}$.

  b If $\frac{a}{b} = \frac{c}{d}$, then $\frac{a}{c} = \frac{b}{d}$.

- **PROPERTY 3** If the denominator is added to or subtracted from the numerator on each side of the proportion, then an equivalent proportion results.

  a If $\frac{a}{b} = \frac{c}{d}$, then $\frac{a+b}{b} = \frac{c+d}{d}$.

  b If $\frac{a}{b} = \frac{c}{d}$, then $\frac{a-b}{b} = \frac{c-d}{d}$.

- **PROPERTY 4** If the product of two nonzero numbers equals the product of another pair of nonzero numbers, then a proportion may be formed by making the factors of one product the extremes, and making the factors of the other product the means. For example, if $R \times S = T \times W$, then we may

  a make R and S the *extremes:* $\frac{R}{T} = \frac{W}{S}$

  or

  b make R and S the *means:* $\frac{T}{R} = \frac{S}{W}$.

Proportions in a Triangle

In Chapter 9 we saw that a line passing through the midpoints of two sides of a triangle was parallel to the third side (and one-half of its length). Suppose we draw a line parallel to a side of a triangle so that it intersects the other two sides, but not necessarily at their midpoints. Many such lines can be drawn, as shown in Figure 10.1.

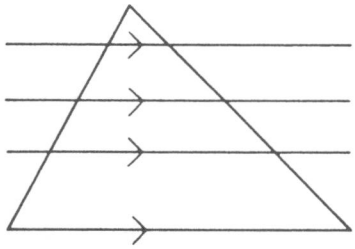

FIGURE 10.1

We will consider one of these lines and the segments that it forms on the sides of the triangle, as shown in Figure 10.2.

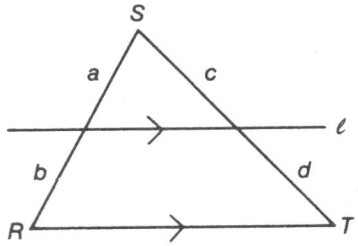

FIGURE 10.2

It will be convenient to postulate that line ℓ divides $\overline{RS}$ and $\overline{ST}$ in such a way that the lengths of corresponding segments on each side have the same ratio:

$$\frac{a}{b} = \frac{c}{d} \quad \text{or} \quad \frac{a}{RS} = \frac{c}{ST} \quad \text{or} \quad \frac{b}{RS} = \frac{d}{ST}$$

If any of the above ratios holds, then the line segments are said to be *divided proportionally*. We notice that each of these ratios has the form

$$\frac{\text{Upper segment of side } \overline{RS}}{\text{Lower segment of side } \overline{RS}} = \frac{\text{upper segment of side } \overline{ST}}{\text{lower segment of side } \overline{ST}}$$

or

$$\frac{\text{Upper segment of side } \overline{RS}}{\text{Whole side}\left(\overline{RS}\right)} = \frac{\text{upper segment of side } \overline{ST}}{\text{whole side}\left(\overline{ST}\right)}$$

or

$$\frac{\text{Lower segment of side } \overline{RS}}{\text{Whole side}\left(\overline{RS}\right)} = \frac{\text{lower segment of side } \overline{ST}}{\text{whole side}\left(\overline{ST}\right)}$$

Keep in mind that the algebraic properties of proportions allow these three proportions to be expressed in equivalent forms. For example, the numerator and denominator of each fraction may be interchanged (that is, each ratio may be *inverted*).

POSTULATE 10.1

A line parallel to one side of a triangle and intersecting the other two sides *divides these sides proportionally.*

EXAMPLE 10.8 In triangle RST, line segment $\overline{EF}$ is parallel to side $\overline{RT}$, intersecting side $\overline{RS}$ at point E and side $\overline{TS}$ at point F.

a. If $SE = 8, ER = 6, FT = 15$, find SF.
b. If $SF = 4, ST = 12, SR = 27$, find SE.
c. If $SE = 6, ER = 4, ST = 20$, find FT.

SOLUTION

a.

$$\frac{SE}{ER} = \frac{SF}{FT}$$
$$\frac{8}{6} = \frac{SF}{15}$$
$$\frac{4}{3} = \frac{SF}{15}$$
$$3(SF) = 60$$
$$SF = \frac{60}{3} = 20$$

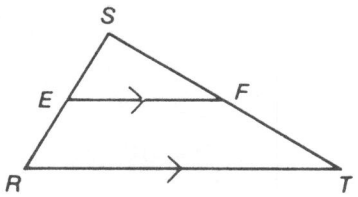

b.

$$\frac{SE}{SR} = \frac{SF}{ST}$$
$$\frac{SE}{27} = \frac{4}{12}$$
$$\frac{SE}{27} = \frac{1}{3}$$
$$3(SE) = 27$$
$$SE = \frac{27}{3} = 9$$

c.

$$\frac{ER}{RS} = \frac{FT}{ST}$$
$$\frac{4}{4+6} = \frac{FT}{20}$$
$$\frac{4}{10} = \frac{FT}{20}$$
$$\frac{2}{5} = \frac{FT}{20}$$
$$5(FT) = 40$$
$$FT = \frac{40}{5} = 8$$

The converse of Postulate 10.1 is also true. If a line is drawn so that the ratio of the segment lengths it cuts off on one side of a triangle is equal to the ratio of the segment lengths it cuts off on a second side of a triangle, then the line must be parallel to the third side of the triangle.

POSTULATE 10.2

A line that divides two sides of a triangle *proportionally* is parallel to the third side of the triangle.

EXAMPLE 10.9 Determine whether $\overline{AB} \parallel \overline{KJ}$ if:

a. $KA = 2, AL = 5, JB = 6$, and $BL = 15$.
b. $AL = 3, KL = 8, JB = 10$, and $JL = 16$.
c. $AL = 5, KA = 9, LB = 10$, and $JB = 15$.

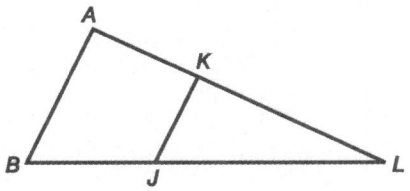

SOLUTION

In each instance, write a tentative proportion and determine whether the proportion is true. If it is true, then $\overline{AB}$ is parallel to $\overline{KJ}$.

a. On the basis of the information provided, use the proportion $\frac{KA}{AL} = \frac{JB}{BL}$ (equivalent proportions may also be formed). Determine whether this proportion is true using the numbers provided:

$$\frac{2}{5} \stackrel{?}{=} \frac{6}{15}$$
$$2 \times 15 = 5 \times 6$$
$$30 = 30$$

Therefore, $\overline{AB} \parallel \overline{KJ}$.

b. Use the proportion $\frac{AL}{KL} = \frac{BL}{JL}$ (other proportions can also be used). Since $BL = 16 - 10 = 6$;

$$\frac{3}{8} \stackrel{?}{=} \frac{6}{16}$$

$\overline{AB}$ is parallel to $\overline{KJ}$ since $3 \times 16 = 8 \times 6$.

c. Use the proportion $\frac{AL}{KA} = \frac{LB}{JB}$:

$$\frac{5}{9} \stackrel{?}{=} \frac{10}{15}$$

$\overline{AB}$ is *not* parallel to $\overline{KJ}$ since $5 \times 15 \neq 9 \times 10$.

EXAMPLE 10.10

GIVEN: Quadrilateral $RSTW$ with $\overline{KJ}$ and $\overline{LM}$ drawn,

$$\frac{RK}{KS} = \frac{RJ}{JW},$$
$$\frac{TL}{LS} = \frac{TM}{MW}.$$

PROVE: $\overline{KJ} \parallel \overline{LM}$.

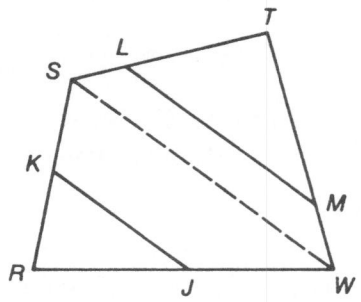

SOLUTION

PLAN: Draw $\overline{SW}$. Applying Postulate 10.2 gives $\overline{KJ} \parallel \overline{SW}$ and $\overline{LM} \parallel \overline{SW}$. Hence, $\overline{KJ} \parallel \overline{LM}$.

PROOF:

Statements	Reasons
1. Quadrilateral $RSTW$ with $\overline{KJ}$ and $\overline{LM}$ drawn.	1. Given.
2. Draw $\overline{SW}$. In $\triangle SRW$:	2. Two points determine a line.
3. $\frac{RK}{KS} = \frac{RJ}{JW}$.	3. Given.
4. $\overline{KJ} \parallel \overline{SW}$.	4. A line that divides two sides of a triangle proportionally is parallel to the third side of the triangle.
In $\triangle WTS$:	
5. $\frac{TL}{LS} = \frac{TM}{MW}$.	5. Given.
6. $\overline{LM} \parallel \overline{SW}$.	6. Same as reason 4.
7. $\overline{KJ} \parallel \overline{LM}$.	7. Two lines parallel to the same line are parallel to each other.

When Are Polygons Similar?

Compare the three triangles in Figure 10.3. Triangles I and III have exactly the same size and shape since they agree in three pairs of angles *and* in three pairs of sides. Triangle I is *congruent* to triangle III.

Triangles II and III have three pairs of congruent corresponding angles, but *each* side of triangle II is twice the length of the corresponding side of triangle III. As a result, triangles II and III have the same shape, although not the same size. Polygons that have the same shape are said to be *similar*.

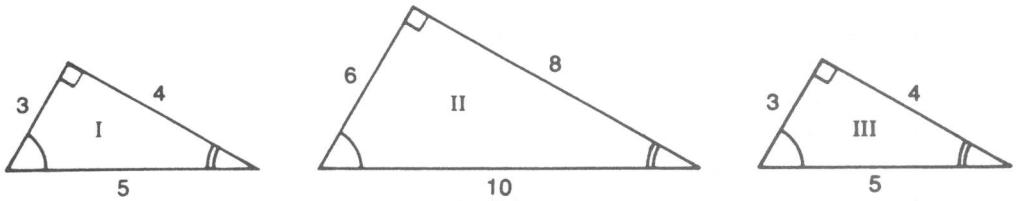

FIGURE 10.3

The concept of similarity is frequently encountered in everyday life. When a photograph is enlarged, the original and enlarged objects are *similar* since they have exactly the same shape. In designing a blueprint, everything must be drawn to scale so that the figures in the blueprint are *similar* to the actual figures.

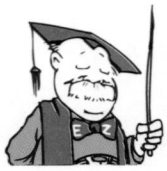

DEFINITION OF SIMILAR POLYGONS

Two polygons are *similar* if their vertices can be paired so that corresponding angles are congruent *and* the ratios of the lengths of all corresponding sides are equal.

The definition of *similar* polygons assumes that there exists a one-to-one correspondence between the vertices of the polygons. This means that each vertex of the first polygon is matched with exactly one vertex of the second polygon and vice versa.

The symbol for similarity is ~. The expression $\triangle ABC \sim \triangle RST$ is read as "Triangle *ABC is similar to* triangle *RST*." If it is known that two polygons are similar, it may be concluded that each pair of corresponding angles are congruent *and* the ratios of the lengths of all pairs of corresponding sides are equal.

REMEMBER
Congruent **polygons have the same size *and* shape, while *similar* polygons have only the same shape.**

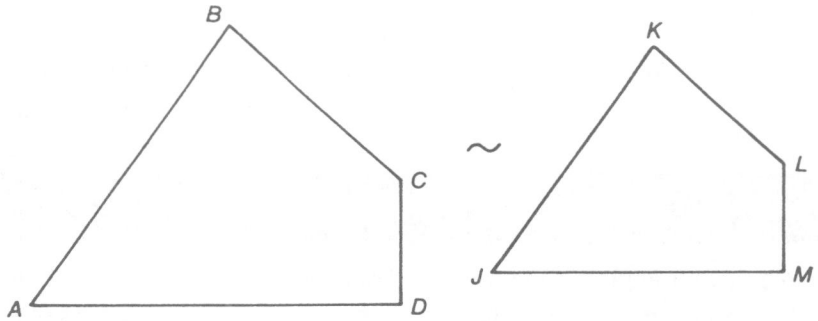

FIGURE 10.4

In Figure 10.4, since quadrilateral *ABCD* is similar to quadrilateral *JKLM*, then the following relationships must hold:

Corresponding Angles are Congruent	Lengths of Corresponding Sides are in Proportion
$\measuredangle A \cong \measuredangle J$ $\measuredangle B \cong \measuredangle K$ $\measuredangle C \cong \measuredangle L$ $\measuredangle D \cong \measuredangle M$	$\frac{AB}{JK} = \frac{BC}{KL} = \frac{CD}{LM} = \frac{AD}{JM}$

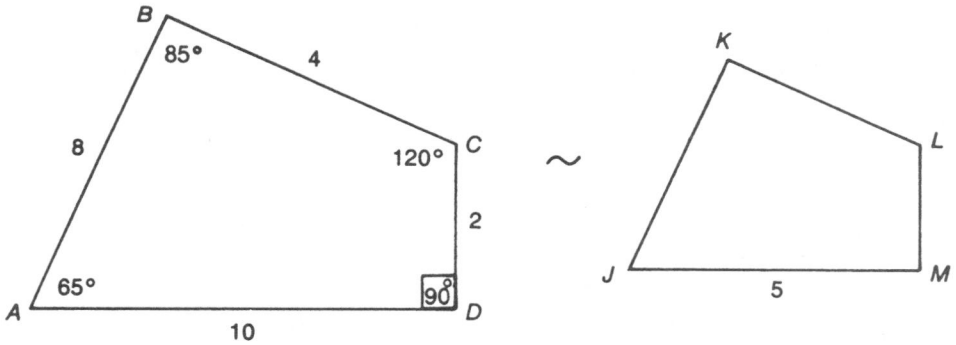

FIGURE 10.5

For example, let's find the measures of the parts of quadrilateral *JKLM* in Figure 10.5. In Figure 10.4 quadrilaterals *ABCD* and *JKLM* were given to be similar. Therefore:

$$m\measuredangle J = m\measuredangle A = 65$$
$$m\measuredangle K = m\measuredangle B = 85$$
$$m\measuredangle L = m\measuredangle C = 120$$
$$m\measuredangle M = m\measuredangle D = 90$$

REMEMBER
In similar polygons, corresponding sides are opposite corresponding congruent angles.

The ratio of the lengths of each pair of corresponding sides must be equal to

$$\frac{JM}{AD} = \frac{5}{10} = \frac{1}{2}$$

The ratio of the length of any side of quadrilateral *JKLM* to the length of the corresponding side of quadrilateral *ABCD* must also be 1 : 2. Hence, $JK = 4$, $KL = 2$, and $LM = 1$.

EXAMPLE 10.11

GIVEN: $\triangle HLX \sim \triangle WKN$.

a. List the pairs of corresponding congruent angles.

b. Write an extended proportion that forms the ratios of the lengths of corresponding sides.

c. If $HL = 8$, $LX = 14$, $HX = 18$, and $WN = 27$, find the lengths of the remaining sides of $\triangle WKN$.

SOLUTION

a. The order in which the vertices of the triangles are written is significant—it defines the pairs of corresponding vertices:

$$H \leftrightarrow W \qquad L \leftrightarrow K \qquad X \leftrightarrow N$$

Hence, $\measuredangle H \cong \measuredangle W$, $\measuredangle L \cong \measuredangle K$, and $\measuredangle X \cong \measuredangle N$.

b. Corresponding sides connect corresponding vertices:

Correspondence	Corresponding Sides
$\triangle HLX \sim \triangle WKN$	$\overline{HL} \leftrightarrow \overline{WK}$
$\triangle HLX \sim \triangle WKN$	$\overline{LX} \leftrightarrow \overline{KN}$
$\triangle HLX \sim \triangle WKN$	$\overline{HX} \leftrightarrow \overline{WN}$

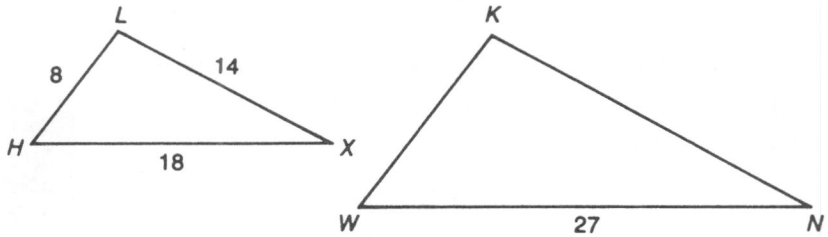

The resulting proportion may be written as follows:

$$\frac{\text{Side of } \triangle HLX}{\text{Corresponding side of } \triangle WKN} = \frac{HL}{WK} = \frac{LX}{KN} = \frac{HX}{WN}$$

After determining the pairs of corresponding vertices, it is probably easier to determine the corresponding sides of two similar figures by drawing a diagram (see the accompanying figure) and then applying the principle that *corresponding sides lie opposite corresponding angles* (vertices). $\overline{HX}$ and $\overline{WN}$ lie opposite corresponding angles L and K; $\overline{HX}$ and $\overline{WN}$ therefore represent a pair of corresponding sides.

c. Since $\overline{HX}$ and $\overline{WN}$ are corresponding sides, the lengths of each pair of corresponding sides must be the same as the ratio of HX to WN:

$$\frac{HX}{WN} = \frac{18}{27} = \frac{2}{3}$$

The length of any side of $\triangle HLX$ to the length of the corresponding side of $\triangle WKN$ must be in the ratio of 2:3. Using this fact, we can find the lengths of the remaining sides of $\triangle WKN$.

To find WK:

$$\frac{2}{3} = \frac{HL}{WK}$$
$$\frac{2}{3} = \frac{8}{WK}$$
$$2(WK) = 24$$
$$WK = \frac{24}{2} = 12$$

To find KN:

$$\frac{2}{3} = \frac{LX}{KN}$$
$$\frac{2}{3} = \frac{14}{KN}$$
$$2(KN) = 42$$
$$KN = \frac{42}{2} = 21$$

EXAMPLE 10.12 In each part, determine whether the pair of polygons are similar.

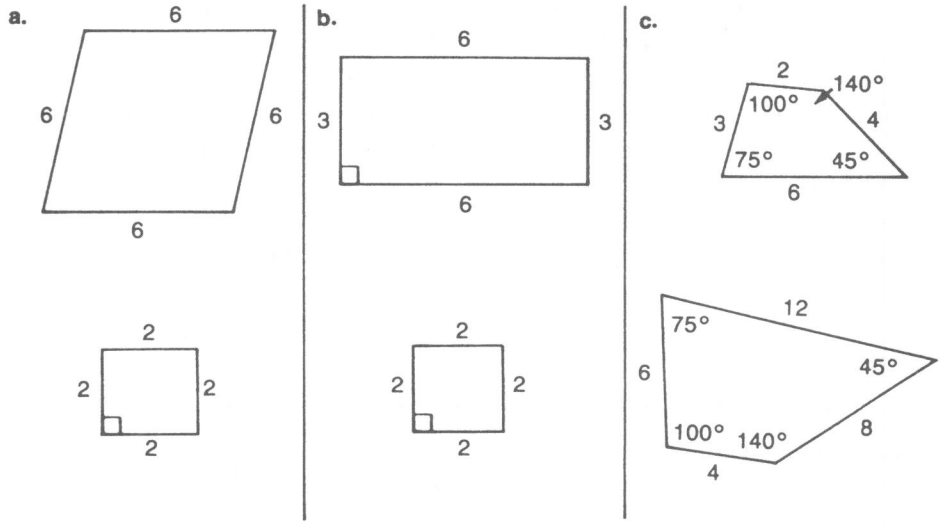

SOLUTION

a. No. The ratios of the lengths of corresponding sides are equal (3 : 1), but corresponding angles are *not* congruent.

b. No. Corresponding angles are congruent (each figure contains four right angles), but the ratios of the lengths of corresponding sides are *not* all equal.

c. Yes. Corresponding angles are congruent and the ratios of the lengths of all pairs of corresponding sides are the same (1 : 2).

If two polygons are similar then the ratio of the lengths of any pair of corresponding sides is called the *ratio of similitude*. If the lengths of a pair of corresponding sides of two similar polygons are 3 and 12, then the ratio of similitude is $\frac{3}{12}$ or 1 : 4.

Are congruent polygons similar? Yes, since they satisfy the two conditions of similarity; all corresponding pairs of angles are congruent and the ratios of the lengths of all pairs of corresponding sides are the same (1 : 1). Are similar polygons also congruent? Generally speaking, no! Similar polygons are congruent only if their ratio of similitude is 1 : 1.

EXAMPLE 10.13 Two quadrilaterals are similar and have a ratio of similitude of 1 : 3. If the lengths of the sides of the smaller quadrilateral are 2, 5, 8, and 12, find the lengths of the sides of the larger quadrilateral.

SOLUTION

Since the ratio of similitude is $1:3$, the length of each side of the larger quadrilateral is 3 times the length of the corresponding side in the smaller quadrilateral. The lengths of the sides of the larger quadrilateral are 6, 15, 24, and 36.

EXAMPLE 10.14 Two quadrilaterals are similar. The length of the sides of the smaller quadrilateral are 4, 6, 12, and 18. The length of the longest side of the larger quadrilateral is 27. Determine each of the following:

a. The ratio of similitude.

b. The lengths of the remaining sides of the larger quadrilateral.

c. The ratio of the perimeters of the two quadrilaterals.

SOLUTION

a. The longest sides of the two quadrilaterals are 27 and 18. The ratio of similitude (larger to smaller quadrilateral) is $\frac{27}{18}$ or $\frac{3}{2}$.

b. Let x = length of a side of the larger quadrilateral. Then:

$$\frac{3}{2} = \frac{x}{4} \quad \text{implies } 2x = 12 \text{ and } x = 6$$

$$\frac{3}{2} = \frac{x}{6} \quad \text{implies } 2x = 18 \text{ and } x = 9$$

$$\frac{3}{2} = \frac{x}{12} \quad \text{implies } 2x = 36 \text{ and } x = 18$$

The sides of the larger quadrilateral are 6, 9, 18, and 27. Alternatively, we could have simply multiplied the length of each side of the smaller quadrilateral by the ratio of similitude $\left(\frac{3}{2}\right)$ in order to obtain the length of the corresponding side in the larger quadrilateral.

c.
$$\frac{\text{Perimeter of larger quadrilateral}}{\text{Perimeter of smaller quadrilateral}} = \frac{6 + 9 + 18 + 27}{4 + 6 + 12 + 18}$$
$$= \frac{60}{40} = \frac{3}{2}$$

Notice that in part **c** of Example 10.14 the ratio of the perimeters of the quadrilaterals is the same as the ratio of the lengths of a pair of corresponding sides.

THEOREM 10.2 RATIO OF PERIMETERS THEOREM

The perimeters of a pair of similar polygons have the same ratio as the lengths of any pair of corresponding sides.

Proving Triangles Similar

To prove triangles *congruent*, we did not have to show that three pairs of angles were congruent *and* three pairs of sides were congruent. Instead, we used shortcut methods that depended on showing that a particular set of *three* parts of one triangle were congruent to the corresponding parts of the second triangle.

The definition of *similarity* requires that, to show two triangles are similar, we need to demonstrate that three pairs of corresponding angles are congruent *and* that the ratios of the lengths of three pairs of corresponding sides are the same. A formidable task! Fortunately, there is shortcut method that can be used to prove a pair of *triangles* similar.

Let's try the following experiment. Suppose we wish to draw a triangle that has at least two angles with the same measures as two angles of triangle ABC (Figure 10.6). We may begin by drawing a line ℓ and choosing *any* two distinct points of line ℓ, say D and E (Figure 10.7). At points D and E we will use a protractor to draw angles having the same measures as angles A and B. The point at which the two rays having D and E as end points meet will be called F.

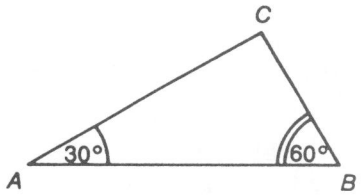

FIGURE 10.6

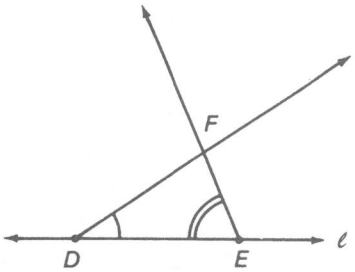

FIGURE 10.7

How do the measures of angles C and F compare? These angles must be congruent since, if two angles of a triangle are congruent to two angles of another triangle, then the third pair of angles must also be congruent (the sum of their measures must equal 180 in each case). Another interesting thing happens. If we actually measured the sides of each triangle and then compared the ratios $\frac{AB}{DE}, \frac{BC}{EF}$, and $\frac{AC}{DF}$, we would find that they are equal! Hence, $\triangle ABC$ must be similar to $\triangle DEF$ since the requirements of the definition of similarity are satisfied. We began by making two angles of $\triangle DEF$ congruent to two angles of $\triangle ABC$. This forced the third pair of angles to be congruent and the lengths of the corresponding sides to be in proportion.

THEOREM 10.3 THE ANGLE-ANGLE (AA) THEOREM OF SIMILARITY

If two angles of one triangle are congruent to two angles of another triangle, then the triangles are similar.

Theorem 10.3 provides a simple method of proving triangles similar. All we need do is show that *two* angles of one triangle are congruent to *two* angles of another triangle. The proof of this theorem is the topic of Exercise 16 at the end of the chapter.

EXAMPLE 10.15

GIVEN: $\overline{CB} \perp \overline{BA}$,
$\overline{CD} \perp \overline{DE}$.

PROVE: $\triangle ABC \sim \triangle EDC$.

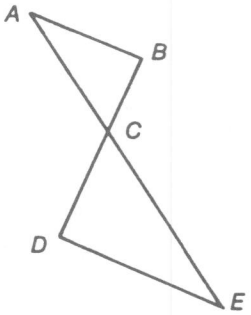

SOLUTION

PLAN: Use the AA Theorem. The two triangles include right and vertical angles that yield two pairs of congruent angles.

PROOF:

Statements	Reasons
1. $\overline{CB} \perp \overline{BA}$ and $\overline{CD} \perp \overline{DE}$.	1. Given.
2. Angles ABC and EDC are right angles.	2. Perpendicular lines intersect to form right angles.
3. $\measuredangle ABC \cong \measuredangle EDC$. (Angle)	3. All right angles are congruent.
4. $\measuredangle ACB \cong \measuredangle ECD$. (Angle)	4. Vertical angles are congruent.
5. $\triangle ABC \sim \triangle EDC$.	5. AA Theorem.

Proving Lengths of Sides of Similar Triangles in Proportion

SIDES OF SIMILAR TRIANGLES

We have learned that congruent triangles may be used to establish that a pair of angles or segments are congruent. In like fashion, we can establish that the lengths of four segments are in proportion by first showing that a pair of triangles that contain these segments as corresponding sides are similar. We may then apply the reverse of the definition of similarity and conclude that the lengths of the segments are in proportion, using as a reason: *The lengths of corresponding sides of similar triangles are in proportion.*

EXAMPLE 10.16

GIVEN: $\overline{AB} \parallel \overline{DE}$.

PROVE: $\dfrac{EC}{BC} = \dfrac{ED}{AB}$.

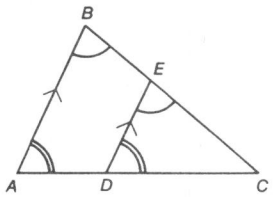

SOLUTION

PLAN: **1** Select the triangles that contain these segments as sides. Read across the proportion:

$$\triangle ECD$$

$$\frac{EC}{BC} = \frac{ED}{AB}$$

$$\triangle BCA$$

2 Mark the diagram with the Given and all pairs of corresponding congruent angle pairs.

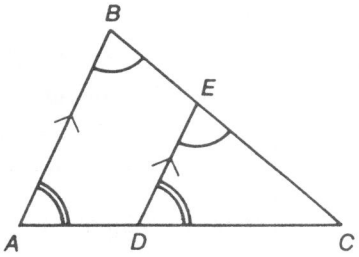

3 Write the proof.

PROOF:

Statements	Reasons
1. $\overline{AB} \parallel \overline{DE}$.	1. Given.
2. $\measuredangle CED \cong \measuredangle CBA$. (Angle) $\measuredangle CDE \cong \measuredangle CAB$. (Angle)	2. If two lines are parallel, then their corresponding angles are congruent.
3. $\triangle ECD \sim \triangle BCA$.	3. AA Theorem.
4. $\dfrac{EC}{BC} = \dfrac{ED}{AB}$.	4. The lengths of corresponding sides of similar triangles are in proportion.

- In this proof notice that the statement "The lengths of corresponding sides of similar triangles are in proportion" is analogous to the statement "Corresponding sides of *congruent* triangles are congruent," which was abbreviated as CPCTC. We will *not* introduce an abbreviation for the expression that appears as reason 4 in the preceding proof.
- Statement 3 of the proof establishes that a *line intersecting two sides of a triangle and parallel to the third side forms two similar triangles*.

EXAMPLE 10.17

GIVEN: $\overline{AC} \perp \overline{CB}, \overline{ED} \perp \overline{AB}$.

PROVE: $EB{:}AB = ED{:}AC$.

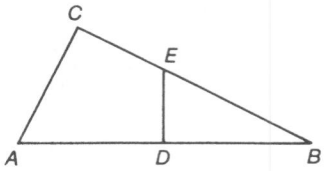

SOLUTION

PLAN: **1** Rewrite the Prove in fractional form: $\dfrac{EB}{AB} = \dfrac{ED}{AC}$.

2 Determine the pair of triangles that must be proved similar.

$$\begin{array}{c} \triangle EBD \\ \downarrow \qquad \downarrow \\ \dfrac{EB}{AB} = \dfrac{ED}{AC} \\ \uparrow \qquad \uparrow \\ \triangle ABC \end{array}$$

3 Mark the diagram with the Given and all pairs of corresponding congruent angle pairs.

4 Write the proof.

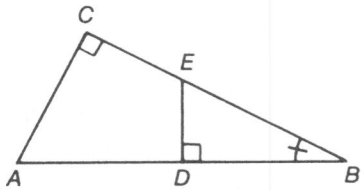

PROOF:

Statements	Reasons
1. $\overline{AC} \perp \overline{CB}, \overline{ED} \perp \overline{AB}$.	1. Given.
2. Angles C and EDB are right angles.	2. Perpendicular lines intersect to form right angles.
3. $\measuredangle C \cong \measuredangle EDB$. (Angle)	3. All right angles are congruent.
4. $\measuredangle B \cong \measuredangle B$. (Angle)	4. Reflexive property of congruence.
5. $\triangle EBD \sim \triangle ABC$.	5. AA Theorem.
6. $EB{:}AB = ED{:}AC$	6. The lengths of corresponding sides of similar triangles are in proportion.

EXAMPLE 10.18

In the accompanying figure, angles D and B are right angles. If $BC = 80$ meters, $DE = 15$ meters, and $BD = 171$ meters, what is the length of $\overline{AB}$?

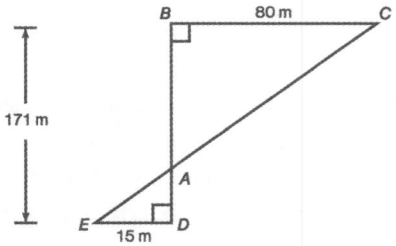

SOLUTION

Look for similar triangles. Angles B and D are right angles, so $\angle B \cong \angle D$. Because angles EAD and CAB are vertical angles, $\angle EAD \cong \angle CAB$. Thus, $\triangle ADE \sim \triangle ABC$.

- Because lengths of corresponding sides of similar triangles are in proportion,

$$\frac{\text{side in } \triangle ADE}{\text{corresponding side in } \triangle ABC} = \frac{AD}{AB} = \frac{DE}{BC}$$

- If $x = AB$, then $AD = 171 - x$:

$$\begin{aligned} \frac{171 - x}{x} &= \frac{15}{80} \\ 15x &= 80(171 - x) \\ 15x &= 13{,}680 - 80x \\ 15x + 80x &= 13{,}680 \\ \frac{95x}{95} &= \frac{13{,}680}{95} \\ x &= 144. \end{aligned}$$

The length of line segment AB is 144 meters.

SUMMARY

To prove that the lengths of segments are in proportion:

1. Use the proportion provided in the Prove to help identify the triangles that contain the desired segments as sides.
2. Mark the diagram with the given as well as with any additional information that may be deduced (for example, vertical angles, right angles, congruent alternate interior angles).
3. Prove the triangles similar.
4. Write the desired proportion using, as a reason: "The lengths of corresponding sides of similar triangles are in proportion."

ALTITUDES AND MEDIANS OF SIMILAR TRIANGLES

In similar triangles the lengths of altitudes or medians drawn to corresponding sides have the same ratio as the lengths of any pair of corresponding sides.

THEOREM 10.4 RATIO OF ALTITUDES AND MEDIANS IN SIMILAR TRIANGLES

If a pair of triangles are similar, then the lengths of a pair of corresponding altitudes or medians have the same ratio as the lengths of any pair of corresponding sides.

EXAMPLE 10.19 $\triangle RST \sim \triangle KLM$. The length of altitude $\overline{SA}$ exceeds the length of altitude $\overline{LB}$ by 5. If $RT = 9$ and $KM = 6$, find the length of each altitude.

SOLUTION

Let x = length of altitude $\overline{LB}$.
Then $x + 5$ = length of altitude $\overline{SA}$.

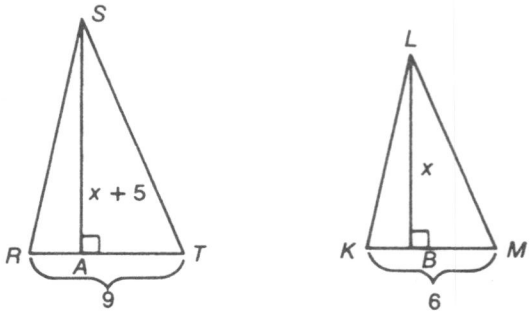

Applying Theorem 10.4 gives

$$\frac{x+5}{x} = \frac{9}{6}$$
$$9x = 6(x+5)$$
$$= 6x + 30$$
$$3x = 30$$
$$x = 10$$
$$\text{Altitude } LB = x = 10$$
$$\text{Altitude } SA = x + 5 = 15$$

SUMMARY

The lengths of corresponding *altitudes*, the lengths of corresponding *medians*, and the *perimeters* of similar triangles have the same ratio as the lengths of any pair of corresponding sides:

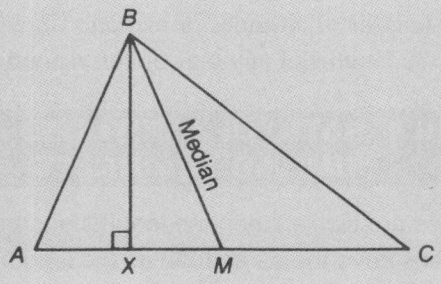

SUMMARY (continued)

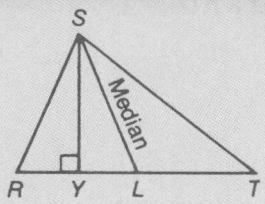

$$\frac{\text{Perimeter of } \triangle ABC}{\text{Perimeter of } \triangle RST} = \frac{BX}{SY} = \frac{BM}{SL} = \frac{AB}{RS} = \frac{BC}{ST} = \frac{AC}{RT}$$

Proving Products of Segment Lengths Equal

We know that, if $\frac{A}{B} = \frac{C}{D}$, then $A \times D = B \times C$. The reason is that in a proportion the product of the means equals the product of the extremes. Instead of generating a product from a proportion, we sometimes need to be able to take a product and determine the related proportion that would yield that product.

Suppose the lengths of four segments are related in such a way that

$$KM \times LB = LM \times KD$$

What proportion gives this result when the products of its means and extremes are set equal to each other? A true proportion may be derived from the product by designating a pair of terms appearing on the same side of the equal sign as the extremes (say, KM and LB). The pair of terms on the opposite side of the equal sign then becomes the means (LM and KD):

$$\frac{KM}{LM} = \frac{KD}{LB}$$

An equivalent proportion results if KM and LB are made the means rather than the extremes.

These ideas are needed in problems in which similar triangles are used to prove products of segment lengths equal. As an illustration, let's look at the following problem.

GIVEN: $\square ABCD$.
PROVE: $KM \times LB = LM \times KD$.

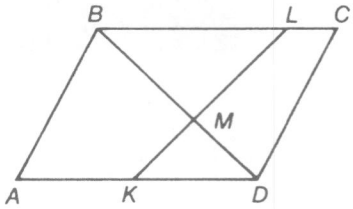

We can identify the needed proportion and the pair of triangles that we will have to prove similar by *working backwards*, from the products in the Prove:

1. Express the product as an equivalent proportion:

$$\frac{KM}{LM} = \frac{KD}{LB}$$

2. From the proportion (and, in some problems, in conjunction with the Given), determine the pair of triangles to be proved similar:

$$\begin{array}{c} \triangle KMD \\ \downarrow \qquad \downarrow \\ \dfrac{KM}{LM} = \dfrac{KD}{LB} \\ \uparrow \qquad \uparrow \\ \triangle LMB \end{array}$$

3. Mark the diagram with the Given and decide how to show that the triangles are similar.
STRATEGY: Use the AA Theorem.

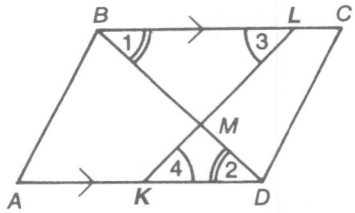

4. Write the formal two-column proof. The steps in the proof should reflect the logic of this analysis, proceeding from step 4 back to step 1 (proving the triangles similar, forming the appropriate proportion, and, lastly, writing the product):

PROOF:

Statements	Reasons
1. $\square ABCD$.	1. Given.
2. $\overline{AD} \parallel \overline{BC}$.	2. Opposite sides of a parallelogram are parallel.
3. $\measuredangle 1 \cong \measuredangle 2$, $\measuredangle 3 \cong \measuredangle 4$.	3. If two lines are parallel, then their alternate interior angles are congruent.
4. $\triangle KMD \sim \triangle LMB$.	4. AA Theorem.

5. $\frac{KM}{LM} = \frac{KD}{LB}$.	5. The lengths of corresponding sides of similar triangles are in proportion.
6. $KM \times LB = LM \times KD$.	6. In a proportion, the product of the means equals the product of the extremes.

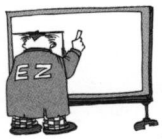

In this proof, notice that:

- Compared with our previous work with similar triangles, the only new step is the last statement/reason of the proof. To prove products of segment lengths equal, we must first prove that a related proportion is true; to prove the proportion is true, we must first establish that the triangles that contain these segments as sides are similar.
- In our analysis of the equality $KM \times LB = LM \times KD$, suppose we formed the proportion

$$\frac{KM}{KD} = \frac{LM}{LB}$$

Reading across the top (K-M-L), we do not find a set of letters that correspond to the vertices of a triangle. When this happens, we switch the terms in either the means or the extremes of the proportion:

$$\frac{KM}{LM} = \frac{KD}{LB}$$

Reading across the top (K-M-D) now gives us the vertices of one of the desired triangles.

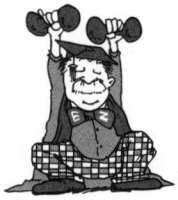

REVIEW EXERCISES FOR CHAPTER 10

1. Find the measure of the largest angle of a triangle if the measures of its interior angles are in the ratio 3:5:7.

2. Find the measure of the vertex angle of an isosceles triangle if the measures of the vertex angle and a base angle have the ratio 4:3.

3. The measures of a pair of consecutive angles of a parallelogram have the ratio 5:7. Find the measure of each angle of the parallelogram.

4. Solve for x.

(a) $\frac{2}{6} = \frac{8}{x}$ (b) $\frac{2}{x} = \frac{x}{50}$ (c) $\frac{2x-5}{3} = \frac{9}{4}$

(d) $\frac{4}{x+3} = \frac{1}{x-3}$ (e) $\frac{3}{x} = \frac{x-4}{7}$

5. Find the mean proportional between each pair of extremes.

(a) 4 and 16 (b) $3e$ and $12e^3$

(c) $\frac{1}{2}$ and $\frac{1}{8}$ (d) 6 and 9

6. In each part, determine whether the pairs of ratios are in proportion:

(a) $\frac{1}{2}$ and $\frac{9}{18}$ (b) $\frac{12}{20}$ and $\frac{3}{5}$

(c) $\frac{4}{9}$ and $\frac{12}{36}$ (d) $\frac{15}{25}$ and $\frac{20}{12}$

7. In $\triangle BAG$, $\overline{LM}$ intersects side $\overline{AB}$ at L and side $\overline{AG}$ at M so that $\overline{LM}$ is parallel to $\overline{BG}$. Write at least three different true proportions. (Do *not* generate equivalent proportions by inverting the numerator and denominator of each ratio.)

8. For each of the following segment lengths, determine whether $\overline{TP} \parallel \overline{BC}$

(a) $AT = 5, TB = 15, AP = 8, PC = 24$.
(b) $TB = 9, AB = 18, AP = 6, PC = 6$.
(c) $AT = 4, AB = 12, AP = 6, AC = 15$.
(d) $AT = 3, TB = 9, PC = 4, AC = 12$.
(e) $AT = 1/3 \cdot AB$ and $PC = 2 \cdot AP$.

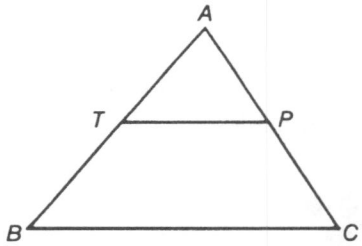

9. If $\overline{KW} \parallel \overline{EG}$, find the length of each indicated segment.

(a) $HE = 20, KE = 12, WG = 9, HG = ?$
(b) $KH = 7, KE = 14, HG = 12, HW = ?$
(c) $HW = 4, WG = 12, HE = 28, KH = ?$
(d) $KH = 9, KE = 12, HG = 42, WG = ?$
(e) $KH = 2x - 15, KE = x, HW = 1, HG = 4$. Find KH and KE.

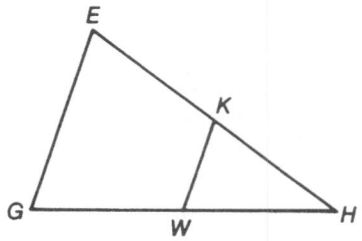

10. $\triangle GAL \sim \triangle SHE$. Name three pairs of congruent angles and three equal ratios.

11. The ratio of similitude of two similar polygons is $3:5$. If the length of the shortest side of the smaller polygon is 24, find the length of the shortest side of the larger polygon.

12. $\triangle ZAP \sim \triangle MYX$. If $ZA = 3$, $AP = 12$, $ZP = 21$, and $YX = 20$, find the lengths of the remaining sides of $\triangle MYX$.

13. Quadrilateral $ABCD \sim$ quadrilateral $RSTW$. The lengths of the sides of quadrilateral $ABCD$ are 3, 6, 9, and 15. If the length of the longest side of quadrilateral $RSTW$ is 20, find the perimeter of $RSTW$.

14. The longest side of a polygon exceeds twice the length of the longest side of a similar polygon by 3. If the ratio of similitude of the polygons is $4:9$, find the length of the longest side of each polygon.

15. $\triangle RST \sim \triangle JKL$.

    (a) $\overline{RA}$ and $\overline{JB}$ are medians to sides $\overline{ST}$ and $\overline{KL}$, respectively. $RS = 10$ and $JK = 15$. If the length of JB exceeds the length of RA by 4, find the lengths of medians $\overline{JB}$ and $\overline{RA}$

    (b) $\overline{SH}$ and $\overline{KO}$ are altitudes to sides $\overline{RT}$ and $\overline{JL}$ respectively. If $SH = 12$, $KO = 15$, $LK = 3x - 2$, and $TS = 2x + 1$, find the lengths of $\overline{LK}$ and $\overline{TS}$.

    (c) The perimeter of $\triangle RST$ is 25 and the perimeter of $\triangle JKL$ is 40. If $\overline{ST} = 3x + 1$ and $\overline{KL} = 4x + 4$, find the lengths of ST and KL.

    (d) The ratio of similitude of $\triangle RST$ to $\triangle JKL$ is $3:x$. The length of altitude $\overline{SU}$ is $x - 4$ and the length of altitude $\overline{KV}$ is 15. Find the length of altitude $\overline{SU}$.

16. Supply the missing reasons in Part I of the following proof of the AA Theorem of Similarity (Theorem 10.3). Supply statements and reasons for Part II of the proof.

GIVEN: $\measuredangle R \cong \measuredangle L$, $\measuredangle T \cong \measuredangle P$.
PROVE: $\triangle RST \sim \triangle LMP$.
PLAN: $\measuredangle S \cong \measuredangle M$. Show sides are in proportion.

PROOF:

Statements	Reasons
Part I. To show that $\dfrac{SR}{ML} = \dfrac{ST}{MP}$	
1. Assume $SR > ML$. On $\overline{SR}$ choose point A such that $SA = ML$.	1. At a given point on a line, a segment may be drawn equal in length to a given segment.

2. Through point A draw a line parallel to $\overline{RT}$, intersecting side $\overline{ST}$ at point B.	2. ?
3. $\dfrac{SR}{SA} = \dfrac{ST}{SB}$.	3. ?
4. $\dfrac{SR}{ML} = \dfrac{ST}{SB}$. Show $\triangle ASB \cong \triangle LMP$ in order to obtain $SB = MP$.	4. ?
5. $\measuredangle SAB \cong \measuredangle R$.	5. ?
6. $\measuredangle R \cong \measuredangle L$ and $\measuredangle T \cong \measuredangle P$.	6. ?
7. $\measuredangle SAB \cong \measuredangle L$.	7. ?
8. $\measuredangle S \cong \measuredangle M$.	8. ?
9. $\triangle ASB \cong \triangle LMP$.	9. ?
10. $SB = MP$.	10. ?
11. $\dfrac{SR}{ML} = \dfrac{ST}{MP}$.	11. ?

Part II. Show that

$$\frac{ST}{MP} = \frac{RT}{LP}.$$

HINT: Locate a point C on side $\overline{RT}$ such that $TC = PL$, and use the procedure followed in statements 2 through 11. Since all pairs of corresponding angles are congruent and $\dfrac{SR}{ML} = \dfrac{ST}{MP} = \dfrac{RT}{LP}$, the reverse of the definition of similar polygons is satisfied. Thus, $\triangle RST \sim \triangle LMP$.

17. GIVEN: $\overline{XW} \cong \overline{XY}$,
$\overline{HA} \perp \overline{WY}, \overline{KB} \perp \overline{WY}$.
PROVE: $\triangle HWA \sim \triangle KYB$.

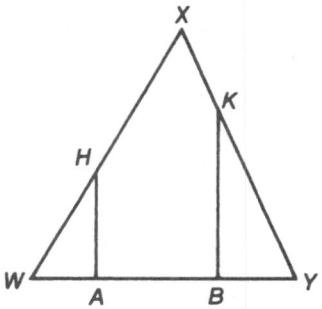

18. GIVEN: $\triangle ABC \sim \triangle RST$,
$\overline{BX}$ bisects $\measuredangle ABC$,
$\overline{SY}$ bisects $\measuredangle RST$.
PROVE: $\triangle BXC \sim \triangle SYT$.

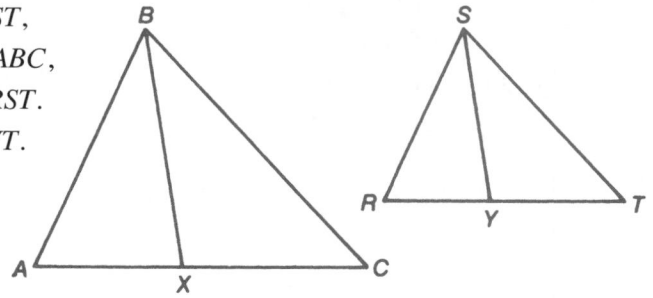

Use the accompanying diagram for Exercises 19 and 20.

19. GIVE: $\overline{AC} \perp \overline{BD}$ and $\overline{DE} \perp \overline{AB}$.
PROVE: $\triangle EFA \sim \triangle CFD$.

20. GIVEN: $\overline{AE} \cong \overline{AF}$ and $\overline{DF} \cong \overline{DC}$.
PROVE: $\dfrac{AF}{EF} = \dfrac{DF}{FC}$.

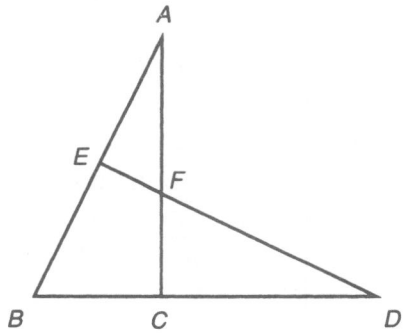

Use the accompanying diagram for Exercises 21 to 23.

21. GIVEN: $\triangle MCT \sim \triangle BAW$,
$\overline{SW}$ bisects $\measuredangle AWM$.
PROVE: $\triangle MCT \sim \triangle BCW$.

22. GIVEN: $\overline{AW} \parallel \overline{ST}$,
$\overline{MS} \cong \overline{MW}$,
$\overline{WC} \cong \overline{WA}$.
PROVE: $\triangle BCW \sim \triangle BTS$.

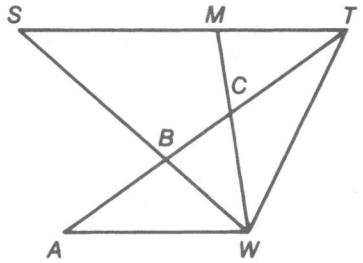

23. GIVEN: $\overline{WB} \cong \overline{WC}$, $\overline{ST} \parallel \overline{AW}$,
$\overline{AT}$ bisects $\measuredangle STW$.
PROVE: $\triangle ABW \sim \triangle TCW$.

24. GIVEN: $\overline{HW} \parallel \overline{TA}$,
$\overline{HY} \parallel \overline{AX}$.
PROVE: $\dfrac{AX}{HY} = \dfrac{AT}{HW}$.

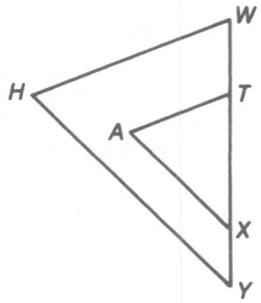

25. GIVEN: $\overline{MN} \parallel \overline{AT}$,
$\measuredangle 1 \cong \measuredangle 2$.
PROVE: $\dfrac{NT}{AT} = \dfrac{RN}{RT}$.

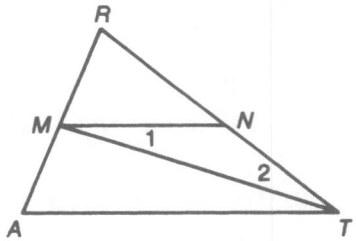

26. GIVEN: $\overline{SR} \cong \overline{SQ}$,
$\overline{RQ}$ bisects $\measuredangle SRW$.
PROVE: $\dfrac{SQ}{RW} = \dfrac{SP}{PW}$.

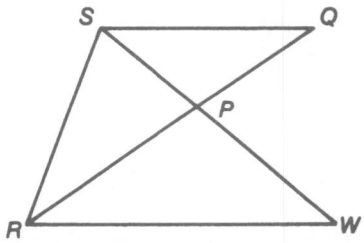

Use the accompanying diagram for Exercises 27 and 28.

27. GIVEN: $\overline{MC} \perp \overline{JK}, \overline{PM} \perp \overline{MQ}$,
$\overline{TP} \cong \overline{TM}$.
PROVE: $\dfrac{PM}{MC} = \dfrac{PQ}{MK}$.

28. GIVEN: T is the midpoint of $\overline{PQ}$,
$\overline{MP} \cong \overline{MQ}, \overline{JK} \parallel \overline{MQ}$.
PROVE: $\dfrac{PM}{JK} = \dfrac{TQ}{JT}$.

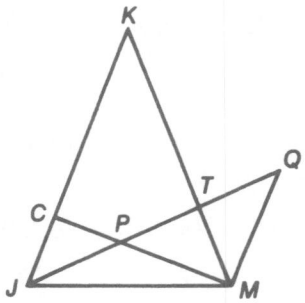

29. GIVEN: $\overline{EF}$ is the median of trapezoid $ABCD$.
PROVE: $EI \times GH = IH \times EF$.

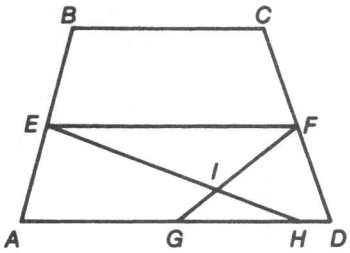

30. GIVEN: $\overline{RS} \perp \overline{ST}$
$\overline{SW} \perp \overline{RT}$.
PROVE: $(ST)^2 = TW \times RT$.

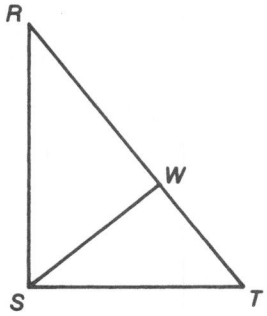

31. GIVEN: $\overline{AF}$ bisects $\measuredangle BAC$,
$\overline{BH}$ bisects $\measuredangle ABC$,
$\overline{BC} \cong \overline{AC}$.
PROVE: $AH \times EF = BF \times EH$.

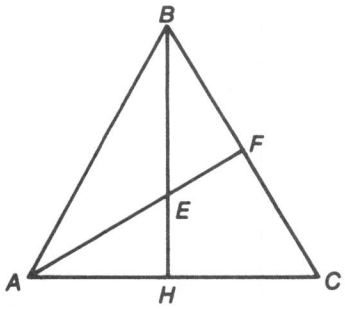

32. GIVEN: $\overline{XY} \parallel \overline{LK}$,
$\overline{XZ} \parallel \overline{JK}$.
PROVE: $JY \times ZL = XZ \times KZ$.

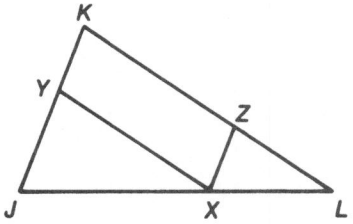

33. GIVEN: Quadrilateral $ABCD$,
$\overline{AB} \perp \overline{BC}$, $\overline{AB} \perp \overline{AD}$,
$\overline{AC} \perp \overline{CD}$.
PROVE: $BC \times AD = (AC)^2$.

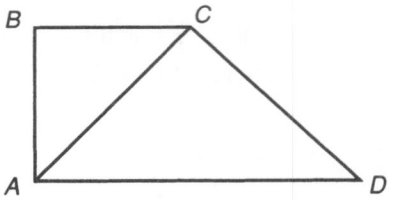

34. GIVEN: $\triangle ABC$ with $\overline{CDA}$, $\overline{CEB}$, $\overline{AFB}$,
$\overline{DE} \parallel \overline{AB}$, $\overline{EF} \parallel \overline{AC}$,
$\overline{CF}$ intersects $\overline{DE}$ at G.
PROVE: (a) $\triangle CAF \sim \triangle FEG$.
(b) $DG \times GF = EG \times GC$.

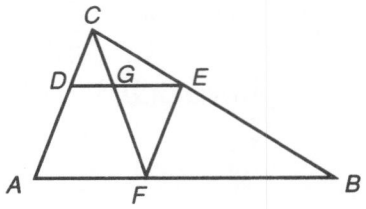

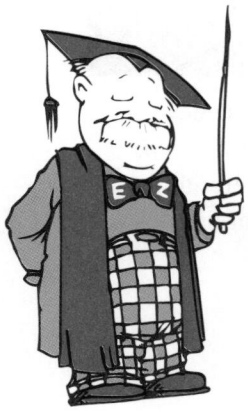

11 The Right Triangle

WHAT YOU WILL LEARN

Right triangles have special properties that are developed by applying the properties of similar triangles. In this chapter you will learn:

- the parts of a right triangle that are in proportion when an altitude is drawn to the hypotenuse of a right triangle;
- the use of the Pythagorean Theorem to find the length of any side of a right triangle when the lengths of the other two sides are known;
- the way to prove the Pythagorean Theorem;
- the special relationships between the sides in a right triangle whose acute angles measure 30 and 60;
- the special relationships between the sides in a right triangle whose acute angles measure 45 and 45;
- the use of trigonometry to find the measures of the angles or sides of a triangle.

SECTIONS IN THIS CHAPTER

- Proportions in a Right Triangle
- The Pythagorean Theorem
- Special Right-Triangle Relationships
- Trigonometric Ratios
- Indirect Measurement in a Right Triangle

Proportions in a Right Triangle

Let's begin by reviewing some important terminology and introducing some new notation. In a right triangle, the side opposite the right angle is called the *hypotenuse*; each of the remaining sides is called a *leg*. It is sometimes convenient to refer to the length of a side of a triangle by using the lowercase letter of the vertex that lies opposite the side. See Figure 11.1.

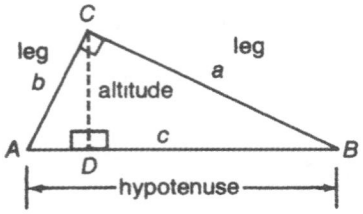

FIGURE 11.1

If we draw an altitude to the hypotenuse of the right triangle shown in Figure 11.1, we notice that altitude $\overline{CD}$ divides hypotenuse $\overline{AB}$ into two segments, $\overline{AD}$ and $\overline{BD}$ Hypotenuse segment $\overline{AD}$ is adjacent to leg $\overline{AC}$, while hypotenuse segment $\overline{BD}$ is adjacent to leg $\overline{BC}$. How many triangles do we see in Figure 11.1? There are *three* different triangles: the original right triangle ($\triangle ACB$) and the two smaller right triangles formed by the altitude on the hypotenuse ($\triangle ADC$ and $\triangle CDB$). As a result of the similarity relationships that exist between these pairs of triangles, a set of proportions can be written. Theorem 11.1 summarizes these relationships.

THEOREM 11.1 PROPORTIONS IN A RIGHT TRIANGLE THEOREM

If in a right triangle the altitude to the hypotenuse is drawn, then:

- The altitude separates the original triangle into two new triangles that are similar to the original triangle and to each other (see Figure 11.1):

$$\triangle ADC \sim \triangle ACB$$
$$\triangle CDB \sim \triangle ACB$$
$$\triangle ADC \sim \triangle CDB$$

- The length of each leg is the mean proportional between the length of the hypotenuse segment adjacent to the leg and the length of the entire hypotenuse.

  Since $\triangle ADC \sim \triangle ACB$,

$$\frac{AD}{AC} = \frac{AC}{AB}$$

Since $\triangle CDB \sim \triangle ACB$,

$$\frac{BD}{BC} = \frac{BC}{AB}$$

- The length of the altitude is the mean proportional between the lengths of the segments it forms on the hypotenuse.

  Since $\triangle ADC \sim \triangle CBD$,

$$\frac{AD}{CD} = \frac{CD}{DB}$$

OUTLINE OF PROOF: $\triangle ADC \sim \triangle ACB$ since $\measuredangle A \cong \measuredangle A$ and $\measuredangle ADC \cong \measuredangle ACB$. $\triangle CDB \sim \triangle ACB$ since $\measuredangle B \cong \measuredangle B$ and $\measuredangle CDB \cong \measuredangle ACB$. $\triangle ADC \sim \triangle CDB$ since, if each triangle is similar to the same triangle, the triangles must be similar to each other. The indicated proportions follow from the principle that the lengths of corresponding sides of similar triangles are in proportion.

EXAMPLE 11.1 Find the value of x.

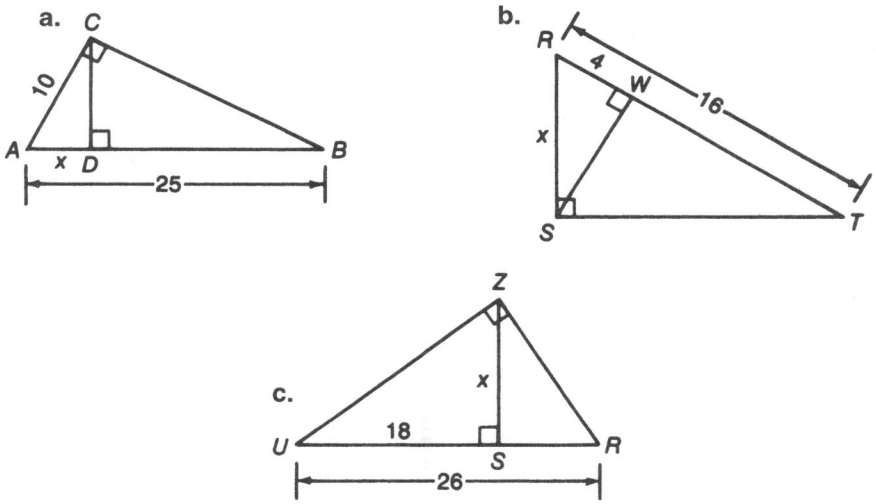

SOLUTION

a. $\dfrac{AD}{AC} = \dfrac{AC}{AB}$ $\left(\dfrac{\text{hyp segment}}{\text{leg}} = \dfrac{\text{leg}}{\text{hyp}}\right)$

$$\frac{x}{10} = \frac{10}{25}$$

$$25x = 100$$

$$x = \frac{100}{25} = 4$$

b. $\dfrac{RW}{RS} = \dfrac{RS}{RT}$ $\left(\dfrac{\text{hyp segment}}{\text{leg}} = \dfrac{\text{leg}}{\text{hyp}}\right)$

$$\frac{4}{x} = \frac{x}{16}$$
$$x^2 = 64$$
$$x = \sqrt{64} = 8$$

c. $\dfrac{US}{ZS} = \dfrac{ZS}{RS}$ $\left(\dfrac{\text{hyp segment 1}}{\text{altitude}} = \dfrac{\text{altitude}}{\text{hyp segment 2}}\right)$

$$\frac{18}{x} = \frac{x}{8}$$

NOTE: $RS = 26 - 18 = 8$

$$x^2 = 144$$
$$x = \sqrt{144} = 12$$

EXAMPLE 11.2 In right triangle JKL, $\measuredangle K$ is a right angle. Altitude $\overline{KH}$ is drawn so that the length of $\overline{JH}$ exceeds the length of $\overline{HL}$ by 5. If $KH = 6$, find the length of the hypotenuse.

SOLUTION

Let x = length of $\overline{LH}$.
Then $x + 5$ = length of $\overline{JH}$.

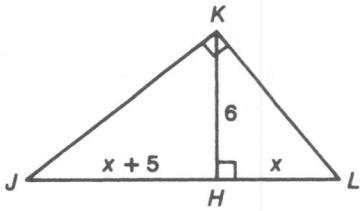

$$\frac{x}{6} = \frac{6}{x+5} \quad \left(\frac{\text{hyp segment 1}}{\text{altitude}} = \frac{\text{altitude}}{\text{hyp segment 2}}\right)$$
$$x(x+5) = 36$$
$$x^2 + 5x = 36$$
$$x^2 + 5x - 36 = 36 - 36$$
$$x^2 + 5x - 36 = 0$$
$$(x+9)(x-4) = 0$$
$$x + 9 = 0 \quad \text{or} \quad x - 4 = 0$$

$x = -9$	$x = 4$
(Reject since a length cannot be a negative number.)	$LH = x = 4$
	$JH = x + 5 = 9$
	$JL = LH + JH = 4 + 9 = 13$

SUMMARY

Proportions in a right triangle

- $\frac{x}{b} = \frac{b}{c}$ and $\frac{y}{a} = \frac{a}{c}$
- $\frac{x}{h} = \frac{h}{y}$

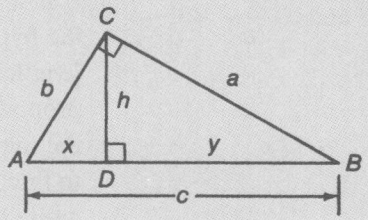

The Pythagorean Theorem

One of the most famous and useful theorems in mathematics provides a means for finding the length of *any* side of a *right* triangle, given the lengths of the other two sides. The sides are related by the equation

$$(\text{Hypotenuse})^2 = (\text{leg } 1)^2 + (\text{leg } 2)^2$$

This relationship, known as the *Pythagorean Theorem*, is named in honor of the Greek mathematician Pythagoras, who is believed to have presented the first proof of this theorem in about 500 B.C. Since that time, many different proofs have been offered. Here are the formal statement of the theorem and one such proof.

THEOREM 11.2 THE PYTHAGOREAN THEOREM

The square of the length of the hypotenuse of a right triangle is equal to the sum of the squares of the lengths of the legs.

GIVEN: Right triangle ACB with right angle C.
PROVE: $c^2 = a^2 + b^2$.

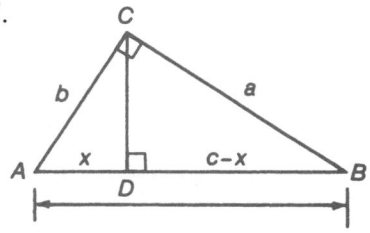

PLAN: Draw the altitude from C to hypotenuse $\overline{AB}$, intersecting $\overline{AB}$ at point D. For convenience, refer to $\overline{AD}$ as x and $\overline{DB}$ as $c - x$. Apply Theorem 11.1.

PROOF:

Statements	Reasons
1. Right $\triangle ACB$ with right angle C.	1. Given.
2. Draw the altitude from vertex C to hypotenuse $\overline{AB}$, intersecting $\overline{AB}$ at point D.	2. From a point not on a line, exactly one perpendicular may be drawn to the line.
Apply Theorem 11.1:	
3. $\frac{x}{b} = \frac{b}{c}$ and $\frac{c-x}{a} = \frac{a}{c}$.	3. If in a right triangle the altitude to the hypotenuse is drawn, then the length of each leg is the mean proportional between the length of the hypotenuse segment adjacent to the leg and the length of the entire hypotenuse.
4. $cx = b^2$ and $c(c-x) = a^2$, which may be written as $c^2 - cx = a^2$.	4. In a proportion, the product of the means equals the product of the extremes.
5. $\begin{array}{rl} c^2 - \cancel{cx} &= a^2 \\ + \quad \cancel{cx} &= b^2 \\ \hline c^2 \quad &= a^2 + b^2 \end{array}$	5. Addition property.

EXAMPLE 11.3 Find the value of x.

a.

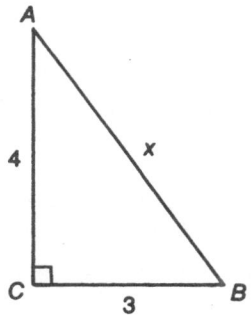

b.

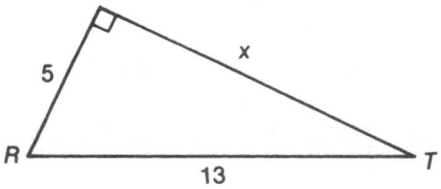

c.

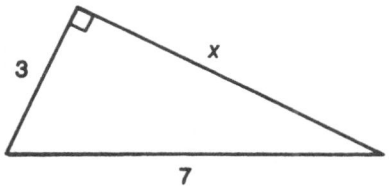

d.

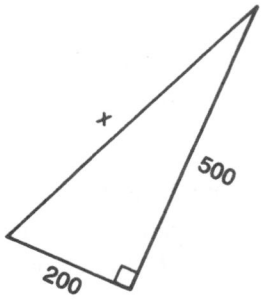

SOLUTION

a. Since x represents the length of the hypotenuse.

$$\begin{aligned} x^2 &= 3^2 + 4^2 \\ &= 9 + 16 \\ &= 25 \\ x &= \sqrt{25} = 5 \end{aligned}$$

b. Since x represents the length of a leg and the hypotenuse is 13,

$$\begin{aligned} 13^2 &= x^2 + 5^2 \\ 169 &= x^2 + 25 \\ 169 - 25 &= x^2 + 25 - 25 \\ 144 &= x^2 \\ \text{or} \qquad x^2 &= 144 \\ x &= \sqrt{144} = 12 \end{aligned}$$

c. Since x represents the length of a leg and the hypotenuse is 7,

$$\begin{aligned} 7^2 &= x^2 + 3^2 \\ 49 &= x^2 + 9 \\ 49 - 9 &= x^2 + 9 - 9 \\ 40 &= x^2 \\ \text{or} \qquad x^2 &= 40 \\ x &= \sqrt{40} \end{aligned}$$

To express a radical in simplest form, write the number underneath the radical sign as the product of two numbers, one of which is the *highest* perfect square factor of the number. Next, distribute the radical sign and simplify:

$$x = \sqrt{40} = \sqrt{4} \times \sqrt{10} = 2\sqrt{10}$$

d. Squaring 200 and 500 would be quite cumbersome. Since each is divisible by 100, work with the numbers 2 and 5. Using these numbers, find the length of the hypotenuse. *Multiply* the answer obtained by 100 to compensate for dividing the original lengths by 100.

$$\begin{aligned} c^2 &= 2^2 + 5^2 \\ &= 4 + 25 \\ &= 29 \\ c &= \sqrt{29} \end{aligned}$$

Find the actual value, x, by multiplying the value obtained by 100:

$$x = 100c = 100\sqrt{29}$$

PYTHAGOREAN TRIPLES

REMEMBER
Memorize **the sets {3, 4, 5}, {5, 12, 13}, and {8, 15, 17} as Pythagorean triples and be able to recognize their multiples.**

In the solutions for Example 11.3, parts **a** and **b**, whole-number values were obtained for the missing sides. Any set of three whole numbers x, y, and z is called a *Pythagorean triple* if the numbers satisfy the equation

$$z^2 = x^2 + y^2$$

The set of numbers $\{3, 4, 5\}$ is an example of a Pythagorean triple. The sets $\{5, 12, 13\}$ and $\{8, 15, 17\}$ are also Pythagorean triples. There are many others.

If $\{x, y, z\}$ is a Pythagorean triple, then so is the set that includes any whole number multiple of *each* member of this set. Thus, the set $\{6, 8, 10\}$ is a Pythagorean triple since each member was obtained by multiplying the corresponding member of the Pythagorean triple $\{3, 4, 5\}$ by 2:

$$\{6, 8, 10\} = \{2 \cdot 3, 2 \cdot 4, 2 \cdot 5\}$$

The following table shows additional examples.

Pythagorean Triple	Multiple of a Pythagorean Triple	Multiplying Factor
{3, 4, 5}	{15, 20, 25}	5
{5, 12, 13}	{10, 24, 26}	2
{8, 15, 17}	{80, 150, 170}	10

EXAMPLE 11.4 The base (length) of a rectangle is 12 and its altitude (height) is 5. Find the length of a diagonal.

SOLUTION

Triangle ABC is a 5-12-13 right triangle. Diagonal $AC = 13$. If you didn't see this pattern, you could apply the Pythagorean Theorem:

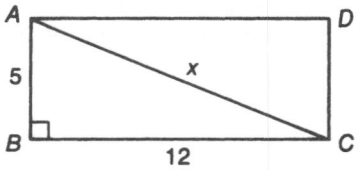

$$\begin{aligned} x^2 &= 5^2 + 12^2 \\ &= 25 + 144 = 169 \\ x &= \sqrt{169} = 13 \end{aligned}$$

EXAMPLE 11.5 The diagonals of a rhombus are 18 and 24. Find the length of a side of the rhombus.

SOLUTION

Recall that the diagonals of a rhombus bisect each other and intersect at right angles.

$\triangle AED$ is a multiple of a 3-4-5 right triangle. Each member of the triple is multiplied by 3. Hence, the length of side $\overline{AD}$ is $3 \bullet 5$ or 15.

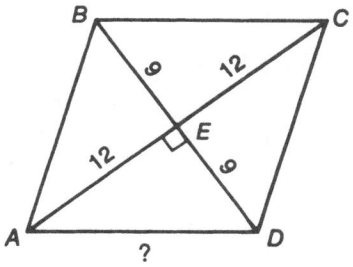

EXAMPLE 11.6 The length of each leg of an isosceles trapezoid is 17. The lengths of its bases are 9 and 39. Find the length of an altitude.

SOLUTION

Drop two altitudes, one from each of the upper vertices.

Quadrilateral $BEFC$ is a rectangle (since $\overline{BE}$ and $\overline{CF}$ are congruent, parallel, and intersect $\overline{AD}$ at right angles). Hence $BC = EF = 9$. Since right $\triangle AEB \cong$ right $\triangle DFC$, $AE = DF = 15$. $\triangle AEB$ is an 8-15-17 right triangle. The length of an altitude is 8.

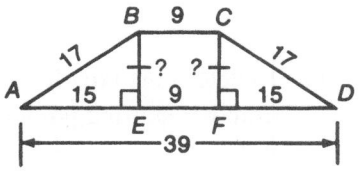

EXAMPLE 11.7 Find the length of a side of a square if a diagonal has a length of 8.

SOLUTION

There are no Pythagorean triples here. Since a square is equilateral, we may represent the lengths of sides $\overline{AB}$ and $\overline{AD}$ as x and apply the Pythagorean Theorem in right $\triangle BAD$:

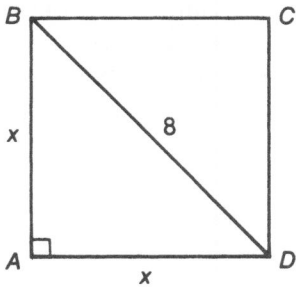

$$x^2 + x^2 = 8^2$$
$$2x^2 = 64$$
$$x^2 = 32$$
$$x = \sqrt{32} = \sqrt{16} \cdot \sqrt{2} = 4\sqrt{2}$$

The length of a side is $4\sqrt{2}$.

EXAMPLE 11.8 Katie hikes 5 miles north, 7 miles east, and then 3 miles north again. To the *nearest tenth of a mile*, how far, in a straight line, is Katie from her starting point?

SOLUTION

The four key points on Katie's trip are labeled A through D in the accompanying diagram.

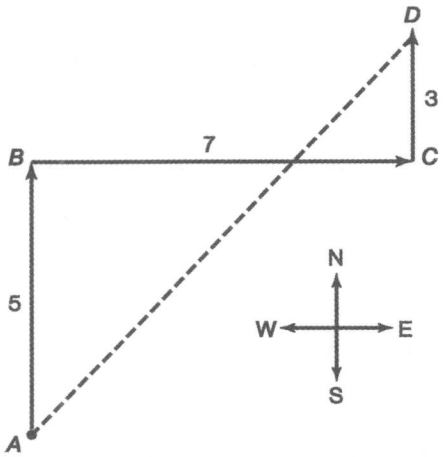

- To determine how far, in a straight line, Katie is from her starting point at A, find the length of $\overline{AD}$.
- Form a right triangle in which $\overline{AD}$ is the hypotenuse by completing rectangle $BCDE$, as shown in the accompanying diagram.
- Because opposite sides of a rectangle have the same length, $ED = BC = 7$, and $BE = CD = 3$. Thus, $AE = 5 + 3 = 8$.
- As $\triangle AED$ is a right triangle, use the Pythagorean theorem to find AD:

$$\begin{aligned}(AD)^2 &= (AF)^2 + (FD)^2 \\ &= 8^2 + 7^2 \\ &= 64 + 49 \\ &= 113 \\ AD &= \sqrt{113} \approx 10.63\end{aligned}$$

Correct to the *nearest tenth of a mile*, Katie is 10.6 miles from her starting point.

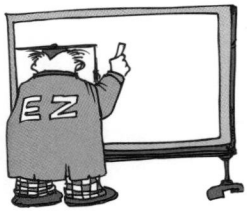

PYTHAGOREAN TRIPLES

Some commonly encountered Pythagorean triples are:

$$\{3n, 4n, 5n\}$$
$$\{5n, 12n, 13n\}$$
$$\{8n, 15n, 17n\}$$

where n is any positive integer ($n = 1, 2, 3, \ldots$). There are many other families of Pythagorean triples, including $\{7, 24, 25\}$ and $\{9, 40, 41\}$.

CONVERSE OF THE PYTHAGOREAN THEOREM

If the lengths of the sides of a triangle are known, then we can determine whether the triangle is a right triangle by applying the *converse* of the Pythagorean Theorem.

THEOREM 11.3 CONVERSE OF THE PYTHAGOREAN THEOREM

If in a triangle the square of the length of a side is equal to the sum of the squares of the lengths of the other two sides, then the triangle is a right triangle.

Suppose a, b, and c represent the lengths of the sides of a triangle and c is the largest of the three numbers. If $c^2 = a^2 + b^2$, then the triangle is a right triangle.

For example, to determine whether the triangle whose sides have lengths of 11, 60, and 61 is a right triangle, we test whether the square of the largest of the three numbers is equal to the sum of the squares of the other two numbers:

$$61^2 \stackrel{?}{=} 11^2 + 60^2$$
$$3721 \stackrel{?}{=} 121 + 3600$$
$$3721 \stackrel{\checkmark}{=} 3721 \quad \text{Hence, the triangle is a right triangle.}$$

Special Right-Triangle Relationships

THE 30-60 RIGHT TRIANGLE

Suppose the length of each side of the equilateral triangle in Figure 11.2 is represented by $2s$. If you drop an altitude from the vertex of this triangle, several interesting relationships materialize:

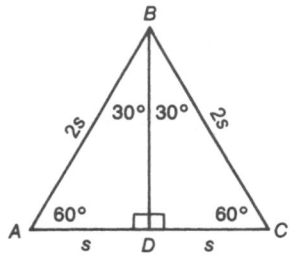

FIGURE 11.2

- Since $\triangle BDA \cong \triangle BDC$, $\overline{BD}$ bisects $\measuredangle ABC$ and also bisects side $\overline{AC}$. Hence, $m\measuredangle ABD = 30$. The acute angles of right triangle ADB measure 60 and 30. Also, $AD = \frac{1}{2}(AC) = \frac{1}{2}(2s) = s$.
- The length of leg $\overline{BD}$ may be found using the Pythagorean Theorem:

$$\begin{aligned}(BD)^2 + (AD)^2 &= (AB)^2\\ (BD)^2 + (s)^2 &= (2s)^2\\ (BD)^2 + s^2 &= 4s^2\\ (BD)^2 &= 3s^2\\ BD &= \sqrt{3s^2} = \sqrt{3}\cdot\sqrt{s^2} = \sqrt{3}\,s\end{aligned}$$

Let's summarize the information we have gathered about $\triangle ADB$. First, $\triangle ADB$ is referred to as a *30-60 right triangle* since these numbers correspond to the measures of its acute angles. In a 30-60 right triangle the following relationships hold:

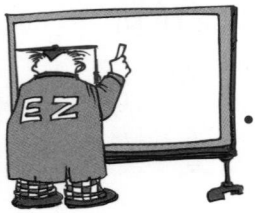

30-60 RIGHT-TRIANGLE SIDE RELATIONSHIPS

- The length of the *shorter* leg (the side opposite the 30 degree angle) is one-half the length of the hypotenuse:

$$AD = \frac{1}{2}AB$$

- The length of the *longer* leg (the side opposite the 60 degree angle) is one-half the length of the hypotenuse multiplied by $\sqrt{3}$:

$$BD = \frac{1}{2} \times AB \times \sqrt{3}$$

- The length of the longer leg is equal to the length of the shorter leg multiplied by $\sqrt{3}$:

$$BD = AD \cdot \sqrt{3}$$

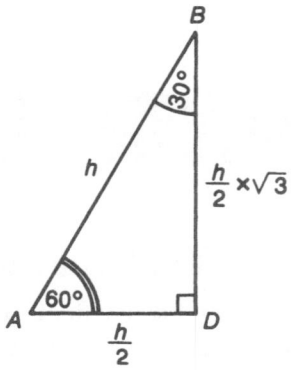

EXAMPLE 11.9 Fill in the following table:

	RS	*ST*	*RT*
a.	?	?	12
b.	4	?	?
c.	?	$7\sqrt{3}$	?

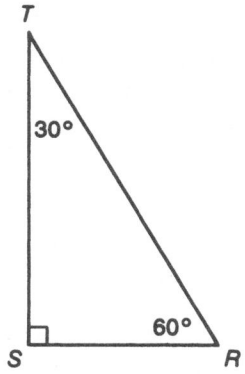

SOLUTION

a. $RS = \frac{1}{2}(RT) = \frac{1}{2}(12) = 6$

$ST = RS(\sqrt{3}) = 6\sqrt{3}$

b. $ST = RS(\sqrt{3}) = 4\sqrt{3}$

$RT = 2(RS) = 2(4) = 8$

c. $RS = 7$

$RT = 2(RS) = 2(7) = 14$

EXAMPLE 11.10 In $\square RSTW$, $m\measuredangle R = 30$ and $RS = 12$. Find the length of an altitude.

SOLUTION

$SH = \frac{1}{2}(12) = 6$

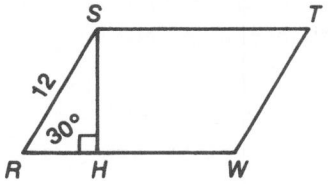

EXAMPLE 11.11 In $\triangle JKL$, $m\measuredangle K = 120$ and $JK = 10$. Find the length of the altitude drawn from vertex J to side $\overline{LK}$ (extended if necessary).

SOLUTION

Since $\measuredangle K$ is obtuse, the altitude falls in the exterior of the triangle, intersecting the extension of $\overline{KL}$, say at point H. $\triangle JHK$ is a 30-60 right triangle. Hence,

$$JH = \frac{1}{2} \cdot 10 \cdot \sqrt{3} = 5\sqrt{3}$$

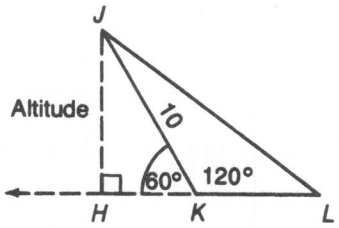

THE 45-45 RIGHT TRIANGLE

Another special right triangle is the isosceles right triangle. Since the legs of an isosceles right triangle are congruent, the angles opposite must also be congruent. This implies that the measure of each acute angle of an isosceles right triangle is 45.

REMEMBER
A *45-45 right triangle* is another name for an isosceles right triangle.

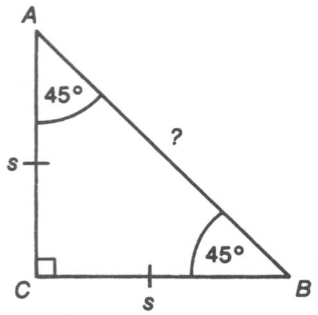

FIGURE 11.3

To determine the relationships between the lengths of the sides of a 45-45 right triangle, we will represent the length of each leg by the letter s (Figure 11.3).

Then the hypotenuse $\overline{AB}$ may be expressed in terms of s by applying the Pythagorean Theorem:

$$\begin{aligned}(AB)^2 &= s^2 + s^2\\ &= 2s^2\\ AB &= \sqrt{2s^2} = \sqrt{2} \times \sqrt{s^2}\\ &= \sqrt{2} \times s\end{aligned}$$

Thus, the length of the hypotenuse is $\sqrt{2}$ times the length of a leg:

$$\text{Hypotenuse} = \sqrt{2} \times \text{leg}$$

By dividing each side of this equation by $\sqrt{2}$ we may find an expression for the length of a leg in terms of the length of the hypotenuse:

$$\text{Leg} = \frac{\text{Hypotenuse}}{\sqrt{2}}$$

Rationalizing the denominator, we obtain:

$$\text{Leg} = \frac{\text{Hypotenuse}}{\sqrt{2}} \times \frac{\sqrt{2}}{\sqrt{2}} = \frac{\text{Hypotenuse}}{2} \times \sqrt{2}$$

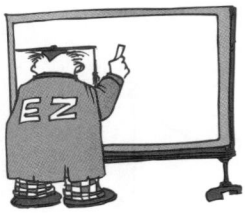

45-45 RIGHT TRIANGLE SIDE RELATIONSHIPS

- The *lengths of the legs* are equal:

$$AC = BC$$

- The *length of the hypotenuse* is equal to the length of either leg multiplied by $\sqrt{2}$:

$$AB = AC \text{ (or } BC) \bullet \sqrt{2}$$

- The *length of either leg* is equal to one-half the length of the hypotenuse multiplied by $\sqrt{2}$:

$$AC \text{ (or } BC) = \frac{1}{2} AB \bullet \sqrt{2}$$

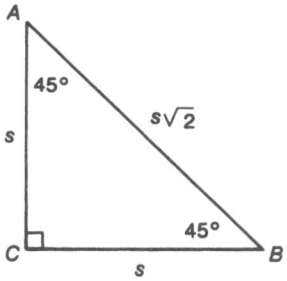

EXAMPLE 11.12 In isosceles trapezoid $ABCD$, the measure of a lower base angle is 45 and the length of upper base $\overline{BC}$ is 5. If the length of an altitude is 7, find the lengths of the legs, $\overline{AB}$ and $\overline{DC}$.

SOLUTION

Drop altitudes from B and C, forming two congruent 45-45 right triangles. $AE = BE = 7$. Also, $FD = 7$. $AB = AE \times \sqrt{2} = 7\sqrt{2}$. The length of each leg is $7\sqrt{2}$.

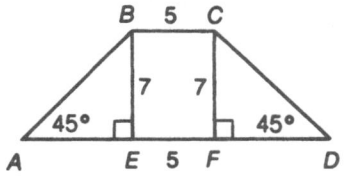

Trigonometric Ratios

DEFINING TRIGONOMETRIC RATIOS

Each of the triangles pictured in Figure 11.4 is a 30-60 right triangle. The lengths of the sides in each of these right triangles must obey the relationships developed in the section on special right-triangle relationships.

In particular, we may write:

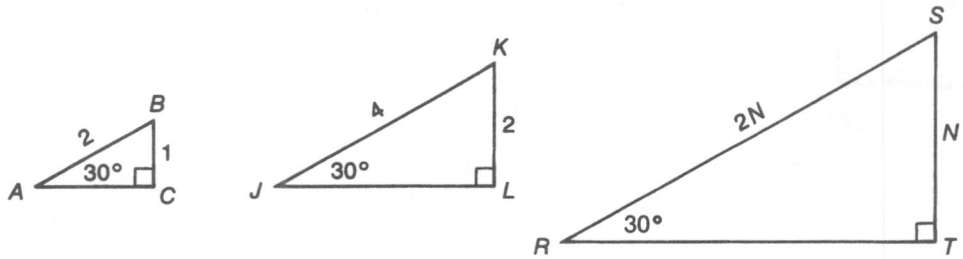

FIGURE 11.4

In $\triangle ABC$	In $\triangle JKL$	In $\triangle RST$
$\frac{BC}{AB} = \frac{1}{2}$	$\frac{KL}{JK} = \frac{2}{4} = \frac{1}{2}$	$\frac{ST}{RS} = \frac{N}{2N} = \frac{1}{2}$

In each of these triangles the ratio of the length of the leg opposite the 30° angle to the length of the hypotenuse is 1 : 2 or 0.5. This ratio will hold in *every* 30-60 right triangle no matter how large or small the triangle is. For other choices of congruent acute angles in right triangles, will this type of ratio also be constant? As the set of triangles in Figure 11.5 illustrates, the answer is yes! In each right triangle, the ratio of the length of the leg opposite the angle whose measure is x degrees to the length of the hypotenuse must be the same.

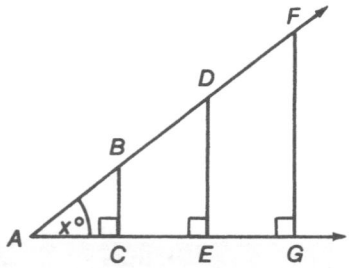

FIGURE 11.5

The ratios $\frac{BC}{AB}, \frac{DE}{AD}, \frac{FG}{AF}$ are equal since they represent the ratios of the lengths of corresponding sides of similar triangles. You should observe that Figure 11.5 includes three right triangles each of which is similar to the other two; that is, $\triangle ABC \sim \triangle ADE \sim \triangle AFG$ by the AA Theorem of similarity since:

- $\measuredangle A \cong \measuredangle A \cong \measuredangle A$.
- Each triangle includes a right angle, and all right angles are congruent.

We are then entitled to write the following extended proportion:

$$\frac{BC}{AB} = \frac{DE}{AD} = \frac{FG}{AF} = \frac{\text{length of leg opposite angle } A}{\text{length of hypotenuse}}$$

For any given acute angle, this ratio is the same regardless of the size of the right triangle. It will be convenient to refer to this type of ratio by a special name—the *sine ratio*. The sine of an acute angle of a right triangle is the ratio formed by taking the length of the leg opposite the angle and dividing it by the length of the hypotenuse. In a right triangle, other ratios may also be formed with each being given a special name. These names are summarized in the following definitions.

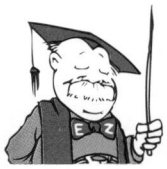

DEFINITIONS OF TRIGONOMETRIC RATIOS

A *trigonometric ratio* is the ratio of the lengths of *any* two sides of a *right* triangle with respect to a given angle. Three commonly formed trigonometric ratios are called the *sine*, *cosine*, and *tangent* ratios:

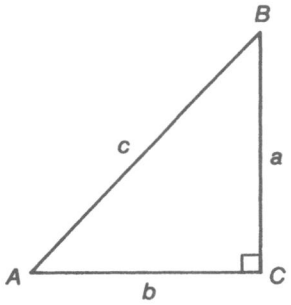

$$\text{Sine of } \measuredangle A = \frac{\text{length of leg opposite } \measuredangle A}{\text{length of hypotenuse}} = \frac{BC}{AB}$$

$$\text{Cosine of } \measuredangle A = \frac{\text{length of leg adjacent } \measuredangle A}{\text{length of hypotenuse}} = \frac{AC}{AB}$$

$$\text{Tangent of } \measuredangle A = \frac{\text{length of leg opposite } \measuredangle A}{\text{length of leg adjacent } \measuredangle A} = \frac{BC}{AC}$$

WORKING WITH TRIGONOMETRIC RATIOS

The Pythagorean Theorem relates the lengths of the three sides of a right triangle, while trigonometric ratios relate the measures of two sides and an acute angle of a right triangle. When working with trigonometric ratios, keep in mind the following:

- Sine, cosine, and tangent may be abbreviated as *sin*, *cos*, and *tan*, respectively.
- The trigonometric ratios may be taken with respect to either of the acute angles of the triangle:

$\sin A = \frac{a}{c}$ $\qquad$ $\sin B = \frac{b}{c}$

$\cos A = \frac{b}{c}$ $\qquad$ $\cos B = \frac{a}{c}$

$\tan A = \frac{a}{b}$ $\qquad$ $\tan B = \frac{b}{a}$

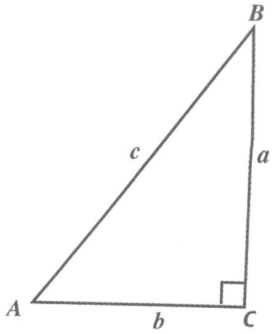

- Since $\sin A$ and $\cos B$ are both equal to $\frac{a}{c}$, they are equal to each other. Similarly, $\cos A$ and $\sin B$ are equal since both are equal to $\frac{b}{c}$. Since angles A and B are complementary, the sine of an angle is equal to the cosine of the angle's complement. For example, $\sin 60° = \cos 30°$.
- The values of the trigonometric ratios do not change in different right triangles in which the measure of acute angle x is the same, as these right triangles are similar.
- Since the hypotenuse is the longest side of a right triangle, the numerators of the sine and cosine ratio are always less than their denominators. Thus, the value of the sine and cosine of an acute angle is always a number between 0 and 1. The tangent ratio may have a value greater than 1.
- Trigonometric *ratios* are also called trigonometric *functions*.

EXAMPLE 11.13 In right triangle ABC, $\measuredangle C$ is a right angle. If $AC = 4$ and $BC = 3$, find the value of $\tan A$, $\sin A$, and $\cos A$.

SOLUTION

Since $\triangle ABC$ is a 3-4-5 right triangle, $AB = 5$.

$$\tan A = \frac{\text{leg opposite } \measuredangle A}{\text{leg adjacent } \measuredangle A} = \frac{3}{4} = 0.75$$

$$\sin A = \frac{\text{leg opposite } \measuredangle A}{\text{hypotenuse}} = \frac{3}{5} = 0.6$$

$$\cos A = \frac{\text{leg adjacent } \measuredangle A}{\text{hypotenuse}} = \frac{4}{5} = 0.8$$

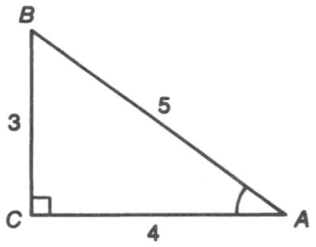

EXAMPLE 11.14 Express the values of $\sin 60°$, $\cos 60°$, and $\tan 60°$ in either radical or decimal form.

SOLUTION

$$\sin 60° = \frac{\sqrt{3} \times n}{2n} = \frac{\sqrt{3}}{2}$$

$$\cos 60° = \frac{n}{2n} = \frac{1}{2} = 0.5$$

$$\tan 60° = \frac{\sqrt{3} \times n}{n} = \sqrt{3}$$

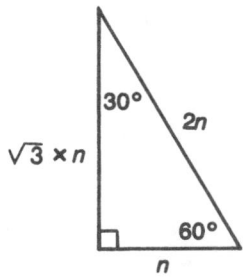

USING A SCIENTIFIC CALCULATOR TO FIND TRIGONOMETRIC VALUES

In Example 11.13 we were able to use special right-triangle relationships to find the exact values of the sine, cosine, and tangent of 60°. Suppose that we need to find the value of the sine, cosine, or tangent of 38°. One approach would be to use a protractor to construct a right triangle with an acute angle of 38° and then, with a ruler, measure the lengths of the three sides of the right triangle. Using these lengths, we could then calculate ratios that would give us an approximate value for the sine, cosine, or tangent of 38°. Unfortunately, this procedure is time consuming, prone to error, and usually does not give results that are sufficiently close to the actual values.

When you first turn on a scientific calculator, the unit of angle measurement is typically set to DEGrees. If you do not see DEG in small type in the display window, press the MODE key or DGR key until DEG is displayed. To evaluate the sine, cosine, or tangent of $n°$, enter n and then press the key labeled with the name of the appropriate trigonometric function. Since most scientific calculators will display a decimal value with at least eight-place decimal accuracy, you will usually need to round off the answer that appears in the display window. As a general practice, you should round off answers correct to four decimal places.

REMEMBER
Scientific calculators provide a fast and easy way to obtain values for the sine, cosine, and tangent of a given angle.

For example, to evaluate sin 38° correct to *four decimal places*, follow this procedure:

STEP 1: Enter 38.
STEP 2: Press the SIN key.
STEP 3: Round off the value that appears in the display window. Thus, correct *to four decimal places,* $\sin 38° = 0.6157$.

Not all calculators work in the same way. For example, some scientific calculators require that you first press the SIN key, enter 38, and then press the = key.

Now consider the opposite situation. Suppose you know that $\cos x = 0.8387$. How can you find the measure of angle x correct to the nearest degree? A scientific calculator has either an INVerse function key, a 2nd function key, or a SHIFT key that allows you to find the degree measure of an angle when the value of a trigonometric function of that angle is known. For example, to find x when $\cos x = 0.8387$, follow these steps:

STEP 1: Enter .8387.
STEP 2: Press the INV or 2nd or SHIFT key, depending on the calculator you are using.
STEP 3: Press the COS key.
STEP 4: Round off the value that appears in the display window to the nearest degree. Thus, correct to the *nearest degree,* $x = 42$.

If this procedure does not work with the calculator you are using, try Steps 2 and 3 first, then enter the angle and press the = key. For example, if your calculator has a SHIFT key, use this key sequence:

SHIFT → COS → .8387 → =

EXAMPLE 11.15 Use your calculator to find the value of each of the following:

a. tan 27° **b.** sin 44° **c.** cos 35°

SOLUTION

a. tan 27° = 0.5095 **b.** sin 44° = 0.6947 **c.** cos 35° = 0.8192

EXAMPLE 11.16 Use your calculator to find the measure of angle x if:

a. $\sin x = 0.7071$ **b.** $\tan x = 0.6009$ **c.** $\cos x = 0.7986$

SOLUTION

a. $x = 45°$ **b.** $x = 31°$ **c.** $x = 37°$

EXAMPLE 11.17 Use your calculator to find the measure of angle A, correct to the *nearest degree*, if:

a. $\tan A = 0.7413$ **b.** $\cos A = 0.8854$ **c.** $\sin A = 0.6599$

SOLUTION

a. $A = 37°$ **b.** $A = 28°$ **c.** $A = 41°$

You should use your calculator to verify that for angles from 0° to 90°:

- As an angle increases in value, its sine and tangent ratios increase, while its cosine ratio decreases.
- The minimum value of sine, cosine, and tangent is 0. (*Note:* $\sin 0° = 0$, $\cos 90° = 0$, $\tan 0° = 0$)
- The maximum value of cosine and sine is 1. (*Note:* $\cos 0° = 1$ and $\sin 90° = 1$.) As the measure of an acute angle of a right triangle approaches 90°, the value of its tangent ratio gets larger and larger, with *no* upper limit. We say that the tangent ratio is *unbounded* or undefined at 90°.

Indirect Measurement in a Right Triangle

Trigonometric ratios may be used to arrive at the measure of a side or angle of a right triangle that may be difficult, if not impossible, to calculate by direct measurement.

For example, consider a plane that takes off from a runway, and climbs while maintaining a constant angle with the horizontal ground. Suppose that, at the instant of time when the plane has traveled 1000 meters, its altitude is 290 meters. Using our knowledge of trigonometry, we can approximate the measure of the angle at which the plane has risen with respect to the horizontal ground.

A right triangle may be used to represent the situation we have just described. The hypotenuse corresponds to the path of the rising plane, the vertical leg of the triangle represents the plane's altitude, and the acute angle formed by the hypotenuse and the horizontal leg (the ground) is the desired angle whose measure we must determine.

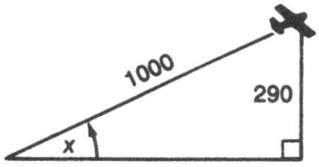

FIGURE 11.6

To find the value of x in Figure 11.6, we must first determine the appropriate trigonometric ratio. The sine ratio relates the three quantities under consideration:

$$\sin x = \frac{\text{leg opposite } \measuredangle}{\text{hypotenuse}}$$
$$= \frac{290}{1000} = 0.2900$$

To find the value of x to the nearest degree, we use a scientific calculator, which gives us $x = 17°$, correct to the *nearest degree*.

EXAMPLE 11.18 Find $m\measuredangle R$ to the *nearest degree*.

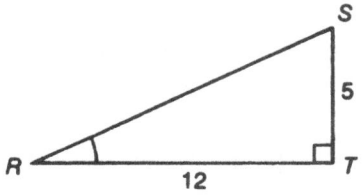

SOLUTION

- Decide which trigonometric ratio to use and write the corresponding equation:

$$\tan R = \frac{\text{leg opposite } \measuredangle R}{\text{leg adjacent } \measuredangle R}$$
$$\tan R = \frac{5}{12}$$

- Express the ratio in decimal form. Using a scientific calculator, you obtain the answer, rounded off to four decimal places: 0.4167. Hence,

$$\tan R = 0.4167$$

- Again use a scientific calculator to find that R measures 23°, correct to the *nearest degree*.

The next two examples illustrate that, given the length of a side of a right triangle and the measure of an acute angle, the length of either of the two remaining sides of the triangle can be found.

EXAMPLE 11.19 Find the value of x to the *nearest tenth*.

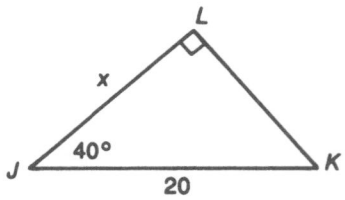

SOLUTION

- Decide which trigonometric ratio to use, and then write the corresponding equation:

$$\cos J = \frac{\text{leg adjacent} \measuredangle j}{\text{hypotenuse}}$$

$$\cos 40° = \frac{x}{20}$$

- Solve for x before evaluating cos 40°:

$$x = 20 \times \cos 40°$$

- Use your calculator to multiply using the value of cos 40°:

$$x = 15.32088886$$

- Round off the answer to the desired accuracy. This example asks for the answer correct to the *nearest tenth*. Hence:

$$x = 15.3$$

EXAMPLE 11.20 To determine the distance across a river, a surveyor marked two points on one riverbank, H and F, 65 meters apart. She also marked one point, K, on the opposite bank such that $\overline{KH} \perp \overline{HF}$, as shown in the accompanying figure. If $\angle K = 54°$, what is the width of the river, to the *nearest tenth of a meter*?

SOLUTION

Represent the width of the river, KH, by x.

- Because the problem involves the sides opposite and adjacent to the given angle, use the tangent ratio:

$$\tan\angle K = \frac{\text{opposite side } (HF)}{\text{adjacent side } (KH)}$$

$$\tan 54^\circ = \frac{65}{x}$$

- Solve for x before evaluating tan 54°:

$$x = 65 \div \tan 54^\circ$$

- Use your calculator to divide, using the value of tan 54°:

$$x = 47.22526432.$$

To the *nearest tenth of a meter*, the width of the river is 47.2 meters.

EXAMPLE 11.21 Find the value of x to the *nearest tenth*.

SOLUTION

$$\tan \measuredangle R = \frac{\text{leg opposite } \measuredangle R}{\text{leg adjacent } \measuredangle R}$$

$$\tan 75^\circ = \frac{28}{x}$$

$$3.7321 = \frac{28}{x}$$

$$3.7321x = 28$$

$$\frac{3.7321x}{3.7321} = \frac{28}{3.7321}$$

$$x = 7.502 = 7.5 \text{ to the } \textit{nearest tenth}$$

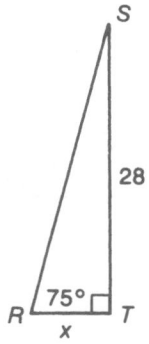

EXAMPLE 11.22 The lengths of the diagonals of a rhombus are 12 and 16. Find to the *nearest degree* the measures of the angles of the rhombus.

SOLUTION

- In $\triangle AED$:

$$\tan x = \frac{\text{side opposite}}{\text{side adjacent}} = \frac{ED}{AE}$$

$$= \frac{6}{8} = 0.7500$$

$$x = 37^\circ \text{ to the } \textit{nearest degree}$$

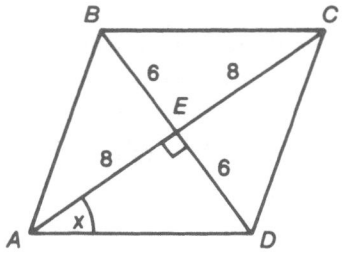

- Since the diagonals of a rhombus bisect its angles,

$$m\measuredangle BAD = 2x = 2(37) = 74$$

- Since consecutive angles of a rhombus are supplementary,

$$\begin{aligned} m\measuredangle ABC &= 180 - m\measuredangle BAD \\ &= 180 - 74 \\ &= 106 \end{aligned}$$

- Since opposite angles of a rhombus are equal in measure:

$$\begin{aligned} m\measuredangle BCD &= m\measuredangle BAD = 74 \\ m\measuredangle ADC &= m\measuredangle ABC = 106 \end{aligned}$$

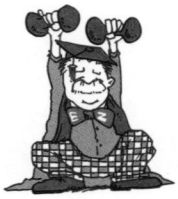

REVIEW EXERCISES FOR CHAPTER 11

In Exercises 1 to 3, find the values of r, s, *and* t.

1.

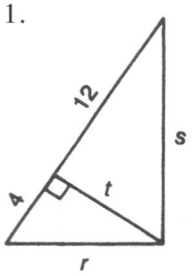

2.

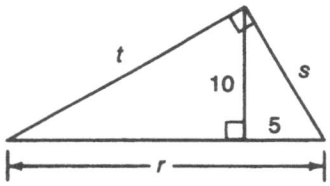

3.

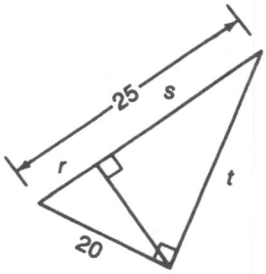

In Exercises 5 to 9, in right triangle JKL, *angle* JKL *is the right angle and* $\overline{KH} \perp \overline{JL}$.

4. The accompanying diagram shows a 24-foot ladder leaning against a building. A steel brace extends from the ladder to the point where the building meets the ground. The brace forms a right angle with the ladder. If the steel brace is connected to the ladder at a point that is 10 feet from the foot of the ladder, find the length, x, of the steel brace to the *nearest tenth* of a foot.

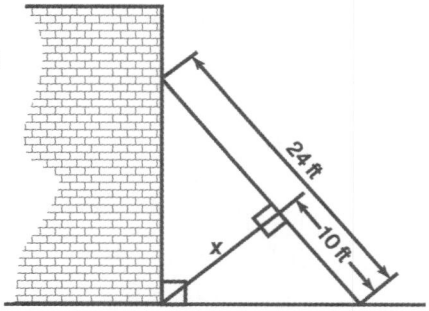

5. If $JH = 5$ and $HL = 4$, find KL.

6. If $JH = 8, JL = 20$, find KH.

7. If $KL = 18, JL = 27$, find JK.

8. If $JK = 14, HL = 21$, find JH.

9. If $KH = 12, JL = 40$, find JK (assume $\overline{JK}$ is the shorter leg of right $\triangle JKL$).

10. The altitude drawn to the hypotenuse of a right triangle divides the hypotenuse into segments such that their lengths are in the ratio of 1:4. If the length of the altitude is 8, find the length of:

    (a) Each segment of the hypotenuse.
    (b) The longer leg of the triangle.

In Exercises 11 to 15, find the value of x.

11.

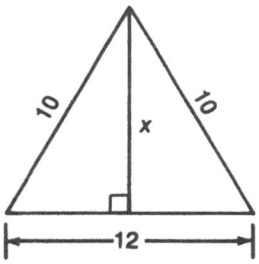

12.

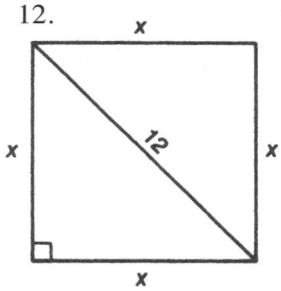

13.

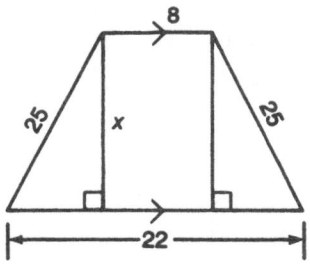

14.

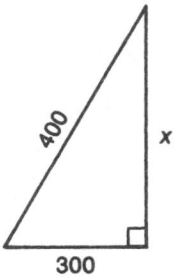

15.

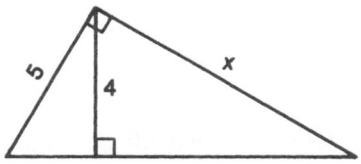

16. If the lengths of the diagonals of a rhombus are 32 and 24, find the perimeter of the rhombus.

17. If the perimeter of a rhombus is 164 and the length of the longer diagonal is 80, find the length of the shorter diagonal.

18. Find the length of the altitude drawn to a side of an equilateral triangle whose perimeter is 30.

19. The length of the base of an isosceles triangle is 14. If the length of the altitude drawn to the base is 5, find the length of each of the legs of the triangle.

20. The measure of the vertex angle of an isosceles triangle is 120 and the length of each leg is 8. Find the length of:

    (a) The altitude drawn to the base.
    (b) The base.

21. If the perimeter of a square is 24, find the length of a diagonal.

22. If the length of a diagonal of a square is 18, find the perimeter of the square.

23. Find the length of the altitude drawn to side $\overline{AC}$ of $\triangle ABC$ if $AB = 8$, $AC = 14$, and $m\measuredangle A$ equals:

    (a) 30
    (b) 120
    (c) 135

24. The lengths of the bases of an isosceles trapezoid are 9 and 25. Find the lengths of the altitude and each of the legs if the measure of each lower base angle is:

    (a) 30
    (b) 45
    (c) 60

25. Find the value of x, y, and z.

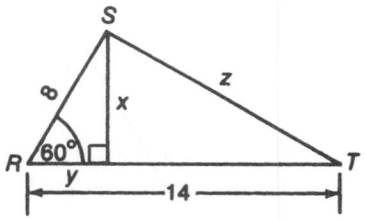

26. The lengths of two adjacent sides of a parallelogram are 6 and 14. If the measure of an included angle is 60, find the length of the shorter diagonal of the parallelogram.

27. The length of each side of a rhombus is 10 and the measure of an angle of the rhombus is 60. Find the length of the longer diagonal of the rhombus.

28. In right $\triangle ABC$, angle C is the right angle, $AC = 7$, and $BC = 24$. Find the value of the sine of the largest acute angle of the triangle.

29. In right $\triangle RST$, angle T is the right angle. If $\sin R = \frac{9}{41}$, find the values of $\cos R$ and $\tan R$.

30. The lengths of a pair of adjacent sides of a rectangle are 10 and 16. Find, correct to the *nearest degree*, the angle a diagonal makes with the longer side.

31. The measure of a vertex angle of an isosceles triangle is 72. If the length of the altitude drawn to the base is 10, find to the *nearest whole number* the length of the base and the length of each leg of the triangle.

32. The shorter diagonal of a rhombus makes an angle of 78° with a side of the rhombus. If the length of the shorter diagonal is 24, find to the *nearest tenth* the lengths of:

    (a) A side of the rhombus.
    (b) The longer diagonal.

33. Find the values of x and y, correct to the *nearest tenth*.

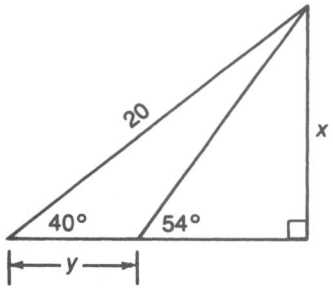

34. $ABCD$ is a trapezoid. If $AB = 14$, $BC = 10$, and $m\measuredangle BCD = 38$, find the lengths of $\overline{AD}$ and $\overline{CD}$, correct to the *nearest tenth*.

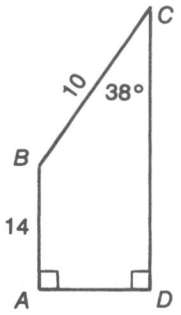

35. Two hikers started at the same location. One traveled 2 miles east and then 1 mile north. The other traveled 1 mile west and then 3 miles south. At the end of their hikes, how many miles apart were the two hikers?

36. At Slippery Ski Resort, the beginner's slope is inclined at an angle of 12.3°, while the advanced slope is inclined at an angle of 26.4°. If Rudy skis 1,000 meters down the advanced slope while Valerie skis the same distance on the

beginner's slope, how much greater was the horizontal distance, to the *nearest tenth* of a meter, that Valerie covered?

37. To get from his high school to his home, Jamal travels 5.0 miles east and then 4.0 miles north. When Sheila goes to her home from the same high school, she travels 8.0 miles east and 2.0 miles south. What is the shortest distance, to the *nearest tenth* of a mile, between Jamal's home and Sheila's home?

38. In the accompanying diagram, the base of a 15-foot ladder rests on the ground 4 feet from a 6-foot fence.

    (a) If the ladder touches the top of the fence and the side of a building, what angle, to the *nearest degree*, does the ladder make with the ground?

    (b) Using the angle found in part *a*, determine how far the top of the ladder reaches up the side of the building, to the *nearest tenth of a foot*.

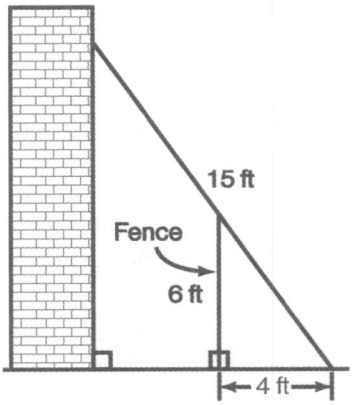

39. In the accompanying diagram of right triangles ABD and DBC, $AB = 5$, $AD = 4$, and $CD = 1$. Find the length of $\overline{BC}$ to the *nearest tenth*.

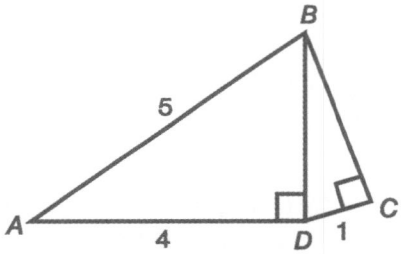

40. The accompanying diagram shows a flagpole that stands on level ground. Two cables, r and s, are attached to the pole at a point 16 feet above the ground. The combined length of the two cables is 50 feet. If cable r is attached to the ground 12 feet from the base of the pole, what is the measure of the angle, x, to the *nearest degree*, that cable s makes with the ground?

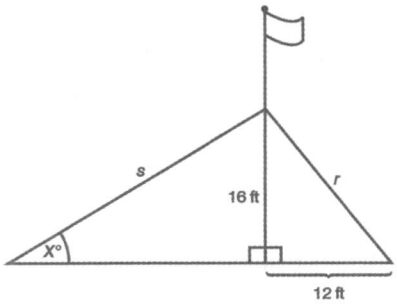

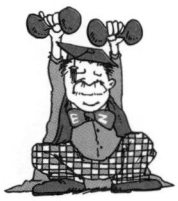

CUMULATIVE REVIEW EXERCISES: CHAPTERS 8–11

1. Which set of numbers may be the lengths of the sides of a triangle?

   (1) {5,4,1} (3) {5,4,9}
   (2) {5,4,6} (4) {5,4,10}

2. In quadrilateral $ABCD$, $\overline{AB} \cong \overline{DC}$ and $\overline{AB} \parallel \overline{DC}$. Which statement *must* be true?

   (1) $\overline{BD} \cong \overline{AC}$ (3) $\overline{AC} \cong \overline{AD}$
   (2) $\overline{AB} \cong \overline{BC}$ (4) $\overline{AD} \cong \overline{BC}$

3. In rhombus $ABCD$, the bisectors of $\measuredangle B$ and $\measuredangle C$ *must* be

   (1) parallel (3) perpendicular
   (2) oblique (4) congruent

4. In $\triangle ABC$, $AB = 8$ and $BC = 6$. Which statement is *always* true?

   (1) $AC > 14$ (3) $AC < 2$ and $AC > 14$
   (2) $AC < 2$ (4) $AC > 2$ and $AC < 14$

5. If the midpoints of the sides of a quadrilateral are joined consecutively, the figure formed *must* be

   (1) equiangular (3) a trapezoid
   (2) equilateral (4) a parallelogram

6. In the accompanying diagram of $\triangle ABC$, $BC > AB$, and $\overline{AB} \cong \overline{BD}$. Which statement is true?

   (1) $BC > BD$
   (2) $AB > BC$
   (3) $m\measuredangle BDA > m\measuredangle BAD$
   (4) $m\measuredangle BDA < m\measuredangle BCA$

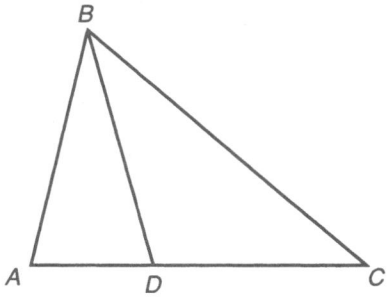

7. Which statement *must* be true?

   (1) If a parallelogram is not a rectangle, then it is not a square.
   (2) If a parallelogram is not a square, then it is not a rectangle.
   (3) All rectangles are squares.
   (4) If a parallelogram is a rectangle, then it is a square.

8. In the accompanying diagram, parallel lines $\overleftrightarrow{AB}$ and $\overleftrightarrow{CD}$ are cut by transveral $\overleftrightarrow{AC}$. Segments $\overline{BC}$ and $\overline{AD}$ intersect at E, and $m\measuredangle BAC < m\measuredangle ACD$. If $\overline{CB}$ bisects $\measuredangle ACD$ and AD bisects $\measuredangle BAC$, then which statement is true?

   (1) $AC = AB$
   (2) $AD < CD$
   (3) $AE < CE$
   (4) $\measuredangle DAC > \measuredangle BCD$

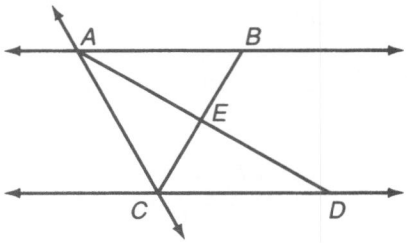

9. In equilateral triangle ABC, $AB = 16$. Find the perimeter of the triangle formed by connecting the midpoints of the sides of triangle ABC.

10. The altitude to the hypotenuse of right triangle ABC divides the hypotenuse into segments whose lengths are represented by $4x$ and x. If the length of the altitude is 2, find x.

11. What is the perimeter of a rhombus whose diagonals are 12 and 16?

12. In $\triangle ABC$, $\overline{AB} \perp \overline{BC}$ and $m\measuredangle CAB = 45$. If $AC = 12$, find AB in radical form.

13. In triangle ABC, $\overline{DE}$ is drawn parallel to $\overline{AB}$ and intersects $\overline{AC}$ and $\overline{BC}$ at D and E, respectively. If $CD = 4$, $DA = 2$, and $BE = 3$, find CE.

14. What is the length of an altitude of an equilateral triangle whose perimeter is 12?

15. The legs of a right triangle have lengths of 5 and 12. Expressed as a fraction, what is the cosine of the larger acute angle?

16. As shown in the accompanying diagram, a person can travel from New York City to Buffalo by going north 170 miles to Albany and then west 280 miles to Buffalo.

(a) If an engineer wants to design a highway to connect New York City directly to Buffalo, at what angle, x, would she need to build the highway? Find the angle to the *nearest tenth* of a degree.

(b) To the *nearest mile*, how many miles would be saved by traveling directly from New York City to Buffalo rather than traveling first to Albany and then from Albany to Buffalo?

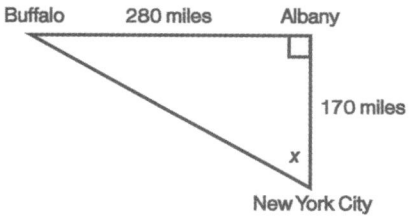

17. In quadrilateral $ABCD$, $\overline{AB} \perp \overline{BC}$, $\overline{AD} \perp \overline{CD}$, $AB = 4$, $BC = 3$, and $AD = 2$.

(a) Find $\angle C$ to the *nearest degree*.

(b) Find the length of $\overline{CD}$ to the *nearest tenth*.

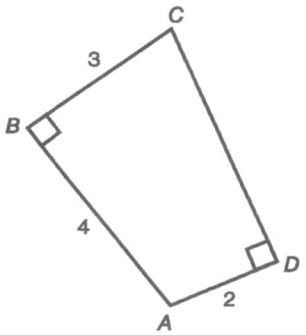

18. GIVEN: $\overline{CD}$ intersects $\overline{EMB}$ at A, $\overline{CB}$ is drawn, $\overline{ED} \perp \overline{EB}$, $\overline{CD} \perp \overline{CB}$, and $\overline{CM}$ is the perpendicular bisector of $\overline{AB}$.

PROVE: $\triangle CMB \sim \triangle DEA$

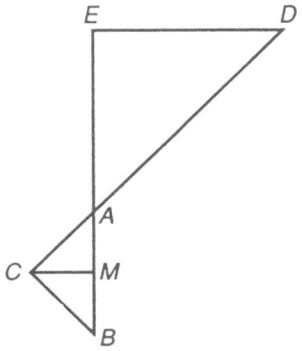

19. GIVEN: $\triangle ABC$, $\overline{CD}$ intersects $\overline{AB}$ at D, and $\overline{CD}$ bisects $\measuredangle BCA$.

PROVE: $CA > DA$

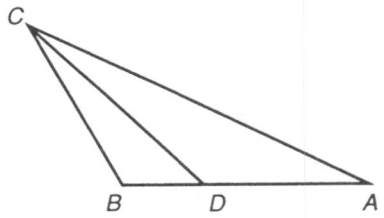

20. GIVEN: $\overline{BC} \parallel \overline{AD}$. $\triangle ABC$ is *not* isosceles.

PROVE: $\overline{AC}$ does *not* bisect $\measuredangle BAD$.

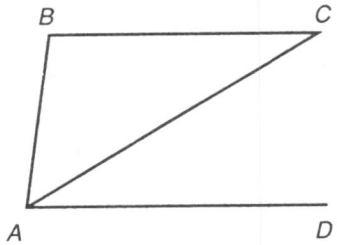

21. GIVEN: $\triangle ABC$ with $\overline{AB} \cong \overline{BC}$. Points D, E, and F are on $\overline{AB}$, $\overline{BC}$, and $\overline{AC}$, respectively, so that $\overline{ED} \perp \overline{DF}$, $\overline{DF} \perp \overline{AC}$, and $\overline{FE} \perp \overline{BC}$.

PROVE: (a) $\triangle FEC \sim \triangle EDF$

(b) $\triangle EDF \sim \triangle DFA$

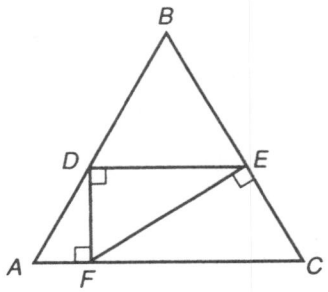

22. GIVEN: Trapezoid $ABCD$ with diagonal $\overline{AFC} \cong$ diagonal $\overline{BFD}$; through C a line is drawn parallel to $\overline{BD}$ and intersecting $\overline{AB}$ at E.

PROVE: (a) $BECD$ is a parallelogram.

(b) $\overline{AC} \cong \overline{CE}$.

(c) $\triangle AFB$ is isosceles.

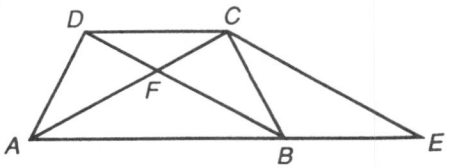

23. GIVEN: Rectangle $ABCD$, $\overline{DFEC}$, $\overline{AGE}$, $\overline{BGF}$, $\overline{DF} \cong \overline{EC}$.

PROVE: (a) $\measuredangle 1 \cong \measuredangle 2$.

(b) $\measuredangle 3 \cong \measuredangle 4$.

(c) $\overline{AG} \cong \overline{GB}$.

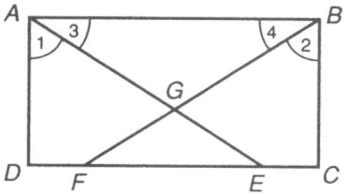

24. GIVEN: Quadrilateral $ABCD$, diagonals $\overline{AC}$ and $\overline{BD}$, $\measuredangle 1 \cong \measuredangle 2$, $\overline{BD}$ bisects $\overline{AC}$ at E.

PROVE: $ABCD$ is a parallelogram.

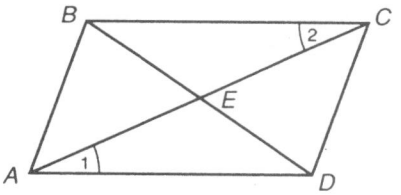

25. GIVEN: V is a point on $\overline{ST}$ such that $\overline{RVW}$ bisects $\measuredangle SRT$, $\overline{TW} \cong \overline{TV}$.

PROVE: $RW \times SV = RV \times TW$.

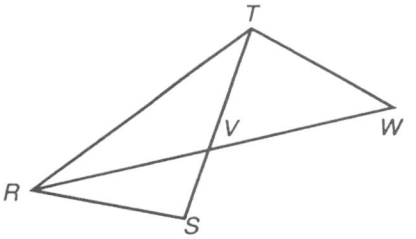

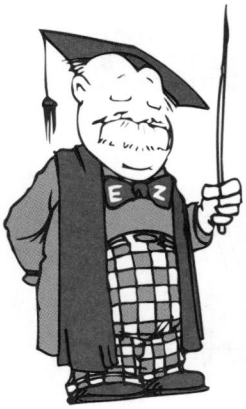

12

Circles and Angle Measurement

WHAT YOU WILL LEARN

Curved figures are encountered in geometry as well as in everyday life. Imagine riding on a bicycle or in an automobile that didn't have circular wheels. It's difficult to think of a machine that doesn't include some circular parts. In this chapter you will learn:

- the names of the various parts of a circle;
- the way in which the measure of an angle whose sides intersect a circle depends on the location of the vertex of the angle;
- the way to find the measure of an angle by using the measure(s) of its intercepted arc(s).

SECTIONS IN THIS CHAPTER

- The Parts of a Circle
- Arcs and Central Angles
- Diameters and Chords
- Tangents and Secants
- Angle Measurement: Vertex on the Circle
- Angle Measurement: Vertex in the Interior of the Circle
- Angle Measurement: Vertex in the Exterior of the Circle
- Using Angle-Measurement Theorems

The Parts of a Circle

If we take a compass and using a fixed setting, draw a closed figure (see Figure 12.1), the resulting figure is a *circle*. The small puncture hole or impression that the metal compass point makes on the paper is called the *center* of the circle. The compass-setting distance is called the *radius* of the circle. Since the compass setting (that is, the radius) remained the same while we drew the circle with the compass, we may define a circle as follows:

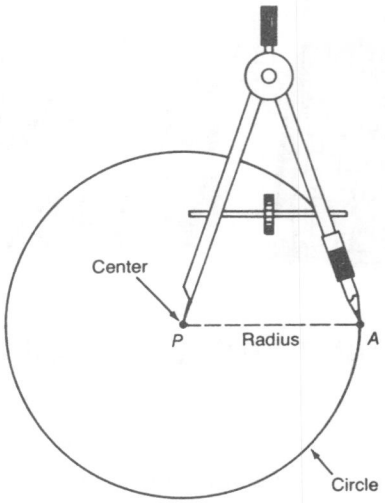

FIGURE 12.1

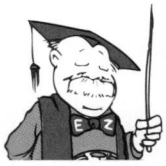

DEFINITION OF A CIRCLE

A *circle* is the set of all points in a plane having the same distance from a fixed point. The fixed point is called the *center* of the circle. The distance between the center of the circle and any point of the circle is called the *radius* (plural: radii) of the circle.

A capital letter is used to designate the center of a circle. If the center of a circle is designated by the letter P, the circle is referred to as *circle* P. The shorthand notation $\odot P$, where the symbol that precedes the letter P is a miniature circle, is commonly used.

From the definition of a circle, it follows that *all radii of the same circle are congruent*. Two different circles are congruent if their radii are congruent. In circle O, shown in Figure 12.2*a*, several radii are drawn. Each of these radii must have the same length so that $\overline{OA} \cong \overline{OB} \cong \overline{OC} \cong \overline{OD} \cong \ldots$. Circle O and circle P (Figure 12.2*b*) are congruent (written as $\odot O \cong \odot P$) if $\overline{OA} \cong \overline{PX}$.

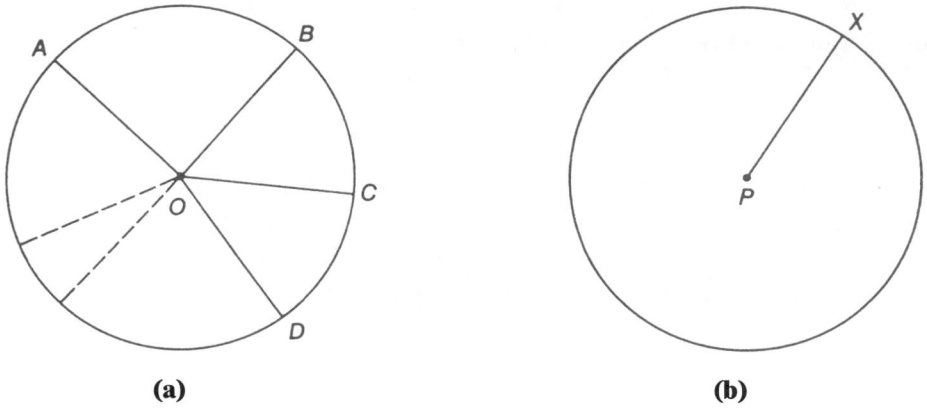

FIGURE 12.2

Several basic terms with which you should be familiar are associated with circles. These terms are defined below and illustrated in Figure 12.3(*a*) and Figure 12.3(*b*).

- A *chord* is a line segment whose end points lie on the circle.
- A *diameter* is a chord that passes through the center of the circle.

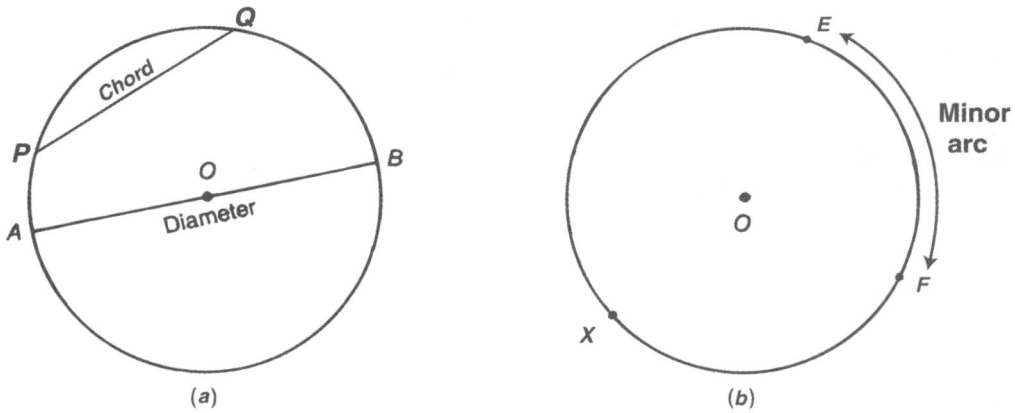

FIGURE 12.3

- An *arc* is a curved $\overline{PQ}$ portion of a circle. In Figure 12.3*a*, diameter $\overline{AB}$ divides the circle into two arcs, each of which is called a *semicircle*. In Figure 12.3*b*, E and F are the end points of two arcs. The shorter path from E to F represents a *minor arc*. A minor arc is less than a semicircle. The notation $\overset{\frown}{EF}$ is read "arc EF." The long way around from E to F is a *major arc*. A major arc is greater than a semicircle. Does $\overset{\frown}{EF}$ name a major arc or a minor arc whose end points are E and F? By using three letters to name a major arc, as in $\overset{\frown}{EXF}$, we can avoid any possible confusion. Thus, $\overset{\frown}{EF}$ names the *minor* arc whose end points are E and F, while $\overset{\frown}{EXF}$ refers to the *major* arc whose end points are E and F.

In Figure 12.3*a*, chord $\overline{PQ}$ cuts off or *determines* an arc of the circle. Referring to Figure 12.4, we can say that chord $\overline{PQ}$ *intercepts* $\overset{\frown}{PQ}$ or *has* $\overline{PQ}$ as its arc.

In Figure 12.3*b* one and only one chord can be drawn that joins the end points of $\overset{\frown}{EF}$. Referring to Figure 12.5, we can say that arc EF has $\overline{EF}$ as its chord or that arc EF *determines* chord $\overline{EF}$.

A diameter of a circle consists of two radii that lie on the same straight line. In Figure 12.3*a* diameter AB = radius OA + radius OB. Hence, the length of a diameter of a circle is twice the length of a radius of the circle. If, for example, the length of a diameter of a circle is 12, then the length of a radius is 6.

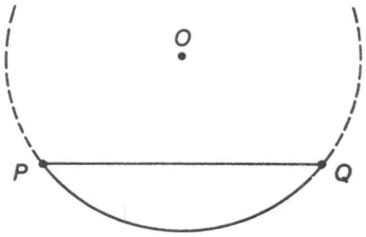

FIGURE 12.4

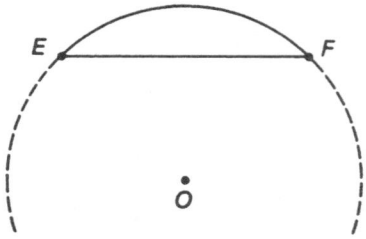

FIGURE 12.5

EXAMPLE 12.1 In circle O, radius $OA = 3n - 10$ and radius $OB = n + 2$. Find the length of a diameter of circle O.

SOLUTION

Since all radii of a circle are congruent,

$$\begin{aligned} OA &= OB \\ 3n - 10 &= n + 2 \\ 3n &= n + 12 \\ 2n &= 12 \\ n &= 6 \\ OB &= OA = n + 2 = 6 + 2 = 8 \end{aligned}$$

The length of a diameter of $\odot O = 2 \times 8 = 16$

EXAMPLE 12.2

GIVEN: $\overline{AB}$ and $\overline{CD}$ are diameters of circle O.
PROVE: $\overline{AD} \cong \overline{BC}$.

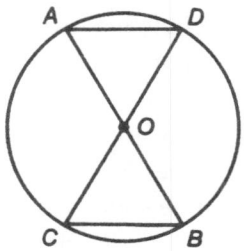

SOLUTION

PLAN: Show $\triangle AOD \cong \triangle BOC$.

PROOF:

Statements		Reasons
1. $\overline{AB}$ and $\overline{CD}$ are diameters of $\odot O$.		1. Given.
2. $\overline{OA} \cong \overline{OB}$.	(S)	2. All radii of a circle are congruent.
3. $\measuredangle AOD \cong \measuredangle BOC$.	(A)	3. Vertical angles are congruent.
4. $\overline{OD} \cong \overline{OC}$.	(S)	4. Same as reason 2.
5. $\triangle AOD \cong \triangle BOC$.		5. SAS Postulate.
6. $\overline{AD} \cong \overline{BC}$.		6. CPCTC.

Consider circle O and a point P. If the length of $\overline{OP}$ is less than the length of the radius of circle O, point P must be located within the boundaries of the circle. If the length of $\overline{OP}$ is greater than the length of the radius of the circle, then point P must fall outside the circle. Suppose the radius of circle O is 5 inches. If the length of $\overline{OP}$ is 6 inches, then point P must lie in the exterior (outside) of the circle. If the length of $\overline{OP}$ is 2 inches, then point P must lie in the interior (inside) of the circle. If the length of $\overline{OP}$ is 5 inches, then point P is a point on the circle.

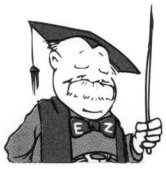

DEFINITIONS OF INTERIOR AND EXTERIOR OF A CIRCLE

- The *interior of a circle* is the set of all points whose distance from the center of the circle is less than the length of the radius of the circle.
- The *exterior of a circle* is the set of all points whose distance from the center of the circle is greater than the length of the radius of the circle.

See Figure 12.6.

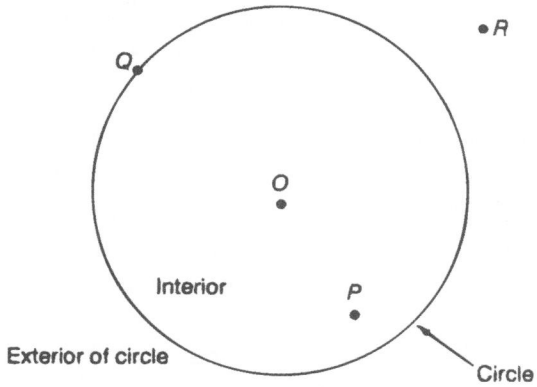

Point	Location
P	interior of $\odot O$
Q	on $\odot O$
R	exterior of $\odot O$

FIGURE 12.6

EXAMPLE 12.3 $\overline{PQ}$ is a diameter of the accompanying circle.

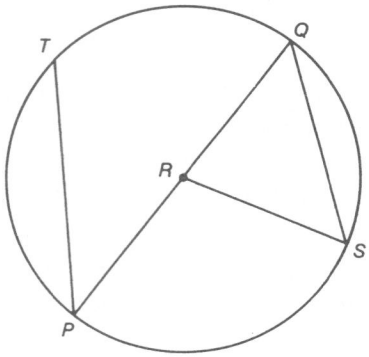

a. What point is the center of the circle?

b. Which segments are chords?

c. Which segments are radii?

d. Name all congruent segments.

e. If $RS = 6$, find PQ.

f. If $PQ = 15$, find RQ.

g. For each of the following conditions, determine whether point X lies in the interior of the circle, the exterior of the circle, or on the circle:

(i) $RQ = 5$, and $RX = 4$.

(ii) $RP = 6$, and $RX = 8$.

(iii) $PQ = 14$, and $RX = 7$.

h. If $m\measuredangle QRS = 60$ and $RQ = 9$, find QS.

i. If $m\measuredangle QRS = 90$ and $QS = 15$, find RQ and RS.

j. If $m\measuredangle QRS = 90$ and $RS = 6$, find RQ and QS.

SOLUTION

a. Point R.

b. $\overline{PT}$, $\overline{PQ}$, and $\overline{QS}$.

c. $\overline{RP}$, $\overline{RQ}$, and $\overline{RS}$.

d. $\overline{RP} \cong \overline{RQ} \cong \overline{RS}$ since all radii of a circle are congruent.

e. Diameter $PQ = 2(6) = 12$.

f. Radius $RQ = \frac{1}{2}(15) = 7.5$.

g. **(i)** X is located in the interior of the circle.

(ii) X is located in the exterior of the circle.

(iii) X is located on the circle.

h. $RQ = RS$, which implies $m\measuredangle RSQ = m\measuredangle RQS$ (base angles of an isosceles triangle are equal in measure). Since the vertex angle has measure 60, each base angle has a measure of one-half of 120 or 60. Triangle RQS is equiangular. An equiangular triangle is also equilateral. Hence, $QS = 9$.

i. Triangle QRS is a 45-45

right triangle.

$$RS = RQ = \frac{\sqrt{2}}{2} \times 15 = 7.5\sqrt{2}$$

j. Triangle QRS is a 45-45 right triangle.

$$RQ = RS = 6$$

$$QS = RS \times \sqrt{2} = 6\sqrt{2}$$

Arcs and Central Angles

The *length* of an arc of a circle is expressed in linear units of measurement such as inches, centimeters, and feet. Degrees are used to represent the *measure* of an arc. A degree is a unit of measurement obtained by dividing one complete revolution (that is, a circle) into 360 equal parts, and then referring to each part as a *degree*. A circle contains 360 degrees, so the measure of an arc of a circle must be some fractional part of 360 degrees.

Since the measure of a circle is 360 degrees, the measure of a semicircle (Figure 12.7*a*) is 180 degrees. An arc whose degree measure is less than 180 is called a *minor arc*. A *major arc* has a degree measure greater than 180 but less than 360. See Figure 12.7*b*.

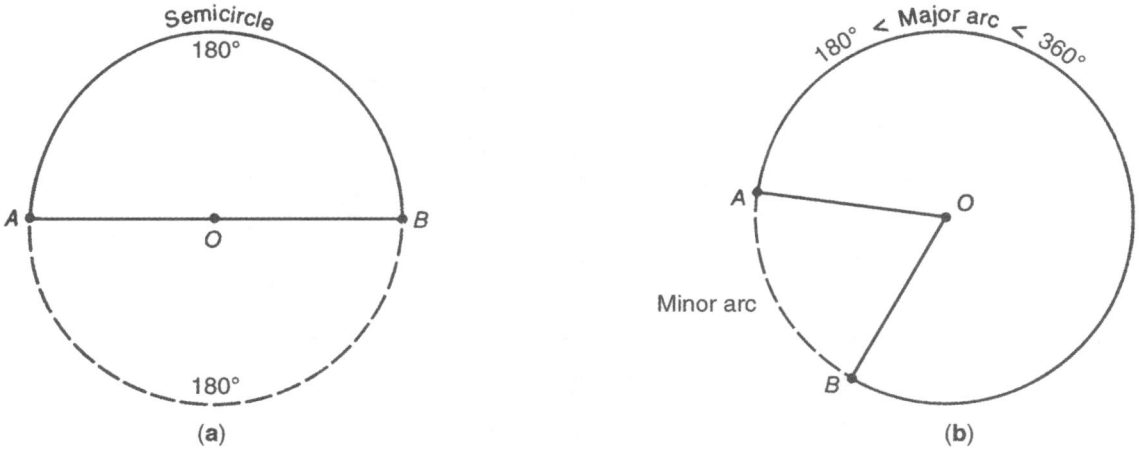

FIGURE 12.7

The measure of an arc of a circle may also be related to an angle whose vertex is the center of the circle and whose sides intercept arcs of the circle. In Figure 12.8 angle AOB is called a *central angle*. The minor arc that lies in the interior of the angle is *defined* to have the same measure as its central angle, $\measuredangle AOB$. For example, if $m\measuredangle AOB = 83$, then $m\overset{\frown}{AB} = 83$.

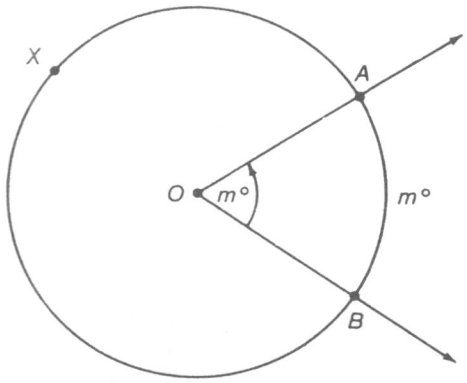

FIGURE 12.8

At times we will need to refer to a central angle and *its intercepted arc*. Although the sides of a central angle determine a minor *and* a major arc, we will refer to the minor arc as *the* intercepted arc.

SUMMARY

- A *semicircle* is an arc of a circle whose end points are the end points of a diameter of the circle; its measure is 180 degrees.
- A *central angle* is an angle whose vertex is at the center of the circle.
- A *minor arc* of a circle is an arc that lies in the interior of the central angle that intercepts the arc. The measure of a minor arc is the same as the measure of its central angle.
- A *major arc* of a circle is an arc that lies in the exterior of the central angle that intercepts the arc. The degree measure of a major arc is equal to 360 minus the measure of the central angle that determines the end points of the arc.
- The *measure* of an arc, as with angle measurement, is denoted by preceding the name of the arc by the letter "m": $m\widehat{AB} = 67$ is read as "the measure of arc *AB* is 67." When this notation is used, the degree symbol is omitted.

EXAMPLE 12.4

a. Name four minor arcs of circle O.

b. Name major $\widehat{TEJ}$ in three different ways. (Use three letters.)

c. Name four different major arcs of circle O.

d. If $m\measuredangle JOT = 73$ and $\overline{NT}$ is a diameter of circle O, find:

(i) $m\widehat{JT}$ **(ii)** $m\widehat{JN}$

(iii) $m\widehat{JET}$ **(iv)** $m\widehat{TEN}$

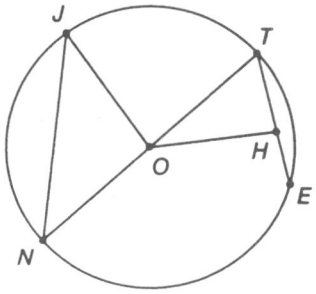

SOLUTION

a. $\widehat{JT}$, $\widehat{JN}$, $\widehat{ET}$, and $\widehat{EN}$.

b. $\widehat{JET}$, $\widehat{TNJ}$, and $\widehat{JNT}$.

c. $\widehat{NTE}$, $\widehat{TNE}$, $\widehat{NTJ}$, and $\widehat{TNJ}$.

d. **(i)** $m\widehat{JT} = m\measuredangle JOT = 73$

(ii) $m\widehat{JN} = m\measuredangle JON = 180 - 73 = 107$

(iii) $m\widehat{JET} = 360 - m\measuredangle JOT = 360 - 73 = 287$

(iv) $m\widehat{TEN} = 180$

In circle O of Figure 12.9, the measures of arcs AC, CB, and AB may be determined as follows:

$$m\widehat{AC} = m\measuredangle AOC = 30$$

$$m\widehat{CB} = m\measuredangle BOC = 40$$

$$m\widehat{AB} = m\measuredangle AOB = 70$$

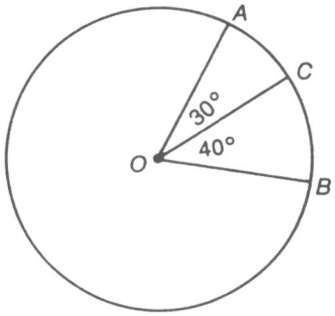

FIGURE 12.9

Notice that $m\widehat{AB} = m\widehat{AC} + m\widehat{CB}$. This result holds provided that arcs AC and CB are consecutive, having exactly one point (C) in common. This concept of arc addition is analogous to the concept of betweenness of points on line segments.

The measures of arcs can be added or subtracted in the same way that measures of angles and line segments are combined.

Are arcs that have the same degree referring to measure congruent (that is, equal in length)? If we are referring to the same circle, the answer is yes. But what if we are comparing two arcs having the same measure in two different circles? We will consider two circles that have the same center (called *concentric* circles).

In Figure 12.10, although $m\widehat{AB} = 40$ and $m\widehat{XY} = 40$, arcs AB and XY are *not* congruent. This observation will help us to frame the definition of congruent arcs.

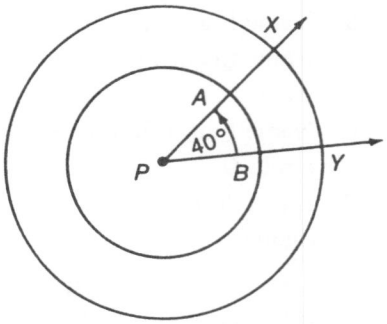

FIGURE 12.10

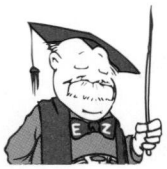

DEFINITION OF CONGRUENT ARCS

Congruent arcs are arcs in the *same circle* or in *congruent circles* that have the same degree measure.

It follows from the definition of congruent arcs that, in the same circle or in congruent circles, *congruent central angles intercept congruent arcs* and *congruent arcs have congruent central angles*. See Figure 12.11.

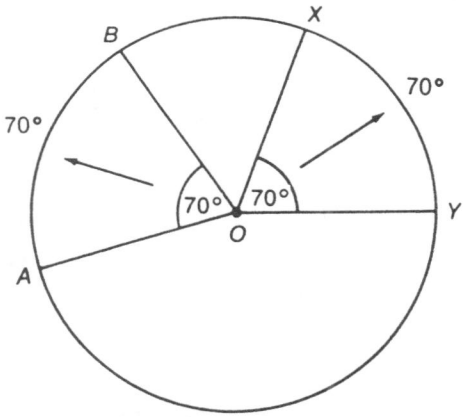

If $\measuredangle AOB \cong \measuredangle XOY$, then $\widehat{AB} \cong \widehat{XY}$.

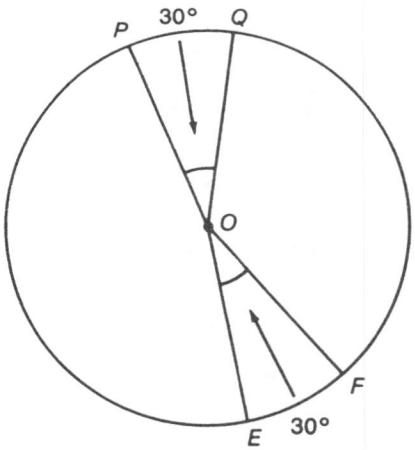

If $\widehat{PQ} \cong \widehat{EF}$, then $\measuredangle POQ \cong \measuredangle EOF$.

FIGURE 12.11

THEOREM 12.1 ≅ CENTRAL ANGLES IMPLIES ≅ ARCS AND VICE VERSA

In the same circle or congruent circles:

- Congruent central angles intercept congruent arcs.
- Congruent arcs have congruent central angles.

If point M is located on $\widehat{AB}$ so that $\widehat{AM} \cong \widehat{MB}$, then M is the midpoint of $\widehat{AB}$, as shown in Figure 12.12.

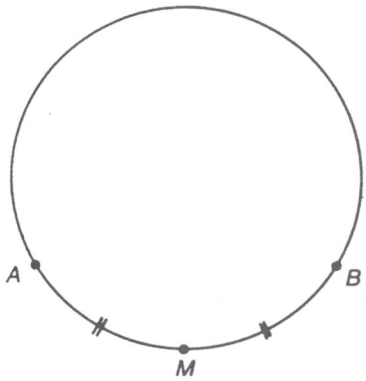

FIGURE 12.12

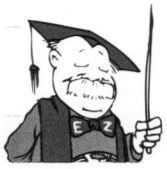

DEFINITION OF MIDPOINT OF AN ARC

The *midpoint of an arc* is the point of the arc that divides the arc into two congruent arcs.

EXAMPLE 12.5

GIVEN: B is the midpoint of $\widehat{AC}$.
PROVE: $\triangle AOB \cong \triangle COB$.

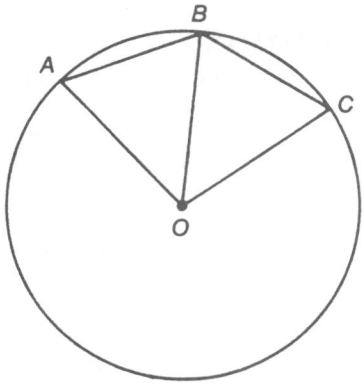

SOLUTION

PLAN: Prove $\triangle AOB \cong \triangle COB$ by SAS.

PROOF:

Statements	Reasons
1. $\overline{OA} \cong \overline{OC}$. (Side)	1. Radii of the same circle are congruent.
2. B is the midpoint of $\widehat{AC}$.	2. Given.
3. $\widehat{AB} \cong \widehat{BC}$.	3. A midpoint of an arc divides the arc into two congruent arcs.
4. $\measuredangle AOB \cong \measuredangle COB$. (Angle)	4. In the same circle, congruent arcs have congruent central angles.
5. $\overline{OB} \cong \overline{OB}$. (Side)	5. Reflexive property of congruence.
6. $\triangle AOB \cong \triangle COB$.	6. SAS Postulate.

EXAMPLE 12.6

GIVEN: $\widehat{AB} \cong \widehat{CD}$.

PROVE: $\overline{AB} \cong \overline{CD}$.

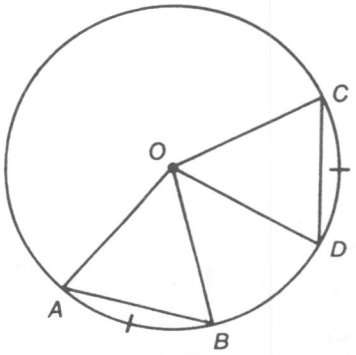

SOLUTION

PLAN: Show $\triangle AOB \cong \triangle DOC$.

PROOF:

Statements	Reasons
1. $\overline{AO} \cong \overline{DO}$. (Side)	1. Radii of the same circle are congruent.
2. $\widehat{AB} \cong \widehat{CD}$.	2. Given.
3. $\measuredangle AOB \cong \measuredangle DOC$. (Angle)	3. In the same circle, congruent arcs have congruent central angles.
4. $\overline{OB} \cong \overline{OC}$. (Side)	4. Same as reason 1.
5. $\triangle AOB \cong \triangle DOC$.	5. SAS Postulate.
6. $\overline{AB} \cong \overline{CD}$.	6. CPCTC.

Example 12.6 establishes the following theorem:

THEOREM 12.2 CONGRUENT ARCS IMPLY CONGRUENT CHORDS

In the same circle or in congruent circles, congruent arcs have congruent chords.

In Figure 12.13, if circle O is congruent to circle P and $\widehat{AB} \cong \widehat{CD}$, then Theorem 12.2 entitles us to conclude that $\overline{AB} \cong \overline{CD}$.

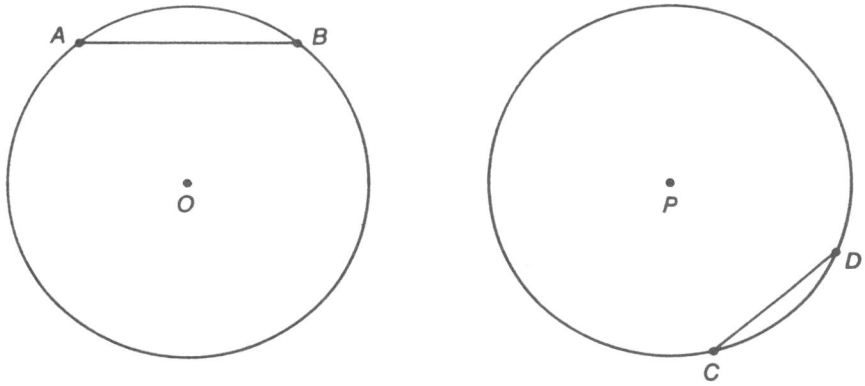

FIGURE 12.13

Theorem 12.2 provides a convenient method for proving a pair of chords are congruent; we show that the chords to be proved congruent cut off congruent arcs in the same circle or in congruent circles. The converse of Theorem 12.2 states that, if we know that a pair of chords are congruent, we may conclude that their intercepted arcs are congruent, provided that we are working in the same circle or in congruent circles.

THEOREM 12.3 CONGRUENT CHORDS IMPLY CONGRUENT ARCS

In the same circle or in congruent circles, congruent chords intercept congruent arcs.

EXAMPLE 12.7

GIVEN: $\overline{CA} \cong \overline{DB}$.

PROVE: $\measuredangle A \cong \measuredangle B$.

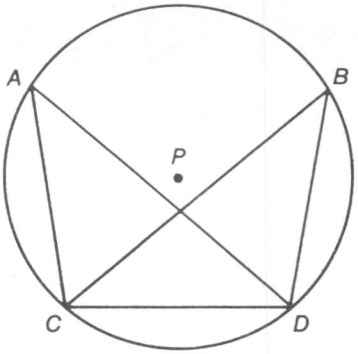

SOLUTION

PLAN: Show $\triangle ACD \cong \triangle BDC$ by SSS Postulate.

PROOF:

Statements	Reasons
1. $\overline{CD} \cong \overline{CD}$. (Side)	1. Reflexive property of congruence.
2. $\overline{CA} \cong \overline{DB}$. (Side)	2. Given.
3. $m\widehat{CA} = m\widehat{DB}$.	3. In the same circle, congruent chords intercept equal arcs.
4. $m\widehat{AB} = m\widehat{AB}$.	4. Reflexive property of equality.
5. $m\widehat{ACD} = m\widehat{BDC}$	5. Arc addition.
6. $\overline{AD} \cong \overline{BC}$. (Side)	6. In the same circle, equal arcs have congruent chords.
7. $\triangle ACD \cong \triangle BDC$.	7. SSS Postulate.
8. $\measuredangle A \cong \measuredangle B$.	8. CPCTC.

EXAMPLE 12.8

Prove that, if two chords are the perpendicular bisectors of each other, then each chord is a diameter.

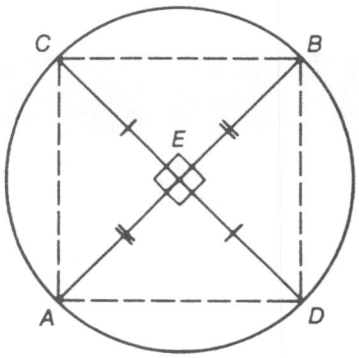

SOLUTION

OUTLINE OF PROOF

GIVEN: Chords $\overline{AB}$ and $\overline{CD}$ are $\perp$ to each other and bisect each other.

PROVE: $\overline{AB}$ and $\overline{CD}$ are diameters.

PLAN: Show that each chord divides the circle into two congruent arcs.

PLAN: To prove that $\overline{AB}$ is a diameter, show that $m\widehat{ACB} = m\widehat{ADB}$.

STEPS:
- Draw $\overline{AC}$, $\overline{AD}$ $\overline{BC}$, and $\overline{BD}$.
- Prove $\triangle AEC \cong \triangle AED$ by *SAS*. $\overline{AC} \cong \overline{AD}$, which implies that $m\widehat{AC} = m\widehat{AD}$.
- Prove $\triangle BEC \cong \triangle BED$ by *SAS*. $\overline{BC} \cong \overline{BD}$, which implies that $m\widehat{BC} = m\widehat{BD}$.
- $$\begin{array}{r} m\widehat{AC} = m\widehat{AD} \\ + \quad m\widehat{BC} = m\widehat{BD} \\ \hline m\widehat{ACB} = m\widehat{ADB} \end{array}$$, which implies that $\overline{AB}$ is a diameter since it divides the circle into two equal arcs.
- Use a similar approach to show that $m\widehat{CAD} = m\widehat{CBD}$, which establishes that $\overline{CD}$ is a diameter.

SUMMARY

- To prove *arcs* are congruent in the same circle or congruent circles, show that one of the following statements is true:
  1. The central angles that intercept the arcs are congruent.
  2. The chords that cut off the arcs are congruent.
- To prove *central angles* are congruent in the same circle or in congruent circles, show that their intercepted arcs are congruent.
- To prove *chords* are congruent in the same circle or in congruent circles, show that the chords have congruent arcs.
- To prove a *chord is a diameter* show that the chord divides the circle into two congruent arcs.

Diameters and Chords

If we draw a chord and then draw a diameter perpendicular to the chord, what appears to be true? The resulting chord segments look as though they are congruent. Also, corresponding pairs of arcs seem to be congruent. These observations are stated in Theorem 12.4.

THEOREM 12.4 DIAMETER ⊥ CHORD THEOREM

In a circle, a diameter drawn perpendicular to a chord bisects the chord and its arcs.

OUTLINE OF PROOF

GIVEN: $\odot O$ with diameter $\overline{AB} \perp \overline{CD}$.

PROVE: **a.** $\overline{CX} \cong \overline{DX}$.

b. $\widehat{CB} \cong \widehat{DB}$.

$\widehat{CA} \cong \widehat{DA}$.

STEPS:

- Draw radii $\overline{OC}$ and $\overline{OD}$.
- Prove $\triangle OXC \cong \triangle OXD$ by Hy-Leg.
- $\overline{CX} \cong \overline{DX}$ and $\measuredangle 1 \cong \measuredangle 2$ by *CPCTC*.
- $\widehat{CB} \cong \widehat{DB}$ since $\cong$ central angles intercept $\cong$ arcs.
- $\measuredangle 3 \cong \measuredangle 4$ since supplements of congruent angles are congruent.
- $\widehat{CA} \cong \widehat{DA}$.

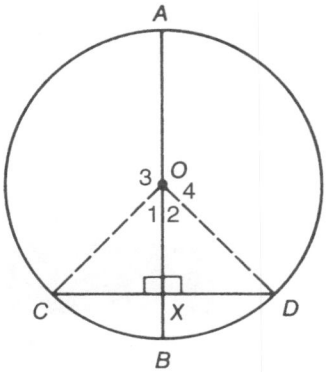

According to Theorem 12.4, any segment that passes through the center of a circle and is perpendicular to a chord bisects the chord.

EXAMPLE 12.9 The length of a diameter of circle O is 20 and the length of chord $\overline{AB}$ is 16. What is the distance between the chord and the center of the circle?

SOLUTION

We recall that the distance from a point (that is, the center of the circle) to a line segment (that is, the chord) is the length of the perpendicular segment from the point to the line segment. We therefore choose to draw a diameter that is perpendicular to chord $\overline{AB}$.

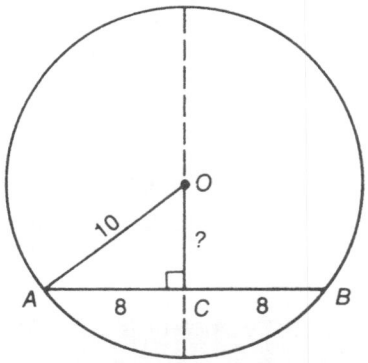

The distance between the chord and the center of the circle is represented by the length of $\overline{OC}$. By drawing radius $\overline{OA}$, we form a right triangle. Since the diameter is 20, the radius is 10. Applying Theorem 12.4, we deduce that the length of $\overline{AC}$ is 8. Hence, $\triangle OCA$ is a right triangle, the lengths of whose sides form a Pythagorean triple (3-4-5). The length of $\overline{OC}$ is 6.

After drawing several pairs of parallel chords in a circle, a pattern emerges, as shown in Figure 12.14.

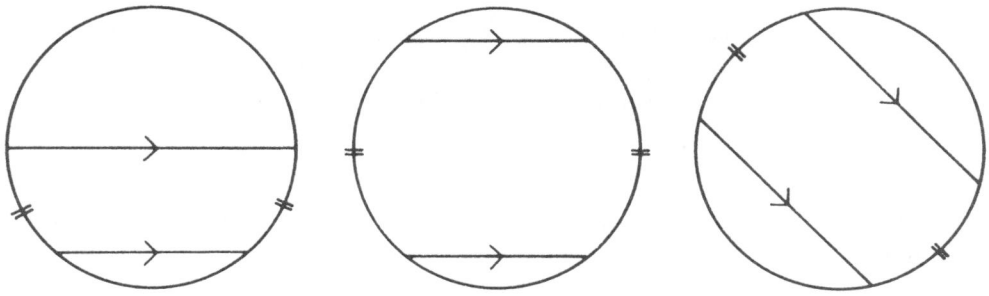

FIGURE 12.14

Notice that the arcs that the parallel chords cut off (and which lie between the chords) appear to be congruent. Here are the formal statement of the corresponding theorem and its proof.

THEOREM 12.5 PARALLEL CHORDS AND CONGRUENT ARCS THEOREM

In a circle, parallel chords cut off equal arcs.

GIVEN: $\odot O$ with $AB \parallel CD$.
PROVE: $m\widehat{AC} = m\widehat{BD}$.
PLAN: Draw diameter $\overline{XY} \perp \overline{AB}$.
$\overline{XY}$ will also be $\perp \overline{CD}$.
Apply Theorem 12.4.

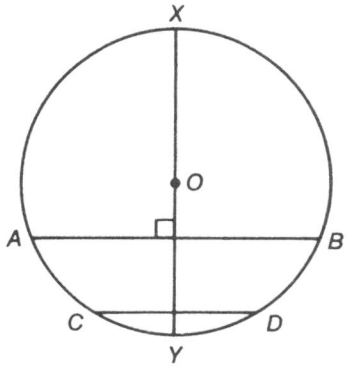

PROOF:

Statements	Reasons
1. $\odot O$ with $\overline{AB} \parallel \overline{CD}$	1. Given.
2. Draw $\overline{XY}$ through point O and perpendicular to $\overline{AB}$.	2. Through a point not on a line, exactly one perpendicular may be drawn from the point to the line.

3. $\overline{XY} \perp \overline{CD}$.	3. A line perpendicular to one of two parallel lines is perpendicular to the other line.
4. $m\widehat{AY} = m\widehat{BY}$ $-m\widehat{CY} = m\widehat{DY}$.	4. A diameter perpendicular to a chord bisects its arcs.
5. $m\widehat{AC} = m\widehat{BD}$.	5. Arc subtraction.

EXAMPLE 12.10 If chord $\overline{CD}$ is parallel to diameter $\overline{AB}$ and $m\overline{CD} = 40$, find the measures of arcs AC and BD.

SOLUTION

Since $\overline{CD} \parallel \overline{AB}$, $m\widehat{AC} = m\widehat{BD} = x$

$$\begin{aligned} x + 40 + x &= 180 \\ 2x + 40 &= 180 \\ 2x &= 140 \\ x &= 70 \\ m\widehat{AC} = m\widehat{BD} &= 70 \end{aligned}$$

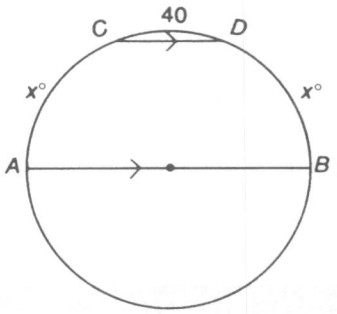

Tangents and Secants

DEFINITIONS AND PROPERTIES OF TANGENTS AND SECANTS

A line may intersect a circle at no points, one point, or two points. See Figure 12.15.

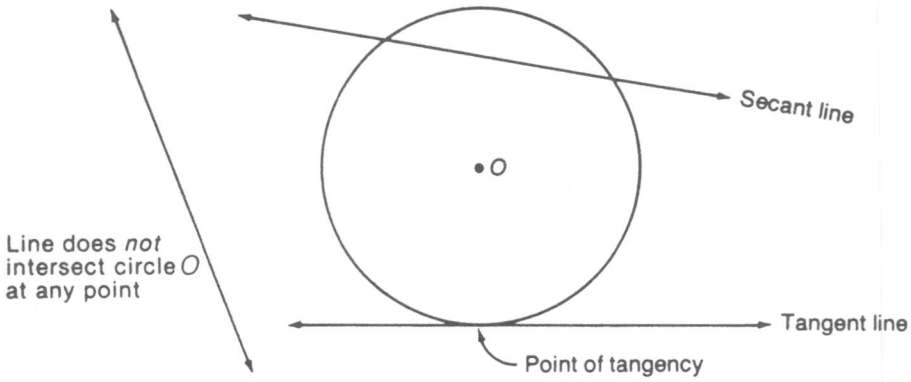

FIGURE 12.15

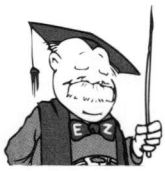

DEFINITIONS OF TANGENT AND SECANT

- A *tangent* line is a line that intersects a circle in exactly one point. The point of contact is called the *point of tangency*.
- A *secant* line is a line that intersects a circle in two different points. (Every secant line includes a chord of the circle.)

In Figure 12.16, lines j, k, and ℓ are tangent to circle O. Radii have been drawn to their points of tangency. In each case the radius appears to be perpendicular to the tangent.

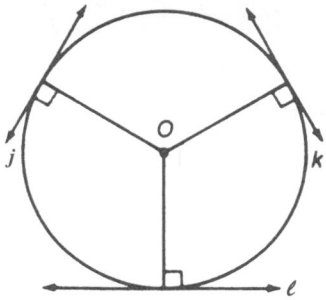

FIGURE 12.16

THEOREM 12.6 RADIUS ⊥ TANGENT THEOREM

A radius drawn to a point of tangency is perpendicular to the tangent.

OUTLINE OF PROOF

GIVEN: $\overleftrightarrow{AB}$ is tangent to $\odot P$ at point A.

PROVE: $PA \perp \overleftrightarrow{AB}$.

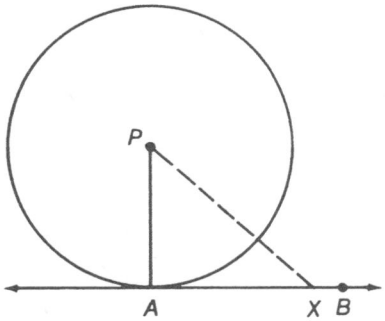

PLAN: Use an indirect proof.

STEPS:
- If we assume $\overline{PA}$ is *not* perpendicular to $\overleftrightarrow{AB}$, then there must be another segment, say PX, which can be drawn from P perpendicular to $\overleftrightarrow{AB}$.
- Since $PX \perp \overleftrightarrow{AB}$, PX represents the *shortest* distance from P to $\overleftrightarrow{AB}$. Thus, PX is *less than* PA.

- But point X lies in the exterior of circle P. Since $\overline{PA}$ is a radius, PX must be *greater than* PA. This contradicts our previous assertion that PX is less than $\overline{PA}$. Our assumption that PA is not perpendicular to $\overleftrightarrow{AB}$ must be false. $\overline{PA}$ is therefore perpendicular to $\overleftrightarrow{AB}$.

The converse of this theorem is given in Theorem 12.7 and may also be proved indirectly.

THEOREM 12.7 RADIUS ⊥ LINE IMPLIES A TANGENT THEOREM

If a radius is perpendicular to a line at the point at which the line intersects a circle, then the line is a tangent.

OUTLINE OF PROOF

GIVEN: $\odot P$ with $\overline{PA} \perp$ line ℓ at point A.

PROVE: ℓ is tangent to $\odot P$.

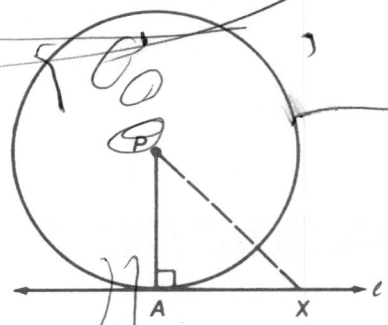

PLAN: Use an indirect proof.

STEPS:
- If ℓ is assumed not to be a tangent line, it must intersect $\odot P$ at another point, say point X.
- Since $\overline{PX}$ is the hypotenuse of right triangle PAX, PX is greater than PA.
- But $\overline{PA}$ is a radius, which implies that point X must lie in the *exterior* of circle P.
- The assumption that there is a second point *on the circle* at which line ℓ intersects is false. Since there is exactly one point at which ℓ intersects the circle, ℓ is a tangent line.

EXAMPLE 12.11 The length of a tangent segment drawn from point P to circle O is 12. If the radius of circle O is 5, find the distance from point P to the center of the circle.

SOLUTION

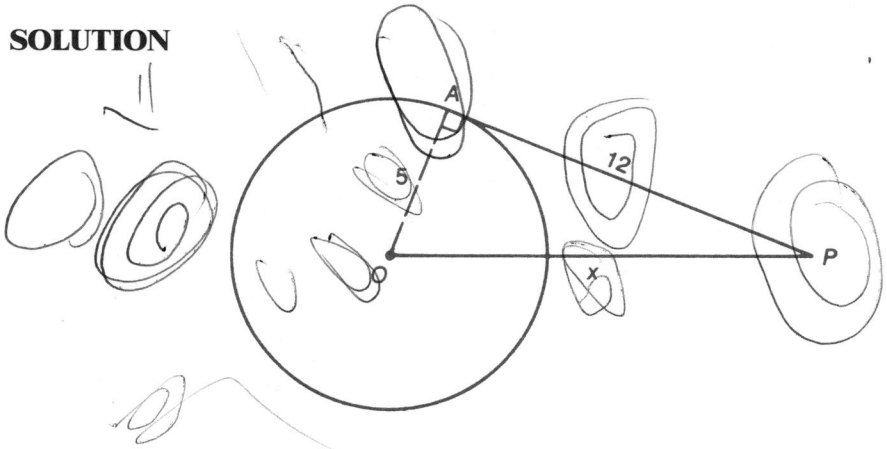

$\overline{PA}$ is a tangent segment since one of its end points, point A, is the point of tangency. Also, PO represents the distance from point P to the center of the circle. Draw a radius to the point of tangency $\overline{OA} \perp \overline{PA}$ and $x = 13$ since a $\triangle OAP$ is a 5-12-13 right triangle.

INSCRIBED AND CIRCUMSCRIBED POLYGONS

Quadrilateral $ABCD$ in Figure 12.17 is *circumscribed about circle O*. Pentagon $ABCDE$ in Figure 12.18 is *inscribed in circle O*.

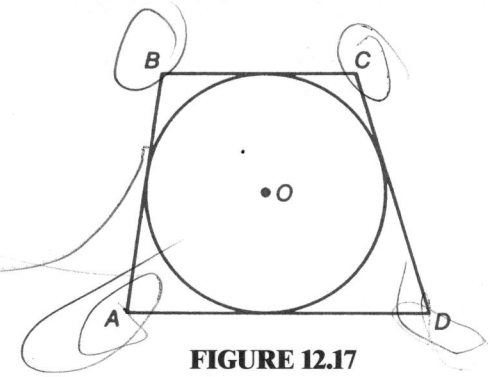

FIGURE 12.17

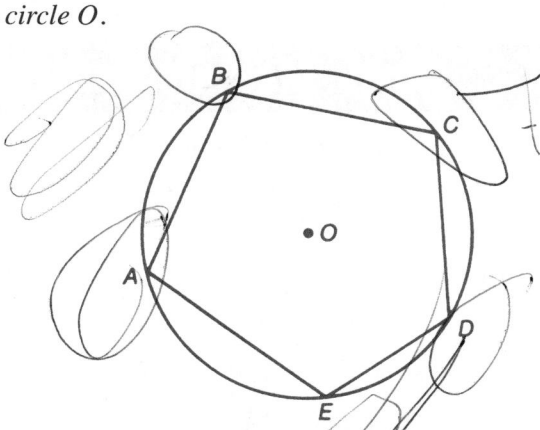

FIGURE 12.18

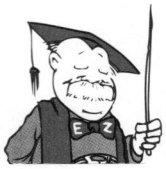

DEFINITIONS OF CIRCUMSCRIBED AND INSCRIBED POLYGONS

A polygon is *circumscribed about* a circle if each of its sides is tangent to the circle.

A polygon is *inscribed in* a circle if each of its vertices lies on the circle.

EXAMPLE 12.12 A quadrilateral is inscribed in a circle in a way that the sides of the quadrilateral divide the circle into arcs whose measures have the ratio 1:2:3:4. How many degrees are there in each arc?

SOLUTION

The sum of the measures of the arcs that comprise a circle must equal 360:

$$\begin{aligned} 1x + 2x + 3x + 4x &= 360° \\ 10x &= 360° \\ x &= 36° \\ 2x &= 72° \\ 3x &= 108° \\ 4x &= 144° \end{aligned}$$

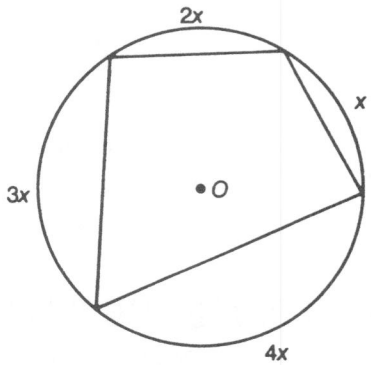

Angle Measurement: Vertex on the Circle

The vertex of an angle whose sides intercept arcs on a circle may be located in any one of the positions shown in Figure 12.19.

At the center of the circle:

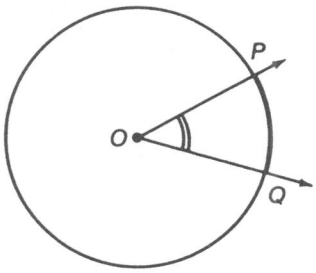

Intercepted arc: $\widehat{PQ}$

On the circle:

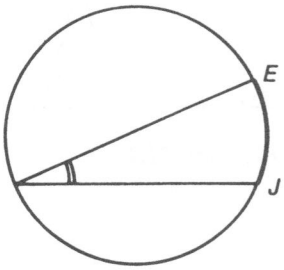

Intercepted arc: $\widehat{EJ}$

In the interior of the circle but not at the center:

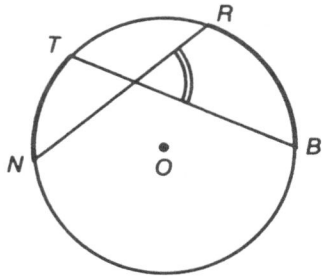

Intercepted arcs: $\widehat{NT}$ and $\widehat{RB}$

In the exterior of the circle:

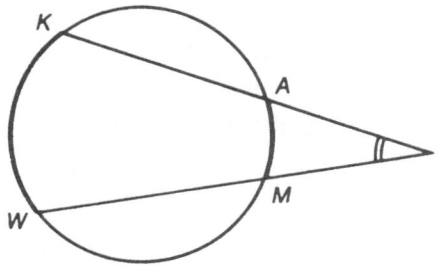

Intercepted arcs: $\widehat{KW}$ and $\widehat{AM}$

FIGURE 12.19

In each of these cases a relationship exists between the measures of the intercepted arc or arcs and the measure of the angle. We have already seen, for example, that the measure of any angle whose vertex is at the center of the circle (a central angle) is equal to the measure of its intercepted arc. Let's now consider the situation in which the vertex of the angle is a point on the circle and the sides of the angle are chords (or secants) of the circle. In Figure 12.20, angle ABC is called an *inscribed angle* and $\widehat{AC}$ is its intercepted arc.

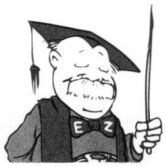

DEFINITION OF INSCRIBED ANGLE

An *inscribed angle* is an angle whose vertex lies on a circle and whose sides are chords (or secants) of the circle.

In Figure 12.21, suppose $m\widehat{AC} = 50$. Then $m\measuredangle AOC = 50$. Since $\measuredangle AOC$ is an exterior angle of $\triangle BOC$, $m\measuredangle AOC = m\measuredangle B + m\measuredangle C$. Since $m\measuredangle B = m\measuredangle C$ (base angles of an isosceles triangle), $m\measuredangle B = 25$. How do the measures of inscribed angle ABC and its intercepted arc ($\widehat{AC}$) compare? Observe that $m\measuredangle ABC = \frac{1}{2}\, m\widehat{AC} = \frac{1}{2}(50) = 25$.

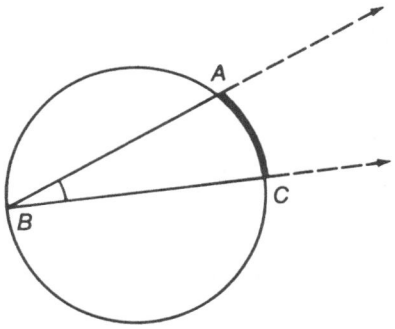

FIGURE 12.20

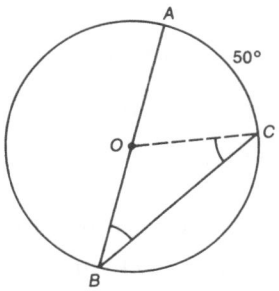

FIGURE 12.21

THEOREM 12.8 INSCRIBED ANGLE THEOREM

The measure of an inscribed angle is equal to one-half the measure of its intercepted arc.

In the accompanying figure,

$$m\measuredangle ABC = \frac{1}{2}x.$$

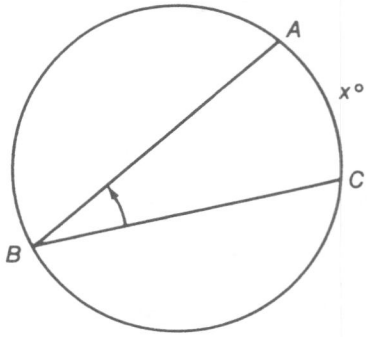

In our example, we assumed that one of the sides of the inscribed angle passed through the center of the circle. In general, this need not be true. The center of the circle may lie on a side of the inscribed angle, in the interior of the angle, or in the exterior of the angle. The proof of the Inscribed Angle Theorem must treat each of these three cases separately.

The three cases of Theorem 12.8 to be proved are shown in Figure 12.22.

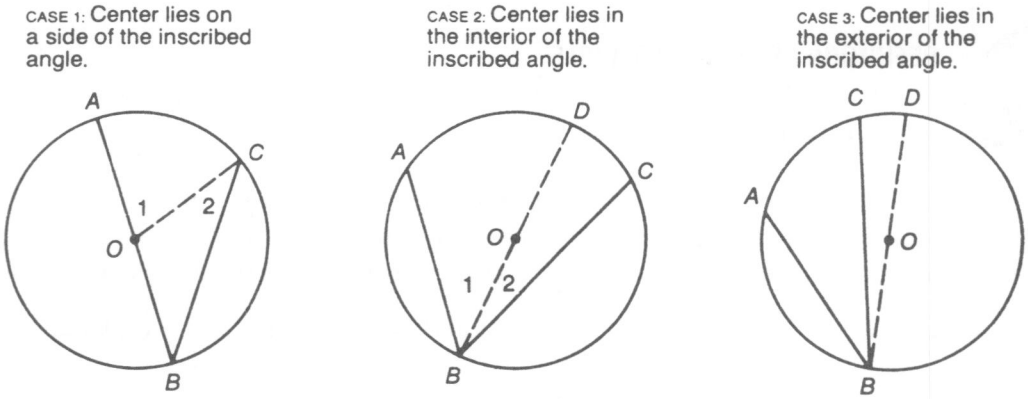

FIGURE 12.22

GIVEN: $\odot O$ with inscribed angle ABC.

PROVE: $m\measuredangle ABC = \frac{1}{2}m\widehat{AC}$.

PROOF OF CASE 1:

PLAN: Use an approach that parallels the solution to the numerical example used to motivate Theorem 12.8.

PROOF:

Statements	Reasons
1. Draw radius OC.	1. Two points determine a line.
2. $m\measuredangle 1 = m\measuredangle ABC + m\measuredangle 2$.	2. The measure of an exterior angle of a triangle is equal to the sum of the measures of the two nonadjacent interior angles of the triangle.
3. $\overline{OB} \cong \overline{OC}$.	3. All radii of the same circle are congruent.
4. $m\measuredangle 2 = m\measuredangle ABC$.	4. If two sides of a triangle are congruent, then the angles opposite these sides are equal in measure.
5. $m\measuredangle 1 = m\measuredangle ABC + m\measuredangle ABC$, or $m\measuredangle 1 = 2\, m\measuredangle ABC$.	5. Substitution.
6. $\frac{1}{2}m\measuredangle 1 = m\measuredangle ABC$ or $m\measuredangle ABC = \frac{1}{2}m\measuredangle 1$.	6. Division property.
7. $m\measuredangle 1 = m\widehat{AC}$.	7. The measure of a central angle is equal to the measure of its intercepted arc.
8. $m\measuredangle ABC = \frac{1}{2}m\widehat{AC}$.	8. Substitution.

OUTLINE OF PROOF FOR CASE 2: Draw diameter $\overline{BD}$.

Apply the result established in Case 1:

$$\begin{aligned} m\measuredangle 1 &= \frac{1}{2}m\widehat{AD} \\ + \quad m\measuredangle 2 &= \frac{1}{2}m\widehat{DC} \\ \hline m\measuredangle 1 + m\measuredangle 2 &= \frac{1}{2}(m\widehat{AD} + m\widehat{DC}) \end{aligned}$$

Use substitution:

$$m\measuredangle ABC = \frac{1}{2}m\widehat{AC}$$

OUTLINE OF PROOF FOR Draw diameter $\overline{BD}$.

Apply the result established in Case 1:

CASE 3:

$$m\measuredangle ABD = \frac{1}{2}m\widehat{ACD}$$

$$-\ m\measuredangle CBD = \frac{1}{2}m\widehat{CD}$$

$$m\measuredangle ABD - m\measuredangle CBD = \frac{1}{2}(m\widehat{ACD} - m\widehat{CD})$$

Use substitution:

$$m\measuredangle ABC = \frac{1}{2}m\widehat{AC}$$

EXAMPLE 12.13 In each of the following, find the value of x.

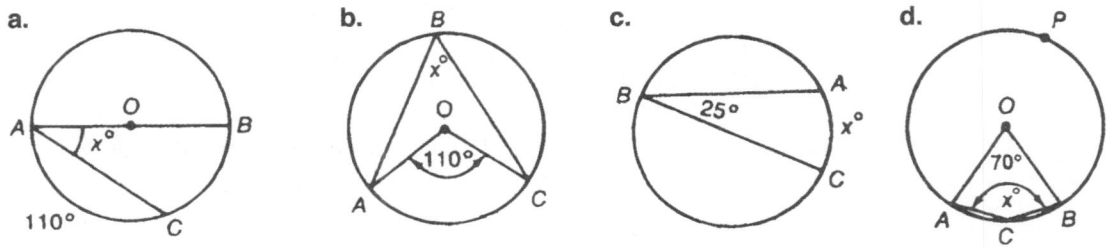

SOLUTION

a. $\widehat{ACB}$ is a semicircle, so $m\widehat{BC} = 180 - 110 = 70$.

$$x = \frac{1}{2}m\widehat{BC} = \frac{1}{2}(70)$$
$$= 35$$

b. $\measuredangle AOC$ is a central angle, so $m\widehat{AC} = 110$.

$$m\measuredangle ABC = \frac{1}{2}m\widehat{AC} = \frac{1}{2}(110)$$
$$x = 55$$

c. The measure of the arc intercepted by an inscribed angle must be twice the measure of the inscribed angle. Hence,

$$x = 2(m\measuredangle ABC) = 2(25)$$
$$= 50$$

d. $\measuredangle AOB$ is a central angle so $m\widehat{ACB} = 70$.

$$m\widehat{APB} = 360 - 70 = 290$$
$$x = \frac{1}{2}m\widehat{APB} = \frac{1}{2}(290)$$
$$= 145$$

EXAMPLE 12.14 A triangle is inscribed in a circle so that its sides divide the circle into arcs whose measures have the ratio of 2:3:7. Find the measure of the largest angle of the triangle.

SOLUTION

First, determine the degree measures of the arcs of the circle.

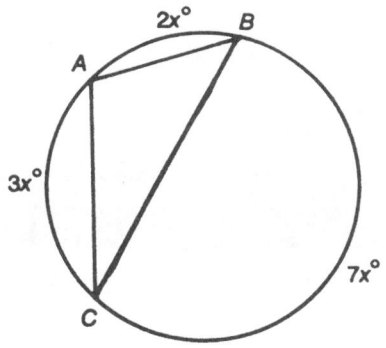

$$\begin{aligned} m\widehat{AB} + m\widehat{AC} + m\widehat{BC} &= 360 \\ 2x + 3x + 7x &= 360 \\ 12x &= 360 \\ x &= 30 \end{aligned}$$

The measures of the arcs of the circle are as follows:

$$\begin{aligned} m\widehat{AB} &= 2x = 60 \\ m\widehat{AC} &= 3x = 90 \\ m\widehat{BC} &= 7x = 210 \end{aligned}$$

Each angle of the triangle is an inscribed angle. The largest angle lies opposite the arc having the greatest measure. Hence,

$$\begin{aligned} m\measuredangle A &= \frac{1}{2}m\widehat{BC} = \frac{1}{2}(210) \\ &= 105 \end{aligned}$$

EXAMPLE 12.15 Quadrilateral $ABCD$ is inscribed in circle O. If $m\measuredangle A = x$, and $m\measuredangle B = y$, express the measures of angles of C and D in terms of x and y.

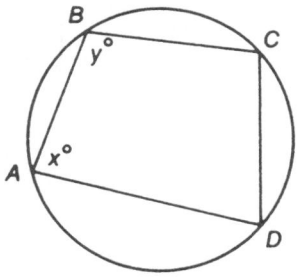

SOLUTION

To find $m\measuredangle C$:	To find $m\measuredangle D$:
$m\widehat{BCD} = 2(x) = 2x$	$m\widehat{ADC} = 2(y) = 2y$
$m\widehat{BAD} = 360 - 2(x)$	$m\widehat{ABC} = 360 - 2y$
$m\measuredangle C = \frac{1}{2}(360 - 2x)$	$m\measuredangle D = \frac{1}{2}(360 - 2y)$
$= 180 - x$	$= 180 - y$

Example 12.15 establishes the following theorem:

THEOREM 12.9 ANGLES OF AN INSCRIBED QUADRILATERAL THEOREM

Opposite angles of an inscribed quadrilateral are supplementary.

$a + c = 180$
$b + d = 180$

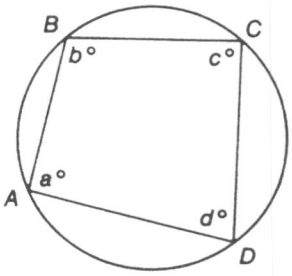

A chord that has as one of its end points the point of tangency of a tangent to a circle forms an angle whose vertex is on the circle. In Figure 12.23, the sides of $\measuredangle ABC$ are chord $\overline{AB}$ and tangent ray $\overrightarrow{BC}$. The end points of the intercepted arc are the vertex of the angle (point B) and the other end point of the chord (point A).

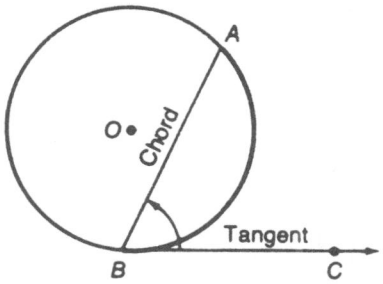

FIGURE 12.23

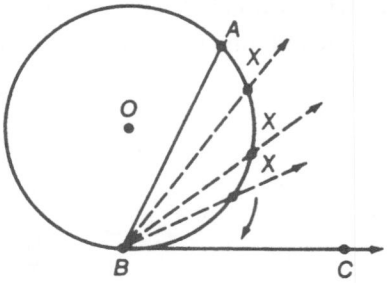

FIGURE 12.24

In Figure 12.24, consider inscribed angle ABX and suppose point X slides down the circle so that $\overrightarrow{BX}$ gets closer and closer to tangent $\overrightarrow{BC}$. At each position, inscribed angle ABX is measured by one-half of its intercepted arc, $\widehat{AX}$. As point X approaches point B, the measure of $\widehat{AX}$ gets closer and closer to the measure of $\widehat{AB}$. Hence, when $\overrightarrow{BX}$ finally coincides with tangent $\overrightarrow{BC}$, the angle formed is measured by one-half the measure of $\widehat{AB}$.

THEOREM 12.10 CHORD-TANGENT ANGLE THEOREM

The measure of an angle formed by a tangent and a chord drawn to the point of tangency is equal to one-half the measure of the intercepted arc.

In the accompanying figure,

$$m\measuredangle ABC = \frac{1}{2}x$$

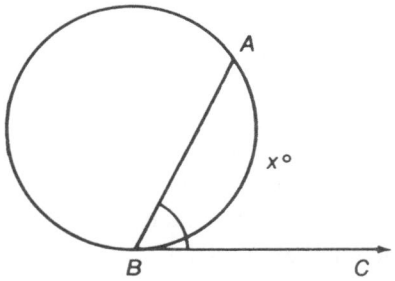

OUTLINE OF PROOF

GIVEN: $\overrightarrow{BC}$ is tangent to $\odot O$ at point B.
$\overline{AB}$ is drawn.

PROVE: $m\measuredangle 1 = \frac{1}{2} m\widehat{AB}$.

PLAN: Through point A draw a chord parallel to $\overline{BC}$, intersecting the circle at point D. Then,

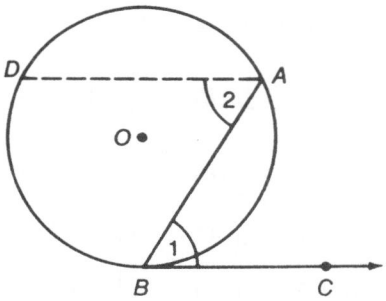

STEPS:
- $m\measuredangle 1 = m\measuredangle 2$ (Alternate interior angles)
- $m\measuredangle 2 = \frac{1}{2} m\widehat{BD}$ (Inscribed Angle Theorem)
- $m\measuredangle 1 = \frac{1}{2} m\widehat{BD}$ (Substitution)
- $m\widehat{BD} = m\widehat{AB}$ (Since $\overline{AD} \parallel \overrightarrow{BC}$)
- $m\measuredangle 1 = \frac{1}{2} m\widehat{AB}$ (Substitution)

EXAMPLE 12.16 Find the value of x if $m\widehat{ACB} = 250$.

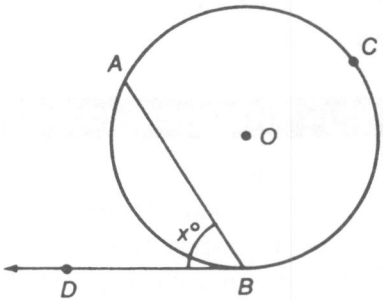

SOLUTION

$$m\widehat{AB} = 360 - 250 = 110$$
$$x = \frac{1}{2}m\widehat{AB} = \frac{1}{2}(110)$$
$$= 55$$

EXAMPLE 12.17 In circle O, $m\widehat{AC} = 110$. Find the measure of each of the numbered angles.

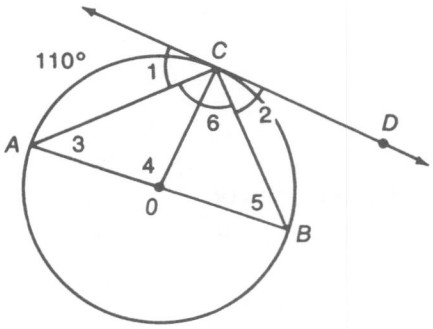

SOLUTION

$$m\measuredangle 1 = \frac{1}{2}m\widehat{AC} = \frac{1}{2}(110) = 55$$

Since $m\widehat{BC} = 180 - 110 = 70$,

$$m\measuredangle 2 = \frac{1}{2}m\widehat{BC} = \frac{1}{2}(70) = 35$$

$$m\measuredangle 3 = \frac{1}{2}m\widehat{BC} = \frac{1}{2}(70) = 35$$

$$m\measuredangle 4 = m\widehat{AC} = 110$$

$$m\measuredangle 5 = \frac{1}{2}m\widehat{AC} = \frac{1}{2}(110) = 55$$

$$m\measuredangle 6 = \frac{1}{2}m\widehat{AB} = \frac{1}{2}(180) = 90$$

NOTE: Angles 1 and 2 are formed by a chord and a tangent; angle 4 is a central angle; and angles 3, 5, and 6 are inscribed angles.

EXAMPLE 12.18 Prove that a tangent drawn parallel to a chord of a circle cuts off congruent arcs between the tangent and the chord.

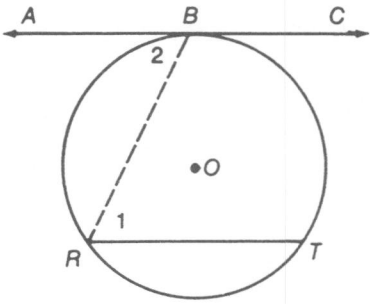

SOLUTION

OUTLINE OF PROOF

GIVEN: $\overleftrightarrow{ABC}$ is tangent to $\odot O$ at point B, $\overleftrightarrow{ABC} \parallel \overline{RT}$.

PROVE: $\widehat{BR} \cong \widehat{BT}$.

STEPS:
- $m\measuredangle 1 = \frac{1}{2}m\widehat{BT}$ and $m\measuredangle 2 = \frac{1}{2}m\widehat{BR}$.
- Since $\overleftrightarrow{ABC} \parallel \overline{RT}$, $m\measuredangle 1 = m\measuredangle 2$ so that $\frac{1}{2}m\widehat{BT} = \frac{1}{2}m\widehat{BR}$.
- Multiplying each side of this equation by 2 yields $m\widehat{BT} = m\widehat{BR}$, or $\widehat{BT} \cong \widehat{BR}$.

Thus, parallel chords, parallel tangents, and a tangent parallel to a chord cut off congruent arcs on a circle.

Angle Measurement: Vertex in the Interior of the Circle

When two chords intersect at a point on the circle, then an inscribed angle is formed. If the chords intersect at a point in the interior of a circle (not at the center), then two pairs of angles whose sides intercept arcs on the circle result, as shown in Figure 12.25. With respect to angle 1 (and its vertical angle 2), the intercepted arcs are the arcs that lie opposite the vertical angle pair: $\widehat{AC}$ *and* $\widehat{BD}$. With respect to the vertical angle pair formed by angles 3 and 4, the intercepted arcs are $\widehat{AD}$ *and* $\widehat{BC}$. For convenience we will refer to angles 1, 2, 3, and 4 as *chord-chord* angles since they are formed by intersecting chords.

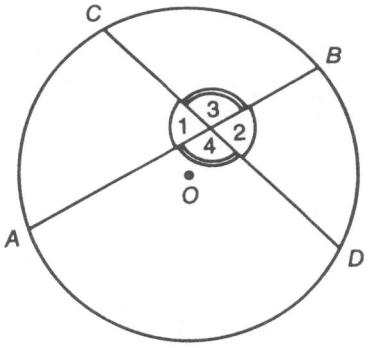

FIGURE 12.25

A chord-chord angle is measured by the *average* of its two intercepted arcs:

$$m\measuredangle 1 = m\measuredangle 2 = \frac{1}{2}(m\widehat{AC} + m\widehat{BD})$$

and

$$m\measuredangle 3 = m\measuredangle 4 = \frac{1}{2}(m\widehat{AD} + m\widehat{BC})$$

THEOREM 12.11 CHORD-CHORD ANGLE THEOREM

The measure of an angle formed by two chords intersecting in the interior of a circle is equal to one-half the sum of the measures of the two intercepted arcs.

In the accompanying figure,

$$m\measuredangle AEC = \frac{1}{2}(x + y).$$

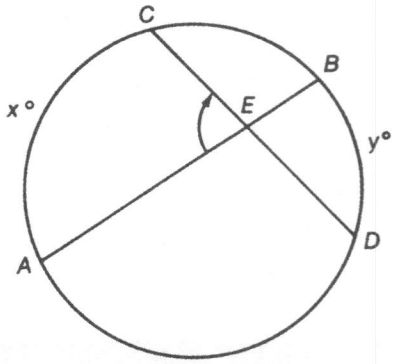

In Theorem 12.11, the sides of the angle may be secants that intersect in the interior of a circle since secants contain chords of the circle.

OUTLINE OF PROOF

GIVEN: In $\odot O$ chords $\overline{AB}$ and $\overline{CD}$ intersect at point E.

PROVE: $m\measuredangle 1 = \frac{1}{2}(m\widehat{AC} + m\widehat{BD})$.

PLAN: Draw $\overline{AD}$. Angle 1 is an exterior angle of $\triangle AED$. Hence,

$$\begin{aligned} m\measuredangle 1 &= (m\measuredangle D + m\measuredangle A) \\ &= \frac{1}{2}m\widehat{AC} + \frac{1}{2}m\widehat{BD} \\ &= \frac{1}{2}(m\widehat{AC} + m\widehat{BD}) \end{aligned}$$

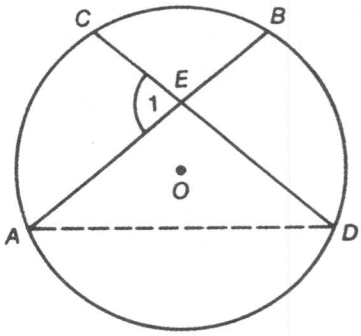

EXAMPLE 12.19 In each of the following, find the value of x.

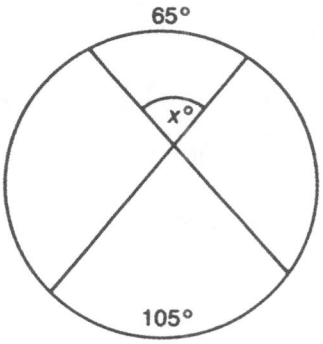

(a)

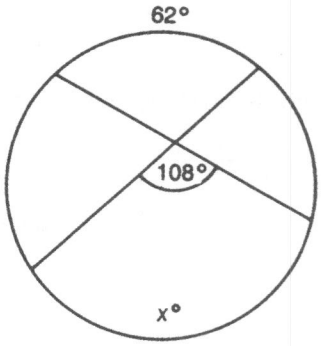

(b)

SOLUTION

a. $x = \frac{1}{2}(65 + 105)$

$= \frac{1}{2}(170) = 85$

b. $108 = \frac{1}{2}(x + 62)$

or

$x + 62 = 2(108) = 216$

$x = 216 - 62 = 154$

Angle Measurement: Vertex in the Exterior of the Circle

If the vertex of an angle whose sides intercept arcs on the circle lies in the exterior of the circle, then the angle may be formed in one of three possible ways, as shown in Figure 12.26.

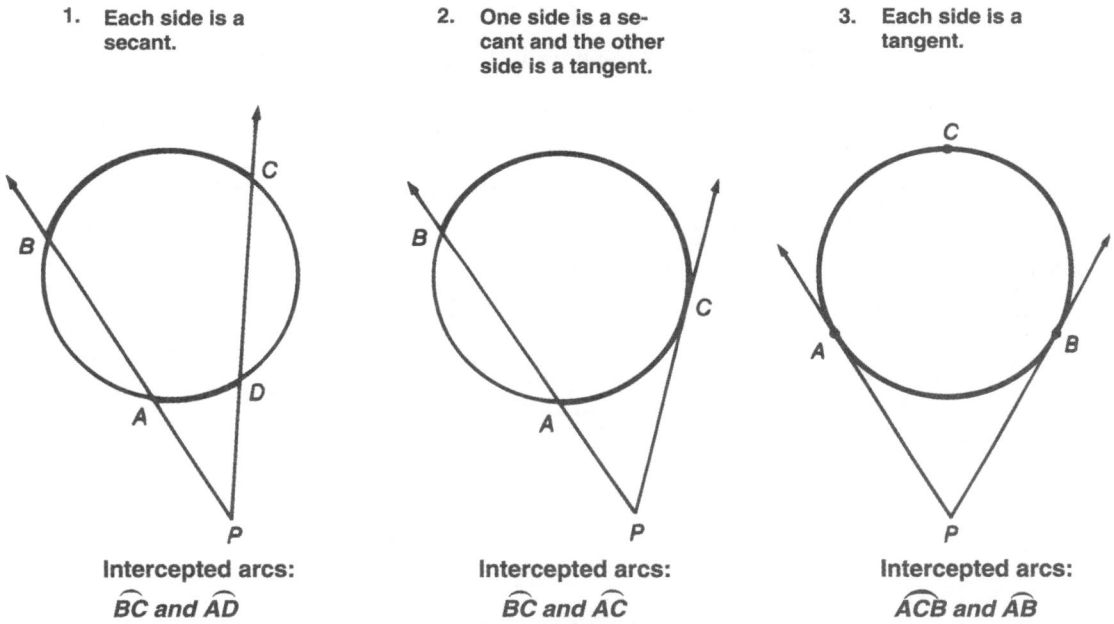

FIGURE 12.26

In each of these three cases, the angle at vertex P is measured by one-half the *difference* of the intercepted arcs.

THEOREM 12.12 SECANT-SECANT, SECANT-TANGENT, TANGENT-TANGENT THEOREM

The measure of an angle formed by two secants, a secant and a tangent, or two tangents intersecting in the exterior of a circle is equal to one-half the difference of the measures of the intercepted arcs.

In the accompanying figure,

$$m\measuredangle P = \frac{1}{2}(x - y).$$

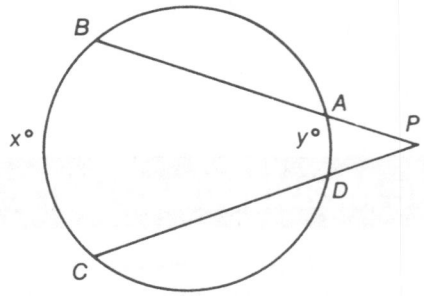

The three cases of Theorem 12.12 to be proved are shown in Figure 12.27.

CASE 1 Secant-Secant

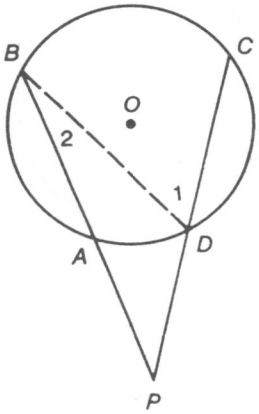

GIVEN: In $\odot O$, secants $\overline{PAB}$ and $\overline{PDC}$ are drawn.

PROVE: $m\measuredangle P = \frac{1}{2}(m\widehat{BC} - m\widehat{AD})$

CASE 2 Secant-Tangent

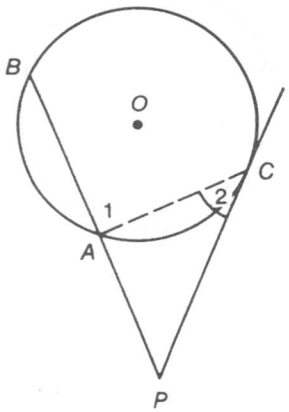

GIVEN: In $\odot O$, secant $\overline{PAB}$ and tangent PC are drawn.

PROVE: $m\measuredangle P = \frac{1}{2}(m\widehat{BC} - m\widehat{AC})$

CASE 3 Tangent-Tangent

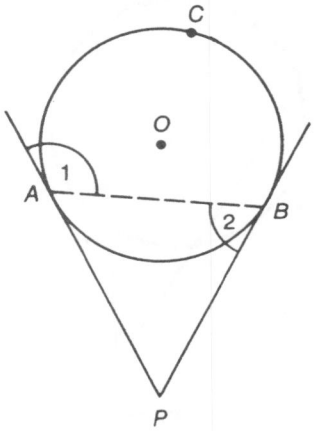

GIVEN: In $\odot O$, tangents $\overline{PA}$ and $\overline{PB}$ drawn.

PROVE: $m\measuredangle P = \frac{1}{2}(m\widehat{ACB} - m\widehat{AB})$

FIGURE 12.27

OUTLINE OF PROOF

OF CASE 1: Draw $\overline{BD}$. Angle 1 is an exterior angle of $\triangle DBP$. Hence,

$$m\measuredangle 1 = m\measuredangle 2 + m\measuredangle P$$

or

$$\begin{aligned} m\measuredangle P &= m\measuredangle 1 - m\measuredangle 2 \\ &= \frac{1}{2} m\widehat{BC} - \frac{1}{2} m\widehat{AD} \\ &= \frac{1}{2}(m\widehat{BC} - m\widehat{AD}) \end{aligned}$$

The proofs of Cases 2 and 3 follow a similar pattern.

SUMMARY OF ANGLE MEASUREMENT IN CIRCLES

Vertex of Angle	Measure of Angle Equals	Model
1. At center (Central Angle)	The measure of the intercepted arc.	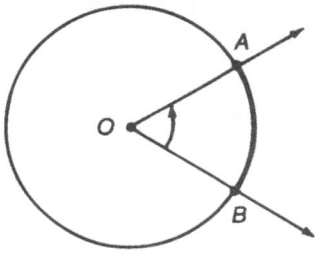 $m\measuredangle AOB = m\widehat{AB}$
2. On circle (Inscribed Angle)	One-half the measure of the intercepted arc.	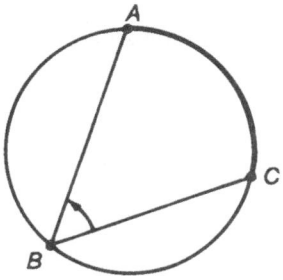 $m\measuredangle ABC = \frac{1}{2} m\widehat{AC}$
3. On circle (Chord-Tangent Angle)	One-half the measure of the intercepted arc.	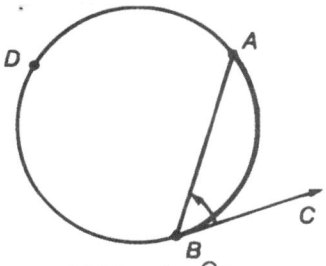 $m\measuredangle ABC = \frac{1}{2} m\widehat{AC}$

SUMMARY OF ANGLE MEASUREMENT IN CIRCLES (*Continued*)

Vertex of Angle	Measure of Angle Equals	Model
4. Interior of Circle (Chord-Chord Angle)	One-half the sum of the measures of the intercepted arcs.	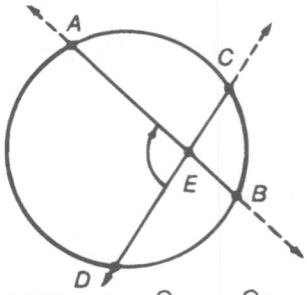 $m\measuredangle AED = \frac{1}{2}(m\widehat{AD} + m\widehat{BC})$
5. Exterior of Circle (Secant-Secant, Secant-Tangent, and Tangent-Tangent Angles)	One-half the difference of the measures of the intercepted arcs.	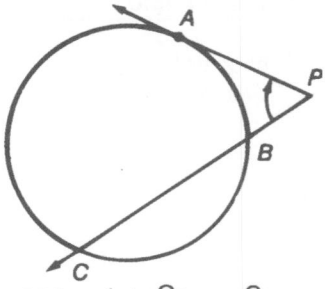 $m\measuredangle APC = \frac{1}{2}(m\widehat{AC} - m\widehat{AB})$

EXAMPLE 12.20 Find the value of x

a.

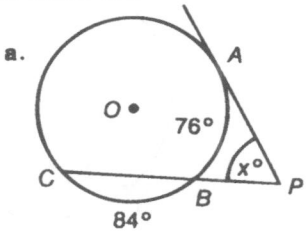

b.

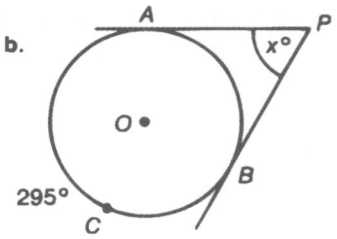

c.

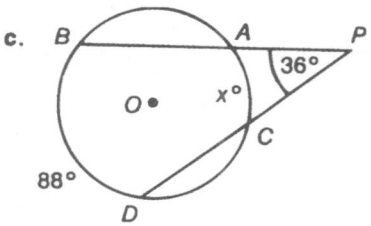

d.

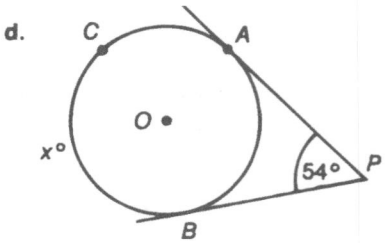

e.

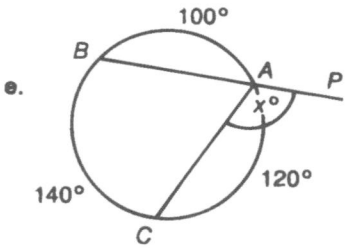

SOLUTION

a.
$$\begin{aligned} m\widehat{AC} &= 360 - (76 + 84) \\ &= 360 - 160 \\ &= 200 \end{aligned}$$

$$\begin{aligned} x &= \frac{1}{2}(m\widehat{AC} - m\widehat{AB}) \\ &= \frac{1}{2}(200 - 76) \\ &= \frac{1}{2}(124) = 62° \end{aligned}$$

b.
$$\begin{aligned} m\widehat{AB} &= 360 - 295 = 65 \\ x &= \frac{1}{2}(m\widehat{ACB} - m\widehat{AB}) \\ &= \frac{1}{2}(295 - 65) \\ &= \frac{1}{2}(230) = 115° \end{aligned}$$

c.
$$\begin{aligned} m\measuredangle P &= \frac{1}{2}(m\widehat{BD} = m\widehat{AC}) \\ 36 &= \frac{1}{2}(88 - x) \end{aligned}$$

Multiplying both sides by 2,
$$\begin{aligned} 72 &= 88 - x \\ x + 72 &= 88 \\ x &= 88 - 72 = 16° \end{aligned}$$

d.
$$\begin{aligned} m\widehat{AB} &= 360 - m\widehat{ACB} \\ m\widehat{AB} &= 360 - x \\ m\measuredangle P &= \frac{1}{2}(m\widehat{ACB} - m\widehat{AB}) \\ 54 &= \frac{1}{2}[x - (360 - x)] \end{aligned}$$

Multiplying both sides by 2,
$$\begin{aligned} 108 &= x - 360 + x \\ 108 &= 2x - 360 \\ 468 &= 2x \\ x &= 234° \end{aligned}$$

e. A common *error* is to find $m\measuredangle CAP$ (that is, x) by taking $\frac{1}{2}m\widehat{AC}$. This is incorrect since the sides of the angle are a chord and *part of a secant*. Since angles CAP and CAB are supplementary, we first determine the measure of inscribed angle CAB:

$$\begin{aligned} m\measuredangle CAB &= \frac{1}{2}m\widehat{BC} = \frac{1}{2}(140) = 70 \\ m\measuredangle CAP &= 180 - 70 \\ x &= 110° \end{aligned}$$

EXAMPLE 12.21 $\overrightarrow{PA}$ is tangent to circle O at point A. Secant $\overline{PBC}$ is drawn. Chords $\overline{CA}$ and $\overline{BD}$ intersect at point E. If $m\widehat{AD} : m\widehat{AB} : m\widehat{DC} : m\widehat{BC} = 2 : 3 : 4 : 6$, find the measure of each of the numbered angles.

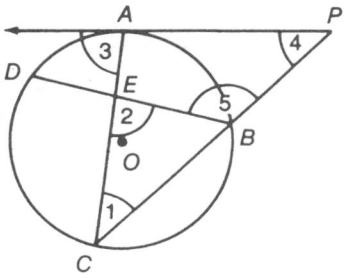

SOLUTION

Let
$$\begin{aligned} m\widehat{AD} &= 2x, \\ m\widehat{AB} &= 3x, \\ m\widehat{DC} &= 4x, \\ m\widehat{BC} &= 6x. \end{aligned}$$

Then $2x + 3x + 4x + 6x = 360$

$$15x = 360$$

$$x = \frac{360}{15} = 24$$

$$m\widehat{AD} = 2x = 2(24) = 48$$
$$m\widehat{AB} = 3x = 3(24) = 72$$
$$m\widehat{DC} = 4x = 4(24) = 96$$
$$m\widehat{BC} = 6x = 6(24) = 144$$

- $m\measuredangle 1 = \frac{1}{2} m\widehat{AB} = \frac{1}{2}(72) = 36$
- $m\measuredangle 2 = \frac{1}{2}(m\widehat{AD} + m\widehat{BC})$
  $= \frac{1}{2}(48 + 144)$
  $= \frac{1}{2}(192)$
  $= 96$
- $m\measuredangle 3 = \frac{1}{2} m\widehat{AC}$
  $= \frac{1}{2}(m\widehat{AD} + m\widehat{DC})$
  $= \frac{1}{2}(48 + 96)$
  $= \frac{1}{2}(144)$
  $= 72$
- $m\measuredangle 4 = \frac{1}{2}(m\widehat{AC} - m\widehat{AB})$
  $= \frac{1}{2}(144 - 72)$
  $= \frac{1}{2}(72)$
  $= 36$
- $m\measuredangle 5 = 180 - m\measuredangle CBD$

Since $m\measuredangle CBD = \frac{1}{2} m\widehat{DC}$
$= \frac{1}{2}(96) = 48,$

$m\measuredangle 5 = 180 - 48$
$= 132$

Using Angle-Measurement Theorems

Angle-measurement theorems can be used to help establish that angles, arcs, and chords of a circle are congruent. First, though, we will formally state some observations that you have probably made while working on numerical applications of these theorems.

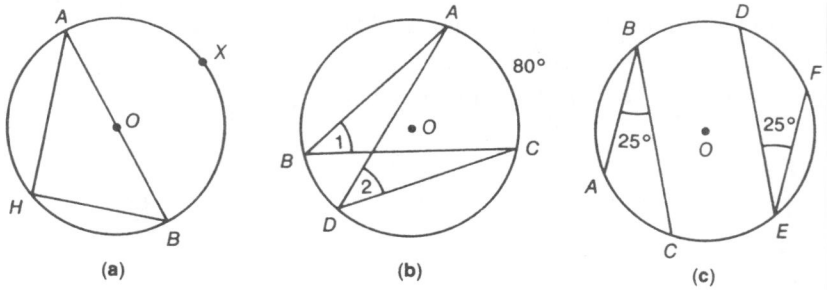

FIGURE 12.28

In Figure 12.28*a* $\overline{AB}$ is a diameter of circle O. Angle AHB is said to be *inscribed in* semicircle $\widehat{AHB}$ since the end points of the chords that form the angle coincide with the end points of a diameter. What is the measure of $\measuredangle AHB$? Angle $\widehat{AHB}$ is an inscribed angle having semicircle $\widehat{AXB}$ as its intercepted arc. Hence, $m\measuredangle AHB = \frac{1}{2}(180) = 90$. *An angle inscribed in a semicircle is a right angle.*

In Figure 12.28*b* the measures of inscribed angles 1 and 2 are each equal to one-half of 80° (or 40°). *Inscribed angles that intercept the same arc (or congruent arcs) are congruent.*

The inscribed angles shown in Figure 12.28*c* are given as congruent (each has a measure of 25). It follows that $m\widehat{AC} = 50$ and $m\widehat{DF} = 50$. *If a pair of inscribed angles are congruent, then their intercepted arcs are congruent.*

Each of the preceding observations was based on a direct application of the Inscribed Angle Theorem (Theorem 12.8) and is therefore referred to as a *corollary*.

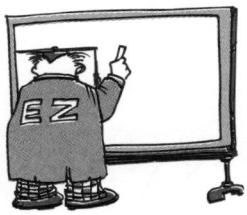

NOTE

COROLLARIES TO THE INSCRIBED ANGLE THEOREM

- An angle inscribed in a semicircle is a right angle.
- If inscribed angles (or angles formed by a tangent and a chord) intercept the same or congruent arcs, then they are congruent.
- If inscribed angles are congruent, then their intercepted arcs are congruent.

EXAMPLE 12.22

GIVEN: In $\odot O$, $\overline{CD} \parallel \overline{AB}$. Chords $\overline{AC}$ and $\overline{BD}$ are extended to meet at point P.

PROVE: $\overline{AP} \cong \overline{BP}$.

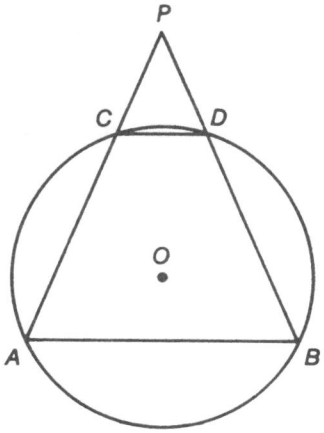

SOLUTION

PLAN: Show that inscribed angles A and B intercept congruent arcs and therefore are congruent. The sides opposite these angles in triangle APB must be congruent by the converse of the Base Angles Theorem.

PROOF:

Statements	Reasons
1. $\overline{CD} \parallel \overline{AB}$.	1. Given.
2. $m\widehat{AC} = m\widehat{BD}$.	2. Parallel chords intercept arcs equal in measure.
3. $m\widehat{ACD} = m\widehat{BDC}$.	3. Arc Sum Postulate.
4. $m\measuredangle B = \frac{1}{2} m\widehat{ACD}$, $m\measuredangle A = \frac{1}{2} m\widehat{BDC}$.	4. The measure of an inscribed angle is one-half the measure of its intercepted arc.
5. $\frac{1}{2} m\widehat{ACD} = \frac{1}{2} m\widehat{BDC}$.	5. Halves of equals are equal.
6. $m\measuredangle B = m\measuredangle A$.	6. Substitution.
7. $\overline{AP} \cong \overline{BP}$.	7. If two angles of a triangle are equal in measure, then the sides opposite are congruent.

EXAMPLE 12.23

GIVEN: In $\odot O$, $\overline{AB}$ is a diameter. $\widehat{AC} \cong \widehat{BD}$.

PROVE: $\triangle ABC \cong \triangle BAD$.

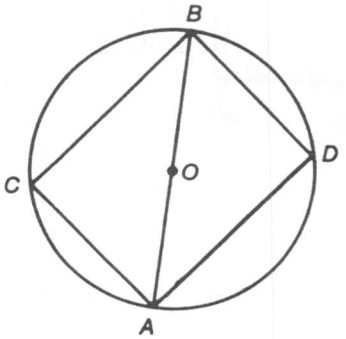

SOLUTION

PLAN: Prove triangles are congruent by Hy-Leg.

PROOF:

Statements	Reasons
1. In $\odot O$, $\overline{AB}$ is a diameter.	1. Given.
2. Angles C and D are right angles.	2. An angle inscribed in a semicircle is a right angle.
3. Triangles ABC and BAD are right triangles.	3. A triangle that contains a right angle is a right triangle.
4. $\overline{AB} \cong \overline{AB}$. (Hy)	4. Reflexive property of congruence.
5. $\widehat{AC} \cong \widehat{BD}$.	5. Given.
6. $\overline{AC} \cong \overline{BD}$. (Leg)	6. In a circle, congruent arcs have congruent chords.
7. $\triangle ACB \cong \triangle BDA$.	7. Hy-Leg.

The AAS theorem could have been used in the proof of Example 12.23.

(A)	$\measuredangle C \cong \measuredangle D$	(Right angles are congruent.)
(A)	$\measuredangle ACB \cong \measuredangle BDA$	(Inscribed angles that intercept congruent arcs are congruent.)
(S)	$\overline{AB} \cong \overline{AB}$	

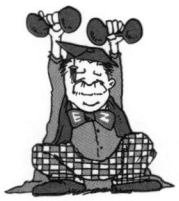

REVIEW EXERCISES FOR CHAPTER 12

In Exercises 1 to 10, find the value of x.

1.

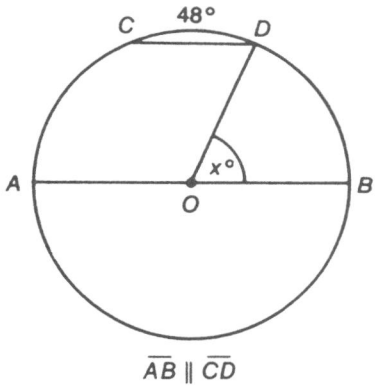

2.

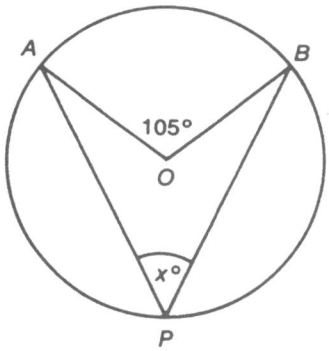

3.

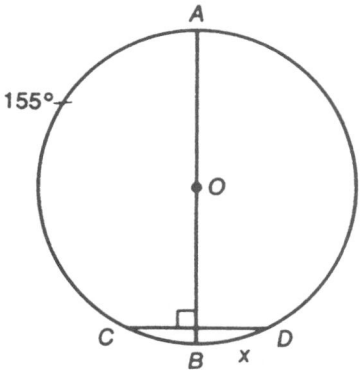

4.

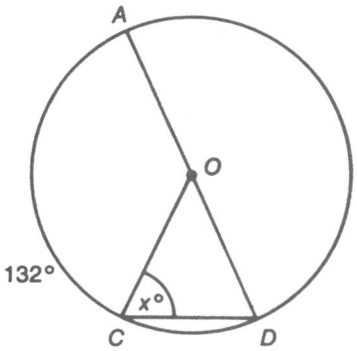

5.

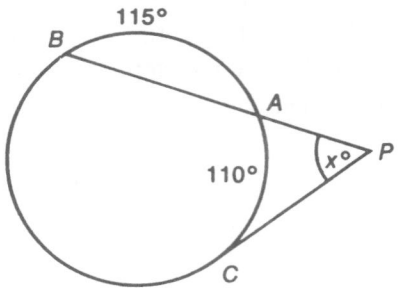

6.

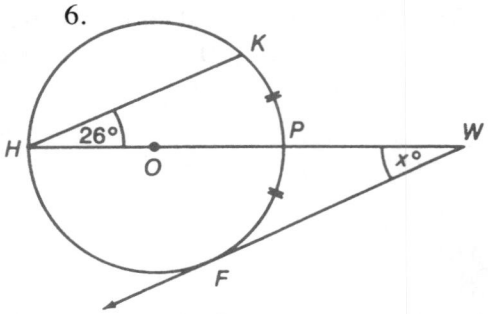

P is the midpoint of $\widehat{KPF}$.

7.

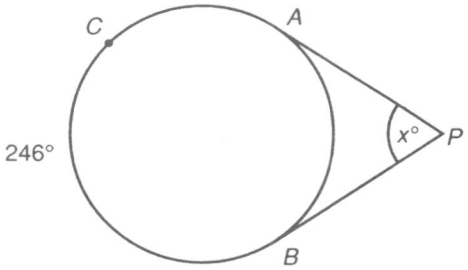

8.

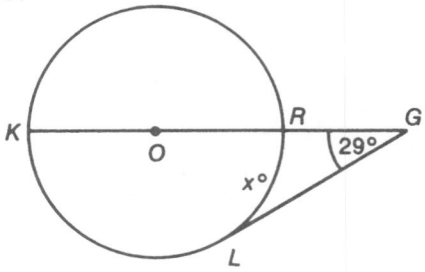

9.

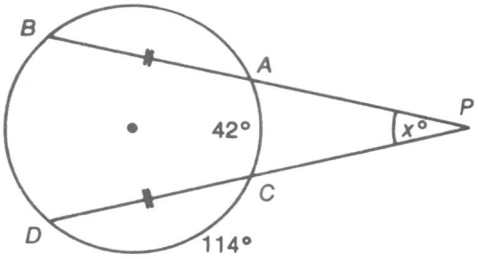

$\overline{AB} \cong \overline{CD}$

10.

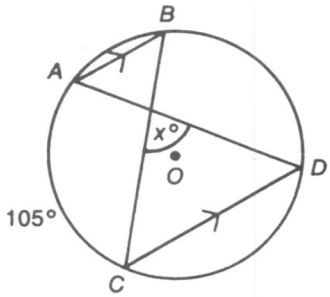

$\overline{AB} \parallel \overline{CD}$

11. Find the value of x.

(a)

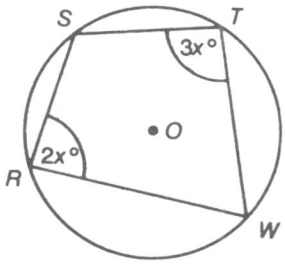

(b)

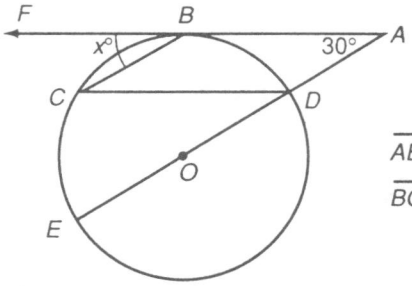

$\overrightarrow{AB} \parallel \overline{CD}$

$\overline{BC} \parallel \overrightarrow{AD}$

12. The length of a chord of a circle is 24 and its distance from the center is 5. Find the length of a diameter of the circle.

13. Tangents $\overline{AX}$ and $\overline{AY}$ are drawn to circle P from an exterior point A. Radii $\overline{PX}$ and $\overline{PY}$ are drawn. If $m\measuredangle XPY = 74$, find $m\measuredangle XAY$.

14. The length of tangent segment $\overline{PA}$ drawn from exterior point P to circle O is 24. If the radius of the circle is 7, find the distance from point P to the center of the circle.

15. In circle P, $\overline{ST}$ is a diameter and $\overleftrightarrow{LAB}$ is a tangent. Point T is the midpoint of $\widehat{ATK}$. If $m\measuredangle AKS = 74$, find each of the following:

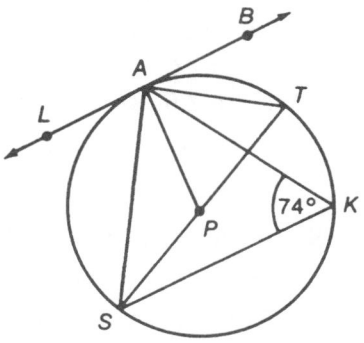

(a) $m\measuredangle ATS$ (b) $m\measuredangle AST$
(c) $m\measuredangle LAS$ (d) $m\measuredangle TAB$
(e) $m\measuredangle APT$ (f) $m\measuredangle SAT$
(g) $m\measuredangle TSK$ (h) $m\measuredangle KAT$
(i) $m\measuredangle TAP$

16. Given circle O with diameter $\overline{GOAL}$, secar $\overline{HUG}$ and $\overline{HTAM}$ intersect at point H; $m\widehat{GM} : m\widehat{ML} : m\widehat{LT} = 7 : 3 : 2$; and $\overline{GU} \cong \overline{UT}$. Find the ratio of $m\angle UGL$ to $m\angle H$.

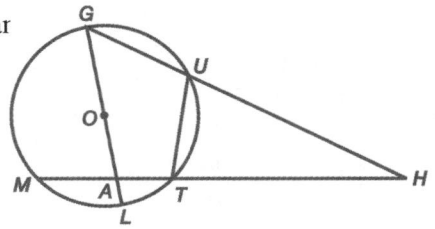

17. Find the values of x, y, and z.

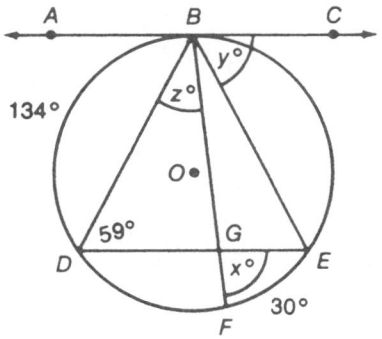

18. In the accompanying diagram, circle O has radius $\overline{OD}$, diameter $\overline{BOHF}$, secant $\overline{CBA}$, and chords $\overline{DHG}$ and $\overline{BD}$. $\overline{CE}$ is tangent to circle O at D; $m\widehat{DF} = 80$; and $m\widehat{BA} : m\widehat{AG} : m\widehat{GF} = 3 : 2 : 1$.

Find:
(a) $m\widehat{GF}$
(b) $m\angle BHD$
(c) $m\angle BDG$
(d) $m\angle GDE$
(e) $m\angle C$
(f) $m\angle BOD$

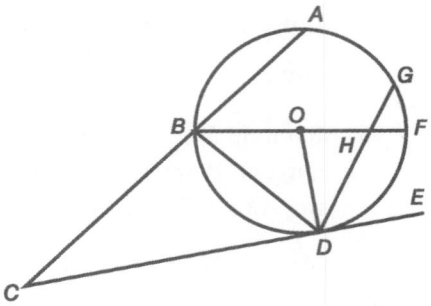

19. In circle O, tangent $\overrightarrow{PW}$ and secant $\overline{PST}$ are drawn. Chord $\overline{WA}$ is parallel to chord $\overline{ST}$. Chords $\overline{AS}$ and $\overline{WT}$ intersect at point B. If $m\widehat{WA} : m\widehat{AT} : m\widehat{ST} = 1:3:5$, find each of the following:

(a) $m\widehat{WA}$, $m\widehat{AT}$, $m\widehat{ST}$, and $m\widehat{SW}$
(b) $m\measuredangle WTS$
(c) $m\measuredangle TBS$
(d) $m\measuredangle TWP$
(e) $m\measuredangle WPT$

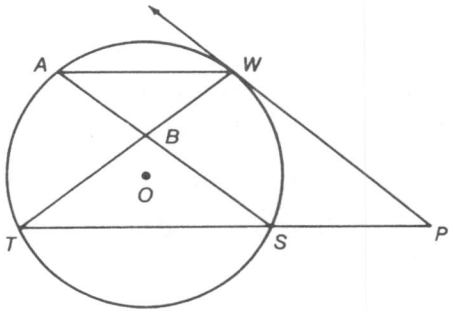

20. In circle P, $\overline{KM}$ is a diameter and $\overline{LFK}$ and $\overline{LHJ}$ are secants. Point F is the midpoint of $\widehat{KFH}$. If $m\widehat{KJ} : m\widehat{JM} = 5:4$ and $m\measuredangle HEM = 64$, find each of the following:

(a) $m\widehat{KF}$, $m\widehat{FM}$, $m\widehat{JK}$, $m\widehat{JM}$, and $m\widehat{HM}$
(b) $m\measuredangle KPF$
(c) $m\measuredangle KJH$
(d) $m\measuredangle KLJ$

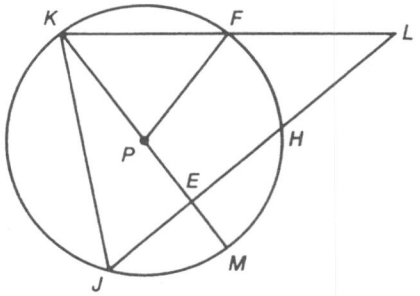

21. GIVEN: In $\odot O$, $\overline{OM} \perp \overline{AB}$.
PROVE: X is the midpoint of $\widehat{AB}$.

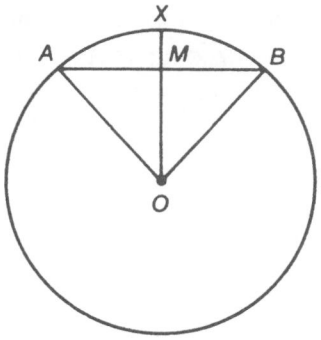

22. GIVEN: In $\odot O$, $\widehat{RST} \cong \widehat{WTS}$.
PROVE: $\measuredangle RTS \cong \measuredangle WST$.

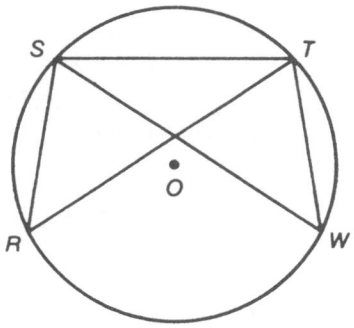

23. In the accompanying diagram of circle O, $\overline{PA}$ is drawn tangent to the circle at A. Assume B is any point on $\overline{PA}$ between P and A. Prove $\overline{OB}$ is *not* perpendicular to $\overline{PA}$.

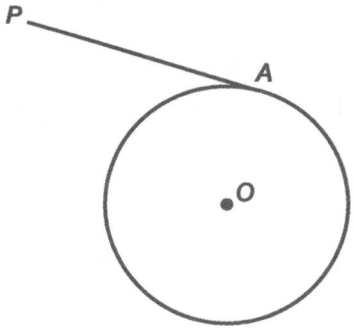

24. GIVEN: In $\odot O$, $OA > AC$.
PROVE: $m\widehat{BC} > m\widehat{AC}$.

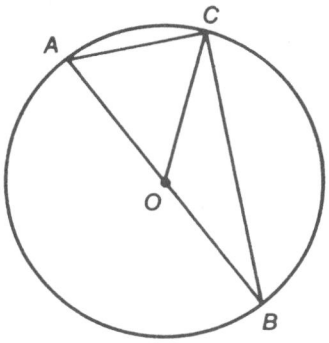

25. GIVEN: In $\odot O$, $m\widehat{BC} > m\widehat{AC}$.
PROVE: $OA > AD$.

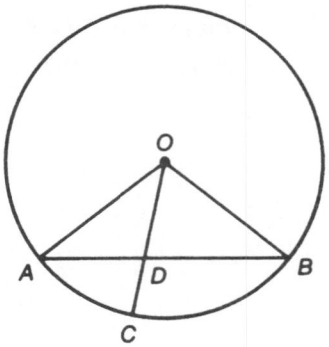

26. GIVEN: $\odot X \cong \odot Y$.
$\overleftrightarrow{PQ}$ is tangent to $\odot X$ at P
and tangent to $\odot Y$ at Q.
PROVE: Point M is the midpoint of $\overline{XY}$.

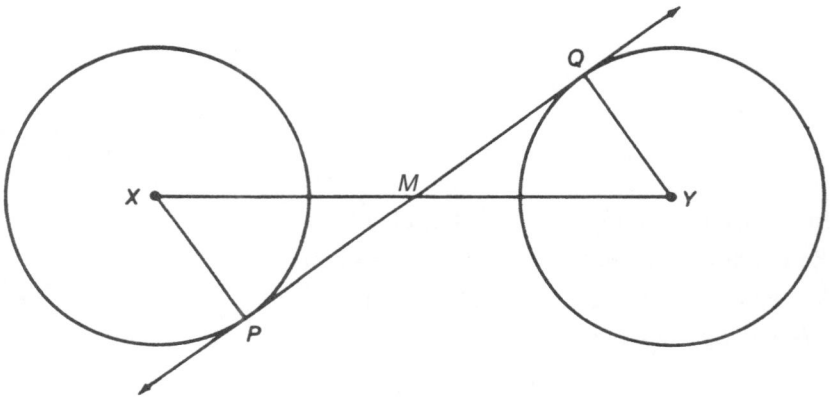

27. GIVEN: $\triangle EFG$ is inscribed in $\odot P$,
$\overleftrightarrow{AB}$ is tangent to F,
$\overline{FE} \cong \overline{FG}$.
PROVE: $\overline{AB} \parallel \overline{EG}$.

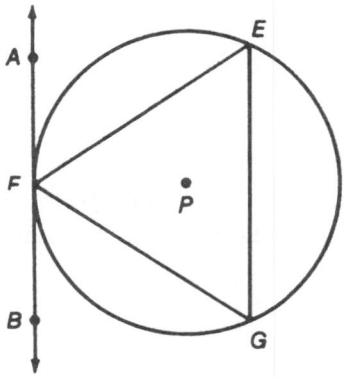

28. GIVEN: In $\odot O$, $\overline{AB}$ is a diameter, point M is the midpoint of $\overline{BD}$, chords $\overline{BM}$ and $\overline{AC}$ are extended to meet at point D.

PROVE: M is the midpoint of $\widehat{BMC}$.

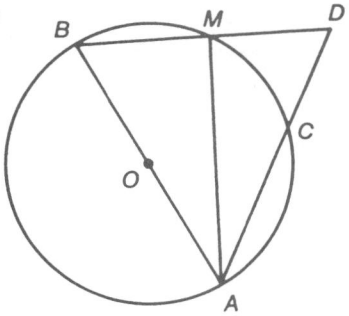

29. GIVEN: $\triangle RST$ is inscribed in $\odot O$ and chords $\overline{SW}$ and $\overline{WK}$ are drawn, $\overline{SW}$ *bisects* $\measuredangle RST$, $\widehat{RW} \cong \widehat{SK}$.

PROVE: (a) $\triangle NWS$ is isosceles.
(b) $\triangle NTK$ is isosceles.

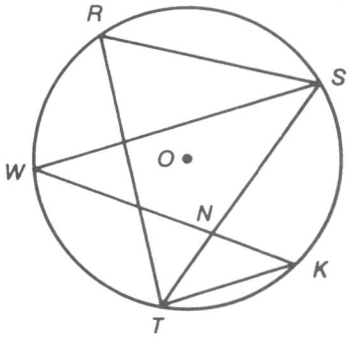

30. GIVEN: Trapezoid $JKLM$ with $\overline{JK} \parallel \overline{LM}$ is inscribed in $\odot O$, $m\widehat{LM} > m\widehat{KL}$.

PROVE: $m\measuredangle MKL > m\measuredangle JKM$.

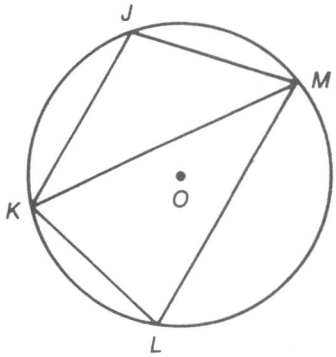

31. GIVEN: $\overline{SR} \cong \overline{TW}$,
$\overline{MA} \cong \overline{MT}$.

PROVE: Quadrilateral $RSTW$ is a parallelogram.

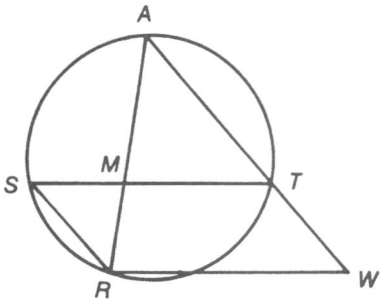

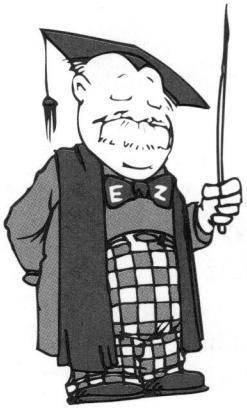

13

Chord, Tangent, and Secant Segments

WHAT YOU WILL LEARN

In Chapter 12 you learned the relationship between the measures of angles and the measures of their intercepted arcs. In this chapter you will learn relationships involving the lengths of chord, tangent, and secant segments. Some of these relationships are not obvious and depend on the properties of similar triangles. In particular, you will learn:

- the conditions under which chords of a circle are congruent and tangent segments drawn to the same circle are congruent;
- the way to draw internally and externally tangent circles;
- the use of angle-measurement theorems in circles to help prove triangles similar;
- the relationship between the four segments formed when two chords intersect inside a circle;
- the relationships between tangent and secant segments drawn to a circle from the same point outside the circle;
- the way to find the circumference of a circle;
- the way to find the length of an arc intercepted by a central angle whose measure is given.

SECTIONS IN THIS CHAPTER

- Equidistant Chords
- Tangents and Circles
- Similar Triangles and Circles
- Tangent- and Secant-Segment Relationships
- Circumference and Arc Length

Equidistant Chords

In a circle, two chords may have the same length or may have different lengths. In this section we will demonstrate that there is a relationship between how two chords compare in length and the distances of these chords from the center of the circle.

In Figure 13.1, $\overline{OC}$ is drawn perpendicular to chord $\overline{AB}$. The length of $\overline{OC}$ represents the *distance* of chord $\overline{AB}$ from the center (point O) of the circle. If the radius of circle O is 5 and the length of $\overline{AB}$ is 8, what is the length of $\overline{OC}$? If a line passes through the center of a circle and is perpendicular to a chord, then it bisects the chord (see Theorem 12.4). Hence, $AC = 4$. If we now draw radius $\overline{OA}$, a 3-4-5 right triangle is formed, where $OC = 3$. (See Figure 13.2.)

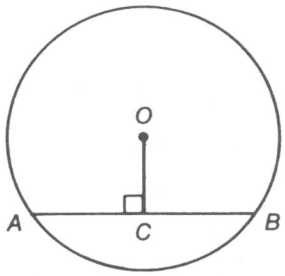

FIGURE 13.1

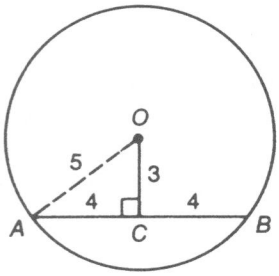

FIGURE 13.2

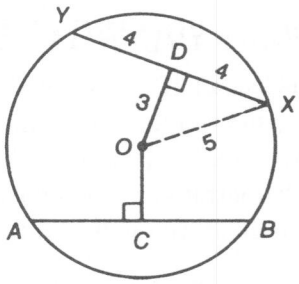

FIGURE 13.3

Suppose we now draw another chord in circle O, say $\overline{XY}$, which also has a length of 8. How far is $\overline{XY}$ from the center of the circle? (See Figure 13.3.)

Using a similar analysis, we find that chord $\overline{XY}$ is also 3 units from the center of the circle. What property do chords $\overline{AB}$ and $\overline{XY}$ have in common? What conclusion follows? The answers are stated in Theorem 13.1.

THEOREM 13.1 CONGRUENT CHORDS AND DISTANCE FROM THE CENTER

In the same circle or in congruent circles, congruent chords are equidistant (the same distance) from the center(s) of the circle(s).

OUTLINE OF PROOF

GIVEN: In $\odot O$, $\overline{AB} \cong \overline{XY}$, $\overline{OC} \perp \overline{AB}$, $\overline{OD} \perp \overline{XY}$.

PROVE: $OC = OD$.

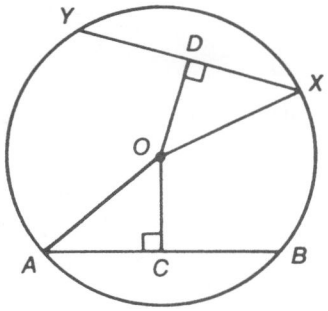

PLAN: Draw $\overline{OA}$ and $\overline{OX}$. Prove $\triangle OCA \cong \triangle ODX$ by Hy-Leg:

STEPS:
- $\overline{OA} \cong \overline{OX}$ (Hy)
- $AC = XD$ (leg) since $AC = \frac{1}{2}AB$ and $XD = \frac{1}{2}XY$, and halves of equals (AB and XY) are equal (AC and XD).
- $\overline{OC} \cong \overline{OD}$ by CPCTC from which it follows that $OC = OD$.

EXAMPLE 13.1

GIVEN: In $\odot P$, $\overline{PB} \perp \overline{DE}$, $\overline{PC} \perp \overline{FG}$, $\overline{DE} \cong \overline{FG}$.

PROVE: $\overline{PA}$ bisects $\measuredangle FAD$.

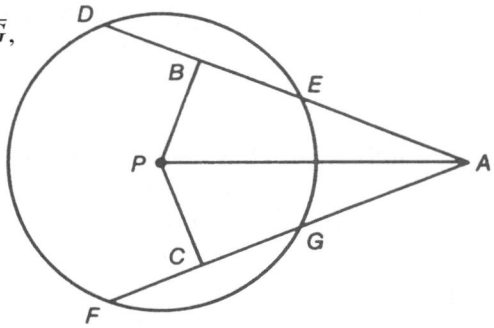

SOLUTION

PLAN: Show right $\triangle PBA \cong$ right $\triangle PCA$.

PROOF:

Statements	Reasons
1. $\overline{PB} \perp \overline{DE}$ and $\overline{PC} \perp \overline{FG}$.	1. Given.
2. $\triangle PBA$ and $\triangle PCA$ are right triangles.	2. Perpendicular lines intersect to form right angles; a triangle that contains a right angle is a right triangle.
3. $\overline{PA} \cong \overline{PA}$. (Hy)	3. Reflexive property of congruence.
4. $\widehat{DE} \cong \widehat{FG}$.	4. Given.
5. $\overline{DE} \cong \overline{FG}$.	5. In a circle, congruent arcs have congruent chords.

6. $PB = PC$.	6. In a circle, congruent chords are equidistant from the center of the circle.
7. $\overline{PB} \cong \overline{PC}$. (Leg)	7. Segments equal in length are congruent.
8. $\triangle PBA \cong \triangle PCA$.	8. Hy-Leg.
9. $\measuredangle PAB \cong \measuredangle PAC$.	9. CPCTC.
10. $\overline{PA}$ bisects $\measuredangle FAD$.	10. If a line divides an angle into two congruent angles, it bisects the angle.

The converse of Theorem 13.1 states that if we know that two chords are the same distance from the center of a circle, then they must be congruent.

THEOREM 13.2 EQUIDISTANT CHORDS

In the same circle or congruent circles, chords equidistant from the center(s) of the circle(s) are congruent.

EXAMPLE 13.2

GIVEN: $\odot O$, with $\overline{OX} \perp \overline{AB}$, $\overline{OY} \perp \overline{CB}$, $\measuredangle OXY \cong \measuredangle OYX$.

PROVE: $\widehat{AB} \cong \widehat{CB}$.

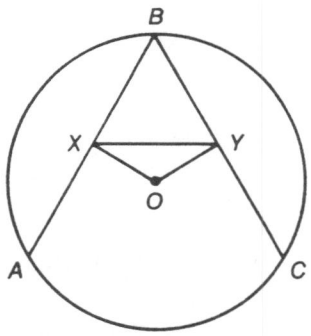

SOLUTION

PLAN: Prove that $\overline{AB}$ and $\overline{CB}$ are the same distance from the center of the circle by showing that $OX = OY$. By Theorem 13.2, $\overline{AB} \cong \overline{CB}$ which implies $\widehat{AB} \cong \widehat{CB}$.

PROOF:

Statements	Reasons
1. $\overline{OX} \perp \overline{AB}$ and $\overline{OY} \perp \overline{CB}$.	1. Given.
2. $\measuredangle OXY \cong \measuredangle OYX$.	2. Given.
3. $OX = OY$	3. If two angles of a triangle are congruent, then the sides opposite are equal in length.
4. $\overline{AB} \cong \overline{CB}$.	4. In the same circle, chords equidistant from the center of the circle are congruent.
5. $\widehat{AB} \cong \widehat{CB}$.	5. In the same circle, congruent chords intercept congruent arcs.

Tangents and Circles

TANGENT SEGMENTS

A tangent segment is a segment whose end points are the point of tangency and a fixed point on the tangent. An infinite number of tangent segments can be drawn to a circle. Under what conditions will the lengths of these tangent segments be equal? If two tangent segments are drawn from the *same* exterior point to a circle, they will have the same length.

THEOREM 13.3 CONGRUENT TANGENT SEGMENTS

If two tangent segments are drawn to a circle from the same exterior point, then they are congruent.

OUTLINE OF PROOF

GIVEN: $\overline{PA}$ and $\overline{PB}$ are tangent to $\odot O$ at points A and B, respectively.

PROVE: $\overline{PA} \cong \overline{PB}$.

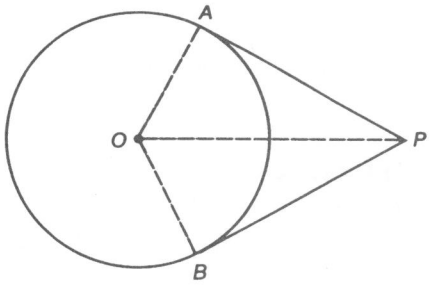

STEPS

- Draw $\overline{OP}$ and radii $\overline{OA}$ and $\overline{OB}$.
  Angles OAP and OBP are right angles.
- Prove $\triangle OAP \cong \triangle OBP$ by Hy-Leg:
  $\overline{OP} \cong \overline{OP}$ (Hy)
  $\overline{OA} \cong \overline{OB}$ (Leg)
- By CPCTC, $\overline{PA} \cong \overline{PB}$.

In the proof of Theorem 13.3, notice that since $\triangle OAP \cong \triangle OBP$, $\measuredangle APO \cong \measuredangle BPO$. In other words, $\overline{OP}$ bisects the angle formed by the two tangent segments.

EXAMPLE 13.3 Find the value of x.

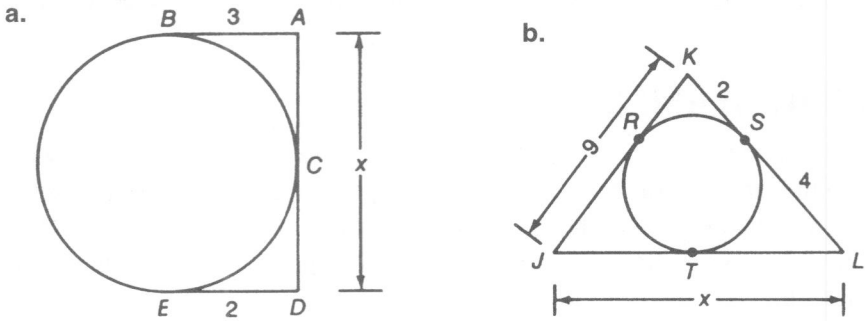

SOLUTION

a. $AC = AB = 3$
$DC = DE = 2$
$x = AC + DC = 5$

b. $KR = KS = 2$
$JR = 9 - KR = 9 - 2 = 7$
$JT = JR = 7$
$LT = LS = 4$
$x = JT + LT = 7 + 4 = 11$

EXAMPLE 13.4 Find the values of x and y.

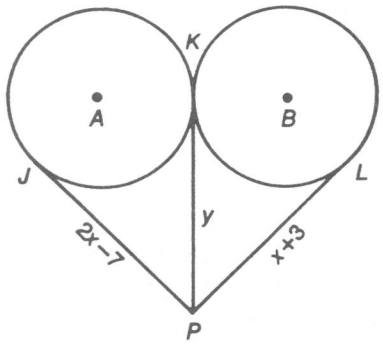

SOLUTION

$PK = PJ$ and $PK = PL$.
Hence, $PJ = PL$.

$$2x - 7 = x + 3$$
$$2x = x + 10$$
$$x = 10$$
$$PL = x + 3 = 13$$
$$y = PK = PL = 13$$

COMMON TANGENTS AND TANGENT CIRCLES

The same line may be tangent to more than one circle (a common tangent), and circles may intersect each other in exactly one point (tangent circles). In order to describe these situations, some terminology must be introduced. In Figure 13.4, line segment $\overline{AB}$ joins the centers of circles A and B and is called the *line of centers* of the two circles. Figure 13.5 illustrates that a line may be tangent to two different circles. A line which has this property is called a *common tangent*. Lines j, k, ℓ, and m are common tangents. In Figure 13.6 two different circles are shown to be tangent to the same line at the same point. These circles are known as *tangent circles*. Circles A and B are tangent to line ℓ at point P. Circles A and C are tangent to line m at point Q.

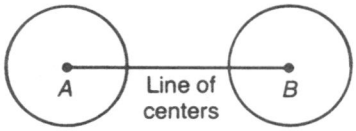

FIGURE 13.4

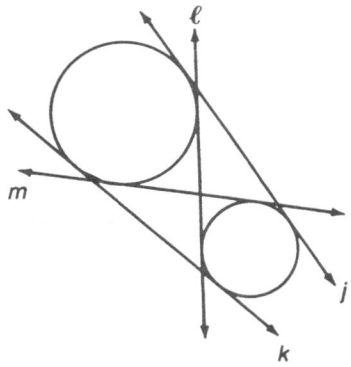

FIGURE 13.5 Common tangents

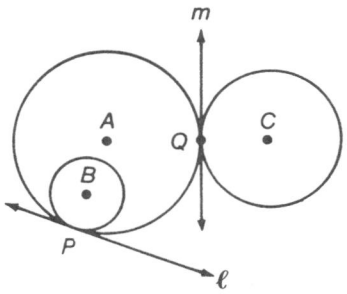

FIGURE 13.6 Tangent circles

We may further distinguish between types of common tangents and types of tangent circles. A common tangent may be either a common *internal* or a common *external* tangent, as shown in Figure 13.7.

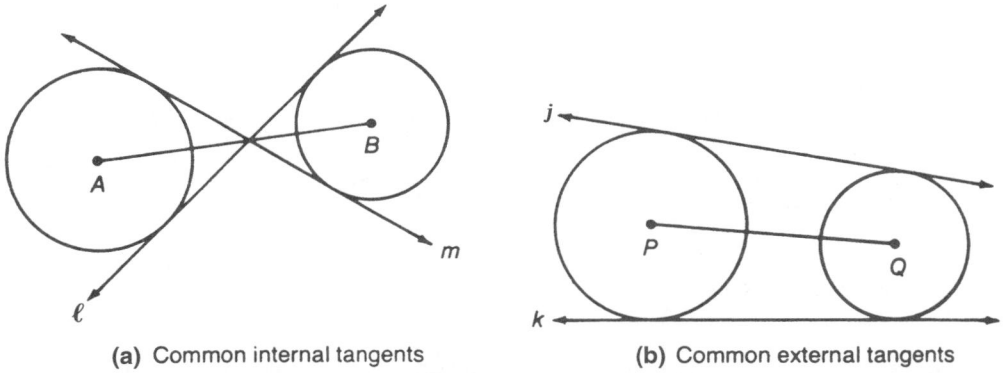

(a) Common internal tangents **(b)** Common external tangents

FIGURE 13.7

In Figure 13.7*a*, lines ℓ and m are common *internal* tangents since each is tangent to both circles and each intersects the line of centers of the two circles. In Figure 13.7*b*, lines j and k are common *external* tangents since each is tangent to both circles and each does *not* intersect the line of centers of the two circles.

Tangent circles may be tangent either *internally* or *externally* to each other, as shown in Figure 13.8. In Figure 13.8*a*, circles A and B are tangent *internally* since they lie on the *same* side of their common tangent. In Figure 13.8*b*, circles P and Q are tangent *externally* since they lie on *opposite* sides of their common tangent.

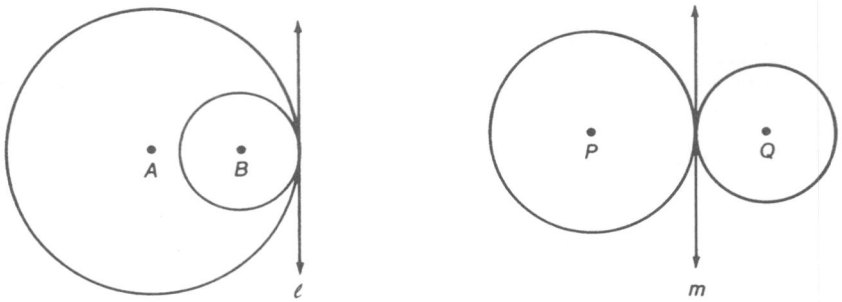

(a) Internally tangent circles (b) Externally tangent circles

FIGURE 13.8

EXAMPLE 13.5 Determine the number of common tangents that can be drawn for each of the following situations:

a. Circle A and circle B intersect in two distinct points.

b. Circle A and circle B are externally tangent circles.

SOLUTION

a.

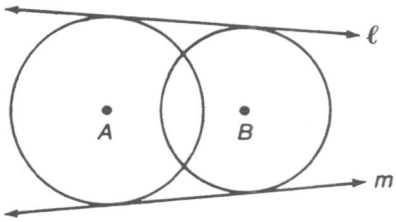

Two common external tangents

b.

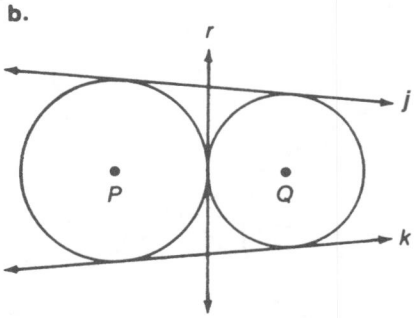

Two common external tangents and one common internal tangent

EXAMPLE 13.6 Prove that common internal tangent segments drawn to two nonintersecting circles are congruent.

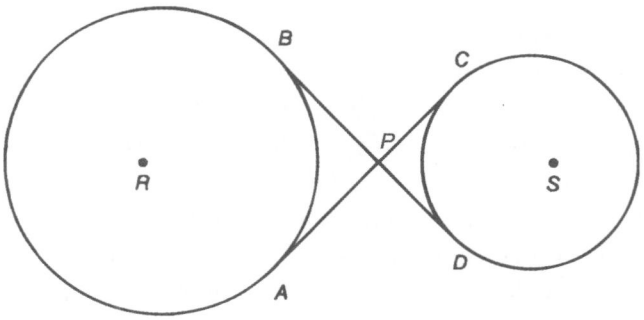

SOLUTION

Draw two nonintersecting circles with their common internal tangent segments shown. Let P represent the point at which the tangent segments intersect.

OUTLINE OF PROOF

GIVEN: Nonintersecting circles R and S, common internal tangent segments $\overline{AC}$ and $\overline{BD}$, intersecting at point P.

PROVE: $\overline{AC} \cong \overline{BD}$.

Apply Theorem 13.3:

$$\begin{aligned} PA &= PB \\ +PC &= PD \\ \hline PA + PC &= PB + PD \end{aligned}$$

Since $AC = PA + PC$ and $BD = PB + PD$, it follows that $\overline{AC} \cong \overline{BD}$.

Similar Triangles and Circles

If two triangles overlap a circle, then some of their angles may have the same arc or congruent arcs. This observation may lead to a pair of congruent angles that can be used to help prove that these triangles are similar.

EXAMPLE 13.7

GIVEN: $\overline{EB}$ is tangent to $\odot O$ at B, $\overline{AB}$ is a diameter, B is the midpoint of $\widehat{CBD}$.

PROVE: $\triangle ABC \sim \triangle AEB$.

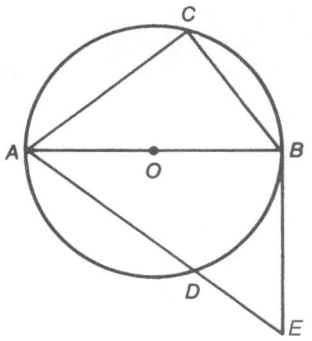

In planning the proof, note in the accompanying figure that:

- Angles ACB and ABE are congruent since they are both right angles.
- Angles CAB and EAB are congruent since they are inscribed angles that intercept congruent arcs ($\widehat{BC} \cong \widehat{BD}$).
- $\triangle ABC \sim \triangle AEB$ by the AA Theorem of Similarity.

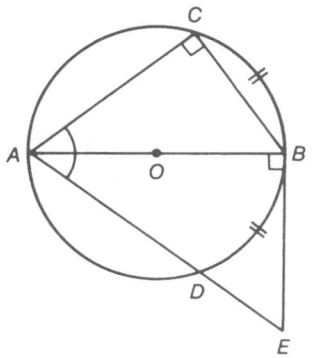

PROOF:

Statements	Reasons
1. B is the midpoint $\widehat{CBD}$.	1. Given
2. $\widehat{BC} \cong \widehat{BD}$.	2. A midpoint of an arc divides the arc into two congruent arcs.
3. $\measuredangle CAB \cong \measuredangle EAB$. (Angle)	3. Inscribed angles of a circle that intercept congruent arcs are congruent.
4. $\measuredangle ACB$ is a right angle.	4. An angle inscribed in a semicircle is a right angle.
5. $\measuredangle ABE$ is a right angle.	5. An angle formed by a radius drawn to the point of tangency is a right angle.
6. $\measuredangle ACB \cong \measuredangle ABE$. (Angle)	6. All right angles are congruent.
7. $\triangle ABC \sim \triangle AEB$.	7. AA Theorem of Similarity.

The properties of similar triangles may be used to establish a special relationship between the segments formed by two chords that intersect in the interior of a circle.

THEOREM 13.4 PRODUCTS OF LENGTHS OF SEGMENTS OF INTERSECTING CHORDS

If two chords intersect in the interior of a circle, then the product of the lengths of the segments of one chord is equal to the product of the lengths of the segments of the other chord.

In the accompanying figure,

$$a \times b = c \times d.$$

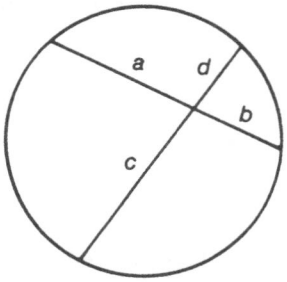

OUTLINE OF PROOF

GIVEN: Chords $\overline{AB}$ and $\overline{CD}$ intersect in the interior of $\odot O$ at point E.

PROVE: $AE \cdot EB = CE \cdot ED$.

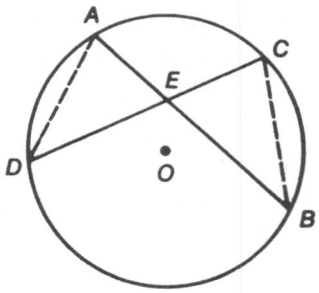

PLAN: Draw $\overline{AD}$ and $\overline{CB}$. Show $\triangle AED \sim \triangle CEB$.

STEPS: $\triangle AED$ is similar to $\triangle CEB$ since:

- $\measuredangle AED \cong \measuredangle BEC$ (Vertical angles)
- $\measuredangle ADC \cong \measuredangle CBA$ (Inscribed angles that intercept the same arc)
- $\dfrac{AE}{CE} = \dfrac{ED}{EB}$ so $AE \cdot EB = CE \cdot ED$

EXAMPLE 13.8 Find the value of x.

a.

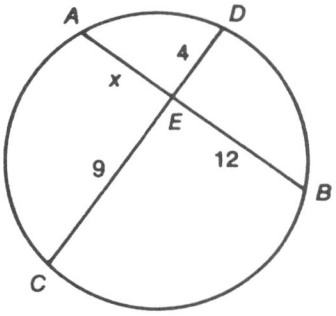

b.

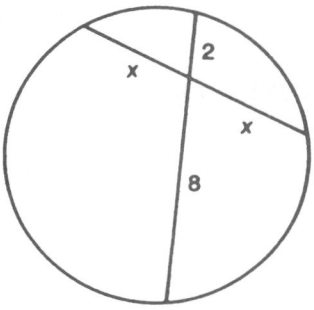

c.

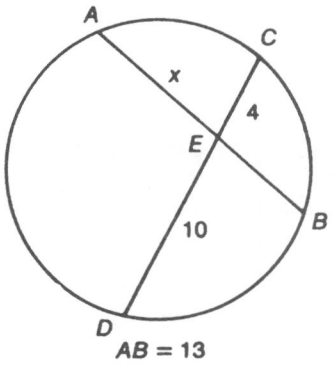

SOLUTION

a. $AE \cdot EB = CE \cdot ED$

$(x)(12) = (4)(9)$

$12x = 36$

$x = \dfrac{36}{12} = 3$

b. $(x)(x) = (8)(2)$

$x^2 = 6$

$x = \sqrt{16} = 4$

c. If $x = AE$, then $13 - x = EB$.

$(x)(13 - x) = (10)(4)$

$13x - x^2 = 40$

Writing the quadratic equation in standard form gives

$$\begin{aligned} x^2 - 13x + 40 &= 0 \\ (x-8)(x-5) &= 0 \\ x - 8 = 0 \quad &or \quad x - 5 = 0 \\ x = 8 \quad &or \quad x = 5 \end{aligned}$$

If $AE = 8$ then $EB = 5$. Alternatively, AE may equal 5, in which case $EB = 8$.

EXAMPLE 13.9 A diameter divides a chord of a circle into two segments whose lengths are 7 and 9. If the length of the shorter segment of the diameter is 3, find the length of a radius of the circle.

SOLUTION

$$\begin{aligned} \text{Let } x &= PB. \\ (3)(x) &= (7)(9) \\ 3x &= 63 \\ x &= \frac{63}{3} = 21 \end{aligned}$$

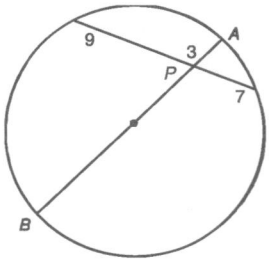

Since diameter $AB = 3 + 21 = 24$, the length of a radius of the circle is $\frac{1}{2}(24)$ or 12.

Tangent- and Secant-Segment Relationships

In Figure 13.9, $\overline{PAB}$ is called a secant *segment*. Its end points are a point in the exterior of the circle (point P) and the point on the circle furthest from point P at which the secant intersects the circle (point B). The circle divides the secant segment into two segments: an *internal* secant segment (chord $\overline{AB}$) and an *external* secant segment ($\overline{AP}$).

When two secant segments are drawn to a circle from the same exterior point, then a special relationship exists between the lengths of the secant segments and the lengths of their external segments.

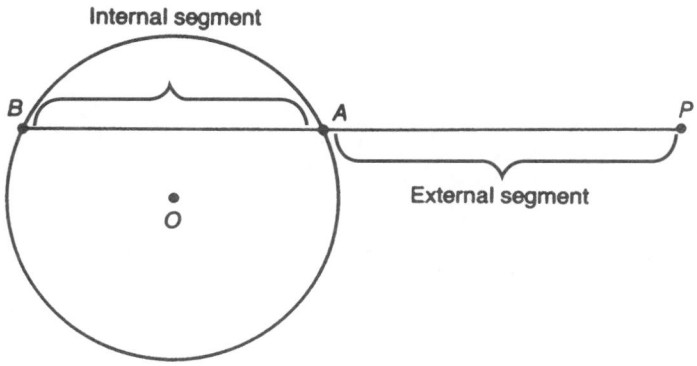

FIGURE 13.9

THEOREM 13.5 SECANT-SECANT SEGMENT PRODUCTS

If two secant segments are drawn to a circle from the same exterior point, then the product of the lengths of one secant segment and its *external* segment is equal to the product of the lengths of the other secant segment and its *external* segment.

In the accompanying figure,

$$s_1 \times e_1 = s_2 \times e_2.$$

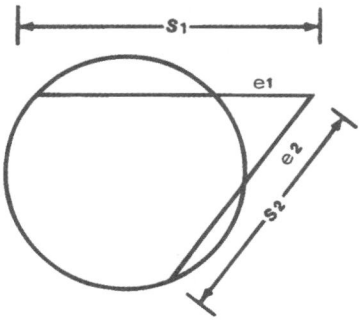

OUTLINE OF PROOF

GIVEN: $\overline{PB}$ and $\overline{PD}$ are secant segments drawn to $\odot O$.

PROVE: $PB \cdot PA = PD \cdot PC$.

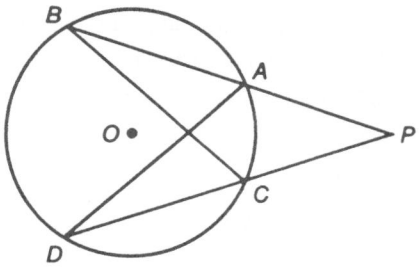

STEPS:

- Draw $\overline{AD}$ and $\overline{CB}$.
- $PB \cdot PA = PD \cdot PC$, so $\dfrac{PB}{PD} = \dfrac{PC}{PA}$. Hence, show $\triangle PBC \sim \triangle PDA$.
- $\triangle PBC \sim \triangle PDA$ since $\measuredangle P \cong \measuredangle P$ and $\measuredangle PBC \cong \measuredangle PDA$

EXAMPLE 13.10 Find the value of x.

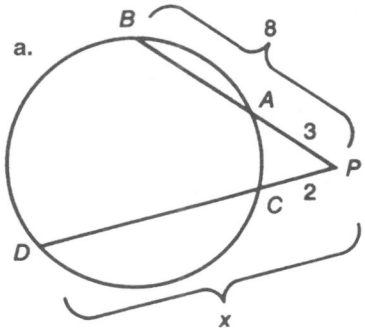

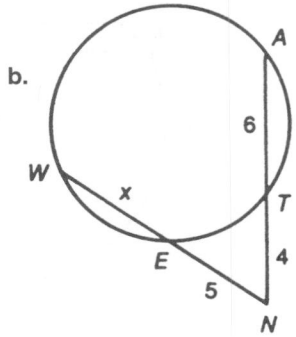

SOLUTION

a. $PA \cdot PB = PC \cdot PD$

$$3 \cdot 8 = 2 \cdot x$$
$$24 = 2x$$
$$12 = x$$

b. $NE \cdot NW = NT \cdot NA$, where

$$NE = 5$$
$$NW = x + 5$$
$$NT = 4$$
$$NA - 6 + 4 = 40$$
$$5(x + 5) = 4 \cdot 10$$
$$5x + 25 = 40$$
$$5x = 15$$
$$x = 3$$

When a *tangent* segment and a secant segment are drawn to a circle from the same exterior point, a relationship analogous to the one stated in Theorem 13.5 results. Figure 13.10*a* illustrates that, when secant $\overline{PAB}$ is rotated clockwise, it will eventually be tangent to the circle as in Figure 13.10*b*. This analysis suggests that we replace the secant-segment length with tangent-segment length $\overline{PA}$ and also replace the external portion of the secant segment with tangent-segment length $\overline{PA}$.

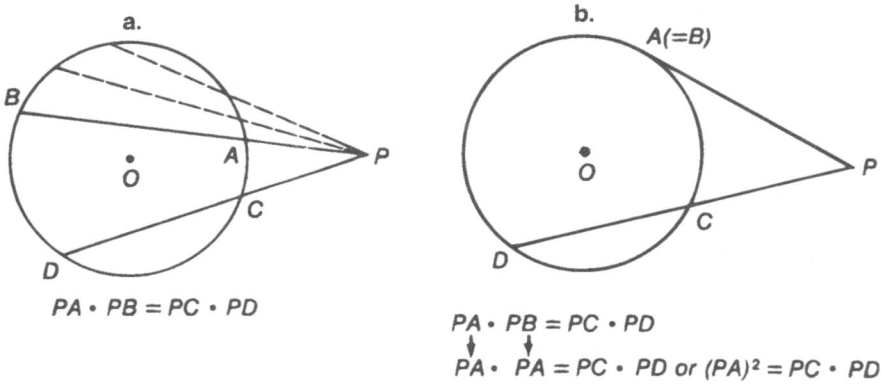

FIGURE 13.10

THEOREM 13.6 TANGENT-SECANT SEGMENT PRODUCTS

If a tangent segment and a secant segment are drawn to a circle from the same exterior point, then the square of the length of the tangent segment is equal to the product of the lengths of the secant segment and its external segment.

In the accompanying figure,

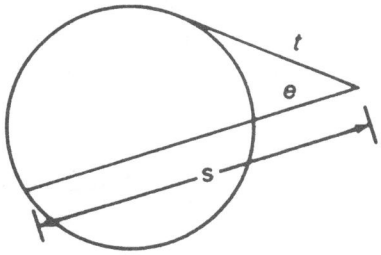

$$t^2 = s \times e.$$

OUTLINE OF PROOF

GIVEN: $\overline{PA}$ is tangent to $\odot O$ at A and $\overline{PC}$ is a secant.

PROVE: $(PA)^2 = PC \cdot PB$.

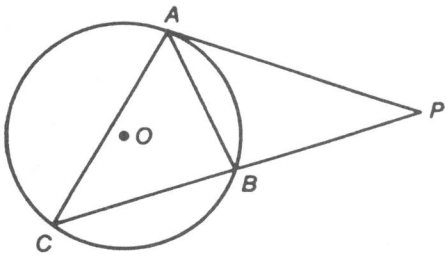

STEPS:

- Draw $\overline{CA}$ and $\overline{BA}$.
- $(PA)^2 = PC \cdot PB$, so $\dfrac{PA}{PC} = \dfrac{PB}{PA}$. Hence, show $\triangle PAB \sim \triangle PCA$.
- $\triangle PAB \sim \triangle PCA$ since:

$$\measuredangle P \cong \measuredangle P$$
$$\measuredangle PCA \cong \measuredangle PAB$$

(NOTE: Both angles are measured by $\frac{1}{2}\, m\widehat{AB}$.)

EXAMPLE 13.11 Find the value of x.

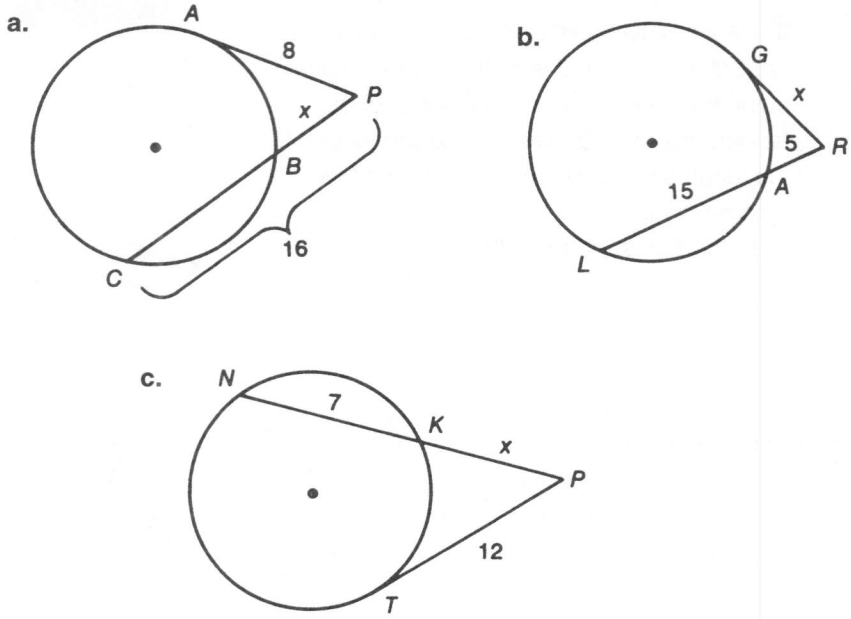

SOLUTION

a.
$$(PA)^2 = PB \cdot PC$$
$$(8)^2 = x \cdot 16$$
$$64 = 16x$$
$$4 = x$$

b.
$$(RG)^2 = RA \cdot RL$$
$$x^2 = 5(15 + 5)$$
$$= 5 \cdot 20$$
$$= 100$$
$$x = \sqrt{100} = 10$$

c.
$$(PT)^2 = PK \cdot PN$$
$$(12)^2 = x(x + 7)$$
$$144 = x^2 + 7x$$

Writing the quadratic equation in standard form gives

$$x^2 + 7x - 144 = 0 \qquad or \qquad x - 9 = 0$$
$$(x + 16)(x - 9) = 0$$
$$x + 16 = 0$$
$$x = -16 \qquad\qquad x = 9$$

Reject since a length cannot be a negative number.

Circumference and Arc Length

The distance around a polygon is called the *perimeter* of the polygon. The perimeter of a circle is given a special name, *circumference*.

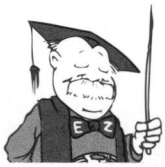

DEFINITION OF CIRCUMFERENCE

The *circumference* of a circle is the distance around the circle, expressed in linear units of measurement (e.g., inches, centimeters, feet).

If a wheel is rolled along a flat surface, as in Figure 13.11, so that the surface is again tangent to the same point on the circle, then the distance the wheel travels along the surface has the same numerical value as the circumference of the circle.

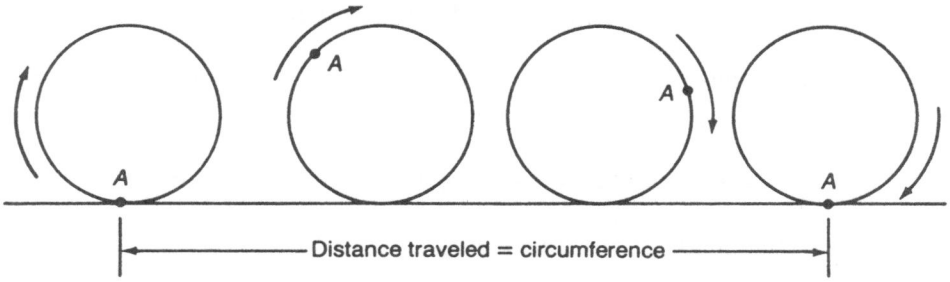

FIGURE 13.11

REMEMBER
$\pi \approx 3.14$, or $\frac{22}{7}$.

The longer the diameter of a circle, the larger the circle's circumference. Interestingly, however, if the circumference of a circle is divided by the length of its diameter, then the value obtained will be the same regardless of the size of the circle:

$$\frac{\text{Circumference}}{\text{Diameter}} = 3.1415926\ldots$$

The three dots that follow the decimal number indicate that this value is a nonterminating (never-ending) decimal number. This constant value is referred to as *pi* and is denoted by the Greek letter π. We may therefore write

$$\frac{\text{Circumference}}{\text{Diameter}} = \pi \quad \text{or} \quad \text{Circumference} = \pi \cdot \text{Diameter}$$

where π is *approximately equal to* 3.14 or *approximately equal to* the improper fraction $\frac{22}{7}$.

THEOREM 13.7 CIRCUMFERENCE OF A CIRCLE

The circumference of a circle is equal to the product of π and the length of its diameter: $C = \pi D$.

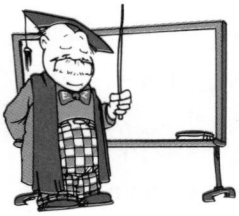

When working with the constant π, note that:

- $\pi \approx 3.14$ and $\pi \approx \frac{22}{7}$, where the symbol $\approx$ is read as "is approximately equal to."
- Since the length of a diameter is numerically equal to twice the length of the radius, we may write

$$\text{Circumference} = \pi \cdot 2 \cdot \text{radius}$$

In writing a formula that includes a number, it is common practice to place the digit (2 in this case) before any symbols. We therefore write

$$\textit{C}\text{ircumference} = \pi \cdot 2 \cdot \textit{r}\text{adius} \qquad \text{or} \qquad C = 2\pi r$$

- The approximation to be used for π (3.14 or $\frac{22}{7}$) is usually specified in the statement of the problem. Frequently, you will be asked *not* to substitute an approximation for π, and instead to express the answer in terms of π, as illustrated in Example 13.12, part **c**.

EXAMPLE 13.12 Find the circumference of a circle if:

a. Diameter = 10 (Use $\pi = 3.14$.)

b. Radius = 14 (Use $\pi = \frac{22}{7}$.)

c. Diameter = 29 (Express answer in terms of π.)

d. Radius = 21 (Round answer to *nearest hundredth*.)

SOLUTION

a. $C = \pi D$
$= 3.14 \times 10 = 31.4$

b. $C = 2\pi r$

$$= 2 \times \frac{22}{\cancel{7}_{1}} \times \frac{\cancel{14}^{2}}{1} = 88$$

c. $C = \pi D$
$= \pi 29 \qquad \text{or} \qquad 29\pi$

d. Use the calculator's stored value of pi by pressing the pi key and multiplying it by 2 times the given value of r:

$\boxed{2}\boxed{\times}\boxed{\pi}\boxed{\times}\boxed{21}\boxed{=}$ $\underbrace{131.9468915}_{\text{Round off}}$

The area, to the *nearest hundredth*, is 131.95.

EXAMPLE 13.13 Rectangle $ABCD$ has a width of 5 and a length of 12 and is inscribed in circle O. Find the circumference of the circle.

SOLUTION

Since $\measuredangle BAD$ is a right angle, diagonal $\overline{BD}$ must coincide with a diameter of the circle. Right $\triangle BAD$ is a 5-12-*13* right triangle where diagonal $BD = 13$. Hence,

$$C = \pi D = 13\pi$$

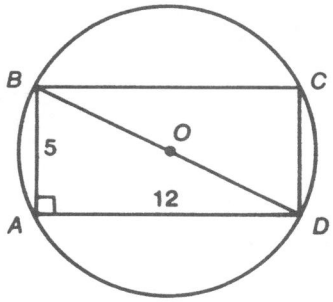

Circumference represents the distance around the *entire* circle. Our next concern is how to determine the length (in linear units) of an *arc* of the circle. Since a circle contains 360°, the circumference of a circle represents the length of a 360° arc of the circle (Figure 13.12). The ratio of the length of an arc to the circumference must be equal to the ratio of the degree measure of the arc to 360°:

$$\frac{\text{Arc measurement}}{\text{Circle measurement}} = \frac{\text{arc length}}{2\pi r} = \frac{n^\circ}{360^0}$$

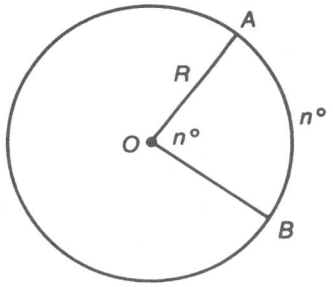

FIGURE 13.12

If you consider the circle to be made of string, then the length of arc AB corresponds to the number arrived at by taking the section of the string from A to B, stretching it out, and then using a ruler to measure its length.

THEOREM 13.8 ARC LENGTH PROPORTION

$$\frac{\text{Length of arc}}{\text{Circumference}} = \frac{\text{degree measure of arc}}{360^\circ}$$

EXAMPLE 13.14 In a circle having a radius of 10, find the length of an arc whose degree measure is 72°. (Leave answer in terms of π.)

SOLUTION

$$\frac{\text{Length of arc}}{\text{Circumference}} = \frac{\text{degree measure of arc}}{360°}$$

$$\frac{L}{2\pi \cdot 10} = \frac{72°}{360°}$$

Simplify each ratio *before* cross-multiplying:

$$\frac{L}{20\pi} = \frac{1}{5}$$

Cross-multiply:

$$5L = 20\pi$$

$$L = 4\pi$$

EXAMPLE 13.15 In a circle a 40° arc has a length of 8π. Determine the radius of the circle.

SOLUTION

$$\frac{\text{Length of arc}}{\text{Circumference}} = \frac{\text{degree measure of arc}}{360°}$$

$$\frac{8\pi}{2\pi r} = \frac{40°}{360°}$$

Simplify each ratio *before* cross-multiplying:

$$\frac{4}{r} = \frac{1}{9}$$

Cross-multiply:

$$r = 36$$

EXAMPLE 13.16 Right triangle ABC is inscribed in circle O so that $\overline{AB}$ is a diameter and has a length of 27. If $m\measuredangle CAB = 50$, find the length of $\widehat{AC}$, expressed in terms of π.

SOLUTION

In order to find the degree measure of $\widehat{AC}$, draw $\widehat{OC}$ and determine the measure of central angle AOC. Since $\overline{OA} \cong \overline{OC}$ $m\measuredangle OCA = m\measuredangle CAB = 50$. The fact that the sum of the angles of a triangle is 180 means that $m\measuredangle AOC = 80$. Since a central angle and its intercepted arc have the same measure, $m\widehat{AC} = 80$.

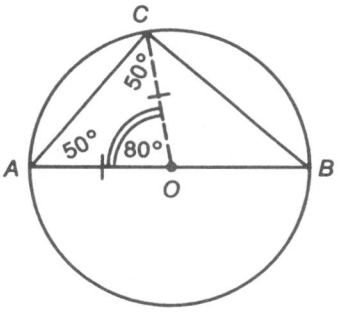

$$\frac{\text{Length of arc}}{\text{Circumference}} = \frac{\text{degree measure of arc}}{360°}$$

$$\frac{L}{27\pi} = \frac{80°}{360°}$$

$$= \frac{2}{9}$$

$$9L = 54\pi$$

$$L = 6\pi$$

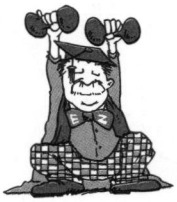

REVIEW EXERCISES FOR CHAPTER 13

1. For each of the given situations, determine the number of common internal tangents and the number of common external tangents that can be drawn.

   (a) Circle P lies in the exterior of circle Q and has no points in common with circle Q.
   (b) Circle P lies in the interior of circle Q and has no points in common with circle Q.
   (c) Circle P and circle Q are tangent internally.

2. Find the value of x.

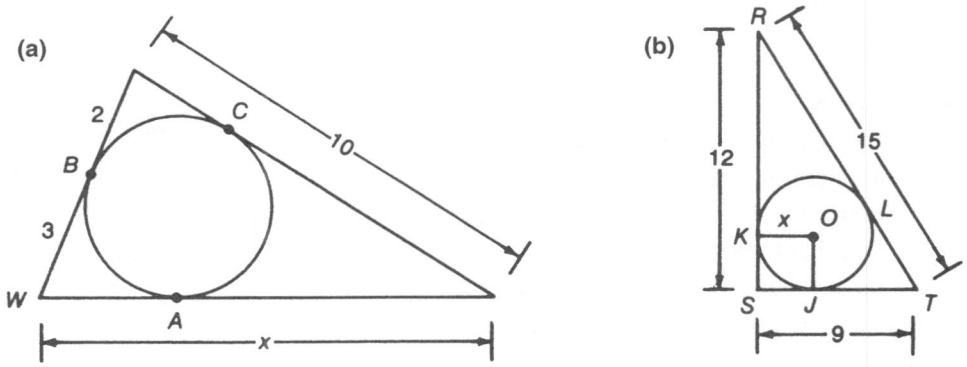

3. In the accompanying diagram of circle O, chords $\overline{AB}$ and $\overline{CD}$ intersect at E. If $AE = 3$, $EB = 4$, $CE = x$, and $ED = x - 4$, what is the value of x?

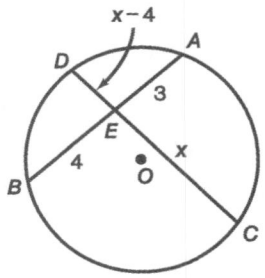

Use the accompanying figure for Exercises 4 to 6.

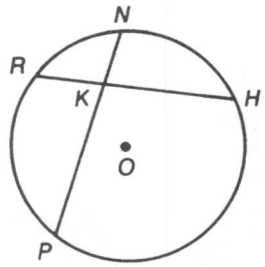

4. If $PK = 27$, $KN = 3$, and K is the midpoint of RH, find RK.

5. If $RH = 16$, $RK = 4$, $PK = 8$, find PN.

6. If $RH = 22$, $PK = 7$, and $KN = 3$, find RK.

Use the accompanying figure for Exercises 7 to 11.

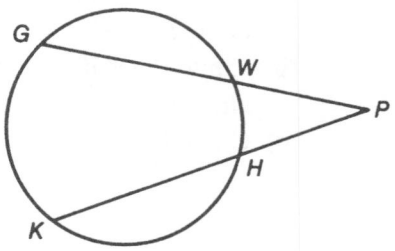

7. If $PW = 5$, $PG = 8$, $PH = 2$, find KH.

8. If $GW = 7$, $PW = 3$, $PK = 15$, find PH.

9. If W is the midpoint of $\overline{GP}$ and $PH = 5$, $KH = 35$, find PG.

10. If $PW = 6$, $WG = 9$, $PH = 9$, find KH.

11. If $GW = 11$, $PH = 8$, $KH = 2$, find PW.

Use the accompanying figure for Exercises 12 to 14.

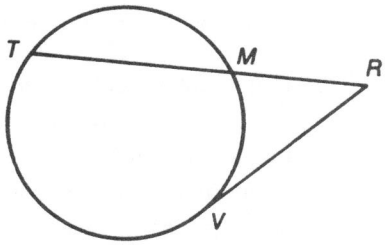

12. If $RV = 9, RM = 3$, find RT.

13. If $MT = 24, RM = 1$, find RV.

14. If $RV = 8, RM = 4$, find MT.

15. $RE = 2, RA = 14.5, ZG = 6, ZF = 8, SF = x$, E is the midpoint of $\overline{BG}$. Find SZ.

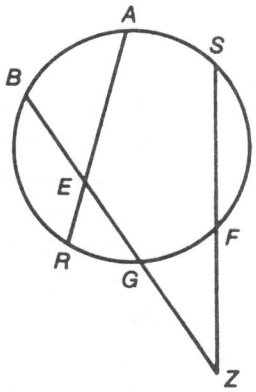

16. (a) If $NA = 4, JA = 8, WA = 11$, find OK.
    (b) If A is the midpoint of $\overline{JW}$, $AK = 32$, $NA = 18$, find JW and OA.
    (c) $JA = 12, AW = 9$, and AK is 3 times the length of $\overline{NA}$. Find OK.

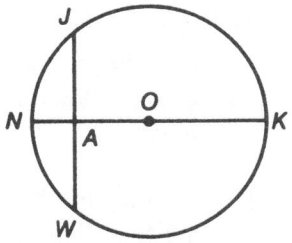

17. A circle divides a secant segment into an internal segment having a length of 8 and an external segment having a length of 2. Find the length of a tangent segment drawn to the circle from the same exterior point.

18. From a point 2 units from a circle, a tangent segment is drawn. If the radius of the circle is 8, find the length of the tangent segment.

Unless otherwise specified, answers to Exercises 19 to 25 may be left in terms of π.

19. Find the circumference of a circle that is inscribed in a square whose side is 9.

20. Find the circumference of a circle that is circumscribed about a square whose perimeter is 36.

21. A square having a side of 12 inches is inscribed in a circle. Find the length of an arc of the circle intercepted by one of the square's sides.

22. Kim wants to determine the radius of a circular pool without getting wet. She is located at point K, which is 4 feet from the pool and 12 feet from the point of tangency, as shown in the accompanying diagram. What is the radius of the pool?

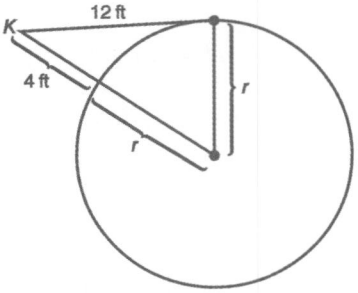

23. The accompanying diagram shows a child's spin toy that is constructed from two chords intersecting in a circle. The curved edge of the larger shaded section is one-quarter of the circumference of the circle, and the curved edge of the smaller shaded section is one-fifth of the circumference of the circle. What is the measure of angle x?

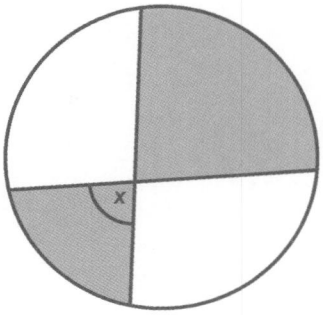

24. A point P on a bicycle wheel is touching the ground. The next time point P hits the ground, the wheel has rolled 3.5 feet in a straight line along the ground. What is the radius of the wheel correct to the *nearest tenth of an inch*?

25. To measure the number of miles in a hiking trail, a worker uses a device with a 2-foot diameter wheel that counts the number of revolutions the wheel makes. If the device reads 0 revolution at the beginning of the trail and 2300 revolutions at the end of the trail, how many miles long, correct to the *nearest tenth of a mile*, is the trail?

26. GIVEN: In $\odot O$, $\overline{MR}$ is a diameter, $\overline{OA} \parallel \overline{MS}$, $\overline{OB} \parallel \overline{MT}$, $\overline{AR} \cong \overline{BR}$.
PROVE: $\overline{SR} \cong \overline{TR}$.

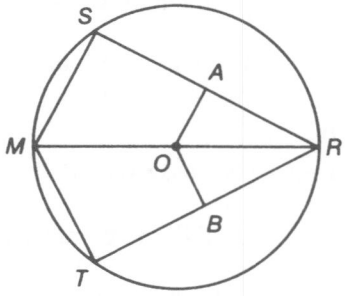

27. In the accompanying diagram of circle O, diameter $\overline{AOB}$ is drawn, tangent $\overline{CB}$ is drawn to the circle at B, E is a point on the circle, and $\overline{BE} \parallel \overline{ADC}$. Prove: $\triangle ABE \sim \triangle CAB$.

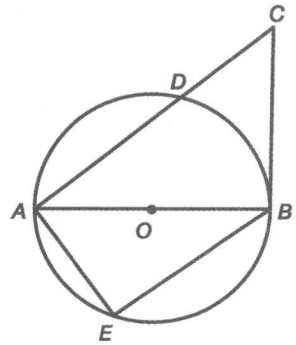

28. GIVEN: In $\odot O$, quadrilateral $OXEY$ is a square.
PROVE: $\widehat{QP} \cong \widehat{JT}$.

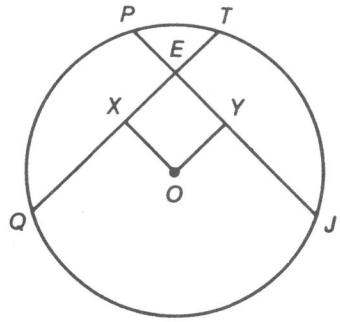

29. GIVEN: $\triangle HBW$ is inscribed in $\odot O$, tangent segment $\overline{AB}$ is tangent at point B, $ABLM$ is a parallelogram.
PROVE: $\dfrac{BL}{BW} = \dfrac{BM}{BH}$.

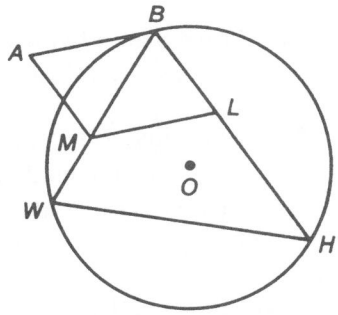

30. GIVEN: In $\odot P$, $\overline{AB}$ is a diameter, $\overline{DB}$ is tangent to $\odot P$ at B, $\overline{CD} \perp \overline{DB}$.
PROVE: BC is the mean proportional between CD and AB.

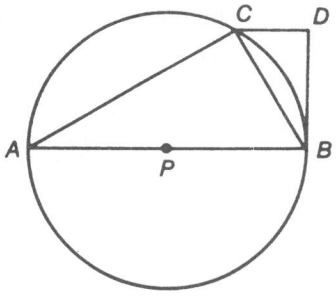

Use the accompanying figure for Exercises 31 and 32.

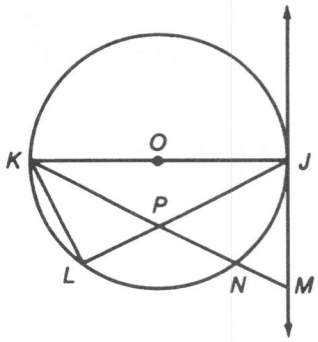

31. GIVEN: In $\odot O$, $\overline{KJ}$ is a diameter, $\overline{MJ}$ is tangent at point J, N is the midpoint of $\widehat{LNJ}$.
PROVE: $KL:KJ = KP:KM$.

32. GIVEN: In $\odot O$, $\overline{KJ}$ is a diameter, $\overline{MJ}$ is tangent at point J, $\overline{JP} \cong \overline{JM}$.
PROVE: $KL \cdot KM = JK \cdot LP$.

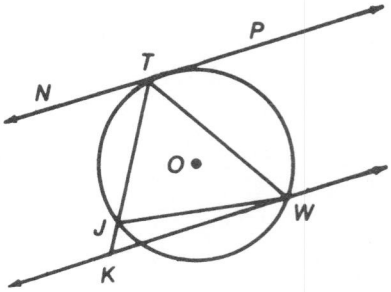

33. GIVEN: $\overline{TK}$ bisects $\measuredangle NTW$, $\overline{WK} \cong \overline{WT}$.
PROVE:
(a) $\overleftrightarrow{NTP} \parallel \overleftrightarrow{KW}$.
(b) $(TW)^2 = JT \cdot TK$.

34. Prove that, if two circles are tangent externally, then the common internal tangent bisects a common external tangent.

35. Prove that, if two circles do not intersect, with one circle lying in the exterior of the other, then their common external tangent segments are congruent. (HINT: Consider as separate cases the situations in which the two circles are congruent and not congruent.)

36. Lines ℓ and m are parallel and each is tangent to circle P. Prove that, if line k is also tangent to circle P, then the segments determined by the points at which line k intersects lines ℓ and m and point P intersect at right angles at P.

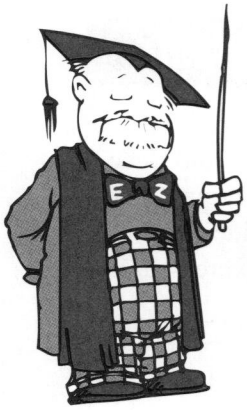

14 Area and Volume

WHAT YOU WILL LEARN

There are convenient formulas for finding the areas and volumes of certain types of plane and solid figures. In this chapter you will learn:

- how to find the areas of triangles, trapezoids, and different types of parallelograms;
- how to compare the areas of similar figures;
- how to find the area of a regular polygon;
- how to find the areas of a circle and a sector;
- how to find the area of a region bounded by a chord and its arc;
- how to find the volumes of familiar types of solid figures.

SECTIONS IN THIS CHAPTER

- Areas of a Rectangle, Square, and Parallelogram
- Areas of a Triangle and Trapezoid
- Comparing Areas
- Area of a Regular Polygon
- Areas of a Circle, Sector, and Segment
- Geometric Solids

Areas of a Rectangle, Square, and Parallelogram

The term *area of a figure* refers to the number of square boxes that the figure can enclose. If a figure can enclose a total of 30 square boxes, then its area is said to be 30 square *units*. If the length of a side of each square box is 1 centimeter (cm), then the area of the figure is 30 square centimeters (cm^2). If the length of a side of each square box is 1 inch, then the area of the figure under consideration is expressed as 30 square inches.

Figure 14.1 shows a rectangle that encloses a total of 36 square boxes. The area of this rectangle is 36 square units.

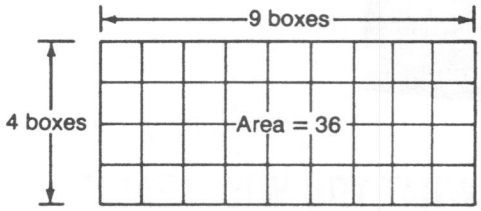

FIGURE 14.1

The value of 36 can be arrived at by multiplying the number of boxes along the length by the number of boxes along the width: Area = $9 \times 4 = 36$. One of our first goals in this chapter is to develop formulas that will enable us to calculate the areas of figures that we have previously investigated: rectangle, square, parallelogram, triangle, rhombus, trapezoid, regular polygon, and circle. We begin by summarizing some fundamental area postulates.

AREA POSTULATES

POSTULATE 14.1 For any given closed region and unit of measurement, there is a positive number that represents the area of the region.

POSTULATE 14.2 If two figures are congruent, then they have equal areas.

POSTULATE 14.3 The area A of a rectangle is equal to the product of its length (ℓ) and width (w).

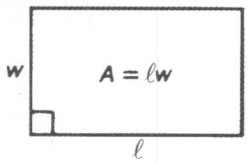

Sometimes one side of a rectangle is referred to as the *base* and an adjacent side (which is perpendicular to the base) is called the *altitude* or *height*. The formula for the area of a rectangle may be expressed as $A = bh$, where b represents the length of the base and h represents the height. The terms *base* and *height* will recur in our investigations of the areas of other figures. In each instance, the height or altitude will always be a segment that is perpendicular to the side that is specified to be the base. Also, when the terms *base* and *altitude* are used in connection with area, we will always understand these terms to mean the *length* of the base and the *length* of the altitude.

EXAMPLE 14.1 Find the area of a rectangle if its base is 12 centimeters and its diagonal has a length of 13 centimeters.

SOLUTION

Triangle *BAD* is a 5-12-13 right triangle where 12 is the base and 5 is the height. Hence,

$$\begin{aligned} A &= bh \\ &= 12 \times 5 \\ &= 60 \text{ cm}^2 \end{aligned}$$

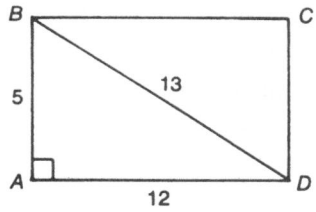

A *square* is a rectangle so that the area formula for to a rectangle also applies to a square. See Figure 14.2. If the length of a side of a square is represented by s, then the area of a square is given by the relationship $A = s \times s$ or $A = s^2$.

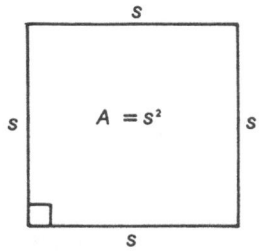

FIGURE 14.2

EXAMPLE 14.2 A certain rectangle and square are *equivalent* (have the same area). The base of the rectangle exceeds three times its altitude by 4. If the length of a side of the square is 8, find the dimensions of the rectangle.

SOLUTION

Let x = altitude of rectangle

Then $3x + 4$ = base of rectangle.

Area of rectangle = Area of square

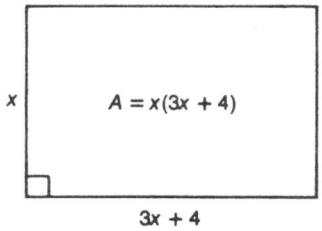

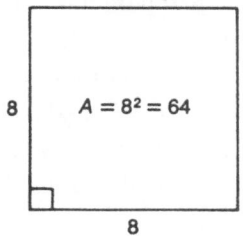

$$\begin{aligned} x(3x+4) &= 64 \\ 3x^2 + 4x &= 64 \\ 3x^2 + 4x - 64 &= 0 \\ (3x+16)(x-4) &= 0 \end{aligned}$$

$3x + 16 = 0$ *or* $x - 4 = 0$

$3x = -16$

$\dfrac{3x}{3} = -\dfrac{16}{3}$

Reject this solution since x (the length of a side) cannot be negative.

x = altitude = 4

$3x + 4$ = base $= 3(4) + 4$

$= 16$

It will sometimes be convenient to subdivide a region into component regions. For example, in Figure 14.3, diagonals $\overline{AC}$ and $\overline{AD}$ separate pentagon $ABCDE$ into three triangular regions such that: Area pentagon $ABCDE$ = area $\triangle ABC$ + area $\triangle ACD$ + area $\triangle ADE$.

The generalization of this notion is presented as Postulate 14.4.

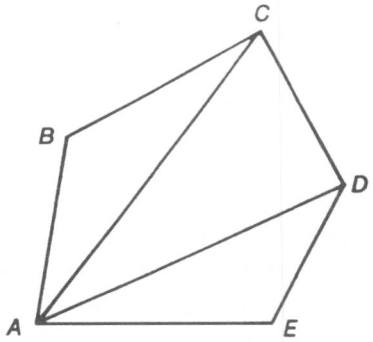

FIGURE 14.3

POSTULATE 14.4 AREA ADDITION

The area of a closed region is equal to the sum of the areas of any non-overlapping divisions of that region.

A rectangle is a special type of parallelogram. It therefore seems reasonable that there may be some relationship between the area formulas for these two figures. We will seek to establish that a parallelogram has the same area as a rectangle having the same base and altitude as the parallelogram. We begin our analysis by relating the concept of base and altitude to a parallelogram.

Any side of a parallelogram may be identified as the *base* of the parallelogram. The *height* or *altitude* of a parallelogram is the length of a segment drawn perpendicular to the base from any point on the side opposite the base.

In Figure 14.4, $\overline{SJ}$, $\overline{AK}$, and $\overline{BL}$ are examples of altitudes, each drawn to base $\overline{RW}$. Since parallel lines are everywhere equidistant, altitudes drawn to a given base of a parallelogram are equal, that is, $SJ = AK = BL = \ldots$.

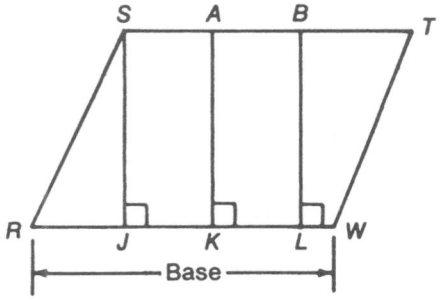

FIGURE 14.4

Next consider ▱$ABCD$ with altitude $\overline{BH}$ drawn to base $\overline{AD}$ (Figure 14.5). If right triangle AHB were cut off the figure and then slid over to the right so that $\overline{AB}$ and $\overline{CD}$ were made to coincide, then the resulting figure would be a rectangle. This is illustrated in Figure 14.6, where right angle DKC corresponds to right angle AHB.

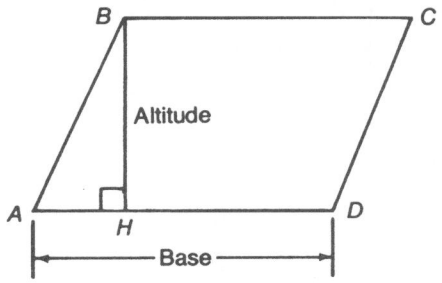

FIGURE 14.5

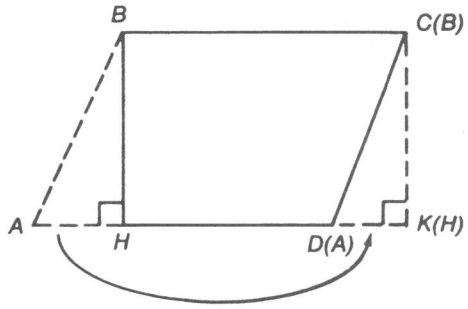

FIGURE 14.6

The area of rectangle $BHKC$ (see Figure 14.6) is given by the relationship $A = HK \times BH$, where $\overline{BH}$ is the altitude and $\overline{HK}$ is the base of the rectangle. Since we have not thrown away or created any additional area, the area of the original parallelogram equals the area of the newly formed rectangle. Hence, the area of ▱$ABCD$ also equals $HK \times BH$. Since $\triangle AHB \cong \triangle DKC$, $\overline{AH} \cong \overline{DK}$. This implies that $AD = HK$. Substituting AD for HK in the area relationship we obtain the following:

$$\begin{aligned} \text{Area of } ▱ABCD &= HK \times BH \\ &= AD \times HK \end{aligned}$$

Note that $\overline{AD}$ is the base of the parallelogram and $\overline{BH}$ is an altitude drawn to that base. This relationship is stated formally in Theorem 14.1.

THEOREM 14.1 AREA OF A PARALLELOGRAM

The area of a parallelogram is equal to the product of the lengths of the base and the altitude drawn to that base.

Thus, $A = bh$.

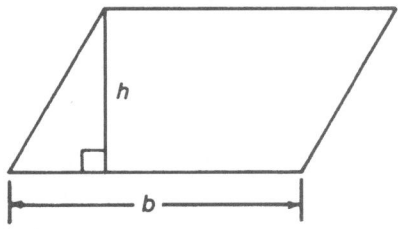

EXAMPLE 14.3 A pair of adjacent sides of a parallelogram are 6 and 10 centimeters in length. If the measure of their included angle is 30, find the area of the parallelogram.

SOLUTION

In ▱$ABCD$, altitude $BH = 3$ since the length of the side opposite a 30° angle in a 30-60 right triangle is one-half the length of the hypotenuse (side $\overline{AB}$).

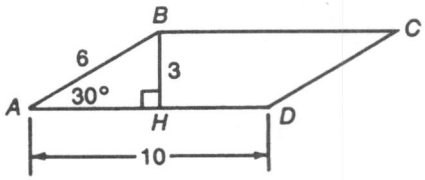

$$\begin{aligned}\text{Area of } ▱ABCD &= bh \\ &= AD \times BH \\ &= 10 \times 3 \\ &= 30 \text{ cm}^2\end{aligned}$$

Areas of a Triangle and Trapezoid

The formula to find the area of a triangle can be derived from the area of a parallelogram relationship. Then, once we know how to find the area of a triangle, we can develop a formula to find the area of a trapezoid.

Recall that a diagonal separates a parallelogram into two congruent triangles. Suppose that the area of a parallelogram is 60 square units. If a diagonal of the parallelogram is drawn (Figure 14.7), what will be the area of each triangle? Since congruent triangles have equal areas, each triangle will have an area of 30 square units. Thus, the area of each triangle is one-half the area of the parallelogram.

$$\begin{aligned}\text{Area of } \triangle ABD &= \frac{1}{2} \text{ area } ▱ABCD \\ &= \frac{1}{2} \text{ base} \times \text{height} \\ &= \frac{1}{2} AD \times BH\end{aligned}$$

This analysis suggests Theorem 14.2.

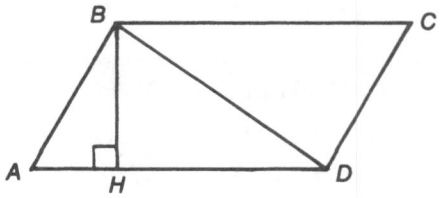

FIGURE 14.7

THEOREM 14.2 AREA OF A TRIANGLE

The area of a triangle is equal to one-half the product of the lengths of the base and the altitude drawn to that base.

Thus, $A = \frac{1}{2}bh.$

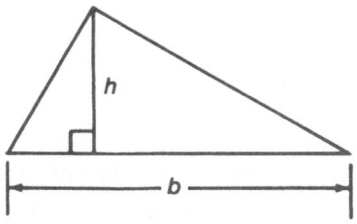

EXAMPLE 14.4 Find the area of each of the following triangles.

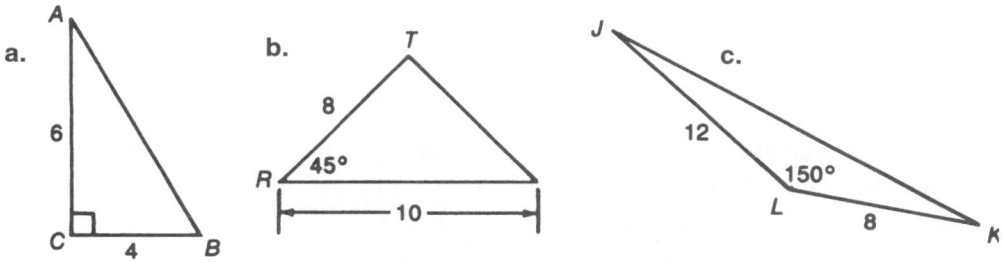

SOLUTION

a. In a right triangle either leg may be considered the base while the other leg then becomes the altitude. The area of a *right* triangle is therefore equal to one-half the product of the lengths of the legs:

$$\begin{aligned} A &= \frac{1}{2}(\text{leg}_1)(\text{leg}_2) \\ &= \frac{1}{2}(6)(4) \\ &= 12 \text{ square units} \end{aligned}$$

b. From vertex T, drop an altitude to base $\overline{RS}$. Since $\overline{TH}$ is the leg opposite a 45° angle in a 45-45 right triangle, the length of $\overline{TH}$ is equal to one-half the length of the hypotenuse ($\overline{RT}$) multiplied by $\sqrt{2}$: $TH = \frac{1}{2}(8)\sqrt{2} = 4\sqrt{2}$.

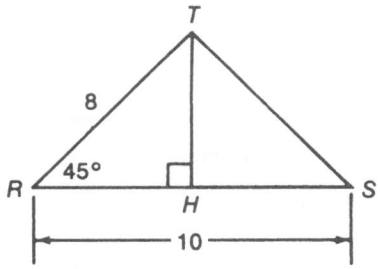

$$\begin{aligned} A &= \frac{1}{2}(RS)(TH) \\ &= \frac{1}{2}(10)(4\sqrt{2}) \\ &= 5(4\sqrt{2}) \\ &= 20\sqrt{2} \text{ square units} \end{aligned}$$

c. From vertex J, drop an altitude to base $\overline{LK}$, extended.

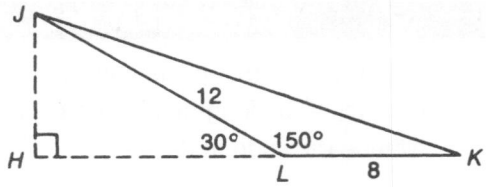

Angle JLH has measure 30 so that the length of altitude $\overline{JH}$ (the leg opposite the 30° angle in a 30-60 right triangle) is equal to one-half the length of the hypotenuse ($\overline{JL}$).

Hence, $JH = \frac{1}{2}(12) = 6$.

$$\begin{aligned} A &= \frac{1}{2}(LK)(JH) \\ &= \frac{1}{2}(8)(6) \\ &= 24 \text{ square units} \end{aligned}$$

To find the area of a *trapezoid*, we draw a diagonal so that the trapezoid is divided into two triangles. The area of the trapezoid is given by the sum of the areas of the two triangles. In Figure 14.8, the lengths of perpendicular segments $\overline{BX}$ and $\overline{DY}$ both represent the distance between bases $\overline{AD}$ and $\overline{BC}$. Since parallel lines are everywhere equidistant, $DY = BX$. For convenience, we will refer to the length of the altitude by the letter h so that $h = DY = BX$. We may then write

$$\begin{aligned} \text{Area trapezoid } ABCD &= area\ \triangle ABD + area\ \triangle BCD \\ &= \frac{1}{2}(AD)(h) \quad + \frac{1}{2}(BC)(h) \\ &= \frac{1}{2}h(AD \quad + BC) \end{aligned}$$

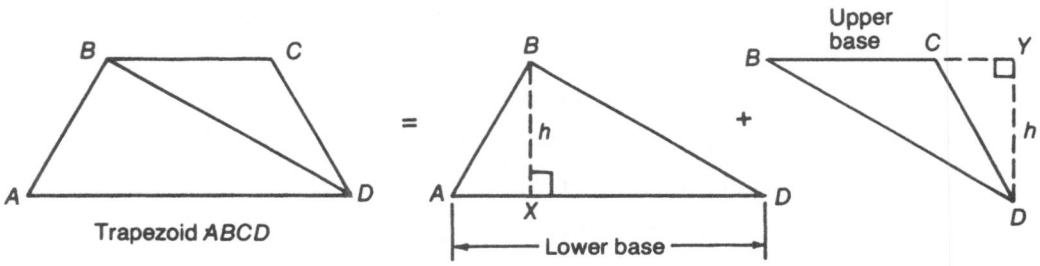

FIGURE 14.8

In this relationship, the terms inside the parentheses represent the lengths of the bases of the trapezoid and the letter h represents the length of an altitude of the trapezoid. This result is stated formally in Theorem 14.3.

THEOREM 14.3 AREA OF A TRAPEZOID

The area of a trapezoid is equal to one-half the product of the length of an altitude and the sum of the lengths of the bases.

Thus, $A = \frac{1}{2}h(b_1 + b_2)$.

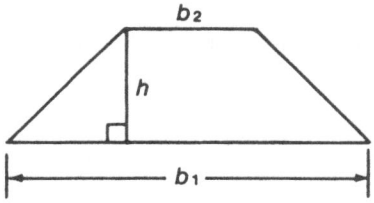

EXAMPLE 14.5 Find the area of an isosceles trapezoid whose bases are 8 and 20 and whose lower base angle has a measure of 45.

SOLUTION

Since $\overline{BH}$ is the leg opposite the 45 degree angle in a 45-45 right triangle, $BH = AH = 6$.

$$\begin{aligned}\text{Area } ABCD &= \frac{1}{2}BH(AD + BC)\\ &= \frac{1}{2}(6)(20 + 8)\\ &= 3(28)\\ &= 84 \text{ square units}\end{aligned}$$

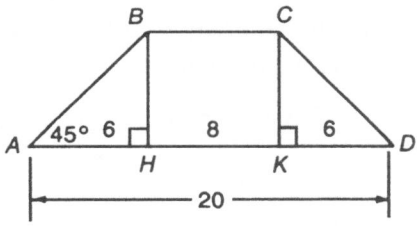

The area of a triangle formula may also be used to derive convenient formulas for finding the areas of equilateral triangles and rhombuses. The formula for the area of an equilateral triangle may be expressed exclusively in terms of the length of a side s of the triangle by representing the length of the altitude in terms of s and then applying the area of a triangle relationship.

Let's illustrate by considering equilateral triangle ABC and drawing altitude $\overline{BD}$ to side $\overline{AC}$ (Figure 14.9).

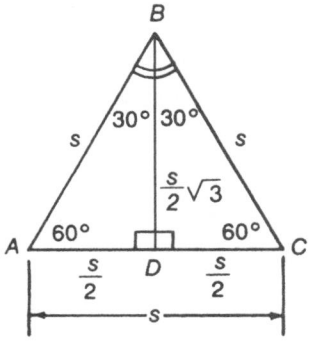

FIGURE 14.9

An equilateral triangle is also equiangular so that altitude $\overline{BD}$ divides the triangle into two 30-60 right triangles. In $\triangle ADB$, the length of $\overline{AD}$ (the side opposite the 30 degree angle) is one-half the length of the hypotenuse ($\overline{AB}$): $AD = \frac{1}{2}AB = \frac{1}{2}s$. The length of altitude $\overline{BD}$ (the leg opposite the 60 degree angle) is equal to one-half the length of the hypotenuse multiplied by $\sqrt{3}$: $BD = \frac{1}{2}AB\sqrt{3} = \frac{1}{2}s\sqrt{3}$.

$$\begin{aligned}\text{Area } \triangle ABC &= \frac{1}{2}(AC)(BD)\\ &= \frac{1}{2}(s)\left(\frac{1}{2}s\sqrt{3}\right)\\ &= \frac{1}{4}s^2\sqrt{3}\end{aligned}$$

THEOREM 14.4 AREA OF AN EQUILATERAL TRIANGLE

The area of an equilateral triangle is equal to one-fourth of the square of the length of a side multiplied by $\sqrt{3}$.

Thus, $A = \frac{s^2}{4}\sqrt{3}$.

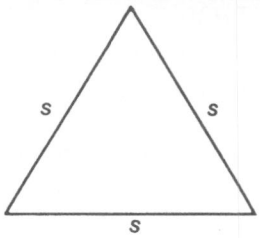

EXAMPLE 14.6 Find the area of an equilateral triangle whose perimeter is 24.

SOLUTION

If the perimeter of the triangle is 24, then the length of each side is 8.

$$\begin{aligned}A &= \frac{s^2}{4}\sqrt{3}\\ &= \frac{8^2}{4}\sqrt{3}\\ &= \frac{64}{4}\sqrt{3}\\ &= 16\sqrt{3} \text{ square units}\end{aligned}$$

EXAMPLE 14.7 Find the length of a side of an equilateral triangle that has an area of $25\sqrt{3}$ square centimeters.

SOLUTION

$$A = \frac{s^2}{4}\sqrt{3}$$
$$25\sqrt{\cancel{3}} = \frac{s^2}{4}\sqrt{\cancel{3}}$$
$$25 = \frac{s^2}{4}$$
$$s^2 = 100$$
$$s = \sqrt{100} = 10 \text{ cm}$$

The diagonals of a rhombus divide the rhombus into four congruent triangles. The area of the rhombus can be obtained by determining the area of one of these triangles and then multiplying it by 4. For convenience, we will refer to the lengths of the two diagonals of a rhombus as d_1 and d_2. It can be shown that the area of any one of the four triangles formed by the diagonals is $\frac{1}{8}(d_1 \cdot d_2)$. To find the area of the rhombus, we multiply $\frac{1}{8}(d_1 \cdot d_2)$ by 4, which gives us $\frac{1}{2}(d_1 \cdot d_2)$.

THEOREM 14.5 AREA OF A RHOMBUS

The area of a rhombus is equal to one-half the product of the lengths of the diagonals.

Thus, $A = \frac{1}{2}(d_1 \cdot d_2)$.

In a square the diagonals are congruent, so

$$\text{Area of square} = \frac{1}{2}d^2$$

where d represents the length of each diagonal.

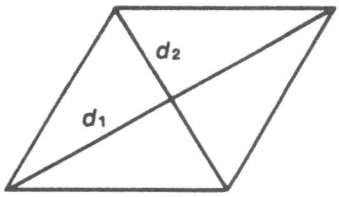

EXAMPLE 14.8 Find the area of a rhombus if the lengths of its diagonals are 10 and 14.

SOLUTION

$$\begin{aligned} A &= \frac{1}{2}(d_1 \cdot d_2) \\ &= \frac{1}{2}(10 \cdot 14) \\ &= \frac{1}{2}(140) \\ &= 70 \text{ square units} \end{aligned}$$

EXAMPLE 14.9 The length of one diagonal of a rhombus is three times the length of the other diagonal. If the area of the rhombus is 54 square units, find the length of each diagonal.

SOLUTION

Let x = length of shorter diagonal.
Then $3x$ = length of longer diagonal.

$$\begin{aligned} A &= \frac{1}{2}(d_1 \cdot d_2) \\ 54 &= \frac{1}{2}(3x \cdot x) \\ 108 &= 3x^2 \\ x^2 &= \frac{108}{3} = 36 \\ x &= \sqrt{36} \\ &= \text{length of shorter diagonal} = 6 \\ 3x &= \text{length of longer diagonal} = 18 \end{aligned}$$

Comparing Areas

COMPARING AREAS OF EQUIVALENT TRIANGLES

Two triangles that are congruent must have the same area. Can two triangles that are *not* congruent have the same area? In the accompanying diagram, *ABCD* is a rectangle. Triangles *AXD* and *AYD* are not necessarily congruent. How do the areas of triangles *AXD* and *AYD* compare? Notice that each of these triangles has the same base ($\overline{AD}$) and a height that is numerically equal to the width of the rectangle. Therefore,

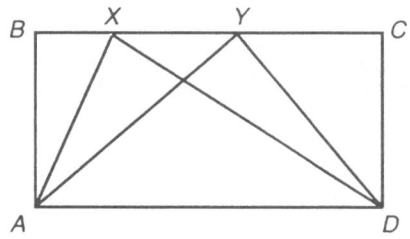

$$\text{Area } \triangle AXD = \text{Area } \triangle AYD = \frac{1}{2}(AD \times AB)$$

THEOREM 14.6 EQUIVALENT TRIANGLES

If two triangles have equal bases and equal altitudes, then their areas are equal.

Since area can be represented by a number, areas can be added or subtracted. For example, in the accompanying diagram *ABCD* is a parallelogram and *E* is the midpoint of diagonal $\overline{BD}$.

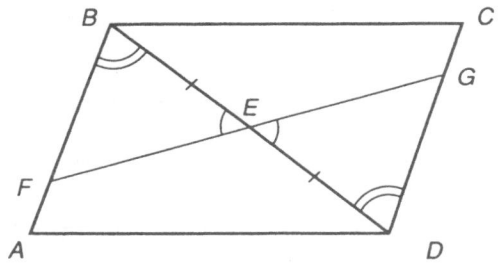

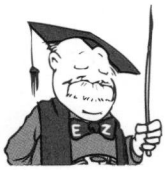

THIS IS THE KEY TO THE METHOD!

- How do the areas of triangles *ABD* and *CBD* compare? Since a diagonal divides a parallelogram into two congruent triangles, the area of triangle *ABD* is equal to the area of triangle *CBD*.
- How do the areas of triangles *FEB* and *GED* compare? Since these triangles are congruent by the ASA postulate, they have the same area.

- How do the areas of quadrilaterals $AFED$ and $CGEB$ compare? By subtracting equal areas ($\triangle FEB$ and $\triangle GED$) from equal areas ($\triangle ABD$ and $\triangle CBD$, respectively), we see that the areas of the two resulting quadrilaterals must also be equal:

$$\begin{array}{r} \text{Area } \triangle ABD = \text{area } \triangle CBD \\ -\text{Area } \triangle FEB = \text{area } \triangle GED \\ \hline \text{Area } AFED = \text{area } CGEB \end{array}$$

COMPARING AREAS OF SIMILAR POLYGONS

Recall that if two triangles are similar, then the ratio of the lengths of any pair of corresponding sides is the same as the ratio of the lengths of any pair of corresponding altitudes (or corresponding medians and angle bisectors). Suppose $\triangle ABC \sim \triangle RST$ and that their ratio of similitude is $3:1$. For example, an interesting relationship arises when we take the ratio of the areas of the two similar triangles in Figure 14.10:

$$\frac{\text{Area } \triangle ABC}{\text{Area } \triangle RST} = \frac{\frac{1}{2}(AC)(BX)}{\frac{1}{2}(RT)(SY)} = \frac{\frac{1}{2}(15)(3)}{\frac{1}{2}(5)(1)} = \frac{9}{1}$$

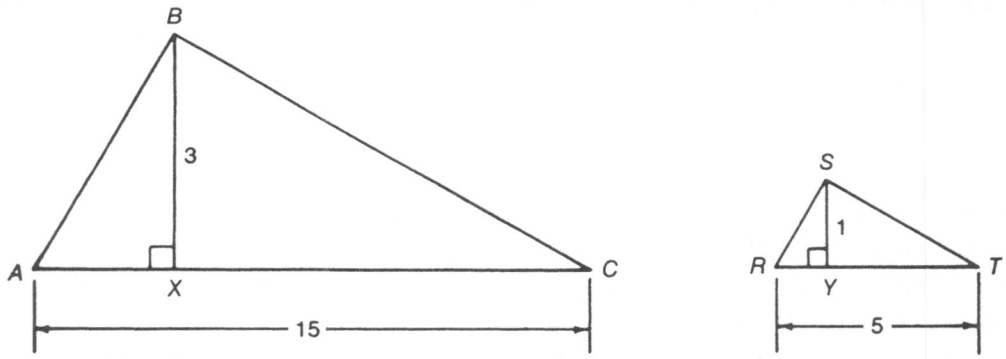

FIGURE 14.10

How does the ratio of the areas of the triangles compare with the ratio of the lengths of a pair of corresponding sides? Since the ratio of similitude is 3:1,

$$\frac{\text{Area } \triangle ABC}{\text{Area } \triangle RST} = \frac{9}{1} = \left(\frac{3}{1}\right)^2 = \left(\frac{\text{length of a side in } \triangle ABC}{\text{length of the corresponding side in } \triangle RST}\right)^2$$

THEOREM 14.7 COMPARING AREAS OF SIMILAR TRIANGLES

If two triangles are similar, then the ratio of their areas is equal to the square of the ratio of the lengths of any pair of corresponding sides:

$$\frac{\text{Area } \triangle \text{I}}{\text{Area } \triangle \text{II}} = \left(\frac{\text{side}_{\text{I}}}{\text{side}_{\text{II}}}\right)^2$$

where $\triangle$I and $\triangle$II refer to a pair of similar triangles and side_{I} and side_{II} represent the lengths of a pair of corresponding sides in triangles I and II, respectively.

Theorem 14.7 may be generalized to apply to any pair of similar *polygons*. Furthermore:

- Since the lengths of corresponding sides, altitudes, medians, and angle bisectors are in proportion, the ratio of the areas of a pair of similar triangles is equal to the square of the ratio of any pair of these corresponding segments.
- Since the perimeters of similar polygons have the same ratio as the lengths of any pair of corresponding sides, the ratio of the areas of similar polygons is equal to the square of the ratio of their perimeters.

EXAMPLE 14.10 The ratio of similitude of two similar triangles is 2:3. If the area of the smaller triangle is 12 square units, find the area of the larger triangle.

SOLUTION

$$\frac{\text{Area of smaller } \triangle}{\text{Area of larger } \triangle} = \left(\frac{2}{3}\right)^2$$

Let x = area of larger triangle. Then:

$$\frac{12}{x} = \frac{4}{9}$$
$$4x = 108$$
$$x = \frac{108}{4} = 27$$

EXAMPLE 14.11 The areas of two similar polygons are 25 and 81 square units. If the length of a side of the larger polygon is 72, find the length of the corresponding side of the smaller polygon.

SOLUTION

Let x = length of the corresponding side of the smaller polygon.

$$\frac{\text{Area of smaller polygon}}{\text{Area of larger polygon}} = \left(\frac{x}{72}\right)^2$$

$$\frac{25}{81} = \left(\frac{x}{72}\right)^2$$

Solve for x by first taking the square root of both sides of the proportion:

$$\sqrt{\frac{25}{81}} = \frac{x}{72}$$

$$\frac{5}{9} = \frac{x}{72}$$

$$9x = 360$$

$$x = \frac{360}{9} = 40$$

Area of a Regular Polygon

We have intentionally restricted our attention to finding the areas of familiar three-sided and four-sided polygons. As a general rule, it is difficult to develop convenient formulas for other types of polygons. There is, however, one notable exception. A formula for the area of a *regular* polygon can be derived by subdividing the regular polygon into a set of congruent triangles and then summing the areas of these triangles. Throughout our analysis we shall assume that *circles having the same center can be inscribed and circumscribed about a regular polygon*. Figures 14.11 and 14.12 identify some important terms relating to regular polygons and their circumscribed and inscribed circles. This information will be needed when we eventually turn our attention to finding the area of a regular polygon.

In Figure 14.11:

- Point O is the center of the circle inscribed and the circle circumscribed about regular pentagon $ABCDE$. Point O is referred to as the *center* of a regular polygon.

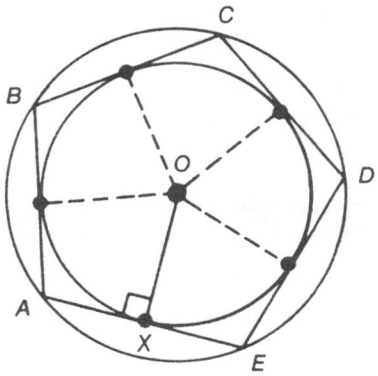

FIGURE 14.11

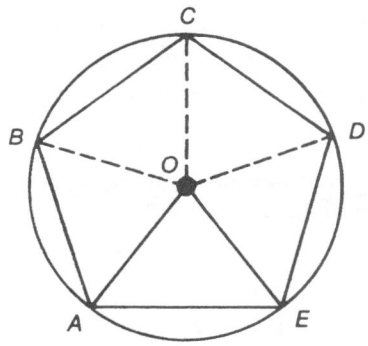

FIGURE 14.12

- $\overline{OX}$ is a segment drawn from the center of the regular pentagon to the point at which the inscribed circle is tangent to side $\overline{AE}$. $\overline{OX}$ is called an *apothem*. Apothems may be drawn to each of the other sides.

In Figure 14.12:

- $\overline{OA}$ and $\overline{OE}$ are *radii* of regular pentagon $ABCDE$. A *radius* of a regular polygon is a segment drawn from the center to any vertex of the polygon.
- Angle AOE is a *central angle* of regular pentagon $ABCDE$. A *central angle* of a regular polygon is formed by drawing two radii to consecutive vertices of the polygon.

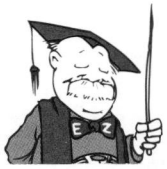

DEFINITIONS RELATING TO REGULAR POLYGONS

- The *center* of a regular polygon is the common center of its inscribed and circumscribed circles.
- An *apothem* of a regular polygon is a segment whose end points are the center of the polygon and a point at which the inscribed circle is tangent to a side. An apothem of a regular polygon is also a radius of the *inscribed* circle.
- A *radius* of a regular polygon is a segment whose end points are the center of the polygon and a vertex of the polygon. A radius of a regular polygon is also a radius of the *circumscribed* circle.
- A *central angle* of a regular polygon is an angle whose vertex is the center of the polygon and whose sides are radii drawn to the end points of the same side of the polygon. An n-sided regular polygon will have n central angles.

In Figure 14.12, the sides of the regular pentagon divide the circle into five congruent arcs. It follows that each central angle must have the same measure. Since there are five central angles, the measure of each central angle may be found by dividing 360 by 5, obtaining a measure of 72 for each of the central angles. In general, the radii of a regular polygon divide the polygon into triangles which may be proven congruent to one another by SAS, where the congruent sides are the radii and the congruent included angles are the central angles. Using the CPCTC principle, we may further conclude that each radius bisects the angle located at the vertex to which it is drawn.

THEOREM 14.8 ANGLES OF A REGULAR POLYGON

- The central angles of a regular polygon are congruent.
- The measure of a central angle of a regular polygon is equal to 360 divided by the number of sides of the polygon.
- The radii of a regular polygon bisect the interior angles of the regular polygon.

EXAMPLE 14.12 The length of a side of a regular hexagon is 14. Find each of the following:

a. The measure of a central angle.
b. The length of a radius of the hexagon.
c. The length of an apothem.

SOLUTION

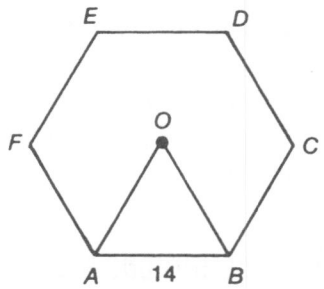

a.
$$m\measuredangle AOB = \frac{360}{n}$$
$$= \frac{360}{6}$$
$$= 60$$

b. Since the central angle has measure 60, and $\overline{OA} \cong \overline{OB}$, angles OAB and OBA must be congruent and be equal to 60. Since $\triangle AOB$ is equiangular, it is also equilateral. Hence, the radius of the hexagon must have the same length as $\overline{AB}$ and is therefore 14.

c. Let us focus on triangle AOB and draw the apothem to side $\overline{AB}$.

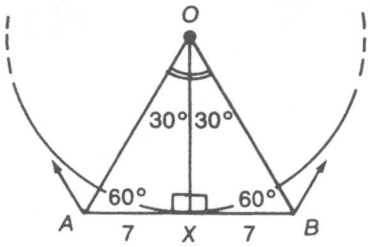

Since an apothem is drawn to a point at which the inscribed circle is tangent to a side of the polygon, the apothem must be perpendicular to the side to which it is drawn. Furthermore, it bisects the side to which it is drawn. As an illustration observe that apothem $\overline{OX}$ divides triangle AOB into two congruent right triangles. $\triangle AOX \cong \triangle BOX$ by Hy-leg since

$$\overline{OA} \cong \overline{OB} \quad \text{(Hy)}$$
$$\overline{OX} \cong \overline{OX} \quad \text{(Leg)}$$

By CPCTC, $\overline{AX} \cong \overline{BX}$. In addition, $\measuredangle AOX \cong \measuredangle BOX$.

For the problem at hand, the apothem $\overline{OX}$ is the leg opposite the 60° angle in a 30-60 right triangle (that is, in $\triangle AOX$):

$$\begin{aligned} OX &= \frac{1}{2} OA\sqrt{3} \\ &= \frac{1}{2}(14)\sqrt{3} \\ &= 7\sqrt{3} \end{aligned}$$

Theorem 14.9 summarizes the properties of an apothem.

THEOREM 14.9 PROPERTIES OF AN APOTHEM

- An apothem of a regular polygon is the perpendicular bisector of the side to which it is drawn.
- An apothem bisects the central angle determined by the side to which it is drawn.

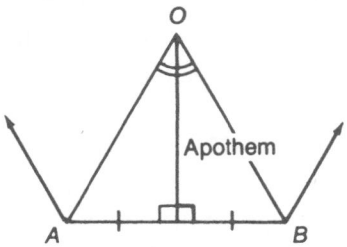

We are now in a position to develop a convenient formula for finding the area of a *regular* polygon. Since an n-sided regular polygon is composed of n identical triangles, the formula for the area of a regular polygon is a generalization of the formula for the area of a triangle. The apothem of a regular polygon corresponds to the altitude of a triangle, and the perimeter of a regular polygon corresponds to the base of a triangle. Therefore, the area of a regular polygon may be found by multiplying one-half the length of the apothem by the perimeter of the regular polygon.

THEOREM 14.10 AREA OF A REGULAR POLYGON

The area (A) of a regular polygon is equal to one-half the product of the length of an apothem (a) and its perimeter (p).

Thus, $A = \frac{1}{2}ap$.

OUTLINE OF PROOF

GIVEN: $ABC\ldots$ is a regular polygon having n sides. Let:
a = length of an apothem,
s = length of each side,
p = perimeter of $ABC\ldots$.

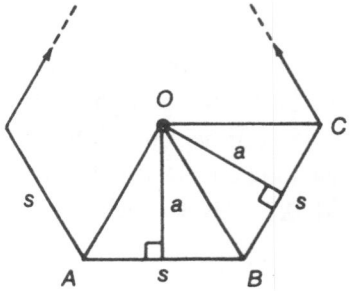

PROVE: Area $ABC\ldots = \frac{1}{2}ap$.

- Draw the radii of the polygon.
- Find the sum of the areas of all the triangles thus formed:

$$\begin{aligned}\text{Area } ABC\ldots &= \text{area } \triangle AOB + \text{area } \triangle BOC + \cdots \\ &= \frac{1}{2}as + \frac{1}{2}as + \cdots + \frac{1}{2}as \\ &= \frac{1}{2}a(s + s + \cdots + s) \\ &= \frac{1}{2}ap\end{aligned}$$

EXAMPLE 14.13 The length of a side of a regular pentagon is 20. Find each of the following:

a. The length of the apothem correct to the *nearest hundredth*.

b. The area correct to the *nearest whole number*.

SOLUTION

First draw a representative triangle of the polygon.

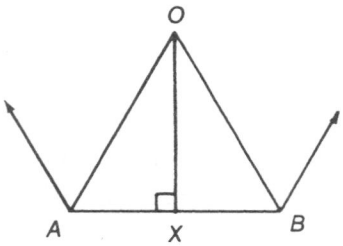

a. Since an apothem is the perpendicular bisector of the side to which it is drawn, $\triangle OAX$ is a right triangle and $AX = \frac{1}{2}(20) = 10$. The measure of central angle $AOB = \frac{360}{5} = 72$. Apothem $\overline{AX}$ bisects central angle AOB, so $m\measuredangle AOX = \frac{1}{2}(72) = 36$. We must find OX by using an appropriate trigonometric ratio. The arithmetic is simplest if we find the tangent of angle OAX. Since $m\measuredangle OAX = 90 - 36 = 54$, we may write the following:

$$\begin{aligned}
\tan 54^\circ &= \frac{\text{side opposite } \measuredangle}{\text{side adjacent } \measuredangle} \\
&= \frac{OX}{AX} \\
1.3764 &= \frac{OX}{10} \\
OX &= 10(1.3764) \\
&= 13.76 \text{ correct to the } \textit{nearest hundredth}
\end{aligned}$$

b. If the length of a side of a regular pentagon is 20, then its perimeter is 5 times 20 or 100.

$$\begin{aligned}
A &= \frac{1}{2}ap \\
&= \frac{1}{2}(13.76)(100) \\
&= \frac{1}{2}(1376) \\
&= 688 \text{ square units correct to the } \textit{nearest whole number}
\end{aligned}$$

Areas of a Circle, Sector, and Segment

Consider what happens if the number of sides, n, of an inscribed regular polygon gets larger, as shown in Figure 14.13.

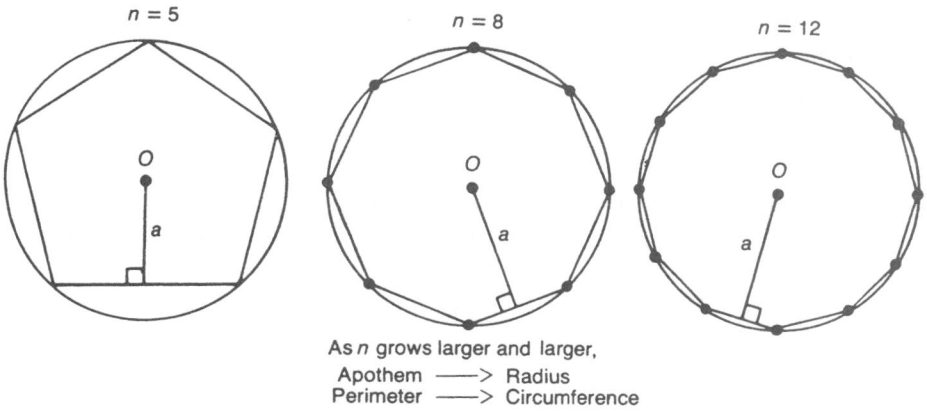

FIGURE 14.13

As the number of sides increases indefinitely without being constrained by an upper bound, the area of the inscribed regular polygon and the area of the circle become indistinguishable. Under these circumstances, we may make the following substitutions in the formula for the area of a regular polygon where r represents the radius of the circle and C its circumference:

As the number of sides becomes infinitely large

$$\text{Area of regular polygon} = \frac{1}{2}\,a\,p$$

$$\text{Area of circumscribed circle} = \frac{1}{2}\,r\,C$$

or

$$= \frac{1}{2}(r)(2\pi r)$$

$$\text{Area of circle} = \pi r^2$$

THEOREM 14.11 AREA OF A CIRCLE

The area (A) of a circle is equal to the product of the constant π and the square of the length of the radius (r) of the circle.

Thus, $A = \pi r^2$.

EXAMPLE 14.14

a. Find the area of circle O.

b. Find the area of the shaded region.

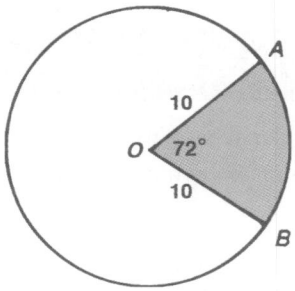

SOLUTION

a. $A = \pi r^2$

$= \pi(10)^2$

$= 100\pi$ square units

b. The shaded region is called a *sector* of the circle. In a manner analogous to determining the length of an arc of a circle, we form the following proportion:

$$\frac{\text{Area of sector}}{\text{Area of circle}} = \frac{\text{degree measure of sector arc}}{360^\circ}$$

Let x = area of the sector.

$$\frac{x}{100\pi} = \frac{72^\circ}{360^\circ}$$
$$\pi = \frac{1}{5}$$
$$5x = 100\pi$$
$$x = 20\pi \text{ square units}$$

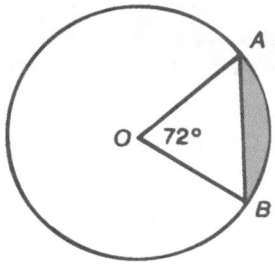

Suppose that in part **b** of Example 14.14 chord $\overline{AB}$ was drawn. Then, if it was known that the area of triangle AOB was 50 square units, what would be the area of the region bounded by chord $\overline{AB}$ and the minor arc that it cuts off? The region described is called a *segment* of the circle. To find the area of this segment, all we need do is to subtract the area of triangle AOB from the area of sector AOB.

$$\begin{aligned} \text{Area segment } \overline{AB} &= \text{area sector } AOB - \text{area } \triangle AOB \\ &= 20\pi - 50 \text{ square units} \end{aligned}$$

Unless otherwise directed, we generally leave the answer in this form.

Let's now summarize the concepts introduced in Example 14.14.

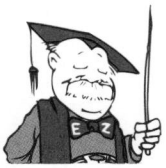

DEFINITIONS OF SECTOR AND SEGMENT OF A CIRCLE

- A *sector of a circle* is a region of a circle bounded by two radii and the minor arc they determine.
- A *segment of a circle* is a region of a circle bounded by a chord and the minor arc that it cuts off.

THEOREM 14.12 AREA OF A SECTOR

The area of a sector of a circle is determined by the following proportion where n represents the degree measure of its minor arc:

$$\frac{\text{Area of sector}}{\text{Area of circle}} = \frac{n}{360}$$

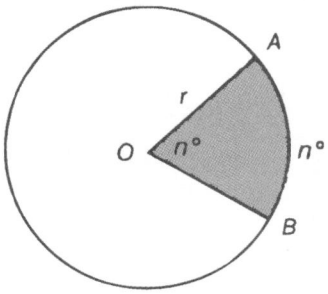

$$\frac{\text{Area sector } AOB}{\pi r^2} = \frac{n}{360}$$

Geometric Solids

The amount of space that a solid encloses is represented by its volume. The *volume* of a solid (Figure 14.14), expressed in cubic units, is the number of cubes having an edge length of 1 unit that the solid can enclose.

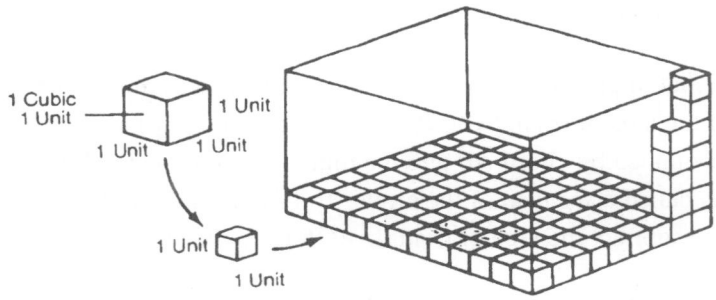

FIGURE 14.14 Volume of a solid

SPECIAL PRISMS

A polygon is "flat" since all its sides lie in the same plane. The figure that is the counterpart of a polygon in space is called a *polyhedron*.

A *prism* (Figure 14.15) is a special type of polyhedron. The sides of a prism are called *faces*. All prisms have these two properties:

- Two of the faces, called *bases*, are congruent polygons lying in parallel planes.
- The faces that are not bases, called *lateral faces*, are parallelograms.

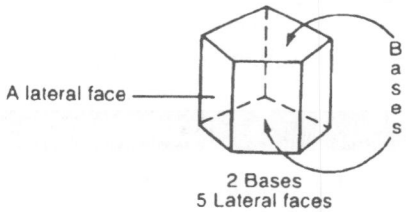

FIGURE 14.15 Prism

A *rectangular solid* (Figure 14.16) is a prism whose two bases and four lateral faces are rectangles. If the bases and lateral faces are squares, the rectangular solid is called a *cube*.

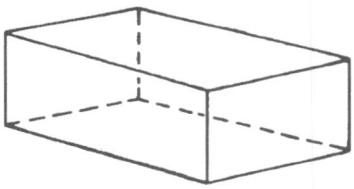

FIGURE 14.16 Rectangular solid

VOLUME OF A RECTANGULAR SOLID

The volume (V) of *any* prism is equal to the product of the area of its base (B) and its height (h):

$$\text{Volume of a prism: } V = Bh.$$

By using this relationship, special formulas can be derived for determining the volumes of a cube (Figure 14.17) and a rectangular solid (Figure 14.18):

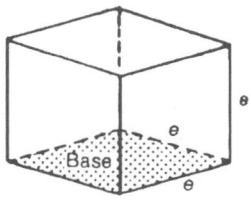

FIGURE 14.17 Cube

Volume of a cube:

$$V = Bh = (e^2)e$$
$$= e^3$$

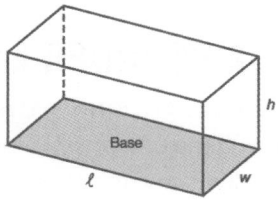

FIGURE 14.18 Rectangular solid

Volume of rectangular solid:

$$V = Bh = (\ell \times w)h$$
$$= \ell wh$$

EXAMPLE 14.15 What is the volume of a rectangular solid having a length, width, and height of 5 centimeters, 4 centimeters, and 3 centimeters, respectively?

SOLUTION

$$V = \ell wh = (5)(4)(3) = 60 \text{ cm}^3$$

EXAMPLE 14.16 A rectangular solid has the same volume as a cube whose edge length is 6. What is the height of the rectangular solid if its length is 8 and its width is 3?

SOLUTION

Volume of rectangular solid = volume of cube

$$\ell wh = e^3$$
$$(8)(3)h = 6^3$$
$$24h = 216$$
$$h = \frac{216}{24} = 9$$

EXAMPLE 14.17 What is the change in volume of a cube whose edge length is doubled?

SOLUTION

In the formula $V = e^3$, replace e with $2e$. The new volume is

$$V = (2e)^3 = 8e^3.$$

The new volume is 8 times as great as the original volume.

VOLUMES OF A CYLINDER AND CONE

A *cylinder* (Figure 14.19) has *two* bases that are congruent circles lying in parallel planes. Like the volume of a prism, a solid that also has two congruent bases, the volume of a cylinder is given by the formula $V = Bh$. In this formula, B represents the area of one of the two identical bases. The area of a cylinder's circular base is πr^2.

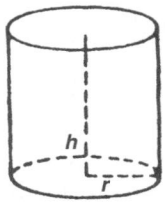

FIGURE 14.19 Cylinder

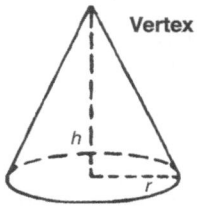

FIGURE 14.20 Cone

Volume of a cylinder:

$$V = \pi r^2 h$$

Volume of a cone:

$$V = \frac{1}{3}\pi r^2 h$$

A *cone* (Figure 14.20), has one base and a volume given by the formula $V = \frac{1}{3}Bh$. Here B represents the area, πr^2, of the circular base of the cone.

EXAMPLE 14.18 The radius of a base of a cylinder is 5 and its height is 6. What is its volume?

SOLUTION

$$\begin{aligned} V &= \pi r^2 h \\ &= \pi(5^2)6 = \pi(25)(6) = 150\,\pi \end{aligned}$$

EXAMPLE 14.19 The *slant height*, ℓ, of a cone is a segment that connects the vertex to a point on the circular base. What is the volume of a cone having a radius of 6 and a slant height of 10?

SOLUTION

The slant height is the hypotenuse of a right triangle whose legs are a radius and the altitude of the cone. The lengths of the sides of this triangle form a 6-8-10 right triangle, where 8 represents the height of the cone. Therefore:

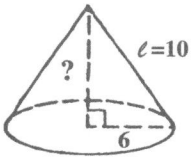

$$\begin{aligned} V &= \frac{1}{3}\pi r^2 h \\ &= \frac{1}{3}\pi(6)^2 \cdot 8 \\ &= \frac{1}{\cancel{3}}\pi(\overset{12}{\cancel{36}}) \cdot 8 \\ &= \overset{1}{96}\pi \end{aligned}$$

EXAMPLE 14.20 The circumference of the base of a cone is 8π centimeters. If the volume of the cone is 16π cubic centimeters, what is the height?

SOLUTION

Since $C = \pi D = 8\pi$, the diameter is 8 cm; therefore, the radius of the base is 4 cm.

$$\begin{aligned} V &= \frac{1}{3}\pi r^2 h \\ 16\pi &= \frac{1}{3}\pi(4^2)h \\ 48\pi &= 16\pi h \\ \frac{48\pi}{16\pi} &= h \\ 3 &= h \end{aligned}$$

The height of the cone is 3 cm.

VOLUME OF A SPHERE

A sphere (Figure 14.21) is a solid that represents the set of all points in space that are at a given distance from a fixed point.

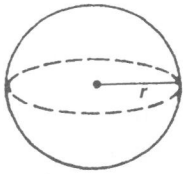

FIGURE 14.21 Sphere

Volume of a sphere:

$$V = \frac{4}{3}\pi r^3$$

EXAMPLE 14.21 Find the volume of a sphere whose diameter is 6 centimeters.

SOLUTION

$$\begin{aligned} \text{Radius} &= \frac{1}{2}(6) = 3 \\ V &= \frac{4}{3}\pi(3)^3 \\ &= \frac{4}{3}\pi(27) \\ &= 4\pi(9) \\ &= 36\pi \text{ cm}^3 \end{aligned}$$

SUMMARY OF VOLUME FORMULAS

Figure	Volume Formula
Rectangular box	$V = \text{length} \times \text{width} \times \text{height}$ $= \ell wh$
Cube	$V = \text{edge} \times \text{edge} \times \text{edge}$ $= e^3$
Cylinder	$V = \pi \times (\text{radius})^2 \times \text{height}$ $= \pi r^2 h$
Cone	$V = \frac{1}{3} \times \pi \times (\text{radius})^2 \times \text{height}$ $= \frac{1}{3}\pi r^2 h$
Sphere	$V = \frac{4}{3} \times \pi \times (\text{radius})^3$ $= \frac{4}{3}\pi r^3$

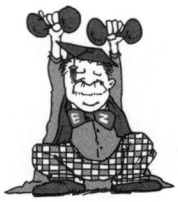

REVIEW EXERCISES FOR CHAPTER 14

1. Which diagram represents the figure with the greatest volume?

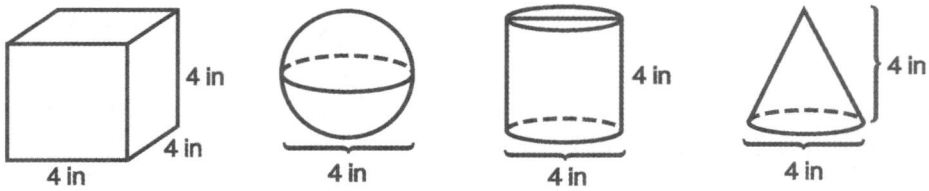

(1) (2) (3) (4)

2. An altitude is drawn from vertex B of rhombus $ABCD$, intersecting $\overline{AD}$ at point H. If $AB = 13$ and $HD = 8$, what is the number of square units in the area of rhombus $ABCD$?

(1) 48 (2) 65 (3) 78 (4) 156

3. What is the length of a side of an equilateral triangle whose area is $16\sqrt{3}$ square units?

(1) 32 (2) 12 (3) 8 (4) 4

4. What is the number of square units in the area of $\triangle ABC$ shown in the accompanying figure?

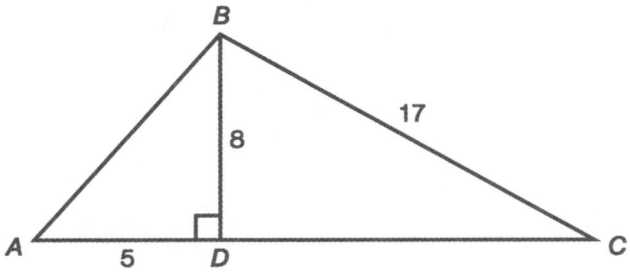

(1) 20 (2) 34 (3) 40 (4) 80

5. Cerise waters her lawn with a sprinkler that sprays water in a circular pattern at a distance of 15 feet from the sprinkler. The sprinkler head rotates through an angle of 300°, as shown by the shaded area in the accompanying diagram. What is the area of the lawn, to the *nearest square foot*, that receives water from this sprinkler?

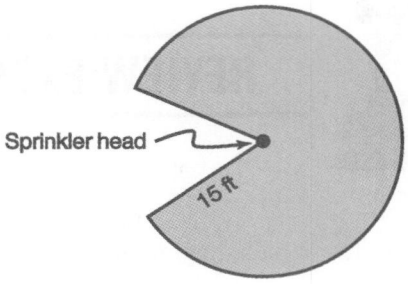

(1) 79 (2) 94 (3) 589 (4) 707

For Exercises 6 to 14, find the area of each of the figures. Whenever appropriate, answers may be left in radical form.

6.

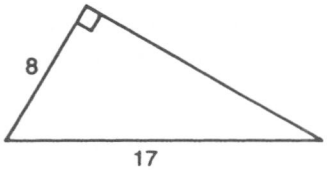

7.

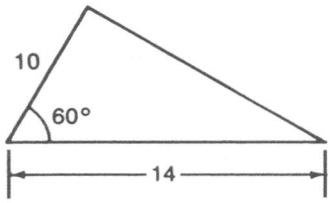

8.

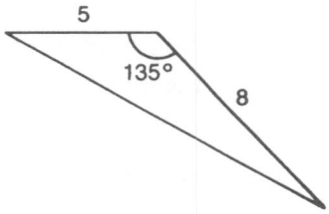

9.

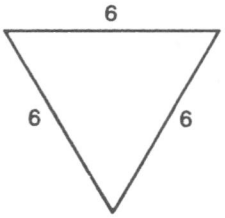

10.

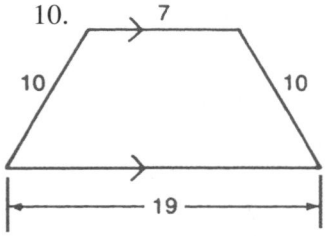

11.

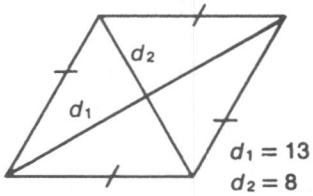

12.

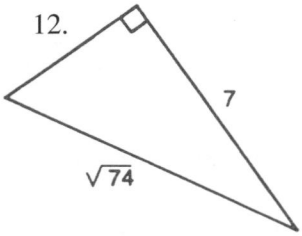

13.

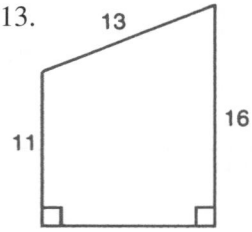

14.

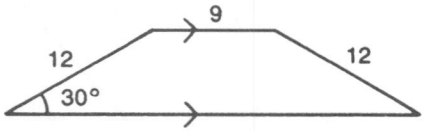

15. In each case, the number of square units in the area of an equilateral triangle is given. Find the length of a side of the triangle.

    (a) $16\sqrt{3}$ (b) $11\sqrt{3}$ (c) 4

16. Find the length of the shorter diagonal of a rhombus if:

    (a) The length of the longer diagonal is 15 and the area is 90.
    (b) The lengths of the diagonals are in the ratio of 2;3 and the area of the rhombus is 147.

17. Find the length of an altitude of a trapezoid if:

    (a) Its area is 72 and the sum of the lengths of the bases is 36.
    (b) Its area is 80 and its median is 16.
    (c) The sum of the lengths of the bases is numerically equal to one-third of the area of the trapezoid.

18. Find the area of a rhombus if its perimeter is 68 and the length of one of its diagonals is 16.

19. Find the area of a triangle if the lengths of a pair of adjacent sides are 6 and 14 and the measure of the included angle is:

    (a) 90 (b) 30 (c) 120 (Leave answer in radical form.)

20. Tamika has a hard rubber ball whose circumference measures 13 inches. She wants to box it for a gift but can only find cube-shaped boxes of sides 3 inches, 4 inches, 5 inches, or 6 inches. What is the smallest box that the ball will fit into with the top on?

21. The areas of two similar polygons are 81 square units and 121 square units.

    (a) Find their ratio of similitude.
    (b) If the perimeter of the smaller polygon is 45, find the perimeter of the larger polygon.

22. The lengths of a pair of corresponding sides of a pair of similar triangles are in the ratio of 5:8. If the area of the smaller triangle is 75, find the area of the larger triangle.

23. Find the measure of a central angle for a regular octagon.

24. If the measure of an interior angle of a regular polygon is 150, find the measure of a central angle.

25. Find the area of a regular hexagon inscribed in a circle having a diameter of 20 centimeters.

26. As shown in the accompanying diagram, the length, width, and height of Richard's fish tank are 24 inches, 16 inches, and 18 inches, respectively. Richard is filling his fish tank with water from a hose at the rate of 500 cubic inches per minute. How long will it take, to the *nearest minute*, to fill the tank to a depth of 15 inches?

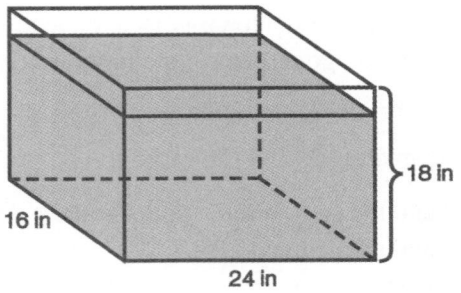

27. A circle having a radius of 6 centimeters is inscribed in a regular hexagon. Another circle whose radius is 6 centimeters is circumscribed about another regular hexagon. Find the ratio of the area of the smaller hexagon to the area of the larger hexagon.

28. The length of a side of a regular decagon is 20 centimeters.

    (a) Find the length of the apothem correct to the *nearest tenth* of a centimeter.
    (b) Using the answer obtained in part **a**, find the area of the decagon.

29. Mr. Gonzalez owns a triangular plot of land, BCD, with $DB = 25$ yards and $BC = 16$ yards. He wishes to purchase the adjacent plot of land in the shape of a right triangle ABD, as shown in the accompanying diagram, with $AD = 15$ yards. If the purchase is made, what will be the total number of square yards in the area of his plot of land, $\triangle ACD$?

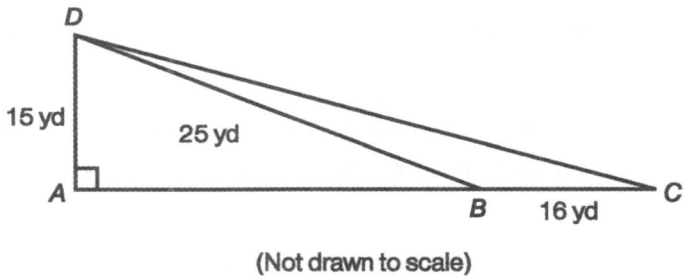

Unless otherwise indicated, answers for exercises 30 to 32 may be left in terms of π.

30. The ratio of the areas of two circles is 1:9. If the radius of the smaller circle is 5 centimeters, find the length of the radius of the larger circle.

31. The ratio of the lengths of the radii of two circles is 4:25. If the area of the smaller circle is 8 square units, find the area of the larger circle.

32. The diameters of two concentric circles are 8 and 12. Find the area of the ring-shaped region (called an *annulus*) bounded by the two circles.

33. In the accompanying diagram, a rectangular container with dimensions 10 inches by 15 inches by 20 inches is to be filled with water, using a cylindrical cup whose radius is 2 inches and whose height is 5 inches. What is the maximum number of full cups of water that can be placed into the container without the water overflowing the container?

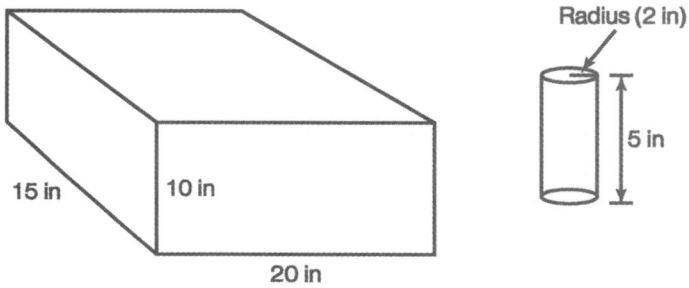

34. In circle O, radii OA and OB are drawn so that OA is equal to the length of chord $\overline{AB}$ If $OA = 12$, find the area of:

    (a) Sector AOB.
    (b) Segment AB.

35. A square whose side is 8 centimeters in length is inscribed in a circle. Find the area of the segment formed by a side of the square and its intercepted arc.

36. What is the ratio of the area of the inscribed and circumscribed circles of a square having a side of 6 centimeters?

For Exercises 37 and 38, find the area of the shaded region in each figure.

37.

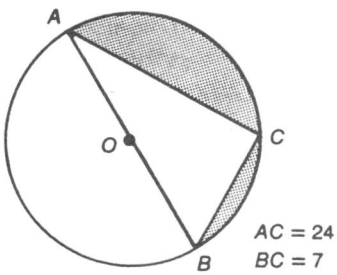

38.

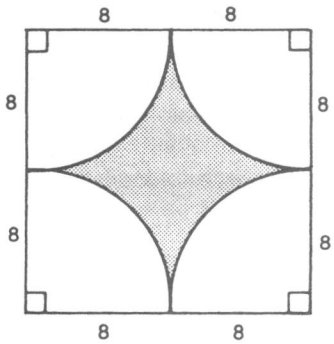

39. A regular pentagon is inscribed in a circle having a radius of 20 centimeters. Find the area of the sector formed by drawing radii to a pair of consecutive vertices of the pentagon.

40. Tracey has two empty cube-shaped containers with sides that measure 5 inches and 7 inches. She fills the smaller container completely with water and then pours all the water from the smaller container into the larger container. How deep, to the *nearest tenth of an inch*, will the water be in the larger container?

41. A parcel of a land is represented by quadrilateral *ABCD* in the accompanying figure, where *DEBC* is a rectangle. If the area of right triangle *AEB* is 600 square feet, find the minimum number of feet of fence needed to completely enclose the entire parcel of land, *ABCD*.

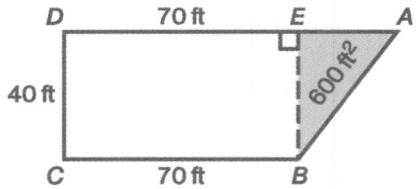

42. A rectangular garden is going to be planted in a person's rectangular backyard, as shown in the accompanying figure. Some dimensions of the back yard and the width of the garden are given. Find the area of the garden to the *nearest square foot*.

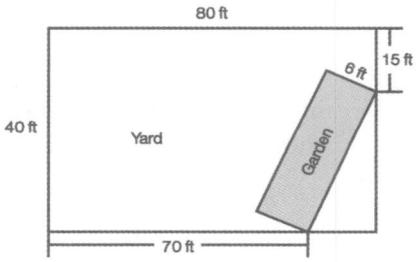

43. Mr. Petri has a rectangular plot of land with length equal to 20 feet and width equal to 10 feet. He wants to design a flower garden in the shape of a circle wit two semicircles at each end of the center circle, as shown in the accompanying diagram. He will fill in the shaded region with wood chips. If one bag of wood chi covers 5 square feet, how many bags mu he buy?

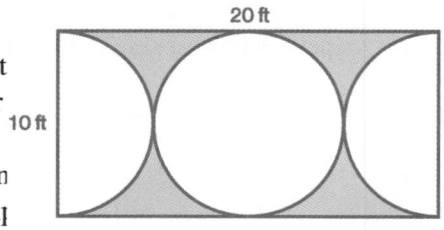

44. The accompanying diagram shows three equal circles aligned in a row such that distance between adjacent circles is 3 inches. The radius of each circle is 5 inches. Line segments AD and BC are diameters of the outside circles, and rectangle $ABCD$ is formed by drawing and $\overline{CD}$.

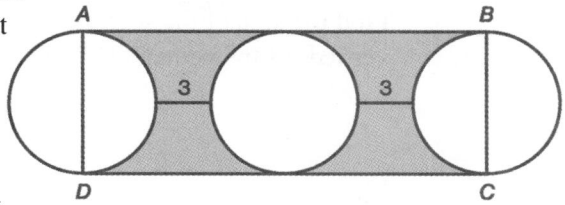

(a) Find the number of square inches in the area of the shaded region to the *nearest integer.*
(b) What percent of the area of rectangle $ABCD$ is shaded, to the *nearest percent*?

45. A regular hexagon is inscribed in a circle. The length of the apothem is $4\sqrt{3}$.

(a) Find the area of the circle.
(b) Find the area of the hexagon.
(c) Find the area of the segment cut off by a side of the hexagon.

46. GIVEN: G is the midpoint of $\overline{CV}$.
PROVE: (a) Area $\triangle CLG$ = area $\triangle VLG$.
(b) Area $\triangle BLC$ = area $\triangle BLV$.

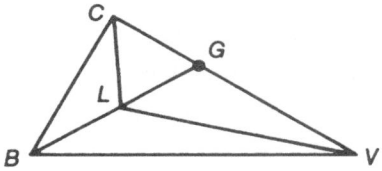

47. In a regular polygon of nine sides, the length of each side is 8.

(a) Find the length of the apothem to the *nearest integer*.
(b) Using the answer obtained in part **a**, find the area of the polygon.

48. GIVEN: $\triangle BCD$, $\overline{BC} \cong \overline{BD}$; points A and E are exterior to $\triangle BCD$. $\overline{AB}$ and $\overline{BE}$ are drawn so that $\overline{AB} \cong \overline{BE}$; $\overline{AC}$ and $\overline{DE}$ are drawn; $\angle ABD \cong \angle EBC$.
PROVE: (a) $\triangle ABC \cong \triangle EBD$.
(b) Area polygon $ABDC$ = area polygon $BEDC$.

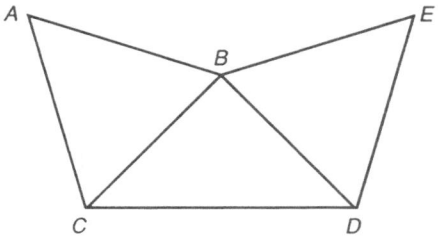

49. GIVEN: $\square ABCD$, $\overline{BE} \cong \overline{FD}$.
PROVE: Area of $\triangle FAC$ = area $\square ABCD$.

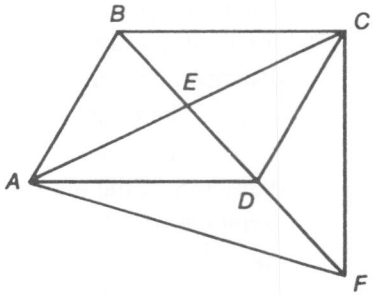

50. Prove, using algebraic methods, that for an equilateral triangle:

    (a) The length of the radius of the inscribed circle is one-third the length of the altitude of the triangle.
    (b) The areas of the inscribed and circumscribed circles are in the ratio of 1:4.

51. The surface area of a rectangular solid is the sum of the areas of its six sides.

    (a) What is the surface area of a cube whose volume is 125 cubic inches?
    (b) What is the surface area of a 3-by-4-by-5 rectangular box?
    (c) What is the volume of a cube whose surface area is 96 square units?

52. A right triangle whose vertical leg measures 8 centimeters and and whose horizontal leg measures 15 centimeters is revolved in space about the shorter leg. What is the volume of the resulting cone?

53. The length, width, and height of a rectangular solid are in the ratio of 3:2:1. If the volume of the box is 48 cubic units, what is the total surface area of the box?

54. If the edge length of a cube is 4, what is the distance from one corner of the cube to the furthest corner on the opposite face of the cube?

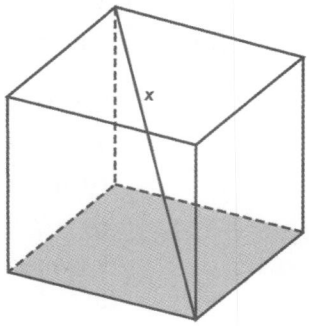

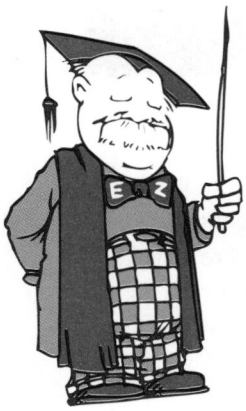

15 Coordinate Geometry

WHAT YOU WILL LEARN

The French mathematician Rene Descartes (1596–1650) developed a method for representing points and equations graphically by drawing a rectangular grid, called a coordinate plane, in which horizontal and vertical number lines intersect at their zero points. In this chapter you will learn:

- the way to plot points and equations in the coordinate plane;
- the way to find areas of figures in the coordinate plane;
- the use of formulas to find the midpoint, length, and slope of a line segment;
- the general form of an equation of a line;
- the general form of an equation of a circle;
- the way to prove geometric relationships using coordinates.

SECTIONS IN THIS CHAPTER

- The Coordinate Plane
- Finding Area Using Coordinates
- The Midpoint and Distance Formulas
- Slope of a Line
- Equation of a Line
- Equation of a Circle
- Proofs Using Coordinates

The Coordinate Plane

You can create a *coordinate plane* by drawing a horizontal number line called the *x-axis* and a vertical number line called the *y-axis*. The x-axis and the y-axis are referred to as the *coordinate axes*, and their point of intersection is called the *origin*. The location of a point in the coordinate plane is determined by two numbers: the *x-coordinate* and the *y-coordinate* of the point. The coordinates of a point are always written as an ordered pair of numbers having the form (x, y). For example, $(2, 3)$ represents the point whose x-coordinate is 2 and whose y-coordinate is 3. This point is named point A in Figure 15.1.

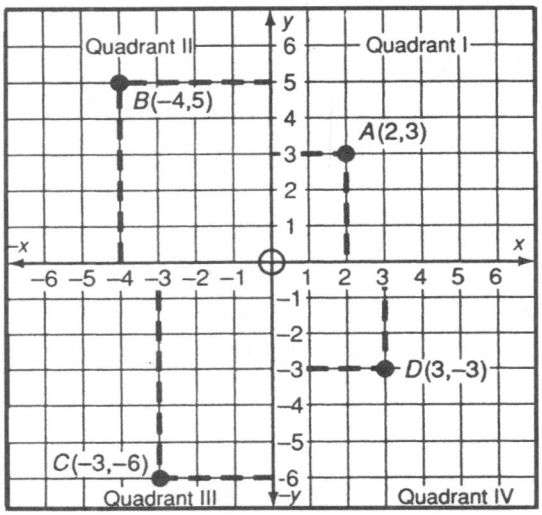

FIGURE 15.1

The x-coordinate of a point is sometimes called the *abscissa*, and the y-coordinate is known as the *ordinate*. Notice in Figure 15.1 that:

- The coordinates of the origin are $(0, 0)$.
- A point is named by writing a single capital letter followed by the coordinates of the point.
- Point A is located by beginning at the origin and moving along the x-axis 2 units to the *right* (since 2 is *positive*), stopping at $x = 2$, and then moving 3 units vertically *up* (since 3 is *positive*).
- The sign of the x-coordinate (positive or negative) tells whether the point is located horizontally to the right of the origin ($x > 0$) or to the left of the origin ($x < 0$). Similarly, the sign of the y-coordinate (positive or negative) tells whether the point is located vertically above the origin ($y > 0$) or below the origin ($y < 0$).

- The coordinate axes divide the coordinate plane into four *quadrants*. Points A, B, C, and D lie in different quadrants. The signs of the x- and y-coordinates of a point determine the quadrant in which the point is located.

Example	Signs of Coordinates	Location of Point
$A(2, 3)$	(+, +)	Quadrant I
$B(-4, 5)$	(–, +)	Quadrant II
$C(-3, -6)$	(–, –)	Quadrant III
$D(3, -3)$	(+, –)	Quadrant IV

Finding Area Using Coordinates

DRAWING AN ALTITUDE TO FIND AREA

If one side of a triangle or quadrilateral is parallel to a coordinate axis, then the area of the figure can be determined by drawing an altitude to this side and using the appropriate area formula.

EXAMPLE 15.1 Graph a parallelogram whose vertices are $A(2, 2)$, $B(5, 6)$, $C(13, 6)$, and $D(10, 2)$, and then find its area.

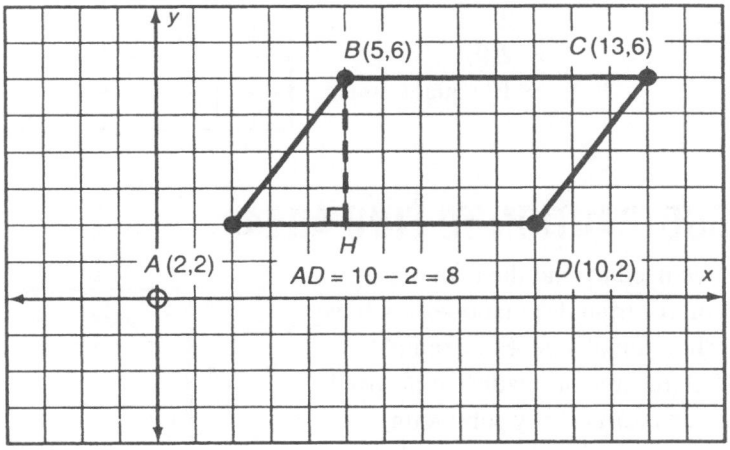

SOLUTION

In the accompanying graph, altitude $\overline{BH}$ has been drawn to $\overline{AD}$. By counting boxes we see that $AD = 8$ and $BH = 4$.
Therefore,

$$\begin{aligned} \text{Area } \square\, ABCD &= bh \\ &= AD \times BH \\ &= 8 \times 4 \\ &= 32 \text{ square units} \end{aligned}$$

EXAMPLE 15.2 Graph the trapezoid whose vertices are $A(-4, 0)$, $B(-4, 3)$, $C(0, 6)$, and $D(0, 0)$, and then find its area.

SOLUTION

In the accompanying graph, $\overline{AB}$ and $\overline{CD}$ are the bases and $\overline{AD}$ an altitude of the trapezoid.

$$\begin{aligned} \text{Area trapezoid } ABCD &= \frac{1}{2} h\,(b_1 + b_2) \\ &= \frac{1}{2}(AD)(AB + CD) \\ &= \frac{1}{2}(4)(3 + 6) \\ &= 2(9) \\ &= 18 \text{ square units} \end{aligned}$$

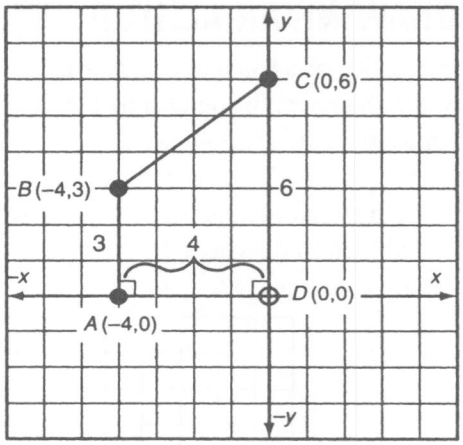

USING SUBTRACTION TO FIND AREA

Sometimes we need to find the area of a quadrilateral (or triangle) that does not have a vertical or horizontal side. For example, in Figure 15.2 the area of quadrilateral $ABCD$ can be obtained *indirectly* by subtracting the sum of the areas of right triangles I, II, III, and IV from the area of rectangle $WXYZ$.

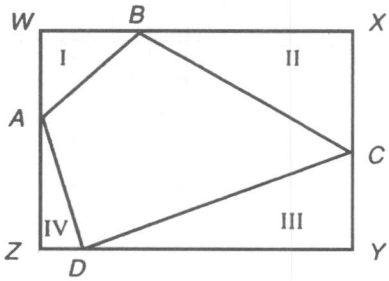

FIGURE 15.2

EXAMPLE 15.3 Find the area of the quadrilateral whose vertices are $A(-2, 2)$, $B(2, 5)$, $C(8, 1)$, and $D(-1, -2)$.

SOLUTION

The area of quadrilateral *ABCD* is calculated *indirectly* as follows:

- Circumscribe a rectangle about quadrilateral *ABCD* by drawing intersecting horizontal and vertical segments through the vertices of the quadrilateral as shown in the accompanying diagram.
- Find the area of the rectangle:

$$\text{Area rectangle } WXYZ = (ZY)(YX) = (10)(7) = 70$$

- Find the sum of the areas of the right triangles in the four corners of the rectangle. Keep in mind that the area of a right triangle is equal to one-half the product of the lengths of the legs of the triangle.

$$\begin{aligned}
\text{Sum} = \text{Area right } \triangle BWA &= \frac{1}{2}(BW)(WA) = \frac{1}{2}(4)(3) = 6 \\
+ \text{ Area right } \triangle BXC &= \frac{1}{2}(BX)(XC) = \frac{1}{2}(6)(4) = 12 \\
+ \text{ Area right } \triangle DYC &= \frac{1}{2}(DY)(YC) = \frac{1}{2}(9)(3) = 13.5 \\
+ \text{ Area right } \triangle DZA &= \frac{1}{2}(DZ)(ZA) = \frac{1}{2}(1)(4) = 2 \\
\hline
\text{Sum of } \triangle \text{ areas} &= 33.5
\end{aligned}$$

- Subtract the sum of the areas of the right triangles from the area of the rectangle.

$$\begin{aligned}
\text{Area quad } ABCD &= \text{Area rect } WXYZ - \text{Sum of areas of right triangles} \\
&= 70 - 33.5 \\
&= 36.5
\end{aligned}$$

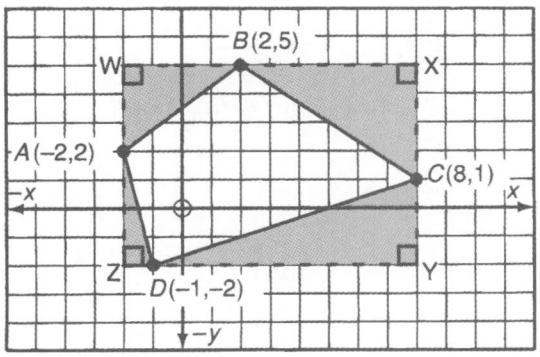

The Midpoint and Distance Formulas

FINDING THE MIDPOINT

In Figure 15.3, what are the coordinates of the *midpoint* of $\overline{AB}$? In general, the coordinates of the midpoint of a segment joining two points are found by taking the average of their x-coordinates and then finding the average of their y-coordinates.

In this case:

$$\frac{2+10}{2} = 6 \quad \text{and} \quad \frac{5+5}{2} = 5$$

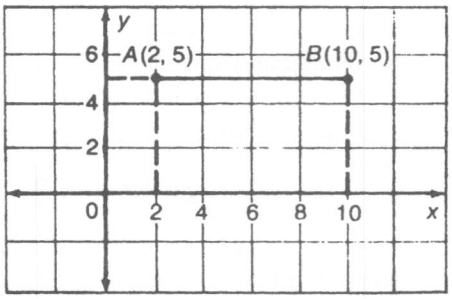

FIGURE 15.3

Hence, the midpoint of $\overline{AB}$ will be located at $M(6, 5)$.

It is sometimes convenient to use *subscript notation* to represent an ordered sequence of two or more related variables. For example, if x_1 refers to the x-coordinate of one point on a line, then x_2 could represent the x-coordinate of a second point on the same line. The subscripted variable x_1 is read "x sub-1," and the subscripted variable x_2 is read "x sub-2." Subscript notation is used in Theorem 15.1.

THEOREM 15.1 THE MIDPOINT FORMULA

The coordinates of the midpoint $M(x_m, y_m)$ of a segment whose end points are $A(x_1, y_1)$ and $B(x_2, y_2)$ may be found using the formulas:

$$x_m = \frac{x_1 + x_2}{2}$$

and

$$y_m = \frac{y_1 + y_2}{2}$$

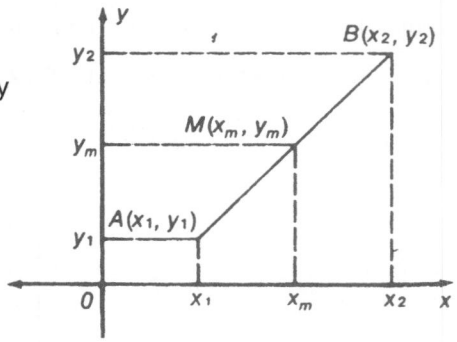

EXAMPLE 15.4 Find the coordinates of the midpoint of a segment whose end points are $H(4, 9)$ and $K(-10, 1)$.

SOLUTION

$x_1 \rightarrow H(4, 9) \leftarrow y_1$ | $x_2 \rightarrow K(-10, 1) \leftarrow y_2$

$$x_m = \frac{x_1 + x_2}{2} = \frac{4 + (-10)}{2} = \frac{-6}{2} = -3$$

$$y_m = \frac{y_1 + y_2}{2} = \frac{9 + 1}{2} = \frac{10}{2} = 5$$

The coordinates of the midpoint of $\overline{HK}$ are $(-3, 5)$.

EXAMPLE 15.5 $M(7, -1)$ is the midpoint of $\overline{WL}$. If the coordinates of end point W are $(5, 4)$, find the coordinates of end point L.

SOLUTION

$$x_m = \frac{5 + x}{2}$$

$W(5, 4)$ $L(x, y)$

$$y_m = \frac{4 + y}{2}$$

$$x_m = 7 = \frac{5 + x}{2}$$

Multiply each side of the equation by 2:

$$2(7) = \cancel{2}\left(\frac{5 + x}{\cancel{2}}\right)$$

$$14 = 5 + x$$

$$9 = x$$

$$y_m = -1 = \frac{4 + y}{2}$$

$$2(-1) = \cancel{2}\left(\frac{4 + y}{\cancel{2}}\right)$$

$$-2 = 4 + y$$

$$-6 = y$$

Hence, the coordinates of point L are $(9, -6)$.

EXAMPLE 15.5 The coordinates of quadrilateral $ABCD$ are $A(-3, 0)$, $B(4, 7)$, $C(9, 2)$, and $D(2, -5)$. Prove that $ABCD$ is a parallelogram.

SOLUTION

If the diagonals of a quadrilateral bisect each other, then the quadrilateral is a parallelogram. In the diagram, diagonals $\overline{AC}$ and $\overline{BD}$ intersect at point E. If the coordinates of point E are the coordinates of the midpoints of $\overline{AC}$ and $\overline{BD}$, then the diagonals bisect each other, and the quadrilateral is a parallelogram.

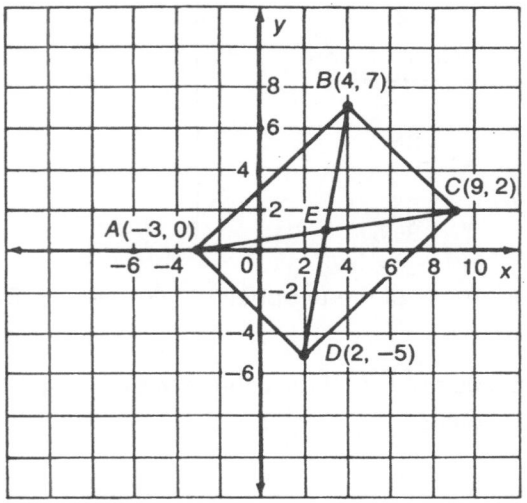

To find the midpoint of $\overline{AC}$, let $(x_1, y_1) = (-3, 0)$ and $(x_2, y_2) = (9, 2)$:

$$x_m = \frac{-3 + 9}{2} = \frac{6}{2} = 3$$

$$y_m = \frac{0 + 2}{2} = \frac{2}{2} = 1$$

Hence, the coordinates of the midpoint of $\overline{AC}$ are $(3, 1)$.

To find the midpoint of $\overline{BD}$, let $(x_1, y_1) = (4, 7)$ and $(x_2, y_2) = (2, -5)$:

$$x_m = \frac{4 + 2}{2} = \frac{6}{2} = 3$$

$$y_m = \frac{7 + (-5)}{2} = \frac{2}{2} = 1$$

The coordinates of the midpoint of $\overline{BD}$ are $(3, 1)$.

Since the diagonals have the same midpoint, they bisect each other, thereby establishing that the quadrilateral is a parallelogram.

FINDING THE DISTANCE BETWEEN TWO POINTS

The distance between two points in the coordinate plane is the length of the segment that joins the two points. Figures 15.4 and 15.5 illustrate how to find the distance between two points that determine a horizontal or a vertical line. In each case, the distance between the two given points is found by subtracting a single pair of coordinates. The length of a *horizontal* segment is obtained by calculating the difference in the x-coordinates of the two points. The length of a *vertical* segment (Figure 15.5) is found by subtracting the y-coordinates of the two points.

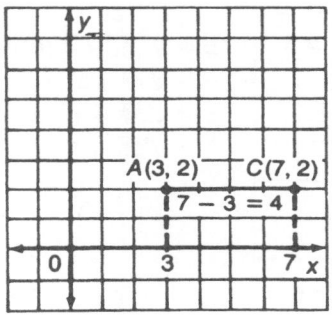

FIGURE 15.4

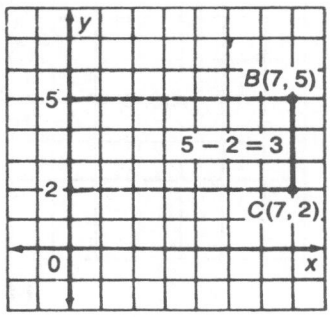

FIGURE 15.5

The Greek letter Δ (delta) is sometimes used to represent the difference or change in a pair of values. In general, if the coordinates of two points are $P(x_1, y_1)$ and $Q(x_2, y_2)$, then $\Delta x = x_2 - x_1$ and $\Delta y = y_2 - y_1$. For example, if the coordinates of P and Q are $P(1, 7)$ and $Q(5, 9)$, then $\Delta x = 5 - 1 = 4$ and $\Delta y = 9 - 7 = 2$.

THEOREM 15.2 LENGTHS OF HORIZONTAL AND VERTICAL SEGMENTS

$PQ = \Delta x = x_2 - x_1 \ (x_2 > x_1)$

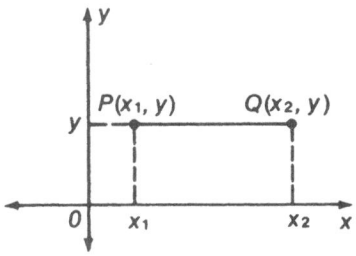

$RW = \Delta y = y_2 - y_1 \ (y_2 > y_1)$

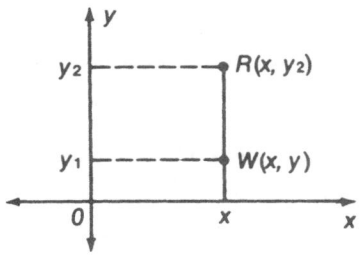

Finding the distance between two points that determine a slanted line is a bit more complicated. Figure 15.6 demonstrates that the distance between points A and B can be found indirectly by first forming a right triangle in which $\overline{AB}$ is the hypotenuse, and horizontal and vertical segments are legs of the triangle. After the lengths of the legs of this right triangle have been determined, a straightforward application of the Pythagorean Theorem leads to the length of $\overline{AB}$, which represents the distance between points A and B. The general formula for determining the distance between any two given points is stated in Theorem 15.3.

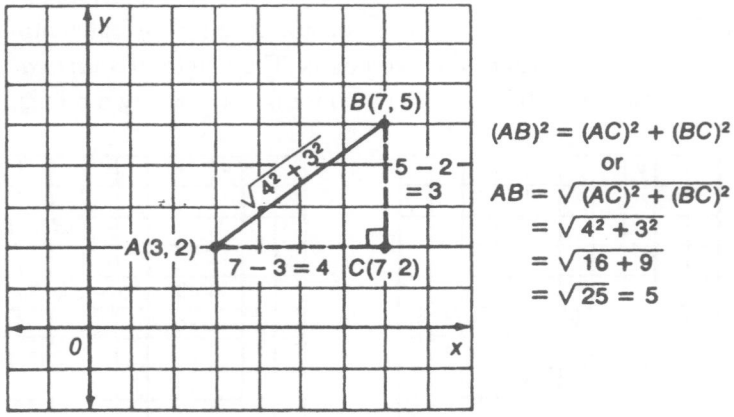

FIGURE 15.6

THEOREM 15.3 THE DISTANCE FORMULA

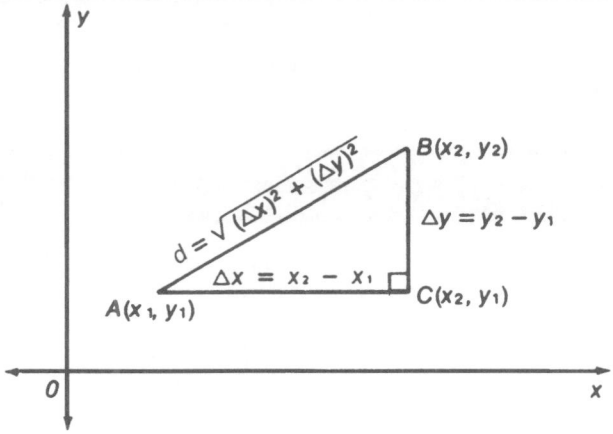

The distance, d, between points $A(x_1, y_1)$ and $B(x_2, y_2)$ may be found using the formula:

$$d = \sqrt{(\Delta x)^2 + (\Delta y)^2}$$

where

$$\Delta x = x_2 - x_1 \quad \text{and} \quad \Delta y = y_2 - y_1$$

EXAMPLE 15.7 Prove that the parallelogram whose coordinates are $A(-3, 0)$, $B(4, 7)$, $C(9, 2)$, and $D(2, -5)$ is a rectangle. (See the diagram of Example 15.6.)

SOLUTION

A parallelogram is a rectangle if its diagonals are equal in length. Let us find and then compare the lengths of diagonals $\overline{AC}$ and $\overline{BD}$.

To find the length of $\overline{AC}$, let $(x_1, y_1) = A\ (-3, 0)$ and $(x_2, y_2) = C(9, 2)$. Therefore:

$$\begin{aligned} \Delta x &= x_2 - x_1 \\ &= 9 - (-3) = 9 + 3 \\ &= 12 \end{aligned} \qquad \begin{aligned} \Delta y &= y_2 - y_1 \\ &= 2 - 0 \\ &= 2 \end{aligned}$$

$$\begin{aligned} AC &= \sqrt{(\Delta x)^2 + (\Delta y)^2} \\ &= \sqrt{(12)^2 + (2)^2} \\ &= \sqrt{144 + 4} \\ &= \sqrt{148} \end{aligned}$$

To find the length of $\overline{BD}$, let $(x_1, y_1) = B(4, 7)$ and $(x_2, y_2) = D(2, -5)$. Therefore:

$$\begin{aligned} \Delta x &= x_2 - x_1 \\ &= 2 - 4 \\ &= -2 \end{aligned} \qquad \begin{aligned} \Delta y &= y_2 - y_1 \\ &= -5 - 7 \\ &= -12 \end{aligned}$$

$$\begin{aligned} BD &= \sqrt{(\Delta x)^2 + (\Delta y)^2} \\ &= \sqrt{(-2)^2 + (-12)^2} \\ &= \sqrt{4 + 144} \\ &= \sqrt{148} \end{aligned}$$

Since AC and BD are each equal to $\sqrt{148}$, they must be equal to each other. Hence, $AC = BD$, and therefore parallelogram $ABCD$ is a rectangle.

Slope of a Line

If you imagine that a line represents a hill, it is obvious that some lines will be more difficult to walk up than other lines.

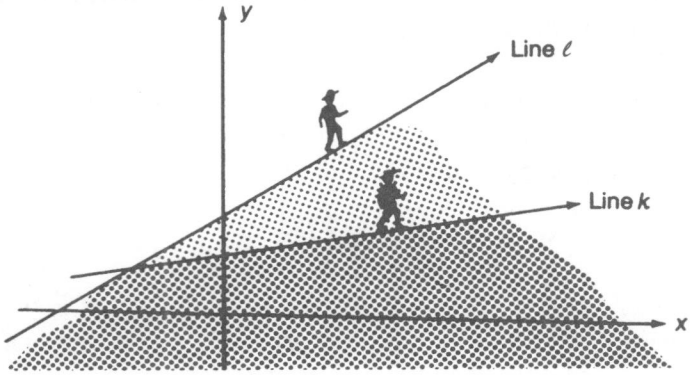

Line ℓ would be more difficult to climb than line k since line ℓ is *steeper* than line k. Another name for steepness is *slope*. The slope of a line may be expressed as a number. To do this, select any two different points on the line. In traveling from one point to the other, compare the change in the vertical distance (Δy) to the change in the horizontal distance (Δx). The ratio of these quantities $\left(\frac{\Delta y}{\Delta x}\right)$ represents the slope of the line.

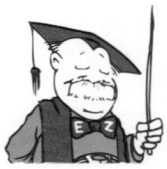

DEFINITION OF SLOPE

The *slope, m*, of a nonvertical line that passes through points $A(x_1, y_1)$ and $B(x_2, y_2)$ is given by the ratio of the change in the values of their y-coordinates to the change in the value of their x-coordinates:

$$\text{Slope} = m = \frac{\Delta y}{\Delta x} = \frac{y_2 - y_1}{x_2 - x_1}$$

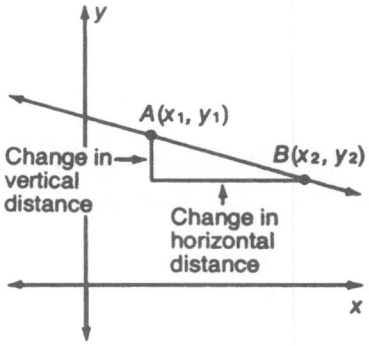

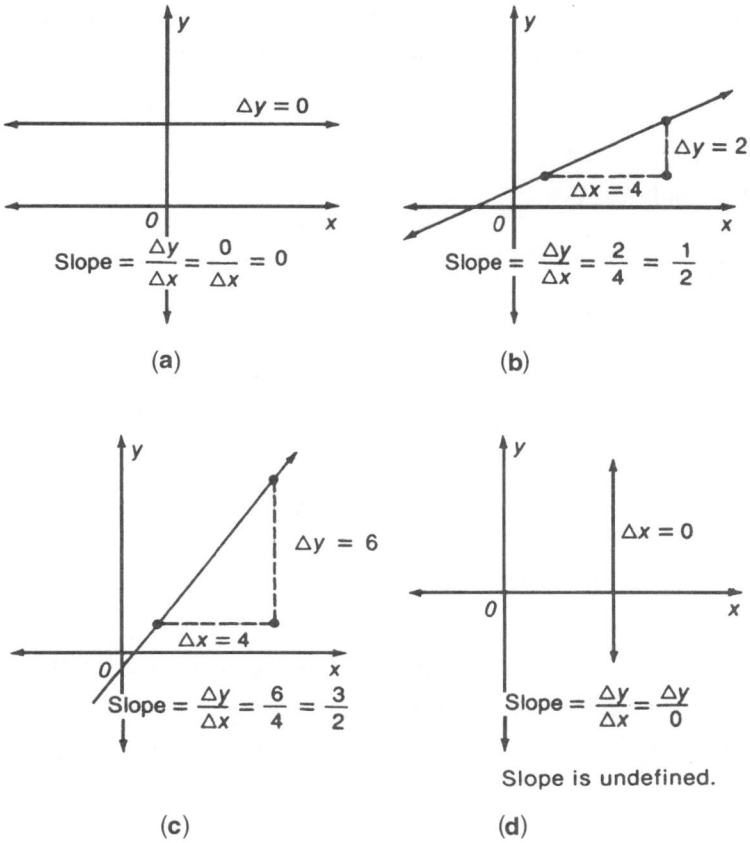

FIGURE 15.7

As the series of graphs in Figure 15.7 illustrates, the steeper the line, the larger the value of its slope. Since a horizontal line has no steepness, its slope has a numerical value of 0. A vertical line is impossible to "walk up," so its slope is not defined. Mathematically speaking:

- The y-coordinate of every point on a *horizontal* line is the same, so $\Delta y = 0$. For any two points on a horizontal line, the slope formula may be expressed as

$$m = \frac{\Delta y}{\Delta x} = \frac{0}{\Delta x}$$

Since 0 divided by any number is 0, the slope of a horizontal line is always 0.

- The x-coordinate of every point on a *vertical* line is the same, so $\Delta x = 0$. For any two points on a vertical line, the slope formula may be expressed as

$$m = \frac{\Delta y}{\Delta x} = \frac{\Delta y}{0}$$

Since division by 0 is not defined, the slope of a vertical line is not defined.

EXAMPLE 15.8 For each of the following pairs of values, plot the points and draw a line through the points. Use the slope formula to determine the slope of each line.

a. $A(4, 2)$ and $B(6, 5)$
b. $P(0, 3)$ and $Q(5, -1)$
c. $J(4, 3)$ and $K(8, 3)$
d. $W(2, 1)$ and $C(2, 7)$

SOLUTION

a. See diagram **a**. To find the slope, observe that:

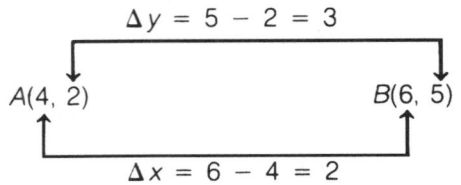

$$\text{Slope of } \overleftrightarrow{AB} = m = \frac{\Delta y}{\Delta x} = \frac{y_2 - y_1}{x_2 - x_1} = \frac{5 - 2}{6 - 4} = \frac{3}{2}$$

The slope of $\overleftrightarrow{AB}$ is $\frac{3}{2}$. The slope of $\overleftrightarrow{AB}$ was calculated assuming that point B was the second point. When using the slope formula either of the two points may be considered the second point. For example, let's repeat the slope calculation considering point A to be the second point (that is, $x_2 = 4$ and $y_2 = 2$):

$$m = \frac{\Delta y}{\Delta x} = \frac{y_2 - y_1}{x_2 - x_1} = \frac{2 - 5}{4 - 6} = \frac{-3}{-2} = \frac{3}{2}$$

The same value, $\frac{3}{2}$, is obtained for the slope of $\overleftrightarrow{AB}$, regardless of whether A is taken as the first or second point.

b. See diagram **b**. Let $(x_1, y_1) = (0, 3)$ and $(x_2, y_2) = (5, -1)$:

$$m = \frac{\Delta y}{\Delta x} = \frac{y_2 - y_1}{x_2 - x_1} = \frac{-1 - 3}{5 - 0} = \frac{-4}{5}$$

The slope of $\overleftrightarrow{PQ}$ is $\frac{-4}{5}$. The slope of $\overleftrightarrow{AB}$ in part **a** was *positive* in value, but the slope of $\overleftrightarrow{PQ}$ has a *negative* value. Compare the directions of these lines.

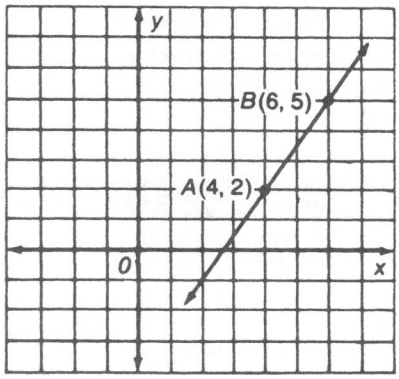

a.

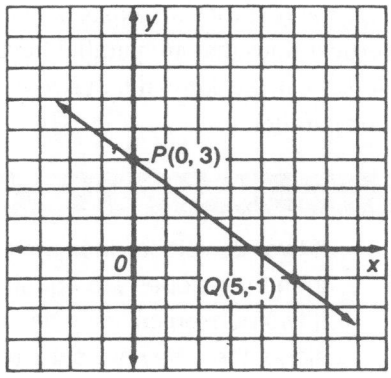

b.

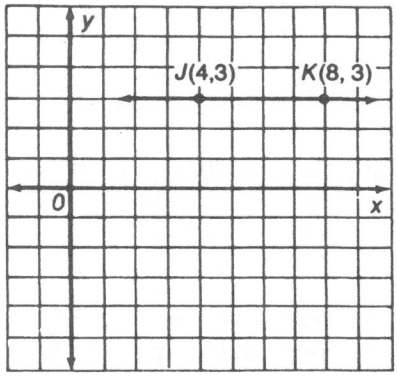

c.

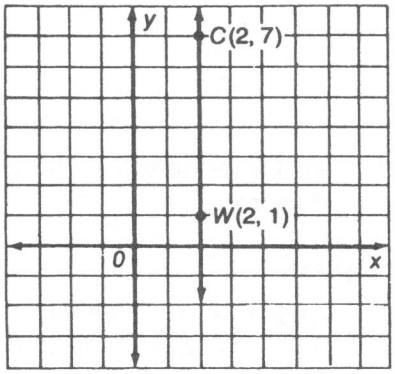

d.

Notice that, if a line has a positive slope, it climbs up *and* to the right; if a line has a negative slope, it falls down *and* to the right. These observations may be restated as follows: If, as you move along a line from left to right, the values of the x-coordinates increase and the values of the y-coordinates also increase, then the line has a positive slope; but if, as the values of the x-coordinates increase, the values of the y-coordinates decrease, then the line has a negative slope.

REMEMBER
The slope of a horizontal line is 0. The slope of a vertical line is undefined.

c. See diagram **c**. Since the line is horizontal, there is no change in y, so $\Delta y = 0$.

$$m = \frac{\Delta y}{\Delta x} = \frac{0}{\Delta x} = 0$$

d. See diagram **d**. Since the line is vertical, there is no change in x, so $\Delta x = 0$. The slope is undefined.

A special relationship exists between the slopes of parallel lines. Since two nonvertical lines that are parallel have the same steepness, their slopes are equal. Conversely, if it is known that two nonvertical lines have the same slope, then the lines must be parallel.

THEOREM 15.4 SLOPES OF PARALLEL LINES

- If two nonvertical lines are parallel, then their slopes are equal.
- If two nonvertical lines have the same slope, then they are parallel.

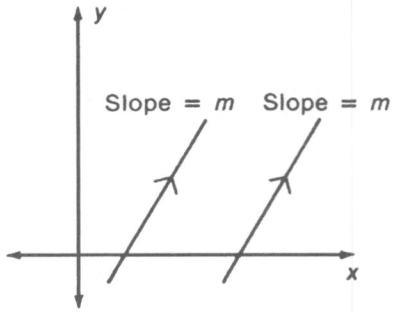

EXAMPLE 15.9 Show that the quadrilateral whose vertices are $A(-3, -8)$, $B(-2, 1)$, $C(2, 5)$, and $D(7, 2)$ is a trapezoid.

SOLUTION

Since a trapezoid is a quadrilateral that has *exactly* one pair of parallel sides, we must show that one pair of sides have the same slope and one pair of sides have different slopes.

- Slope of $\overline{AB} = \dfrac{\Delta y}{\Delta x} = \dfrac{1-(-8)}{-2-(-3)} = \dfrac{1+8}{-2+3} = \dfrac{9}{1} = 9$
- Slope of $\overline{BC} = \dfrac{\Delta y}{\Delta x} = \dfrac{5-1}{2-(-2)} = \dfrac{4}{2+2} = \dfrac{4}{4} = 1$
- Slope of $\overline{CD} = \dfrac{\Delta y}{\Delta x} = \dfrac{2-5}{7-2} = \dfrac{-3}{5} = -\dfrac{3}{5}$
- Slope of $\overline{AD} = \dfrac{\Delta y}{\Delta x} = \dfrac{2-(-8)}{7-(-3)} = \dfrac{2+8}{7+3} = \dfrac{10}{10} = 1$

Since the slope of $\overline{BC}$ = the slope of $\overline{AD}$, $\overline{BC} \parallel \overline{AD}$ and $\overline{BC}$ and $\overline{AD}$ are the bases of the trapezoid.

Since the slope of $\overline{AB} \neq$ slope of $\overline{CD}$, $\overline{AB}$ and $\overline{CD}$ are the legs or nonparallel sides of the trapezoid.

A less obvious relationship exists between the slopes of *perpendicular* lines. If two lines are perpendicular, the slope of one line will be the negative reciprocal of the slope of the other line. For example, suppose line ℓ is perpendicular to line k. If the slope of line ℓ is $\frac{2}{3}$, then the slope of line k is $-\frac{3}{2}$. Conversely, if it is known that the slopes of two lines are negative reciprocals of one another (or, equivalently, that the product of their slopes is negative), then the lines must be perpendicular.

THEOREM 15.5 SLOPES OF PERPENDICULAR LINES

- If two lines are perpendicular, then their slopes are negative reciprocals of one another.
- If the slopes of two lines are negative reciprocals of one another, then the lines are perpendicular.

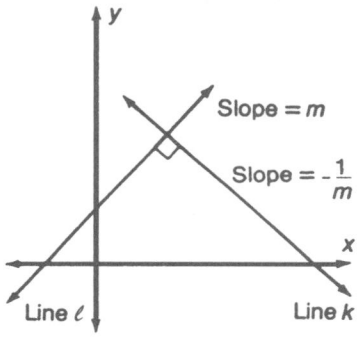

EXAMPLE 15.10 Show by means of slope that the triangle whose vertices are $A(2, 0)$, $B(11, 8)$, and $C(6, 10)$ is a right triangle.

SOLUTION

A triangle is a right triangle if two sides are perpendicular. Let us find and then compare the slopes of the three sides of the triangle.

- Slope of $\overline{AB} = \frac{\Delta y}{\Delta x} = \frac{8-0}{11-2} = \frac{8}{9}$
- Slope of $\overline{AC} = \frac{\Delta y}{\Delta x} = \frac{10-0}{6-2} = \frac{10}{4} = \frac{5}{2}$
- Slope of $\overline{BC} = \frac{\Delta y}{\Delta x} = \frac{10-8}{6-11} = \frac{2}{-5} = -\frac{2}{5}$

REMEMBER
The slopes of two perpendicular lines are negative reciprocals of one another.

Notice that the slopes of $\overline{AC}$ and $\overline{BC}$ are negative reciprocals. Hence, $\overline{AC} \perp \overline{BC}$, and angle C of $\triangle ABC$ is a right angle.

Equation of a Line

For each point that lies on a given line, the same relationship between the values of the x- and y-coordinates of these points must hold. For example, the coordinates of three representative points on line ℓ in Figure 15.8 are (0, 1), (1, 3), and (2, 5). In comparing the x- and y-coordinates of these points, do you see a pattern? In each case the y-coordinate is arrived at by multiplying the corresponding x-coordinate by 2 and then adding 1.

A general rule that expresses this relationship is the equation $y = 2x + 1$, which is referred to as *an equation of line* ℓ. The x- and y-coordinates for *every* point on line ℓ in Figure 15.8 must satisfy this equation. Without looking at the figure, can you predict the y-coordinate of a point that lies of line ℓ if its x-coordinate is –2? All you need do is substitute –2 for x in the equation $y = 2x + 1$:

$$\begin{aligned} y &= 2x + 1 \\ &= 2(-2) + 1 \\ &= -4 + 1 \\ &= -3 \end{aligned}$$

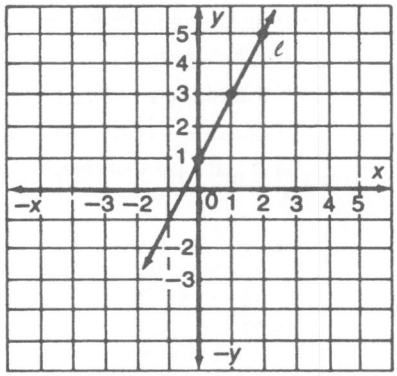

FIGURE 15.8

Now check Figure 15.8. Notice that the point whose coordinates are (–2, –3) lies on line ℓ.

In the equation $y = 2x + 1$, let's see if any special significance can be attached to the numbers 2 and 1. Referring to Figure 15.8, notice that the line crosses the y-axis at the y-value of 1. We call this point the *y-intercept* of a line. The slope of the line can be found by considering any two points on the line, say (1, 3) and (2, 5). For these points, if we let $x_1 = 1, y_1 = 3, x_2 = 2$, and $y^2 = 5$, we have

$$m = \frac{\Delta y}{\Delta x} = \frac{5-3}{2-1} = \frac{2}{1} = 2$$

The slope of the line is 2. The coefficient of x in the equation $y = 2x + 1$ is also 2. Hence,

$$y = 2x + 1$$

slope of the line

y-intercept of the line

In general, whenever an equation of a nonvertical line is written in the form

$$y = mx + b,$$

the equation is said to be written in *slope-intercept* form since m represents the slope of the line and b is its y-intercept.

EXAMPLE 15.11 Line p is parallel to the line $y + 4x = 3$, and line q is perpendicular to the line $y + 4x = 3$.

a. What is the slope of line p?

b. What is the slope of the line q?

SOLUTION

Write an equation of the given line in $y = mx + b$ form $y = -4x + 3$. The slope of the line is $m = -4$.

a. Since parallel lines have the same slope, the slope of line p is -4.

b. Since perpendicular lines have slopes that are negative reciprocals, the slope of line q is $\frac{1}{4}$.

EXAMPLE 15.12 The line whose equation is $y - 3x = b$ passes through point $A(2, 5)$.

a. Find the slope of the line.

b. Find the y-intercept of the line.

SOLUTION

a. To find the slope of the line, put the equation of the line in $y = mx + b$ form. The equation $y - 3x = b$ may be written as $y = 3x + b$, which means that the slope of the line is 3.

b. In the equation $y = 3x + b$, b represents the y-intercept of the line. Since point A lies on the line, its coordinates must satisfy the equation. To find the value of b, replace x by 2 and y by 5 in the original equation.

$$\begin{aligned} y - 3x &= b \\ 5 - 3 \cdot 2 &= b \\ 5 - 6 &= b \\ -1 &= b \end{aligned}$$

The y-intercept is -1.

WRITING EQUATIONS OF LINES

If a line has a slope of 3 ($m = 3$) and a y-intercept of -4($b = -4$), then an equation of this line may be written by replacing m and b in $y = mx + b$ with their numerical values: $y = 3x - 4$.

EXAMPLE 15.13 Write an equation of the line whose y-intercept is 1 and that is perpendicular to the line $y = \frac{1}{2}x - 4$.

SOLUTION

The slope of the given line is $\frac{1}{2}$. Since the lines are perpendicular, the slope m of the desired line is the negative reciprocal of $\frac{1}{2}$, which is -2. An equation of the desired line is $y = -2x + 1$.

If (1) the slope m of a line and the coordinates of a point $P(a, b)$ on the line, or (2) the coordinates of two points on the line are known, then it is usually more convenient to write an equation of the line using the *point-slope* form:

$$y - b = m(x - a).$$

EXAMPLE 15.14 Write an equation of a line that is parallel to the line $y + 2x = 5$ and passes through point $(1, 4)$.

SOLUTION

Since the lines are parallel, the slope m of the desired line is equal to the slope of the line $y + 2x = 5$. The line $y + 2x = 5$ may be written in the slope-intercept form as $y = -2x + 5$. Therefore, its slope is -2, so $m = -2$.

Method 1 (Slope-Intercept Form)	Method 2 (Point-Slope Form)
Find b by replacing m by -2, x by 1, and y by 4: $y = mx + b$ $4 = -2(1) + b$ $4 = -2 + b$ $6 = b$ Therefore, $y = -2x + 6$	Since $(a, b) = (1, 4)$, replace m by -2, a by 1, and b by 4: $y - b = m(x - a)$ $y - 4 = -2(x - 1)$

Keep in mind that the slope-intercept and point-slope forms of an equation of a line are equivalent. For example, the equation $y - 4 = -2(x - 1)$ may be expressed in $y = mx + b$ form as follows:

$$\begin{aligned} y - 4 &= -2(x - 1) \\ &= -2x + 2 \\ y &= -2x + 2 + 4 \\ &= -2x + 6 \end{aligned}$$

EXAMPLE 15.15 Write an equation of the line that passes through the origin and is perpendicular to the line whose equation is $y = 3x - 6$.

SOLUTION

The slope of the line whose equation is $y = 3x - 6$ is 3, so the slope m of a perpendicular line is $-\frac{1}{3}$ (the negative reciprocal of 3). Since the desired line passes through the origin, $(a, b) = (0, 0)$. Therefore,

$$y - b = m(x - a)$$

$$y - 0 = -\frac{1}{3}(x - 0)$$

or

$$y = -\frac{1}{3}x$$

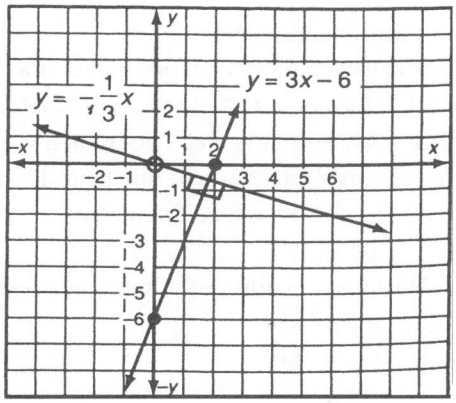

EXAMPLE 15.16 Write an equation of the line that contains points $A(6, 0)$ and $B(2, -6)$.

SOLUTION

Use the point-slope form. First, find the slope m of $\overleftrightarrow{AB}$.

$$m = \frac{-6-0}{2-6} = \frac{-6}{-4} = \frac{3}{2}$$

Next, choose either point and then use its x- and y-coordinates. Consider point A. Then $(a, b) = (6, 0)$.

$$y - b = m(x - a)$$
$$y - 0 = \frac{3}{2}(x - 6)$$
$$y = \frac{3}{2}(x - 6)$$

Multiplying each term inside the parentheses by $\frac{3}{2}$ gives

$$y = \frac{3}{2}x - 9$$

EQUATIONS OF HORIZONTAL AND VERTICAL LINES

Until now we have been considering only *oblique* or *slanted* lines. Oblique lines have nonzero slopes. The slope of a horizontal line is 0, while the slope of a vertical line is undefined. The forms of the equations of horizontal and vertical lines, therefore, require special consideration.

Since each point on a horizontal line has the same y-coordinate, call it b, its equation has the form $y = b$. Note that b is also the y-intercept of the horizontal line, extended, if necessary. Similarly, each point on a vertical line has the same x-coordinate, say a, so that equation of a vertical line takes the form $x = a$. Note that a is the x-intercept of the vertical line, extended, if necessary.

EXAMPLE 15.17 Write an equation of a line that passes through point $A(-1, 2)$ and is parallel to

a. the x-axis

b. the y-axis

SOLUTION

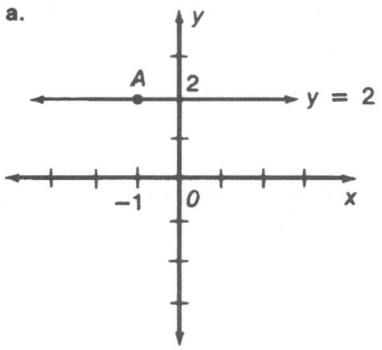

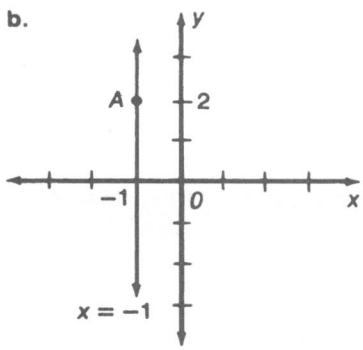

General Form of Equation	What it Means
$y = mx + b$	Equation of a line with m = slope of line; b = y-intercept
$y - b = m(x - a)$	Form to use when the slope m of the line and the coordinates (a, b) of a point on the line are known. Also use when two points on the line are given, after first using their coordinates to calculate m.
$x = a$	Vertical line (parallel to y-axis) that intersects the x-axis at a.
$y = b$	Horizontal line (parallel to x-axis) that intersects the y-axis at b.

Equation of a Circle

The set of all points that are 4 units from point $A(2, 3)$ is a circle having A as its center and a radius of 4 units. (See Figure 15.9.) Suppose we wish to find the *equation* that describes this situation. We can begin by choosing a representative point, $P(x, y)$, that satisfies the stated condition. The distance between point A and point P must therefore be 4. A straightforward application of the distance formula, where $\Delta x = x - 2$ and $\Delta y = y - 3$, yields the following result:

$$PA = \sqrt{(\Delta x)^2 + (\Delta y)^2}$$
$$= \sqrt{(x-2)^2 + (y-3)^2}$$

Since $PA = 4$, we may write:

$$\sqrt{(x-2)^2 + (y-3)^2} = 4$$

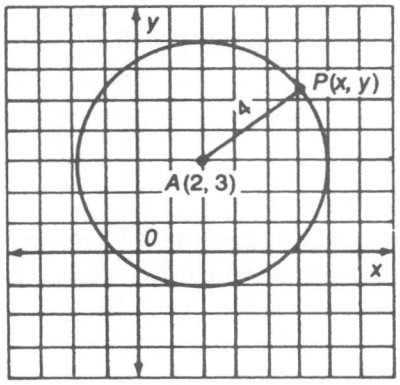

FIGURE 15.9

By squaring both sides of this equation, we obtain

$$(x - 2)^2 + (y - 3)^2 = 16$$

coordinates of the center of the circle

the square of the radius

THEOREM 15.6 EQUATION OF A CIRCLE

A circle whose center is located at $A(p, q)$ and that has a radius r units in length is defined by the equation:

$$(x - p)^2 + (y - q)^2 = r^2$$

If the center of the circle having radius r is the origin, then its equation is

$$x^2 + y^2 = r^2$$

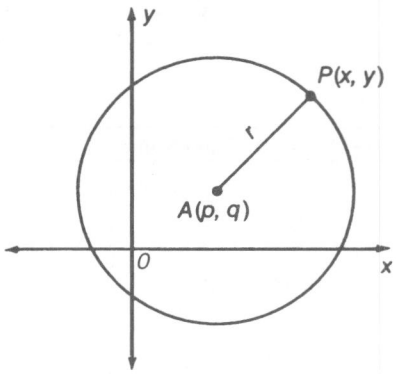

EXAMPLE 15.18 Write an equation for each of the following circles:

a. The center is at $(3, -5)$, and the radius is 7 units in length.

b. The center is at the origin, and the radius is 5 units in length.

SOLUTION

a. Here, $p = 3$, $q = -5$, and $r = 7$.

$$\begin{aligned}(x - p)^2 + (y - q)^2 &= r^2\\ (x - 3)^2 + y - (-5)^2 &= 7^2\\ (x - 3)^2 + (y + 5)^2 &= 49\end{aligned}$$

b. At the origin, $p = 0$ and $q = 0$. Since $r = 5$,

$$\begin{aligned}(x - p)^2 + (y - q)^2 &= r^2\\ (x - 0)^2 + (y - 0)^2 &= 5^2\\ x^2 + y^2 &= 25\end{aligned}$$

Notice that in part **b** of Example 15.18 an equation of the form $x^2 + y^2 = r^2$ defines a circle whose center is at the origin and that has a radius r units in length.

EXAMPLE 15.19 Determine the coordinates of the center of a circle and the length of its radius given the following equations:

a. $x^2 + y^2 = 100$

b. $(x + 1)^2 + (y - 4)^2 = 39$

SOLUTION

a. *Center:* origin. *Radius:* $\sqrt{100} = 10$.

b. Rewrite the equation as follows:

$$[x - (-1)]^2 + (y - 4)^2 = 39$$

Center: $(-1, 4)$ *Radius:* $\sqrt{39}$

Proofs Using Coordinates

By placing a polygon in a convenient position in a coordinate plane, and then representing the coordinates of its vertices using letters and zeros, we can easily prove many theorems. We will illustrate this fact by showing how the methods of coordinate geometry can be used to establish that the diagonals of a rectangle are congruent.

Our first concern is to decide where in the coordinate plane a representative rectangle should be placed. Figure 15.10 shows three possible positionings of a rectangle in a coordinate plane. One of these has a clear advantage over the other two placements. We generally look to position a polygon so that:

- The origin is a vertex of the polygon.
- At least one of the sides of the polygon coincides with a coordinate axis.

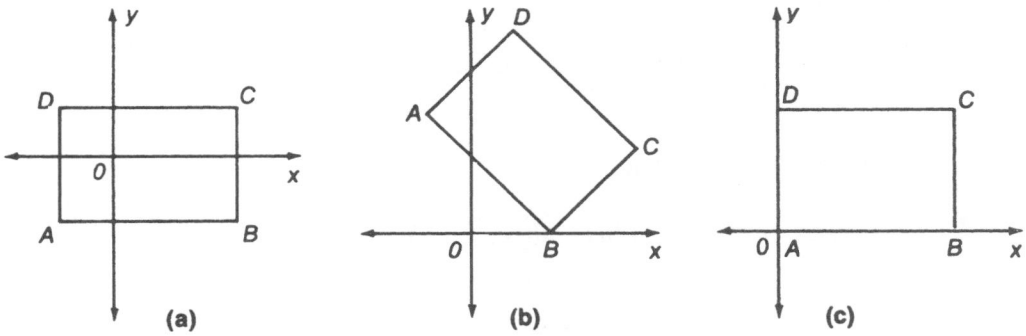

FIGURE 15.10

As we shall see, this positioning tends to simplify matters because it introduces zeros for some of the coordinates of the vertices. Figure 15.10*c* conforms to the guidelines given above. Since the sides of a rectangle are perpendicular to each other, a pair of adjacent sides of the rectangle can be made to coincide with the coordinate axes so that the origin is a vertex of the rectangle. The actual proof follows.

GIVEN: Rectangle $ABCD$ with diagonals $\overline{AC}$ and $\overline{BD}$.

PROVE: $AC = BD$.

PLAN: Use the distance formula to show $AC = BD$.

PROOF: STEP 1 Find AC.

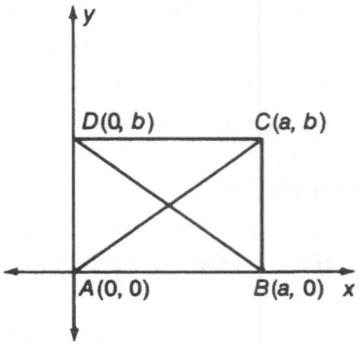

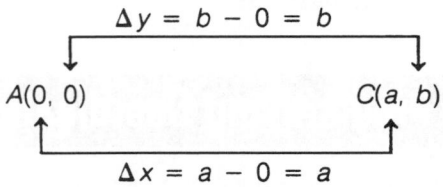

Use the distance formula:

$$\begin{aligned} AC &= \sqrt{(\Delta x)^2 + (\Delta y)^2} \\ &= \sqrt{(a)^2 + (b)^2} \\ &= \sqrt{a^2 + b^2} \end{aligned}$$

STEP 2 Find BD.

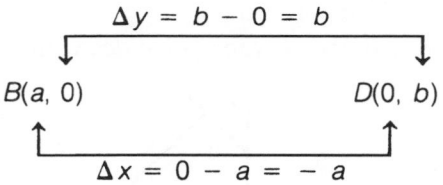

Use the distance formula:

$$\begin{aligned} BD &= \sqrt{(\Delta x)^2 + (\Delta y)^2} \\ &= \sqrt{(-a)^2 + (b)^2} \\ &= \sqrt{a^2 + b^2} \end{aligned}$$

Since AC and BD are both equal to $\sqrt{a^2 + b^2}$, they are equal to each other. Since $AC = BD$, $\overline{AC} \cong \overline{BD}$.

EXAMPLE 15.20 Prove the diagonals of a parallelogram bisect each other.

SOLUTION

Place parallelogram $QRST$ in the coordinate plane so that side $\overline{RS}$ coincides with the x-axis and one of the end points, point R, is the origin, as shown in the diagram. Next, draw $\overline{QT}$ parallel to $\overline{RS}$. Since $\overline{QT}$ is a horizontal line segment, the y-coordinates of points Q and T must be the same, say b. Represent the x-coordinate of point Q by a. Since the lengths of $\overline{QT}$ and $\overline{RS}$ must be the same, the x-coordinate of point T must then be represented by $a + c$.

PLAN: To prove the diagonals of parallelogram $QRST$ bisect each other, find and then compare the midpoints of RT and QS.

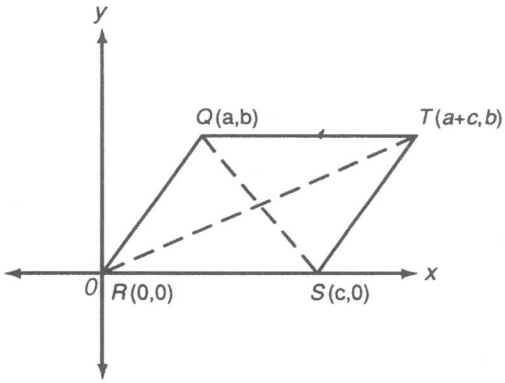

Midpoint of $\overline{RT}$:

$$x_m = \frac{0 + (a + c)}{2} = \frac{a + c}{2}; \qquad y_m = \frac{0 + b}{2} = \frac{b}{2}$$

The midpoint of diagonal $\overline{RT}$ is $\left(\frac{a + c}{2}, \frac{b}{2}\right)$.

Midpoint of $\overline{QT}$:

$$x_m = \frac{a + c}{2}; \qquad y_m = \frac{b + 0}{2} = \frac{b}{2}$$

The midpoint of diagonal $\overline{QT}$ is $\left(\frac{a + c}{2}, \frac{b}{2}\right)$.

Since the diagonals have the same midpoint, they bisect each other.

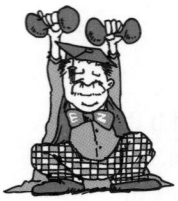

REVIEW EXERCISES FOR CHAPTER 15

In Exercises 1 to 4, draw the triangles whose vertices are given and then find their areas.

1. $A(0, 0)$, $B(5, 0)$, and $C(0, 8)$.

2. $R(-6, 0)$, $S(0, 0)$, and $T(-6, -4)$.

3. $A(-5, 2)$, $B(-3, 6)$, and $C(3, 1)$.

4. $J(2, -7)$, $K(11, 7)$, and $L(8, -3)$.

5. What is an equation of the circle shown in the accompanying figure?

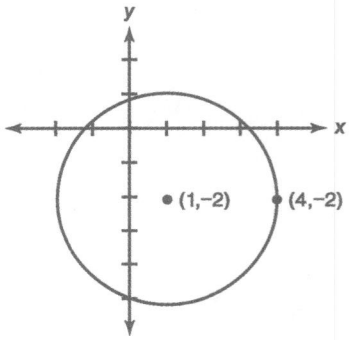

6. $P(2, -3)$ and $Q(-4, 5)$ are the end points of diameter $\overline{PQ}$ of circle O.

   (a) Find the coordinates of the center of the circle
   (b) Find the length of a radius of the circle.
   (c) Determine whether the point $K(-5, -2)$ lies on the circle.

7. The coordinates of the vertices of quadrilateral $MATH$ are $M(-3, 2)$, $A(4, 8)$, $T(15, 5)$, and $H(8, y)$. If $MATH$ is a parallelogram, find the value of y:

   (a) by using midpoint relationships
   (b) by using slope relationships

8. The coordinates of the vertices of $\triangle RST$ are $R(4, -4)$, $S(-1, 5)$, and $T(13, 9)$. Find the length of the median drawn to side $\overline{ST}$.

9. The coordinates of the vertices of quadrilateral $JKLM$ are $J(-7, 6)$, $K(-4, 8)$, $L(-2, 5)$, and $M(-5, 3)$. Prove that $JKLM$ is a square. (*Hint:* Show that $JKLM$ is a rectangle having a pair of congruent adjacent sides.)

10. The coordinates of the vertices of a triangle are $J(-10, -6)$, $K(0, 6)$, and $L(6, 1)$.

    (a) By means of slope, show that $\triangle JKL$ is a right triangle.
    (b) Verify your conclusion by showing that the lengths of the sides of $\triangle JKL$ satisfy the converse of the Pythagorean Theorem.
    (c) Find the area of $\triangle JKL$.

11. The coordinates of the vertices of a triangle are $A(2, -3)$, $B(5, 5)$, and $C(11, 3)$.

    (a) Find the slope of a line that is parallel to $\overline{AB}$.
    (b) Find the slope of the altitude drawn to side $\overline{AC}$.
    (c) Find the slope of the median drawn to side $\overline{BC}$.

12. Three points are *collinear* if the slopes of two of the segments determined by selecting any two pairs of points are the same. For example, points A, B, and C are collinear if any one of the following relationships is true: (1) slope of $\overline{AB}$ = slope of $\overline{BC}$; or (2) slope of $\overline{AB}$ = slope of $\overline{AC}$; or (3) slope of $\overline{BC}$ = slope of $\overline{AC}$.

    Determine in each case whether the three points are collinear.

    (a) $A(-4, -5)$, $B(0, -2)$, $C(8, 4)$.
    (b) $R(2, 1)$, $S(10, 7)$, $T(-4, -6)$.
    (c) $J(1, 2)$, $K(5, 8)$, $L(-3, -4)$.

13. Show by means of slope that the quadrilateral whose vertices are $S(-3, 2)$, $T(2, 10)$, $A(10, 5)$, and $R(5, -3)$ is a:

    (a) parallelogram
    (b) rectangle

14. Write the equation of the line that passes through the point $A(-4, 3)$ and that is parallel to the line $2y - x = 6$.

15. Write the equation of the line that passes through the point $(3, 1)$ and that is perpendicular to the line $3y + 2x = 15$.

16. Write the equation of the line that passes through points A and B with the given coordinates.

    (a) $A(-2, 7)$ and $B(4, 7)$
    (b) $A(0, 3)$ and $B(2, 1)$
    (c) $A(6, 8)$ and $B(6, -3)$
    (d) $A(1, -3)$ and $B(-1, 5)$

17. Write the equation of the line that contains point $(-5, 2)$ and that is parallel to:

    (a) the x-axis
    (b) the y-axis

18. The coordinates of the vertices of a triangle are $A(-1, -4)$, $B(7, 8)$, and $C(9, 6)$. Write the equation of each of the following lines:

    (a) The line that passes through point B and is parallel to $\overline{AC}$.
    (b) The line that contains the median drawn to side $\overline{BC}$ from vertex A.
    (c) The line that contains the altitude drawn to side $\overline{AB}$ from vertex C.

19. Write the equation of the perpendicular bisector of the segment that joins the points $A(3, -7)$ and $B(5, 1)$.

20. The equation of the line that contains the perpendicular bisector of the segment that joins the points $R(-8, t)$ and $S(2, -3)$ is $2y + 5x = k$. Find the values of k and t.

21. (a) On the same set of axes, graph and label the following lines: $y = 5, x = -4$, and $y = \frac{5}{4}x + 5$.

    (b) Find the number of square units in the area of the triangle whose vertices are the three points of intersection of the lines graphed in part **a**.

22. The line whose equation is $x = 5$ is tangent to a circle whose center is at the origin. Write the equation of the circle.

23. The center of a circle is located at $O(1, h)$. The line whose equation is $y = kx + 1$ is tangent to circle O at the point $P(3, 6)$. Find the values of k and h.

24. Two circles are tangent externally at point P. The equation of one of the circles is $x^2 + y^2 = 16$. If the other circle has its center on the positive y-axis and has a radius of 5 units, find its equation.

25. Jim is experimenting with a new drawing program on his computer. He created quadrilateral $TEAM$ with coordinates $T(-2,3)$, $E(-5,-4)$, $A(2,-1)$, and $M(5,6)$. Jim believes that he has created a rhombus but not a square. Prove that Jim is correct.

26. Find the area of the quadrilateral whose vertices are given.

    (a) $A(-2, 2)$, $B(2, 5)$, $C(8, 1)$, and $D(-1, -2)$.
    (b) $T(-4, -4)$, $R(-1, 5)$, $A(6, 5)$, and $P(9, -4)$.
    (c) $A(-4, -2)$, $B(0, 5)$, $C(9, 3)$, and $D(7, -4)$.
    (d) $M(-3, 5)$, $A(4, 8)$, $T(15, 3)$, and $H(8, -1)$.

27. Given: points $A(1, -1)$, $B(5, 7)$, $C(0, 4)$ and $D(3, k)$.

    (a) If $\overline{AB}$ is parallel to $\overline{CD}$, find k.
    (b) Write an equation of $\overleftrightarrow{CD}$.
    (c) Find the area of quadrilateral $ABDC$.

28. In $\triangle ABC$, the coordinates of A are $(-6, -8)$, of B are $(6, 4)$, and of C are $(-6, 10)$.

    (a) Write an equation of the altitude of $\triangle ABC$ from C to $\overline{AB}$.
    (b) Write an equation of the altitude of $\triangle ABC$ from B to $\overline{AC}$.
    (c) Find the x-coordinate of the point of intersection of the two altitudes in parts **a** and **b**.
    (d) Write an equation of the circle whose center is at the origin and which is tangent to the altitude in part **b**.

29. Quadrilateral $JAKE$ has coordinates $J(0, 3a)$, $A(3a, 3a)$, $K(4a, 0)$, $E(-a, 0)$. Prove by coordinate geometry that quadrilateral $JAKE$ is an isosceles trapezoid.

30. Quadrilateral $ABCD$ has vertices $A(-2,2)$, $B(6,5)$, $C(4,0)$, and $D(-4,-3)$. Prove that $ABCD$ is a parallelogram but *not* a rectangle.

31. Quadrilateral $KATE$ has vertices $K(1,5)$, $A(4,7)$, $T(7,3)$, and $E(1,-1)$.

    (a) Prove that $KATE$ is a trapezoid.
    (b) Prove that $KATE$ is *not* an isosceles trapezoid.

In Exercises 32 to 35, prove each theorem using the methods of coordinate geometry.

32. The opposite sides of a parallelogram are congruent.

33. The length of the median drawn to the hypotenuse of a right triangle is equal to one-half the length of the hypotenuse.

34. The segment joining the midpoints of two sides of a triangle is parallel to the third side and one-half of its length.

35. The diagonals of an isosceles trapezoid are congruent.

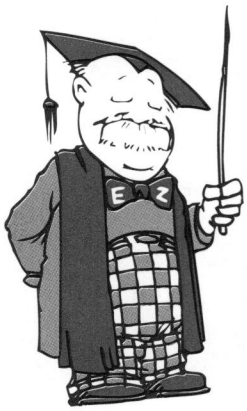

16

Locus and Constructions

WHAT YOU WILL LEARN

Points that satisfy one or more given conditions can be located by drawing a sketch based on geometric principles or by performing a construction using a straightedge to make straight lines and a compass to mark off equal distances. In this chapter you will learn:

- how to locate and describe the set of all points that fit one or more conditions;
- how to use a compass and straightedge to copy angles and segments, to bisect angles and segments, and to construct a line that is perpendicular or parallel to another line.

SECTIONS IN THIS CHAPTER

- Describing Points That Fit One Condition
- Describing Points That Fit More Than One Condition
- Locus and Coordinates
- Basic Constructions

Describing Points That Fit One Condition

Draw a point on a piece of paper and label it K. Take a ruler and locate all points that are 3 inches from point K. How many such points can you find? How would you concisely describe the set of *all* points that satisfy the condition that the point be 3 inches from point K?

The set of all points that are 3 inches from point K forms a *circle* having K as its center and a radius of 3 inches. The circle drawn is said to represent the *locus* of all points 3 inches from point K.

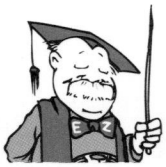

DEFINITION OF LOCUS

A *locus* (plural: *loci*) is the set of all points, and only those points, that satisfy one or more stated conditions.

Think of a locus as a path consisting of one or more points such that each point along the path conforms to the given condition(s). In our example, the given condition was that the points had to be 3 inches from point K. The circular path having point K as a center and a radius of 3 inches is the locus.

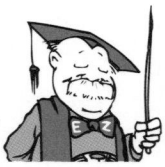

THIS IS THE KEY TO THE METHOD!

STEP 1: Draw a diagram that includes all the given information (point K) and several representative points that satisfy the given condition (that the point be 3 inches from point K).

STEP 2: Keep drawing points until you discover a pattern. Connect the points by using a broken curve or line.

STEP 3: Describe the locus in a complete sentence.

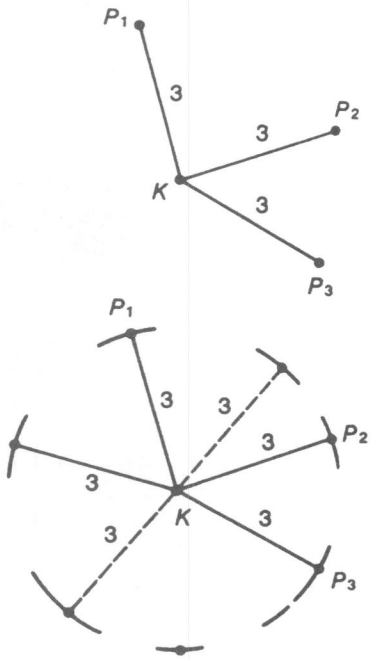

The locus is a circle having point K as its center and a radius of 3 inches.

FIGURE 16.1

Figure 16.1 uses our example to summarize the steps to be followed in determining a locus. (Note that the representative points in Figure 16.1 have been labeled P_1, P_2, and P_3. As discussed in Chapter 15, it is a common practice in mathematics to use *subscripts* to name a related set of points. The symbol P_1, for example, is read "*P* one" or "*P* sub-one." The symbol P_2 names the point "*P* two" and so on.)

Some additional examples of conditions for determining loci are provided in Table 16.1.

SUMMARY

- The locus of all points at a given distance from a point is a circle having the point as its center and the given distance as the length of its radius.
- The locus of all points equidistant from the sides of an angle is the ray that bisects the angle.
- The loci of all points at a given distance from a line are two lines parallel to the original line, on opposite sides of the original line, and each at the given distance from the line.
- The locus of all points equidistant from two parallel lines is a line parallel to the two lines and halfway between them.
- The locus of all points equidistant from the end points of a line segment is the perpendicular bisector of the segment.
- The loci of all points d units from a circle having a radius of r units are two concentric circles with radii of $r - d$ and $r + d$ units.
- The locus of all points equidistant from two concentric circles is a third concentric circle whose radius is the average of the radii of the given circles.

Describing Points That Fit More Than One Condition

Tree *A* stands 5 meters away from tree *B*. A map indicates that treasure is buried 2 meters from tree *A* (condition 1) *and* 4 meters from tree *B* (condition 2). Where would you dig for the treasure? You can narrow down the possibilities by determining the points at which the loci of the individual conditions intersect. Since the sum of the lengths of the radii of the two circles (6 meters) is greater than the distance between the trees (5 meters), the circles that define the loci for the two conditions will intersect at two different points, as illustrated in Figure 16.2. The treasure is buried at either point *X or* point *Y*.

Suppose the trees were 6 meters apart. Where would you dig? Since the circles that define the loci for the two conditions will be tangent to each other, the desired locus is a single point that corresponds to the point of tangency.

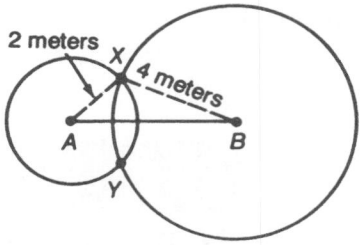

Note: $AB = 5$ meters

FIGURE 16.2

TABLE 16.1

Condition	Diagram	Locus
1. All points that are equidistant from the sides of an angle.	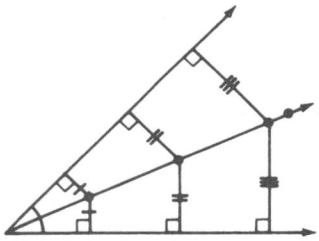	The ray that bisects the angle.
2. All points 4 cm from line ℓ.	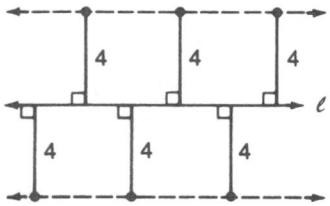	Two lines parallel to line ℓ, on opposite sides of ℓ, and a distance of 4 cm from line ℓ.
3. All points that are equidistant from two parallel lines.	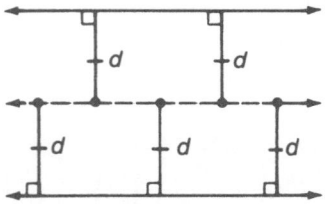	A line parallel to the two lines and halfway between them.

TABLE 16.1 ***(continued)***

Condition	Diagram	Locus
4. All points equidistant from the end points of a line segment.	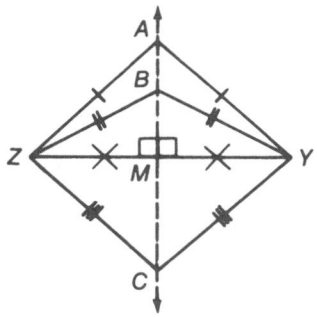	The perpendicular bisector of the segment ($ZA = YA$, for example, since $\triangle ZAM \cong \triangle YAM$ by SAS.)
5. All points d units from a circle having a radius of r units.	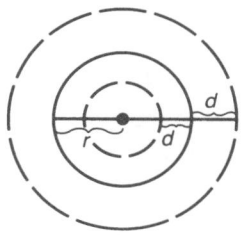	Two concentric circles having the same center as the original circle, the smaller circle having a radius of $r - d$ units and the larger circle having a radius of $r + d$ units.
6. All points equidistant from two concentric circles having radii of p and q units.	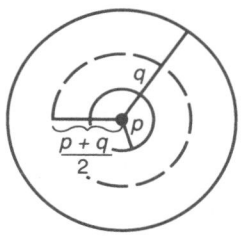	A circle having the same center as the given circles and a radius of $\frac{p+q}{2}$ units.

In general, the loci that satisfy two or more conditions are found by following these steps:

STEP 1: Identify the Given and the conditions.
STEP 2: Draw a diagram that reflects the given information.
STEP 3: Determine the locus that satisfies the first condition.
STEP 4: Determine the locus for each of the remaining conditions, using the same diagram.
STEP 5: Note the points of intersection of the loci (if any).

EXAMPLE 16.1 Two parallel lines are 8 inches apart. Point A is located on one of the lines. Find the number of points that are the same distance from each of the parallel lines *and* that are also at a distance from point A of:

a. 5 inches **b.** 4 inches **c.** 3 inches

SOLUTION

STEP 1: Identify the Given and the conditions.

GIVEN: Two parallel lines 8 inches apart and a point A located on one of them.

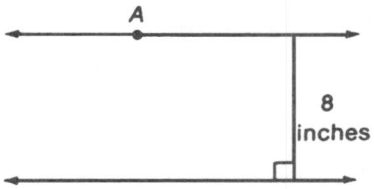

CONDITIONS

a. Points must be:

1. Equidistant from the two lines.
2. 5 inches from point A

STEP: 2: Draw a diagram that reflects the given information.

STEP 3: The locus of all points equidistant from two parallel lines is a line parallel to the original lines and midway between them.

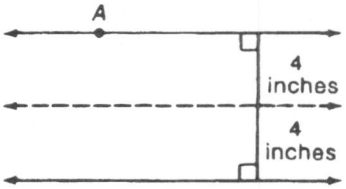

STEP 4: The locus of all points 5 inches from point A is a circle having point A as its center and a radius of 5 inches.

STEP 5: Since the length of the radius of the circle is greater than the distance of the middle line from the line that contains point A, the circle intersects this line in *two* distinct points

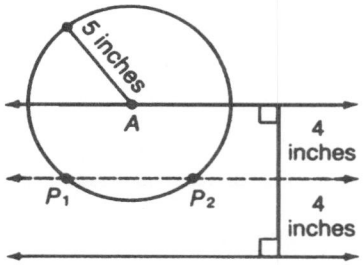

Thus, there are two points, P_1 and P_2, that are the same distance from each of the parallel lines and that are 5 inches from point A.

b. Since the circle has a radius of 4 inches, it is tangent to the line that is midway between the original pair of parallel lines. The loci intersect at point P. There is one point, P, that satisfies both conditions.

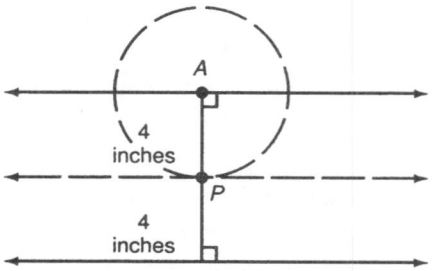

c. Since the circle has a radius of 3 inches, it does *not* intersect the line that is midway between the original pair of parallel lines. Since the loci do *not* intersect, there is no point that satisfies both conditions.

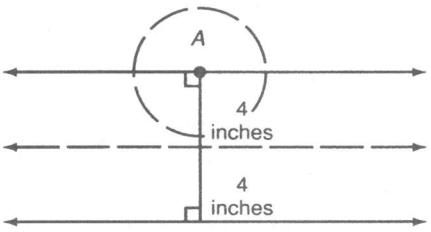

EXAMPLE 16.2 Point P on line ℓ. Find the loci of points that are 2 inches from line ℓ and also 2 inches from point P.

SOLUTION

- *Locus condition 1:* All points 2 inches from line ℓ. The desired locus is a pair of parallel lines on either side of line ℓ, each line at a distance of 2 inches from line ℓ.
- *Locus condition 2:* All points 2 inches from point P. The desired locus is a circle that has P as its center and has a radius of 2 inches. See the accompanying diagram.
- The required loci are the two points, A and B, that satisfy both conditions.

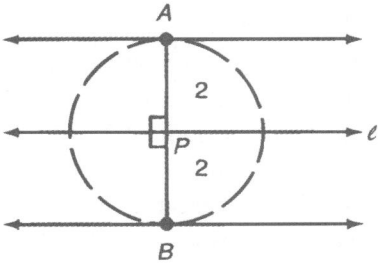

Locus and Coordinates

Points that satisfy one or more conditions can also be located in the coordinate plane.

EXAMPLE 16.3 Write the equation of the locus of all points whose ordinates exceed their abscissas by 3.

SOLUTION

The locus is a line. The y-coordinate (ordinate) of each point on this line is 3 more than the corresponding x-coordinate (abscissa). Therefore, the equation $y = x + 3$ completely describes the required locus.

EXAMPLE 16.4 Find the loci of all points 3 units from the y-axis.

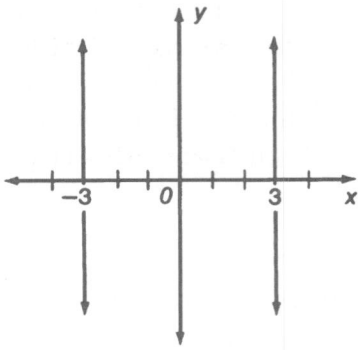

SOLUTION

The loci of all points 3 units from the y-axis (a line) may be determined by drawing two vertical lines on either side of the y-axis and 3 units from the y-axis. The loci of all points 3 units from the y-axis are the lines whose equations are $x = -3$ and $x = 3$.

EXAMPLE 16.5 Find an equation of a line that describes the locus of points equidistant from the lines whose equations are $y = 3x - 1$ and $y = 3x + 5$.

SOLUTION

The lines given by the equations $y = 3x - 1$ and $y = 3x + 5$ are parallel since the slope of each line is 3. The line parallel to these lines and midway between them must have a slope of 3 and a y-intercept of 2 since 2 is midway between (the average of) the y-intercepts of the original pair of lines. Hence, the locus of points equidistant from the lines whose equations are $y = 3x - 1$ and $y = 3x + 5$ is the line whose equation is $y = 3x + 2$.

EXAMPLE 16.6 How many points are 3 units from the origin and 2 units from the y-axis?

SOLUTION

- *Locus condition 1:* All points 3 units from the origin. The desired locus is a circle with the origin as its center and a radius of 3 units. Note in the accompanying diagram that the circle intersects each coordinate axis at 3 and –3.

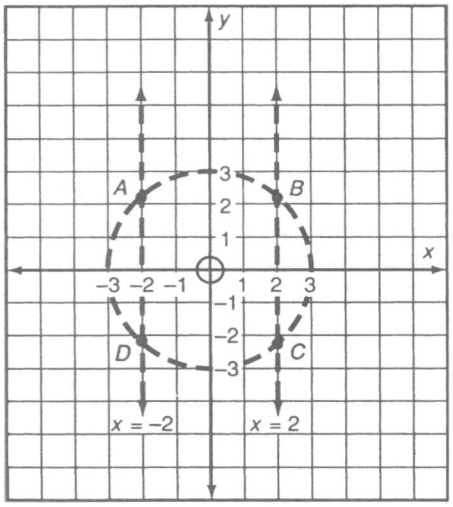

- *Locus condition 2*: All points 2 units from the y-axis. The desired loci are a pair of parallel lines; one line is 2 units to the right of the y-axis ($x = 2$), and the other line is 2 units to the left of the y-axis ($x = -2$). See the accompanying diagram.
- Since the loci intersect at points A, B, C, and D, there are four points that satisfy both conditions.

EXAMPLE 16.7

GIVEN: Points $A(2, 7)$ and $B(6, 7)$.

a. Write an equation of $\overline{AB}$.

b. Describe the locus or loci of points equidistant from:

(i) points A and B **(ii)** the x- and y-axes

c. How many points satisfy both conditions obtained in part **b**?

SOLUTIONS

a. Since the y-coordinates of points A and B are the same, points A and B determine a horizontal line, $\overline{AB}$, whose equation is $y = 7$.

b. **(i)** The locus of points equidistant from two points is the perpendicular bisector of the line segment determined by the given points. The midpoint of $\overline{AB}$ is $M(4, 7)$. Since $\overline{AB}$ is a horizontal line, a vertical line that contains $(4, 7)$ will be the perpendicular bisector of $\overline{AB}$ An equation of this line is $x = 4$.

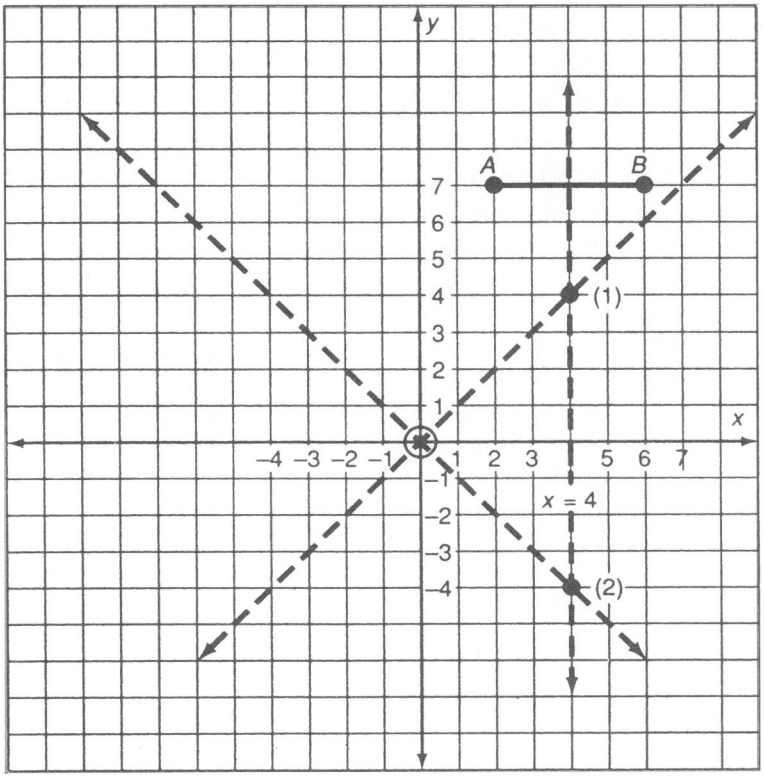

(ii) The loci of points equidistant from the x- and y-axes are the pair of lines that bisect the pairs of vertical angles formed by the intersecting coordinate axes.

c. Two points satisfy the two conditions in part **b**.

Basic Constructions

Geometric constructions, unlike *drawings*, are made only with a straightedge (for example, an unmarked ruler) and compass. The point at which the pivot point of the compass is placed is sometimes referred to as the *center*, while the fixed compass setting that is used is called the *radius length*.

COPYING SEGMENTS AND ANGLES

Given a line segment or angle, it is possible to construct another line segment or angle that is congruent to the original segment or angle without using a ruler or protractor.

CONSTRUCTION 1 Given line segment $\overline{AB}$, construct a congruent segment.

STEP 1: Using a compass, measure $\overline{AB}$ by placing the compass point on A and the pencil point on B.

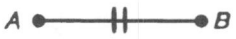

STEP 2: Draw any line, and choose any convenient point on it. Label the line as ℓ and the point as C.

STEP 3: Using the same compass setting, place the compass point on C and draw an arc that intersects line ℓ. Label the point of intersection as D.

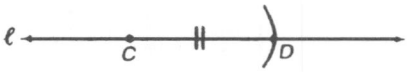

Conclusion: $\overline{AB} \cong \overline{CD}$.

CONSTRUCTION 2 Given $\angle ABC$, construct a congruent angle.

STEP 1: Using any convenient compass setting, place the compass point on B and draw an arc intersecting $\overrightarrow{BC}$ at X and $\overrightarrow{BA}$ at Y.

STEP 2: Draw any line and choose any point on it. Label the line as ℓ and the point as S.

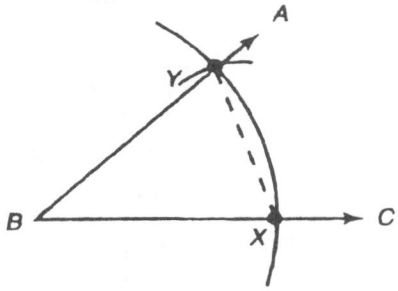

STEP 3: Using the same compass setting, place the compass point at S and draw arc WT, intersecting line ℓ at T.

STEP 4: Adjust the compass setting to measure the line segment determined by points X and Y by placing the compass point at X and the pencil at Y.

STEP 5: Using the same compass setting, place the compass point at T and construct an arc intersecting arc WT at point R.

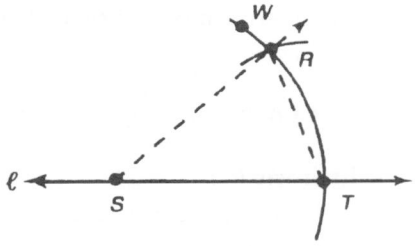

STEP 6: Using a straightedge, draw $\overrightarrow{SR}$.

Conclusion: $\angle ABC \cong \angle RST$.

Why it works:

- The arcs are constructed in such a way that $\overline{BX} \cong \overline{ST}$, $\overline{BY} \cong \overline{SR}$, and $\overline{XY} \cong \overline{TR}$.
- Therefore, $\triangle XYB \cong \triangle TRS$ by the SSS postulate.
- By CPCTC, $\angle ABC \cong \angle RST$.

CONSTRUCTION 3 Given two points, construct the perpendicular bisector of the segment determined by the two points.

STEP 1: Label points A and B, and draw $\overline{AB}$. Choose any compass setting (radius length) that is more than one-half the length of $\overline{AB}$.

STEP 2: Using this compass setting, and points A and B as centers, construct a pair of arcs above and below $\overline{AB}$. Label the points at which the pairs of arcs intersect as P and Q.

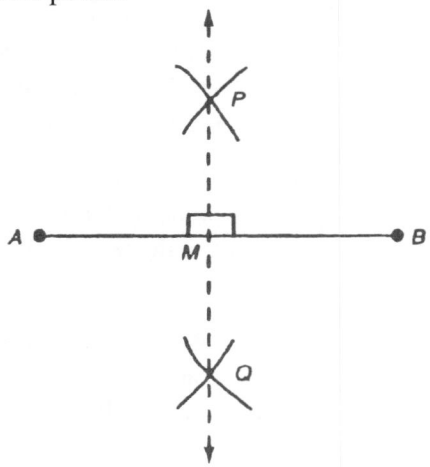

STEP 3: Draw $\overleftrightarrow{PQ}$ and label the point of intersection of $\overleftrightarrow{PQ}$ and $\overline{AB}$ as M.

Conclusion: $\overleftrightarrow{PQ}$ is the perpendicular bisector of $\overline{AB}$.

Why it works:

- The arcs are constructed in such a way that $AP = BP = AQ = BQ$.
- Since quadrilateral $APBQ$ is equilateral, it is a rhombus.
- Since the diagonals of a rhombus are perpendicular bisectors, $\overline{AM} \cong \overline{BM}$ and $\overline{PQ} \perp \overline{AB}$

CONSTRUCTION 4 Given an angle, construct the bisector of the angle.

STEP 1: Designate the angle as $\angle ABC$. Using B as a center, construct an arc, using any convenient radius length, that intersects $\overrightarrow{BA}$ at point P and $\overrightarrow{BC}$ at point Q.

STEP 2: Using points P and Q as centers and the same radius length, draw a pair of arcs that intersect. Label the point at which the arcs intersect as D.

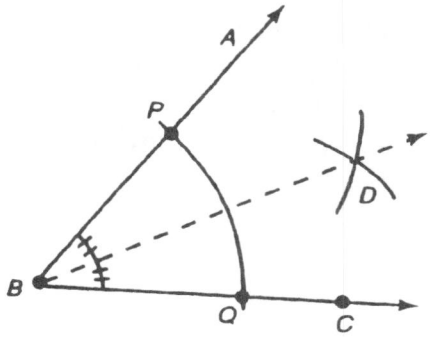

STEP 3: Draw $\overrightarrow{BD}$.

Conclusion: $\overrightarrow{BD}$ is the bisector of $\angle ABC$.

Why it works: Draw $\overline{PD}$ and $\overline{QD}$. Then:

- $\triangle BPD \cong \triangle BQD$ by SSS.
- By CPCTC, $\angle ABD \cong \angle CBD$.

CONSTRUCTING PERPENDICULAR LINES

A line can be constructed perpendicular to a given line at a given point on the line or through a given point not on the line.

CONSTRUCTION 5 Given a line ℓ and a point P *not* on the line, construct a line through P and perpendicular to line ℓ.

STEP 1: Using P as a center and any convenient radius length, construct an arc that intersects line ℓ at two points. Label these points as A and B.

STEP 2: Choose a radius length greater than one-half the length of $\overline{AB}$. Using points A and B as centers, construct a pair of arcs that intersect at point Q.

STEP 3: Draw $\overleftrightarrow{PQ}$, intersecting line ℓ at point M.

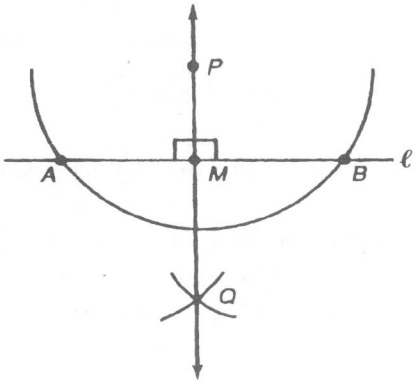

Conclusion: $\overleftrightarrow{PQ}$ is perpendicular to line ℓ at point M.

Why it works: Draw $\underline{PA}$ $\underline{PB}$, $\underline{AQ}$, and $\underline{BQ}$. Then:

- $\triangle PAQ \cong \triangle PBQ$ by SSS.
- By CPCTC, $\angle APM \cong \angle BPM$.
- $\triangle PMA \cong \triangle PMB$ by SAS.
- By CPCTC, $\angle AMP \cong \angle BMP$ so $\underline{PQ} \perp \ell$.

CONSTRUCTION 6 Given a line ℓ and a point P on line ℓ, construct a line through P and perpendicular to line ℓ.

STEP 1: Using P as a center and any convenient radius length, construct an arc that intersects line ℓ at two points. Label these points as A and B.

STEP 2: Choose a radius length greater than one-half the length of $\overline{AB}$. Using points A and B as centers, construct a pair of arcs on either side of line ℓ that intersect at point Q.

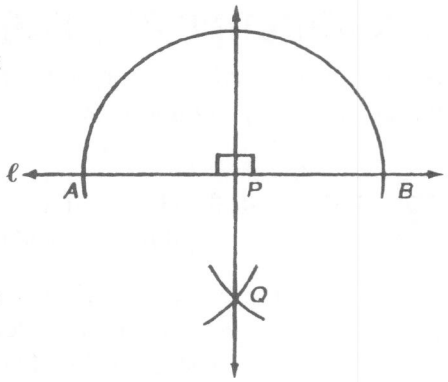

STEP 3: Draw $\overleftrightarrow{PQ}$.

Conclusion: $\overleftrightarrow{PQ}$ is perpendicular to line ℓ at point P.

Why it works: Draw $\overline{AQ}$ and $\overline{BQ}$. Then:

- $\triangle PAQ \cong \triangle PBQ$ by SSS.
- By CPCTC, $\angle APQ \cong \angle BPQ$ so $PQ \perp \ell$.

CONSTRUCTING PARALLEL LINES

A line can be constructed parallel to a given line and through a given point not on the line by drawing any convenient transversal through the point and then constructing a congruent corresponding angle, using this point as a vertex.

CONSTRUCTION 7 Given a line $\overleftrightarrow{AB}$ and a point P *not* on the line, construct a line through P and parallel to $\overleftrightarrow{AB}$.

STEP 1: Through P draw any convenient line, extending it so that it intersects $\overleftrightarrow{AB}$. Label the point of intersection as Q.

STEP 2: Using P and Q as centers, draw arcs having the same radius length. Label the point at which the arc intersects the ray opposite $\overrightarrow{PQ}$ as R.

STEP 3: Construct at P an angle, one of whose sides is PR, congruent to $\angle PQB$.

STEP 4: Draw $\overleftrightarrow{PS}$.

Conclusion: $\overrightarrow{PS}$ is parallel to $\overleftrightarrow{AB}$.

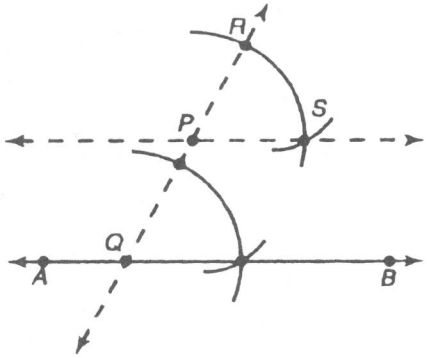

Why it works: $\angle PQB$ and $\angle RPS$ are congruent corresponding angles.

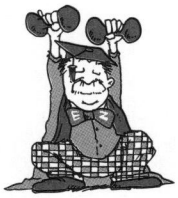

REVIEW EXERCISES FOR CHAPTER 16

1. How many points are equidistant from two parallel lines and also equidistant from two points on one of the lines?

   (1) 1 (2) 2 (3) 3 (4) 4

2. Which graph represents the locus of points 3 units from the origin?

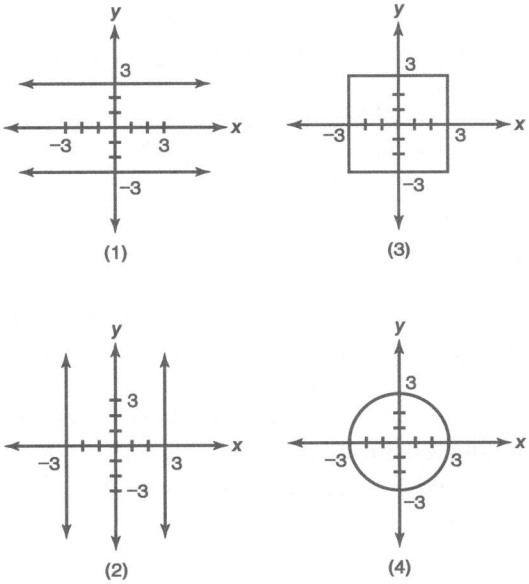

3. What is the total number of points equidistant from two intersecting straight roads and also 300 feet from the traffic light at the center of the intersection?

   (1) 1 (2) 2 (3) 3 (4) 4

4. What is the total number of points 2 inches from a given line and also 3 inches from a given point on the line?

   (1) 1 (2) 2 (3) 3 (4) 4

5. Lines ℓ and m are parallel lines 8 centimeters apart, and point P is on line m. What is the total number of points that are equidistant from lines ℓ and m and 5 centimeters from P?

   (1) 1 (2) 2 (3) 0 (4) 4

In Exercises 6 to 12, find the loci.

6. The locus of all points 1 centimeter from a circle whose radius is 8 centimeters.

7. The locus of the centers of all circles tangent to each of two parallel lines that are 10 inches apart.

8. The locus of the center of a circle that rolls along a flat surface.

9. The locus of all points equidistant from two concentric circles having radii of 7 centimeters and 11 centimeters.

10. The locus of the vertices of all isosceles triangles having a common base.

11. The loci of all points equidistant from points A and B and 4 inches from $\overleftrightarrow{AB}$.

12. The loci of all points equidistant from two intersecting lines and 3 inches from their point of intersection.

13. Point A is 4 inches from line k. What is the loci of all points 1 inch from line k and 2 inches from point A?

14. Point X is the midpoint of $\overline{PQ}$. If $PQ = 10$ centimeters, what is the loci of all points equidistant from points P and Q and 5 centimeters from point X?

15. Lines p and q are parallel and 10 centimeters apart. Point A is between lines p and q and 2 centimeters from line q. What is the loci of all points equidistant from lines p and q and d centimeters from point A, if d equals the given number of centimeters?

    (a) 6 cm (b) 1 cm (c) 3 cm

16. Points A and B are d units apart. What is the number of points equidistant from points A and B and r units from point A given the following conditions?

    (a) $r < \frac{d}{2}$ (b) $r = \frac{d}{2}$ (c) $r > \frac{d}{2}$

17. Write the equation(s) that describe the following loci.

    (a) The locus of all points that are 5 units from the x-axis.
    (b) The locus of all points that are 4 units from the y-axis.
    (c) The locus of all points such that their ordinates are twice their abscissas.
    (d) The locus of all points such that the sum of twice their ordinates and three times their abscissas is 6.
    (e) The locus of all points such that their ordinates exceed their abscissas by 5.

18. How many points are 3 inches from line ℓ and also 3 inches from a point on line ℓ?

19. Two parallel lines are 12 inches apart. Point P is located on one of the lines. How many points are equidistant from the parallel lines and are also at a distance of 5 inches from point P?

20. Find the number of points that are 3 inches from point A and 5 inches from point B if:

    (a) $AB = 8$
    (b) $AB = 6$
    (c) $AB = 10$

21. How many points are equidistant from two parallel lines and also equidistant from two points on one of the lines?

22. Give an equation or equations that describe the locus of points:

    (a) equidistant from the lines $x = -1$ and $x = 5$
    (b) equidistant from the lines $y = -3$ and $y = -7$
    (c) equidistant from the lines $y = 3x + 1$ and $y = 3x + 9$

23. Find the number of points that are 3 units from the origin and also 2 units from the x-axis.

24. Find the number of points that are equidistant from points $(-1, 0)$ and $(3, 0)$ and are also 2 units from the origin.

25. The coordinates of point P are $(3, 5)$.

    (a) Describe fully the locus of points at a distance of:
    1 d units from P **2** 1 unit from the y-axis

    (b) How many points satisfy the conditions in part **a** simultaneously for the following values of d?
    1 $d = 2$ **2** $d = 4$ **3** $d = 5$

26. The total number of points in a plane that are distance d from a given straight line and are also distance r from a certain point on the line can *not* be

    (1) 0 (2) 2 (3) 3 (4) 4

27. Draw obtuse triangle ABC, where $\angle B$ is obtuse. Construct:

    (a) the altitude to side $\overline{BC}$ (extended, if necessary)
    (b) the median to side $\overline{BC}$
    (c) a line through A and parallel to $\overline{BC}$

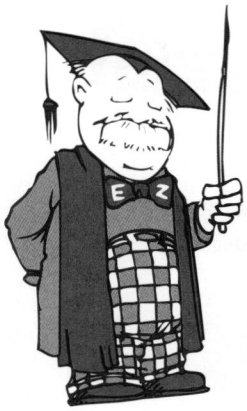

17 Transformation Geometry

WHAT YOU WILL LEARN

In this chapter you will learn:

- that a geometric transformation may change the position, orientation, or size of an object;
- transformations may produce images that are congruent or similar to the original figure;
- different types of symmetry;
- how to apply transformations using coordinates;
- the single transformations that are equivalent to performing consecutive reflections either over two parallel lines or over two intersecting lines.

SECTIONS IN THIS CHAPTER

- Terms and Notation
- Congruence Transformations
- Classifying Isometries
- Size Transformations
- Types of Symmetry
- Transformations in the Coordinate Plane
- Composing Transformations

Terms and Notation

ONE-TO-ONE MAPPINGS

A *mapping* between two sets is a pairing of their elements. Suppose the members of set X are three married women, and the three members of set Y are their husbands:

$$X = \{\text{Alice, Barbara, Carol}\} \text{ and } Y = \{\text{Ray, Mike, Bob}\}.$$

Matching each wife with her husband, as in

$$\{(\text{Alice, Bob}\}, (\text{Barbara, Ray}), (\text{Carol, Mike})\},$$

represents a *one-to-one mapping* between sets X and Y, as each woman is paired with exactly one man and each man is paired with exactly one woman.

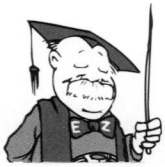

DEFINITION OF ONE-TO-ONE MAPPING

A mapping between sets X and Y is *one-to-one (1–1)* if each element of set X is paired with exactly one element of set Y, and each element of set Y is matched with exactly one element of set X. Pairs of elements that are matched together are *corresponding* elements.

In our earlier work with congruent and similar triangles, the idea of a 1–1 mapping arose naturally when matching the vertices of one triangle with the corresponding vertices of the congruent or similar triangle.

TRANSFORMATIONS

Figure 17.1 shows one possible transformation of $\triangle ABC$. Triangle $A'B'C'$, called the *image* of $\triangle ABC$, is determined by using some rule that tells how to map the points of $\triangle ABC$ onto it.

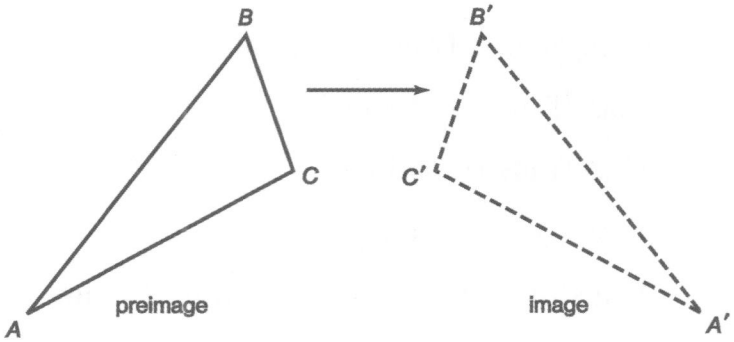

FIGURE 17.1 Transformation of $\triangle ABC$ onto $\triangle A'B'C''$.

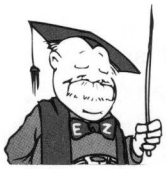

DEFINITION OF TRANSFORMATION

A *transformation* is a one-to-one mapping between the elements of two sets where the elements are points. Under a transformation, each point of an object is mapped onto exactly one point called its *image*. Each image point corresponds to exactly one point of the original object called the *preimage*.

It is customary to use the same capital letter to represent a preimage point and its image. To distinguish between the two points, the letter representing the image point is followed by a prime mark (′). In Figure 17.1, A is mapped onto A' (read "A prime"), B is mapped onto B', and C is mapped onto C'. This type of correspondence is usually represented using "arrow" notation. For example, $A \rightarrow A'$ indicates that point A is mapped onto point A' or, equivalently, that point A' is the image of point A under the given transformation. Similarly, the notation $\triangle ABC \rightarrow \triangle A'B'C'$ means that $\triangle ABC$ is mapped onto $\triangle A'B'C'$ so $\triangle A'B'C'$ is the image of $\triangle ABC$ under the given transformation.

Congruence Transformations

An *isometry* is a transformation that preserves the distance between points. Because an isometry produces an image congruent to the original figure, it is sometimes referred to as a *congruence transformation*. Figure 17.2 shows three simple types of isometries: reflection, translation, and rotation.

Reflection: "Flip" **Translation: "Slide"** **Rotation: "Turn"**

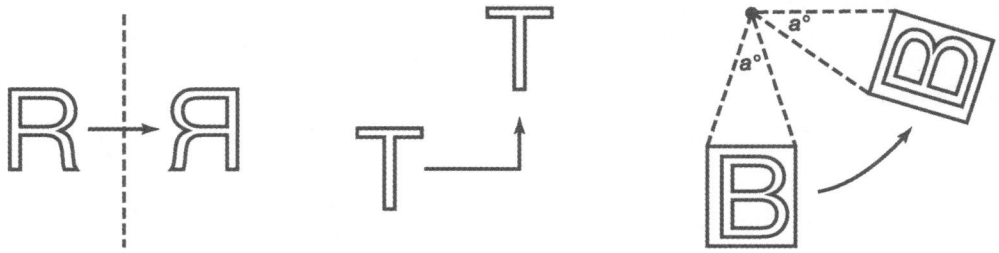

FIGURE 17.2 Transformations that are isometries.

For any isometry:

- collinearity and betweenness of points are preserved. In Figure 17.3, A' and B' are the image points of A and B under some isometry. If C is between A and B, then C' is between A' and B'.
- the image of a line segment is a line segment of the same length. In Figure 17.3, if $AB = 6$, then $A'B' = 6$.
- the image of an angle is an angle of the same measure, as illustrated in Figure 17.3.

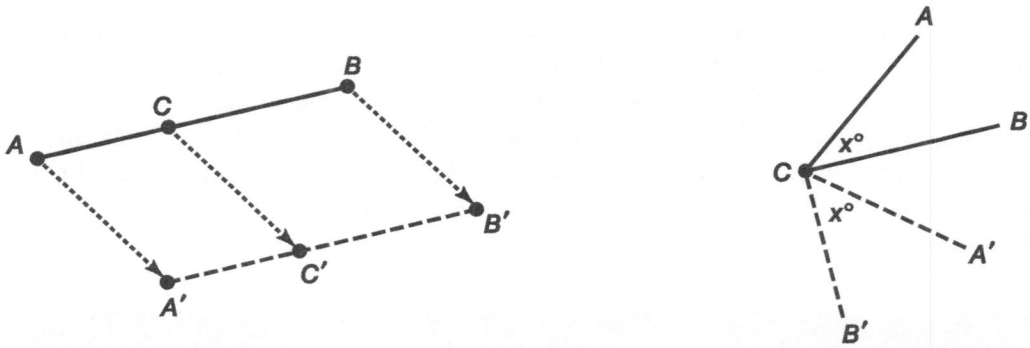

FIGURE 17.3 Properties of isometries.

REFLECTIONS

A line *reflection* "flips" an object over the line so that the image appears "backwards," much like how the reflected image of that object would appear in a mirror. The transformation represented in Figure 17.1 is the reflection of $\triangle ABC$ over a vertical line (not drawn) midway between $\triangle ABC$ and its reflected image, $\triangle A'B'C'$. Figure 17.4 shows how to determine the reflected image of point A over line ℓ:

- Draw a line segment from A perpendicular to line ℓ.
- Extend that segment its own length to A'.

The shorthand notation $r_\ell(A) = A'$ indicates the reflection of point A over line ℓ is A'.

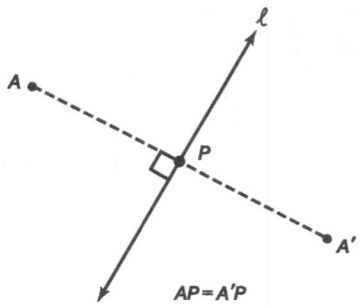

FIGURE 17.4 Reflecting point A over line ℓ.

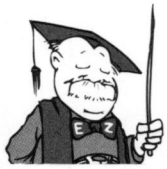

DEFINITION OF A LINE REFLECTION

A reflection over a line is an isometry that maps all points of a figure such that each image point is on the opposite side of the reflecting line and the same distance from it as its preimage. If a point of the figure is on the reflecting line, then its image is the point itself.

A polygon can be reflected over a line simply by reflecting its vertices and then connecting the reflected image points, as shown in Figure 17.5.

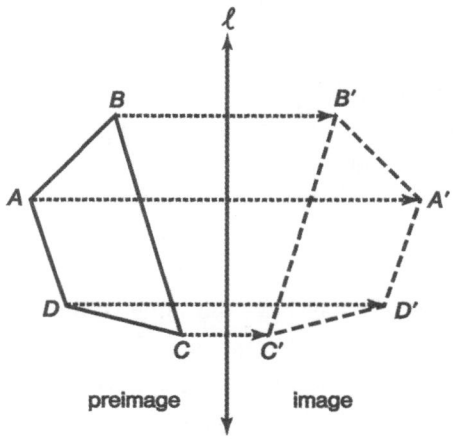

FIGURE 17.5 Reflecting trapezoid *ABCD* over a line.

ORIENTATION

REMEMBER
A reflection makes the reflected image appear backwards.

Every convex polygon has two orientations: clockwise and counterclockwise. The orientation assigned to a polygon depends on the path traced when the vertices are read in a given order. In Figure 17.5, when the lettered vertices of trapezoid *ABCD* are traced from *A* to *B* to *C* to *D*, they are being traced in a *clockwise* direction so the orientation of the trapezoid is *clockwise*. When the lettered vertices of the reflected image are read in corresponding order from *A′* to *B′* to *C′* and *D′*, the vertices are now being traced in a *counterclockwise* direction. Trapezoid *A′B′C′D′* has *counterclockwise* orientation. A reflection, therefore, *reverses* orientation. It is this property of a reflection that makes the reflected image appear "backwards."

TRANSLATIONS

A *translation* slides a figure to the left or to the right, up or down, or both horizontally and vertically, as in Figure 17.6. A translation preserves orientation.

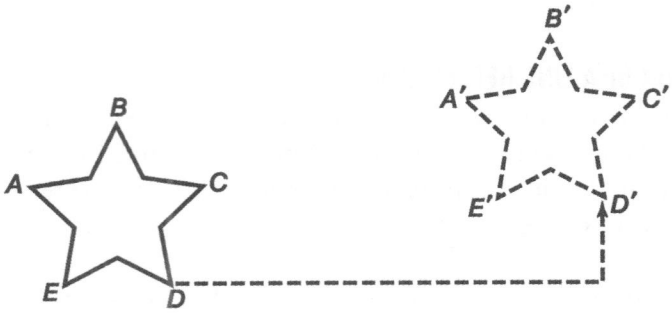

FIGURE 17.6 A translation or "slide" of an object.

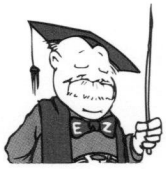

DEFINITION OF TRANSLATION

A *translation* is an isometry that "slides" each point of a figure the same distance in the same direction.

ROTATIONS

Suppose two identical pieces of paper have the same smiley face drawn in the same location. A pin is pushed through the papers when their edges are aligned. A rotation of the smiley face can be modeled by holding one paper fixed and turning the other paper, as illustrated in Figure 17.7. The pin represents the *center of rotation*.

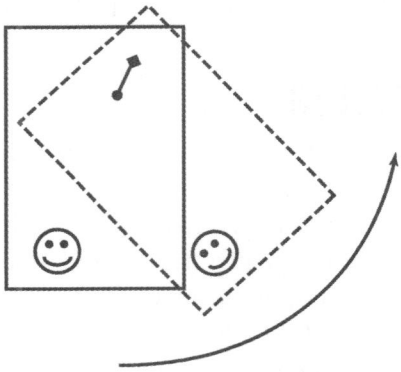

FIGURE 17.7 Modeling a rotation.

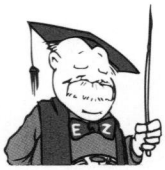

DEFINITION OF ROTATION

A *rotation* is an isometry that "turns" a figure a specified number of degrees in a given direction (clockwise or counterclockwise) about some fixed point called the *center of rotation*.

Unless otherwise indicated, rotations are performed counterclockwise. Figure 17.8 shows a counterclockwise rotation of $\triangle ABC$ $x°$ about point O. The images of points A, B, and C are determined so that corresponding sides of the figure and its image have the same lengths and

$$m\angle AOA' = m\angle BOB' = m\angle COC' = x.$$

The shorthand notation $R_{x°}(A) = A'$ indicates that the rotated image of point A after a counterclockwise rotation of $x°$ is point A'. You should verify that the rotation preserves orientation.

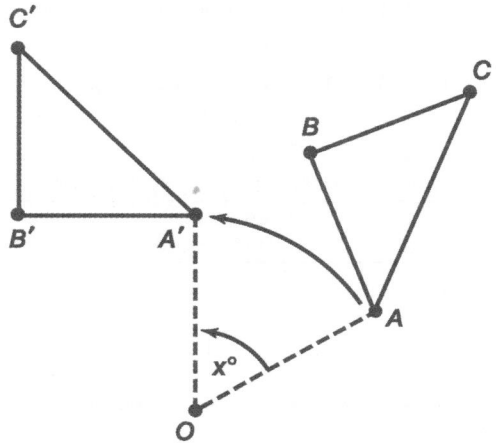

FIGURE 17.8 Counterclockwise rotation of $\triangle ABC$ $x°$ about point O.

GLIDE REFLECTION

There are only four types of isometries: reflection, translation, rotation, and glide reflection. A *glide reflection* combines a reflection over a line and a translation or "glide" in the direction parallel to the reflecting line, as in Figure 17.9. The line reflection and translation may be performed in either order. Because this transformation includes a reflection, a glide reflection reverses orientation.

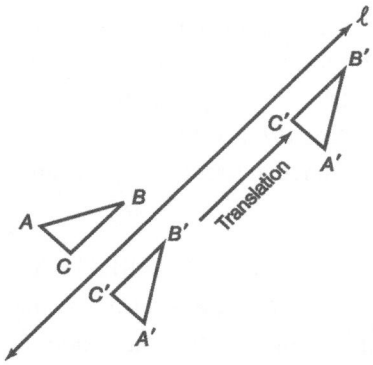

FIGURE 17.9
Glide reflection.

Classifying Isometries

Isometries that reverse and maintain orientation are given special names.

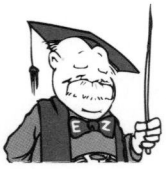

DEFINITIONS OF OPPOSITE AND DIRECT ISOMETRIES

- A *direct isometry* is an isometry that preserves orientation. Translations and rotations are direct isometries.
- An *opposite isometry* is an isometry that reverses orientation. A line reflection and glide reflection are opposite isometries.

Table 17.1 summarizes some key properties of congruence transformations.

TABLE 17.1
PROPERTIES OF CONGRUENCE TRANSFORMATIONS

Congruence Transformation	Preserves Distance	Preserves Angle Measure	Preserves Orientation	Type of Isometry
Reflection	√	√	✗	Opposite
Translation	√	√	√	Direct
Rotation	√	√	√	Direct
Glide Reflection	√	√	✗	Opposite

Size Transformations

Not all transformations produce congruent images. A dilation is a size transformation that produces a similar rather than a congruent image. When the image size setting of an office copying machine is set to a value that is lower or higher than 100%, the copy machine *dilates* the original by shrinking or enlarging it without changing its shape. The image size setting represents the *scale factor* of the dilation.

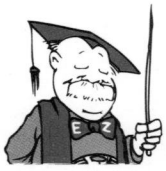

DEFINITION OF DILATION

A *dilation* is a transformation that changes the size of a figure by mapping each point to its image such that the distance from the center of the dilation to the image is c times the distance from the center to the preimage. The multiplying factor c is called the *scale factor* or *constant of dilation*.

The scale factor of a dilation determines whether the image is larger or smaller than the original object:

- If $c > 1$, the dilation enlarges the figure, as in Figure 17.10.
- If $0 < c < 1$, the dilatation shrinks the figure, as in Figure 17.11.

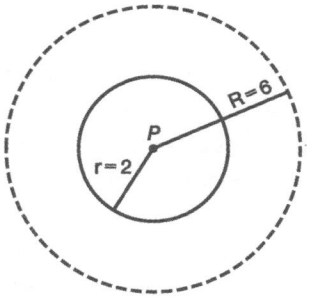

FIGURE 17.10 Dilation with center P of a circle with radius 2, using a scale factor of 3.

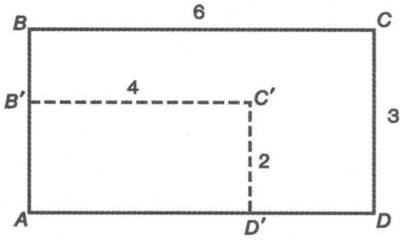

FIGURE 17.11 Dilation with center A of rectangle $ABCD$, using a scale factor of $\frac{2}{3}$.

EXAMPLE 17.1 Which transformation appears to represent an isometry?

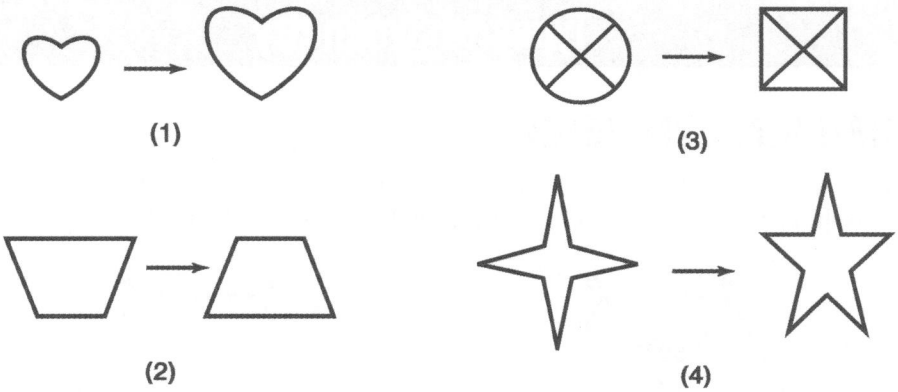

SOLUTION

Choice (2).

A transformation is an isometry if the preimage (original figure) and image are congruent. In choice (2), the pair of figures appear congruent, although the image is turned upside-down.

EXAMPLE 17.2 Under what type of transformation, shown in the accompanying figure, is $\triangle AB'C'$ the image of $\triangle ABC$?

(1) dilation
(2) translation
(3) rotation about point A
(4) reflection in line ℓ

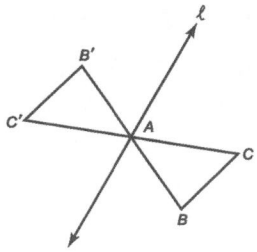

SOLUTION

Choice (3).

Consider each choice in turn:

- Choice (1): A dilation changes the size of the original figure. Since $\triangle AB'C'$ and $\triangle ABC$ are the same size, the figure does *not* represent a dilation.
- Choice (2): Since $\triangle AB'C'$ cannot be obtained by "sliding" $\triangle ABC$ in the horizontal ("sideways") or vertical (up and down) direction, or in both directions, the figure does *not* represent a translation.
- Choice (3): A rotation about a fixed point "turns" a figure about that point. Since angles BAB' and CAC' are straight angles, $\triangle AB'C'$ is the image of $\triangle ABC$ after a rotation of 180° about point A.
- Choice (4): Since line ℓ is *not* the perpendicular bisector of $\overline{BB'}$ and $\overline{CC'}$, points B' and C' are not the reflected images of points B and C, respectively. Hence, the figure does *not* represent a reflection.

Types of Symmetry

ROTATIONAL SYMMETRY

After a clockwise rotation of 60° about its center O, a regular hexagon will coincide with itself, as indicated in Figure 17.12. Regular hexagon $ABCDEF$ has 60° rotational

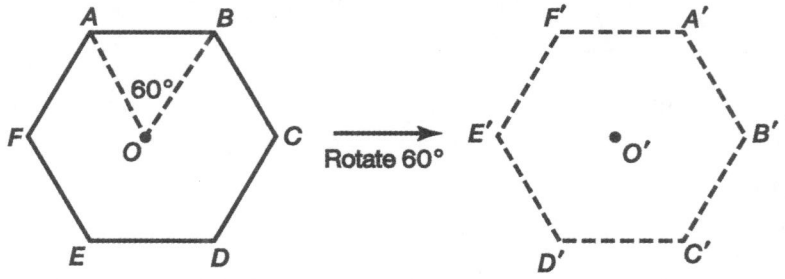

FIGURE 17.12 Rotational symmetry with a 60° angle of rotation.

symmetry. A figure has *rotational symmetry* if it coincides with its image for some rotation of 180° or less. Every regular polygon enjoys rotational symmetry about its center for an angle of rotation of $\frac{360°}{n}$, where n is the number of sides of the polygon. For an equilateral triangle, square, regular pentagon, and regular octagon, these angles of rotation are 120°, 90°, 72°, and 45°, respectively.

POINT SYMMETRY

A figure has *point symmetry* about a point when it has 180° rotational symmetry about that point, as in Figure 17.13.

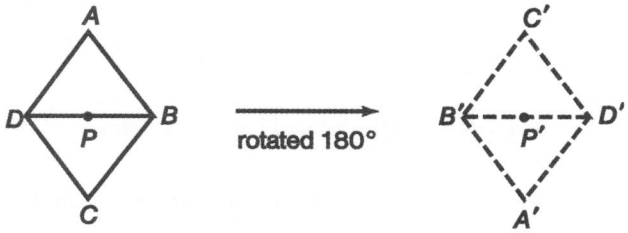

FIGURE 17.13 Point symmetry.

If you are not sure whether a figure has point symmetry, turn the page on which the figure is drawn upside down. Now compare the rotated figure with the original. If they look exactly the same, the figure has point symmetry.

LINE SYMMETRY

A figure has *line symmetry* if a line can be drawn that divides the figure into two "mirror image" parts that coincide when folded along the line. The line of symmetry may be a horizontal line, a vertical line, or neither. The objects in Figure 17.14 have horizontal line symmetry.

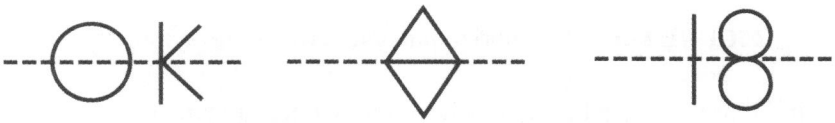

FIGURE 17.14 Horizontal line symmetry.

The objects in Figure 17.15 have vertical line symmetry.

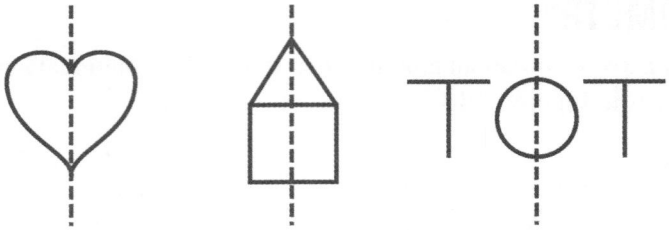

FIGURE 17.15 Vertical line symmetry.

Figure 17.16 illustrates that a figure may have both a horizontal and a vertical line of symmetry.

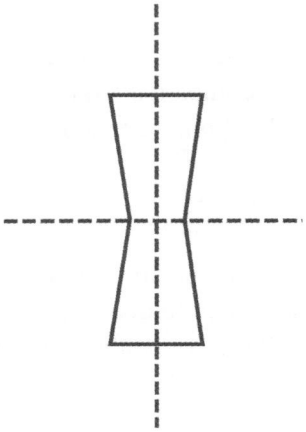

FIGURE 17.16 Both horizontal and vertical line symmetry.

As shown in Figure 17.17, a figure may have many lines of symmetry or may have no line of symmetry.

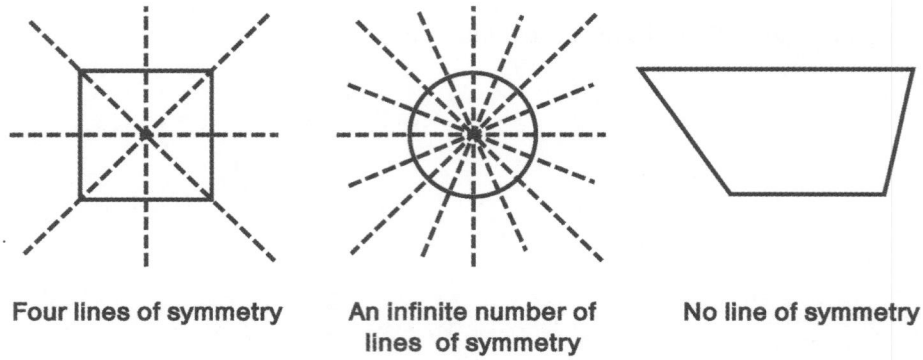

FIGURE 17.17 Figures with many or no lines of symmetry.

Transformations in the Coordinate Plane

Transformations can also be performed in the coordinate plane.

REFLECTIONS USING COORDINATES

To reflect a *point* over a coordinate axis, flip it over the axis so that the image is the same distance from the reflecting line as the original point. Figure 17.18 shows the reflection of (2,4) over each coordinate axis and in the origin. The notation $r_{x\text{-}axis}(2,4) = (2,-4)$ indicates that the reflected image of point (2,4) over the x-axis is (2,–4). In general,

- $r_{x\text{-}axis}(x,y) = (x,-y)$.
- $r_{y\text{-}axis}(x,y) = (-x,y)$.
- $r_{origin}(x,y) = (-x,-y)$.

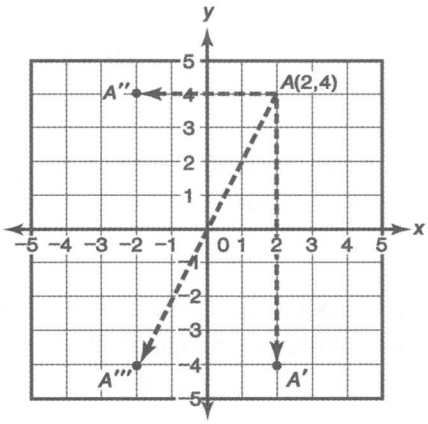

FIGURE 17.18 Reflecting over an axis and over the origin.

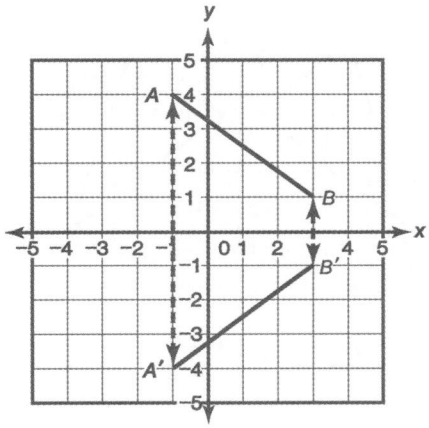

FIGURE 17.19 Reflecting a line segment.

To reflect a *line segment* over a coordinate axis, flip it over the axis by reflecting each endpoint of that segment. Then connect the two image points. If the endpoints of $\overline{AB}$ are $A(-1,4)$ and $B(3,1)$, then, after a reflection of $\overline{AB}$ over the x-axis, the image is $\overline{A'B'}$, with endpoints $A'(-1,-4)$ and $B'(3,-1)$, as shown in Figure 17.19.

To reflect a point over the line $y = x$ or over the line $y = -x$, as illustrated in Figure 17.20, use these rules:

- $r_{y=x}\, A\,(x, y) = A'\,(y, x)$
- $r_{y=-x}\, A\,(x, y) = A''\,(-y, -x)$

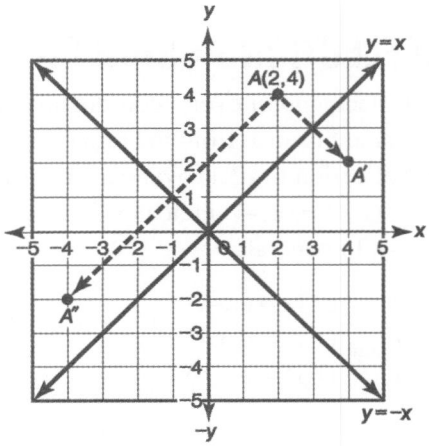

FIGURE 17.20 Reflecting over $y = \pm x$.

EXAMPLE 17.3 Graph $\triangle ABC$ with coordinates $A(1,3)$, $B(5,7)$, and $C(8,-3)$. On the same set of axes graph $\triangle A'B'C'$, the reflection of $\triangle ABC$ over the y-axis.

SOLUTION

After graphing $\triangle ABC$, reflect points A, B, and C over the y-axis, as shown in the accompanying figure. Then connect the image points A', B', and C' with line segments.

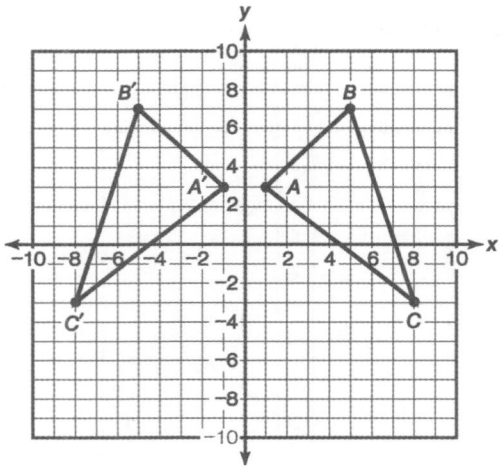

TRANSLATIONS USING COORDINATES

Sliding a point $P(x, y)$ horizontally h units and then vertically k units places the image at $P'(x + h, y + k)$. Figure 17.21 illustrates a translation in which both h and k stand for positive numbers. The signs of h and k indicate direction. For example, $P'(x + 1, y - 2)$ is a translation of $P(x, y)$. Since h is +1 and k is –2, point P is shifted 1 unit to the *right* ($h > 0$) and then 2 units *down* ($k < 0$).

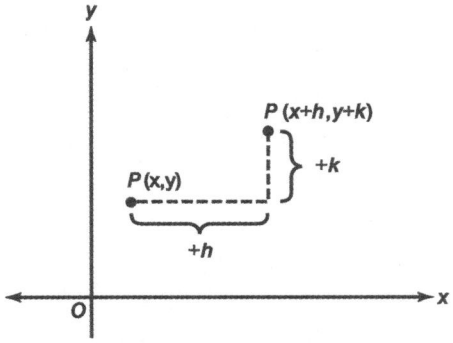

FIGURE 17.21 Translation of a point.

The shorthand notation $T_{h,k}$ is sometimes used to represent a translation of an object h units horizontally and k units vertically. For example:

$$T_{2,-3}\, A(3, 4) = A'(3 + 2, 4 + (-3)) = A'(5, 1).$$

Thus, the image of $A(3,4)$ under the translation $T_{2,-3}$ is $A'(5,1)$. In general,

$$T_{h,k}\,(x, y) = (x + h, y + k).$$

EXAMPLE 17.4 The coordinates of the vertices of $\triangle ABC$ are $A(2,-3)$, $B(0,4)$ and $C(-1, 5)$. If the image of point A under a translation is point $A'(0, 0)$, find the images of points B and C under the same translation.

SOLUTION

In general, after a translation of h units in the horizontal direction and k units in the vertical direction, the image of $P(x, y)$ is $P'(x + h, y + k)$. Since

$$A(2, -3) \rightarrow A'(2 + h, -3 + k) = A'(0, 0)$$

it follows that

$$2 + h = 0 \text{ and } h = -2,$$
$$-3 + k = 0 \text{ and } k = 3$$

Therefore:

$$B(0,4) \rightarrow B'(0 + [-2], 4 + 3) = B'(-2,7),$$
$$C(-1,5) \rightarrow C'(-1 + [-2], 5 + 3) = C'(-3,8).$$

ROTATIONS USING COORDINATES

The notation $R_{x°}(x, y)$ represents the counterclockwise rotation of point (x, y) through an angle of $x°$. Unless otherwise indicated, the center of rotation is the origin. In Figure 17.22, rectangle $AB'C'D'$ is the image of rectangle $ABCD$ under a 90° counterclockwise rotation about the origin. The vertices of rectangle $ABCD$ are mapped as follows:

$$R_{90°}\,A(0, 0) = A(0, 0)$$
$$R_{90°}\,B(6, 0) = B'(0, 6)$$
$$R_{90°}\,C(6, 3) = C'(-3, 6)$$
$$R_{90°}\,D(0, 3) = D'(-3, 0)$$

In general,

$$R_{90°}\,(x, y) = (-y, x).$$

The coordinates of the images of points rotated about the origin through angles of 180° and 270° can be determined using these rules:
$R_{180°}\,(x, y) = (-x, -y)$ and $R_{270°}\,(x, y) = (y, -x)$.

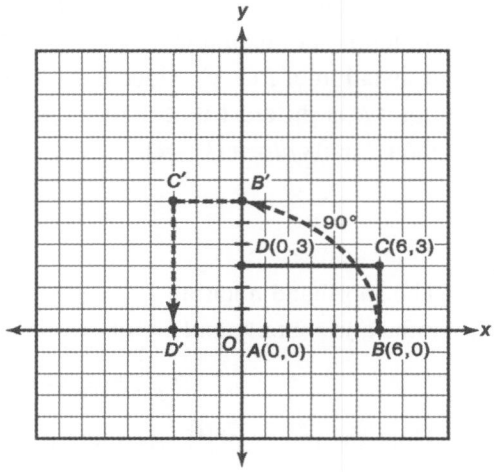

FIGURE 17.22 Rotation of a rectangle 90°.

EXAMPLE 17.5 In the accompanying figure, each grid box is 1 unit. Describe a transformation that maps:

a. $\triangle JCK$ onto $\triangle BCQ$
b. $\triangle WST$ onto $\triangle JCL$
c. $\triangle WST$ onto $\triangle JCK$
d. $\triangle JCK$ onto $\triangle CAP$

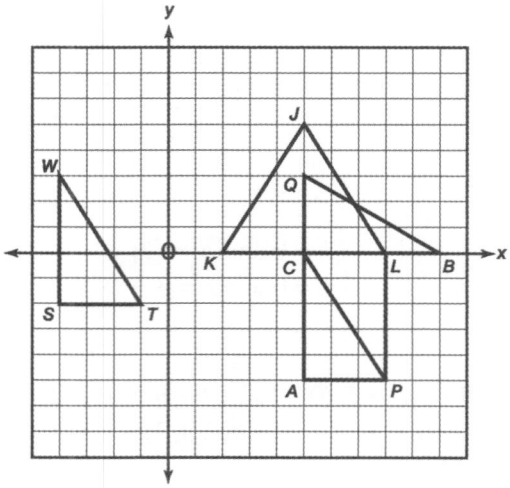

SOLUTION

a. A *rotation* of $\triangle JCK$ 270° counterclockwise about point C produces an image that coincides with $\triangle BCQ$. Thus, using C as the center of rotation,

$$R_{270°}\ \triangle JCK = \triangle BCQ.$$

b. A *translation* of 9 units horizontally and 2 units vertically will shift the vertices of $\triangle WST$ such that $W \rightarrow J$, $S \rightarrow C$, and $T \rightarrow L$. Thus,

$$T_{9,2}\ \triangle WST = \triangle JCL.$$

c. A *glide reflection* comprised of a reflection of $\triangle WST$ in the y-axis, followed by a horizontal translation of the reflected triangle 1 unit to the right and 2 units up:

$$r_{y\text{-}axis}\ \triangle WST = \triangle W'S'T' \text{ followed by } T_{1,2}\ \triangle W'S'T' = \triangle JCK.$$

d. A *glide reflection* comprised of a reflection of $\triangle JCK$ in vertical segment $\overline{JCA}$, followed by a vertical translation of its image, $\triangle JCL$, 5 units down:

$$r_{\overline{JCA}}\ \triangle JCK = \triangle JCL \text{ followed by } T_{0,-5}\ \triangle JCL = \triangle CAP.$$

DILATIONS USING COORDINATES

A dilation with a nonzero scale factor of k maps $P(x,y)$ onto $P'(kx, ky)$ where the origin is the center of the dilation. This transformation can be expressed using the notation

$$D_k\,(x, y) = (kx, ky).$$

Figure 17.23 illustrates a dilation in which $k > 1$ so that $OP' > OP$. To illustrate further, assume the coordinates of the endpoints of $\overline{AB}$ are $A(2,4)$ and $B\left(\frac{1}{3},1\right)$. The coordinates of the endpoints of $\overline{A'B'}$, the image of $\overline{AB}$ under a dilation with a scale factor of 3, are $A'(6,12)$ and $B'(1,3)$.

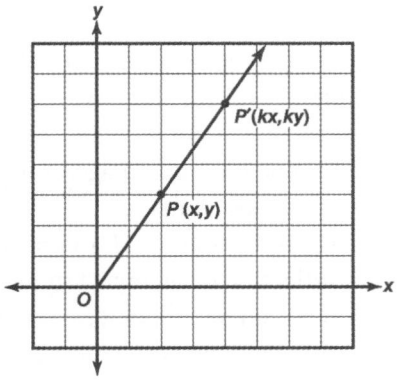

FIGURE 17.23 Dilation of point P.

EXAMPLE 17.6 After a dilation with respect to the origin, the image of $A(2,3)$ is $A'(4,6)$. What are the coordinates of the point that is the image of $B(1,5)$ after the same dilation?

SOLUTION

Determine the constant of dilation. The constant of dilation is 2, since

$$A(2,3) \rightarrow A'(\underline{2}\times 2,\underline{2}\times 3) = A'(4,6).$$

Under the same dilation, the x- and y-coordinates of point B are also multiplied by 2:

$$D_2(1,5) \rightarrow B'(\underline{2}\times 1,\underline{2}\times 5) = B'(2,10).$$

Composing Transformations

A glide reflection is an example of a *composite transformation*, as it combines two other transformations to form a new transformation.

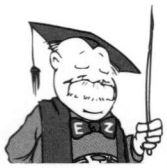

DEFINITION OF COMPOSITE TRANSFORMATION

A *composite transformation* is a series of two or more transformations in which each transformation after the first is performed on the image of the transformation that was applied before it.

REMEMBER

When performing a composite transformation, the order in which the transformations are performed matters.

EXAMPLE 17.7 In the accompanying figure, p and q are lines of symmetry for regular hexagon $ABCDEF$ intersecting at point O, the center of the hexagon. Determine the final image of the composite transformation of:

a. the reflection of $\overline{AB}$ in line q followed by a reflection of its image in line p.

b. a 60° counterclockwise rotation of point E followed by a reflection of its image in line q.

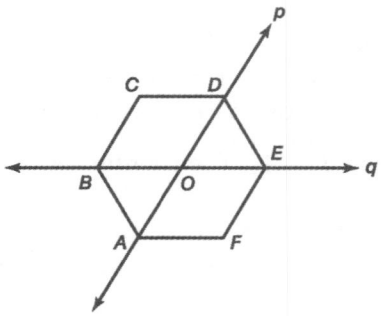

SOLUTION

a. $r_{\text{line } q}\left(\overline{AB}\right) = \overline{CB}$ and $r_{\text{line } p}\left(\overline{CB}\right) = \overline{EF}$.

b. $R_{60°}(E) = D$ and $r_{\text{line } q}(D) = F$.

Sometimes a special shorthand notation is used to indicate a composite transformation. The notation $r_{\text{line } q} \circ R_{60°}(E)$ represents the composite transformation of a 60° counterclockwise rotation of point E followed by a reflection of the rotated image point in line q. The transformation on the right side of the small centered circle (∘) is always performed first.

EXAMPLE 17.8 The coordinates of the vertices of $\triangle ABC$ are $A(2,0)$, $B(1,7)$, and $C(5,1)$.

a. Graph $\triangle A'B'C'$, the reflection of $\triangle ABC$ over the y-axis, and graph $\triangle A''B''C''$, the reflection of $\triangle A'B'C'$ over the x-axis.

b. What single type of transformation maps $\triangle ABC$ onto $\triangle A''B''C''$?

SOLUTION

a. Under the given composite transformation:

$$A(2,0) \rightarrow A'(-2,0) \rightarrow A''(-2,0).$$
$$B(1,7) \rightarrow B'(-1,7) \rightarrow B''(-1,-7).$$
$$C(5,1) \rightarrow C'(-5,1) \rightarrow C''(-5,-1).$$

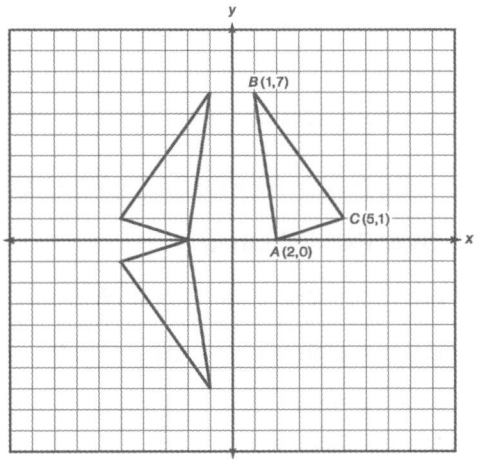

b. $R_{180°}\triangle ABC = \triangle A''B''C''$.

COMPOSING REFLECTIONS OVER TWO LINES

The composition of two rotations with the same center is a *rotation*. The composition of two translations is a *translation*. The composition of two reflections, however, is *not* a reflection. There are two possibilities to consider: composing the reflections over parallel lines and composing the reflections over intersecting lines.

- Composing two reflections over parallel lines *translates* the original figure, as shown in Figure 17.24.
- Composing two reflections over intersecting lines *rotates* the original figure, as shown in Figure 17.25.

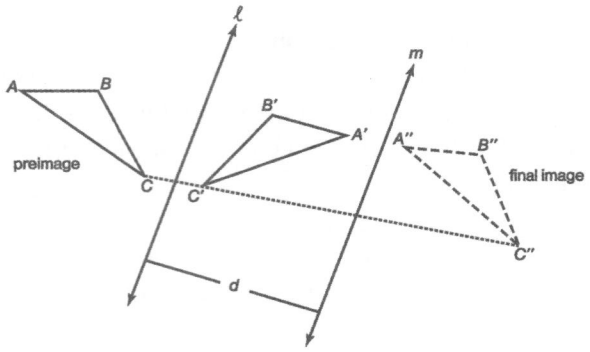

FIGURE 17.24 Composing reflections over two parallel lines.

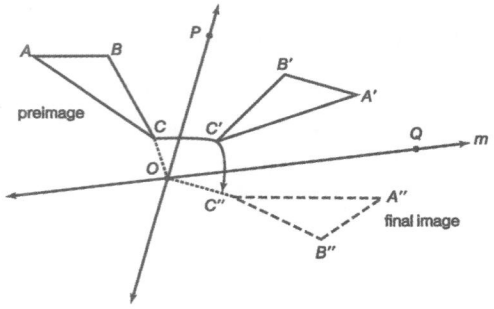

FIGURE 17.25 Composing reflections over two lines intersecting at point O.

REFLECTION–REFLECTION THEOREM

Case 1: Reflecting Over Parallel Lines

The composition of two reflections over two parallel lines is a translation.

- The direction of the translation is perpendicular to the reflecting lines.
- The distance between the final image and the preimage is two times the distance between the parallel lines.

Case 2: Reflecting Over Intersecting Lines

The composition of two reflections over two intersecting lines is a rotation about their point of intersection. The angle of rotation is equal to two times the measure of the angle formed by the reflecting lines at their point of intersection.

Based on the Reflection–Reflection theorem, we can generalize that any translation or rotation can be expressed as the composition of two reflections.

COMPOSITE TRANSFORMATIONS AND CONGRUENT FIGURES

Here are a couple of other observations about composite transformations that you should know:

- If two figures are congruent, there exists a transformation that maps one figure onto the other.
- In a plane, one of two congruent figures can be mapped onto the other by a composition of at most three reflections.

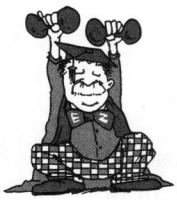

REVIEW EXERCISES FOR CHAPTER 17

1. Which figure best represents a line reflection?

   (1) P q (2) P p (3) P P (4) P P

2. One function of a movie projector is to enlarge the image on the film. This procedure is an example of a

   (1) dilation (2) reflection (3) rotation (4) translation

3. A reflection does *not* preserve

   (1) collinearity (3) orientation
   (2) segment measure (4) angle measure

4. Which letter has both point and line symmetry?

   (1) Z (2) T (3) C (4) H

5. Which letter has point symmetry but no line symmetry?

   (1) E (2) S (3) W (4) I

6. Which letter has line symmetry but no point symmetry?

   (1) O (2) X (3) N (4) M

7. In the accompanying diagram, $\triangle A'B'C'$ is the image of $\triangle ABC$ under a transformation in which $\triangle ABC \cong \triangle A'B'C'$. This transformation is an example of a

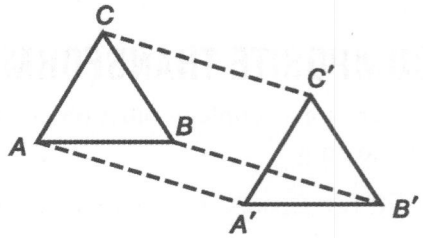

(1) line reflection
(2) rotation
(3) translation
(4) dilation

8. The composition of two equal glide reflections is equivalent to

(1) a translation that is twice the distance of a single glide reflection
(2) a dilation with a scale factor of 2
(3) a rotation
(4) a reflection in a line perpendicular to the direction of the translation

9. Ms. Brewer's art class is drawing reflected images. She wants her students to draw images reflected over a line. Which diagram represents a correctly drawn image?

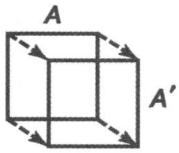

(1)

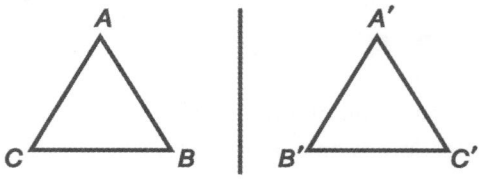

(3)

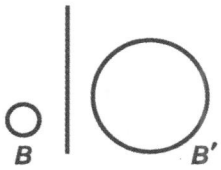

(2)

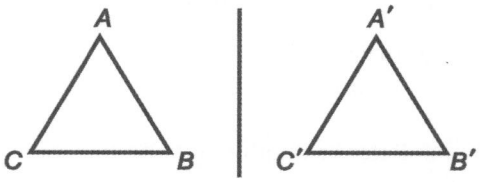

(4)

10. The image of point A after a dilation with a scale factor of 3 is (6,15). What was the original location of point A?

(1) (2,5) (2) (3,12) (3) (9,18) (4) (18,45)

11. What is the image of point (–3,4) under the translation that shifts (x,y) to $(x - 3, y + 2)$?

(1) (0,6) (2) (6,6) (3) (–6,8) (4) (–6,6)

Use the accompanying diagram for exercises 12 and 13. In the diagram of regular octagon ABCDEFGH, *lines* ℓ *and* m *are lines of symmetry.*

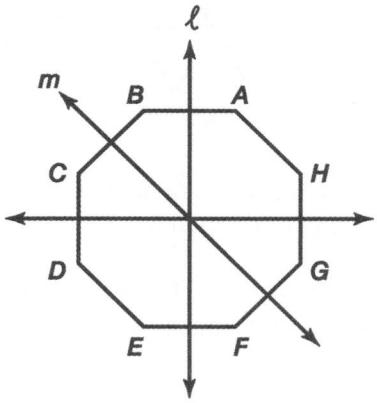

12. What is the final image of the composite transformation of the reflection of $\overline{AB}$ over line m followed by a reflection of the image over line ℓ?

(1) $\overline{CD}$ (2) $\overline{AH}$ (3) $\overline{HG}$ (4) $\overline{FG}$

13. What is the final image of the composite transformation of a counterclockwise rotation of point H 135° followed by a reflection of the image over line m?

(1) B (2) D (3) E (4) G

14. The image of point (–2,3) after a certain translation is (3,–1). What is the image of point (4,2) after the same translation?

(1) (–1,6) (2) (0,7) (3) (5,4) (4) (9,–2)

15. What is image of point (–3,–1) after a rotation of 90° about the origin?

(1) (3,1) (2) (1,–3) (3) (3,–1) (4) (1,3)

16. In the accompanying diagram, K is the image of A after a translation. Under the same translation, which point is the image of J?

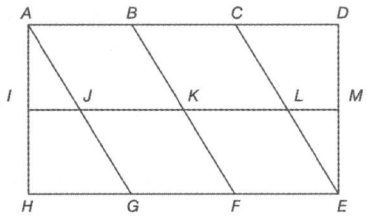

(1) B (3) E
(2) C (4) F

17. Which figures, if any, have *both* point symmetry and line symmetry?

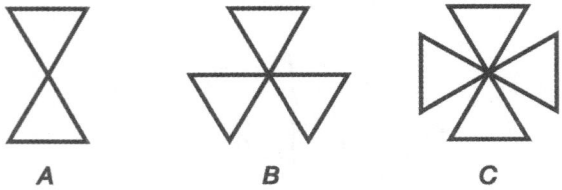

(1) A and C, only (3) none of the figures
(2) B and C, only (4) all of the figures

18. The coordinates of $\triangle JRB$ are $J(1,-2)$, $R(-3,6)$, and $B(4,5)$. What are the coordinates of the vertices of its image after the transformation $T_{2,-1} \circ r_{y\text{-axis}}$?

(1) $(3,1), (-1,-7), (6,-6)$ (3) $(1,-3), (5,5), (-2,4)$
(2) $(3,-3), (-1,5), (6,4)$ (4) $(-1,-2), (3,6), (-4,5)$

19. Which mapping rule does *not* represent an isometry in the coordinate plane?

(1) $(x,y) \rightarrow (2x,2x)$ (3) $(x,y) \rightarrow (-x,y)$
(2) $(x,y) \rightarrow (x+2, y+2)$ (4) $(x,y) \rightarrow (x,-y)$

20. The composite transformation that reflects point P through the origin, the x-axis, and the line $y = x$, in the order given, is equivalent to which rotation of point P about the origin?

(1) $R_{90°}$ (2) $R_{180°}$ (3) $R_{270°}$ (4) $R_{360°}$

Use the accompanying figure for exercises 21 to 24. In the figure, each grid box is 1 unit. Identify each of the given transformations as either a reflection, translation, rotation, dilation, or glide reflection. State the reflection line, translation rule, center and angle of rotation, or the reflecting line and translation for a glide reflection.

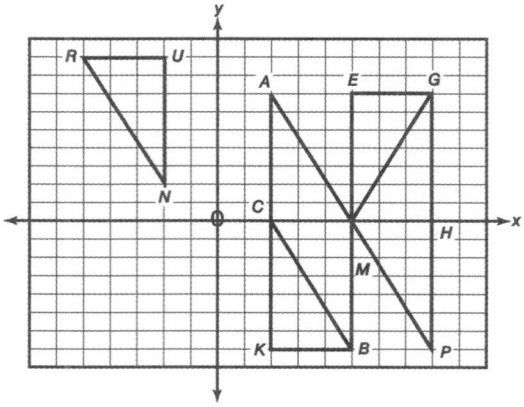

21. $\triangle GHM \rightarrow \triangle ACM$

22. $\triangle BMC \rightarrow \triangle NUR$

23. $\triangle BKC \rightarrow \triangle GEM$

24. $\triangle ACM \rightarrow \triangle PHM$

25. Using the same figure as for Exercises 21–24, describe the composite transformation that maps $\triangle RUN$ onto $\triangle GEM$.

26. Carson is a decorator. He often sketches his room designs on the coordinate plane. He has graphed a square table on his grid so that its corners are at the coordinates $A(2,6)$, $B(7,8)$, $C(9,3)$, and $D(4,1)$. To graph a second identical table, he reflects $ABCD$ over the y-axis.

    (a) Using graph paper, sketch and label $ABCD$ and its image $A'B'C'D'$, which show the locations of the two tables.
    (b) Find the number of square units in the area of $ABCD$.

27. If point A is the image of $(-1,4)$ after a reflection over the line $y = x$, and point B is the image of $(3,-5)$ after a reflection over the line $y = -x$, determine an equation of $\overleftrightarrow{AB}$.

28. (a) On graph paper, graph and label the triangle whose vertices are $A(0,0)$, $B(8,1)$, and $C(8,4)$. Then graph and state the coordinates of $\triangle A''B''C''$, the final image under the composite transformation of the reflection of $\triangle ABC$, over the line $y = x$ followed by a reflection over the y-axis.
    (b) Which single transformation maps $\triangle ABC$ onto $\triangle A''B''C''$?

29. In the accompanying diagram of regular hexagon $ABCDEF$ with center O, L, and P are lines of symmetry. Identify the final image under each composite transformation.

    (a) A 120° counterclockwise rotation of point C followed by a reflection of its image over line P.
    (b) A reflection of $\overline{AB}$ in line P followed by a reflection of its image over line L.
    (c) A reflection of point A in point O followed by a counterclockwise 60° rotation of its image.

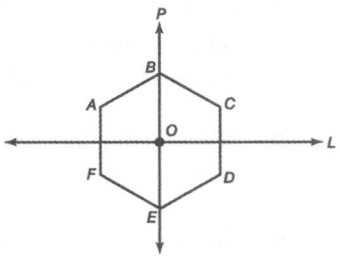

30. The engineering office in the village of Kingsboro has a map of the village that is laid out on a rectangular coordinate system. A traffic circle located on the map is represented by the equation $(x + 4)^2 + (y - 2)^2 = 81$. The village planning commission wants to expand the traffic circle. A new traffic circle is designed by applying the transformation D_2 to the original traffic circle, where the center of dilation is at the origin. Find the center and radius of the new traffic circle.

31. In parallelogram $ABCD$, diagonals $\overline{AC}$ and $\overline{BD}$ intersect at point E.

    (a) Describe the isometry that can be used to map $\triangle AED$ onto $\triangle CEB$.
    (b) Use the properties of a parallelogram to prove that $\triangle CEB$ is the image of $\triangle AED$ under that isometry.

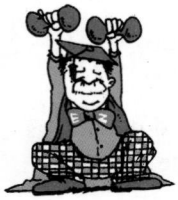

CUMULATIVE REVIEW EXERCISES: CHAPTERS 12–17

1. Given the points $A(-2, 4)$ and $B(4, -4)$. Find:

    (a) the slope of $\overleftrightarrow{AB}$.
    (b) the length of $\overline{AB}$.

2. If the diagonals of a rhombus have lengths 4 and 5, what is the area of the rhombus?

3. Determine an equation of the line that contains the origin and the image of $T_{2,3}$ $(1,-4)$.

4. From the same external point, two tangents are drawn to a circle. If the tangents intercept arcs whose degree measures are 300° and 60°, what is the measure of the angle formed by the two tangents?

5. What is the perimeter of a regular hexagon inscribed in a circle whose radius is 5?

6. Tangents $\overline{PA}$ and $\overline{PB}$ are drawn to circle O from external point P, and chord $\overline{AB}$ is drawn. If m $\measuredangle APB = 60$ and $AB = 8$, what is the length of $\overline{PB}$?

7. If the number of square centimeters in the area of a circle is equal to the number of centimeters in its circumference, what is the length (in centimeters) of the radius of the circle?

8. What is an equation for the locus of points that are equidistant from:

   (a) $(1, 0)$ and $(7, 0)$?
   (b) $(-2, -1)$ and $(-2, 5)$?

9. The degree measure of the central angle of a sector of a circle is $120°$. If the area of the sector is 27π square units, what is the *circumference* of the circle?

10. In trapezoid $ABCD$, $\overline{AB} \parallel \overline{DC}$, $DC = 4$, $AB = 20$, $AD = 12$, and $m\measuredangle A = 30$. What is the area of trapezoid $ABCD$?

11. In circle O, chords $\overline{PQ}$ and $\overline{RS}$ intersect at T. If $PT = 4$, $TQ = 4$, and $ST = 8$, what is TR?

12. Determine an equation of the line that contains the reflected images of $(-1,4)$ and $(3,-4)$ in the line $y = x$.

13. In a circle, an arc measures $120°$. If the length of the arc is 8π, find the circumference of the circle.

14. In the accompanying diagram, triangles AEB, EBD, and BDC are equilateral. If the perimeter of $ACDE$ is 20, then what is the area of $ACDE$?

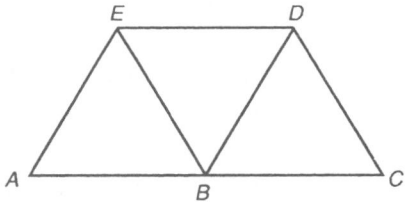

15. In circle, O, chords $\overline{AB}$ and $\overline{CD}$ intersect at point E. If $AE = EB$, $CE = 4$, and $ED = 9$, find the length of $\overline{AB}$.

16. Point M is the midpoint of $\overline{AB}$. If the coordinates of A are $(7, -3)$ and the coordinates of M are $(7, 7)$, what are the coordinates of B?

17. If P' represents the image of point P after a reflection in line ℓ, what are the coordinates of $r_{y=x}(P')$?

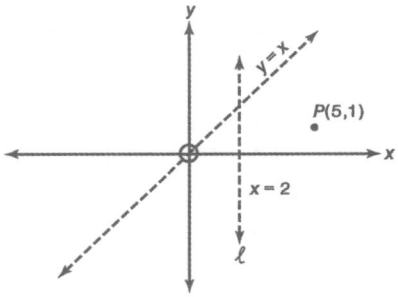

18. What is the slope of the line that is parallel to the line whose equation is $2y = 3x - 10$?

19. A regular hexagon with an apothem of 3 centimeters circumscribes a circle. What is the area of the circle in square centimeters?

20. What is an equation of the line that passes through the point (0, 0) and has a slope of $-\frac{2}{3}$?

21. Secant $\overline{ADB}$ and tangent $\overline{AC}$ are drawn to circle O from external point A. If $AD = 3$ and $DB = 9$, what is the length of $\overline{AC}$?

22. If two circles with radii 2 and 5 are drawn so that the distance between their centers is 6, what is the maximum number of common tangents they can have?

23. As shown in the diagram, two semicircles of radius 2 are drawn in a square. If the length of a side of the square is 4, what is the area of the shaded portion of the figure?

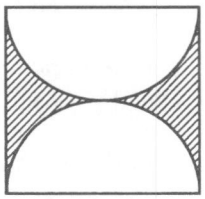

24. In the accompanying diagram, circle O is inscribed in quadrilateral $ABCD$ and E, F, G, and H are the points of tangency of the sides. If $AH = 6$, $DG = 4$, $CF = 2$, and $BF = 3$, what is the perimeter of quadrilateral $ABCD$?

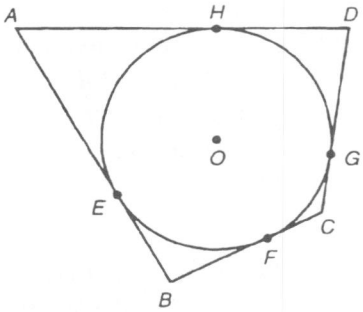

25. Quadrilateral $TRAP$ has vertices $T(0, 0)$, $R(0, 5)$, $A(9, 8)$, and $P(12, 4)$. Prove by coordinate geometry that quadrilateral $TRAP$ is an isosceles trapezoid.

26. In the accompanying figure, point $S(-3, 4)$ lies on circle O with center (0, 0). Line $\overleftrightarrow{ASB}$ and radius $\overline{OS}$ are drawn.

(a) Find the length of $\overline{OS}$.
(b) Write an equation of circle O.
(c) If $\overleftrightarrow{AB} \perp \overline{OS}$, find the slope of $\overleftrightarrow{AB}$.
(d) Write an equation of line $\overleftrightarrow{ASB}$.
(e) Find the coordinates of any point on $\overleftrightarrow{AB}$ other than S.

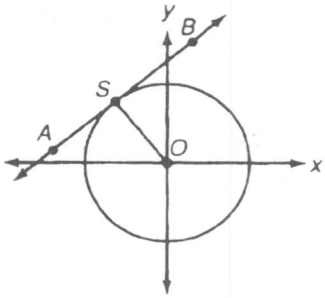

27. Quadrilateral $ABCD$ has vertices $A(-1, 0)$, $B(3, 3)$, $C(6, -1)$, and $D(2, -4)$. Prove that quadrilateral $ABCD$ is a square.

28. The vertices of $\triangle ABC$ are $A(-4, -2)$, $B(2, 6)$, and $C(2, -2)$.

    (a) Write an equation of the locus of points equidistant from vertex B and vertex C.
    (b) Write an equation of the line parallel to $\overline{BC}$ and passing through vertex A.
    (c) Find the coordinates of the point of intersection of the locus in part **a** and the line determined in part **b**.
    (d) Write an equation of the locus of points that are 4 units from vertex C.
    (e) What is the total number of points that satisfy the loci described in parts **a** and **d**?

29. In a trapezoid, the length of one base is 5 times the length of the other base. The height of the trapezoid is 1 less than the length of the shorter base. If the area of the trapezoid is 90 square units, find the length of the *shorter* base.

30. Isosceles trapezoid $ABCD$ with bases $\overline{AB}$ and $\overline{DC}$ is inscribed in circle O. Secant $\overline{RCT}$ intersects $\overline{AB}$ at S, $\overline{RE}$ is tangent to the circle at B, chord $\overline{TB}$ is drawn, $m\measuredangle AST = 50$, $m\measuredangle ABR = 110$, and $m\widehat{AD} = 80$.

    Find: (a) $m\widehat{DC}$
    (b) $m\widehat{AT}$
    (c) $m\measuredangle TBE$
    (d) $m\measuredangle R$
    (e) $m\measuredangle ADC$

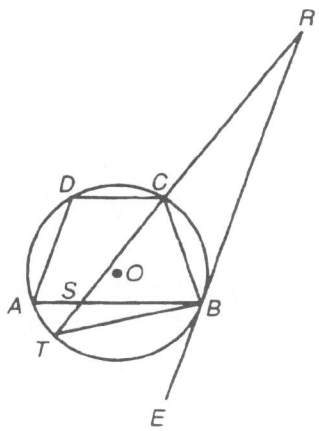

31. The vertices of $\triangle ABC$ are $A(-2, 3)$, $B(0, -3)$, and $C(4, 1)$. Prove, by means of coordinate geometry, that:

    (a) $\triangle ABC$ is isosceles
    (b) The median to side $\overline{BC}$ is also the altitude to side $\overline{BC}$

32. The equation of line $\overleftrightarrow{AB}$ is $x = 2$.

    (a) Describe fully the locus of points d units from line $\overleftrightarrow{AB}$.
    (b) Describe fully the locus of points 1 unit from the origin.
    (c) How many points satisfy the conditions in parts **a** and **b** simultaneously for the following values of d?
    1 $d = 2$ **2** $d = 3$ **3** $d = 4$

33. GIVEN: $\overline{RN}$ tangent to circle T at N, and diameter $\overline{AN}$.

PROVE: $(AN)^2 = AR \times AB$

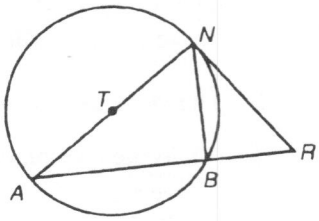

34. GIVEN: Secants $\overline{PAB}$ and $\overline{PCD}$ are drawn to circle O, $\overline{PAB} \cong \overline{PCD}$, and $\overline{AE}$ and $\overline{CF}$ are perpendicular to $\overline{BD}$ at points E and F, respectively.

PROVE: $\overline{AE} \cong \overline{CF}$

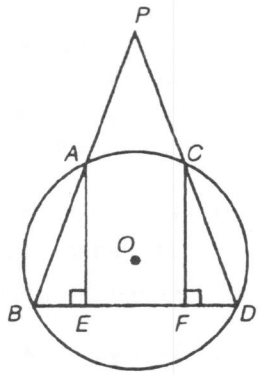

35. In the accompanying diagram, $\triangle LRS$ is inscribed in circle O. The length of diameter $\overline{LOS}$ is 10. Altitude $\overline{RP}$ is drawn so that $PO = 3$.

(a) Find $m\measuredangle LRS$

(b) Find RP

(c) Using the result from part **b**, find

1 $m\measuredangle L$ to the *nearest degree*

2 $m\measuredangle PRS$ to the *nearest degree*

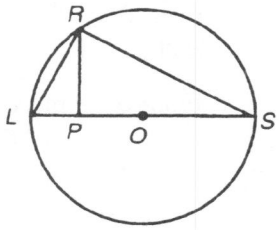

36. Regular pentagon $ABCDE$ is inscribed in circle O. Diameter $\overline{AHJ} \perp \overline{DC}$. Segments $\overline{AHJ}$ and $\overline{ED}$ are extended and intersect at external point G.

Find: (a) $m\widehat{BC}$
(b) $m\measuredangle BDC$
(c) $m\measuredangle AHB$
(d) $m\measuredangle G$
(e) $m\measuredangle CDG$

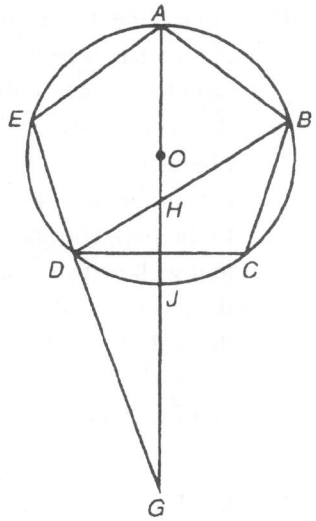

37. Lines $\overleftrightarrow{AB}$ and $\overleftrightarrow{CD}$ are parallel and 4 units apart. Point P is on $\overleftrightarrow{AB}$.

(a) Describe fully the locus of points equidistant from $\overleftrightarrow{AB}$ and $\overleftrightarrow{CD}$.
(b) Describe fully the locus of points at a distance of k units from P.
(c) How many points will satisfy the conditions of both parts **a** and **b** if
1 $k = 2$ **2** $k = 4$ **3** $k = 1$

38. The vertices of $\triangle ABC$ are $A(-1, -2)$, $B(3, 1)$, and $C(0, 5)$.

(a) Show, by means of coordinate geometry, that $\triangle ABC$ is a right triangle and state a reason for your conclusion.
(b) Find the area of $\triangle ABC$.

39. Find the area of quadrilateral $ABCD$ with vertices $A(5, 2)$, $B(0, 5)$, $C(-2, -2)$, and $D(0, 0)$.

40. GIVEN: $\overline{CEB}$ is tangent to circle O at B, diameter $\overline{BA} \parallel \overline{CD}$. Secants $\overline{DFB}$ and $\overline{EFA}$ intersect circle O at F.

PROVE: $\dfrac{BD}{AE} = \dfrac{CD}{BE}$

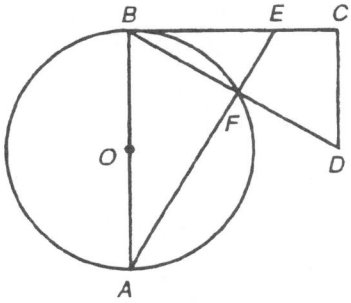

41. An architect is designing a park with an entrance represented by point C and a circular garden with center O, as shown in the accompanying diagram. The architect plans to connect three points on the circumference of the garden, A, B, and D, to the park entrance, C, with walkways so that walkways $\overline{CA}$ and $\overline{CB}$ are tangent to the garden, walkway $\overline{DOEC}$ is a path through the center of the garden $m\widehat{ADB} : m\widehat{AEB} = 3 : 2$, $BC = 60$ meters, and $EC = 43.6$ meters.

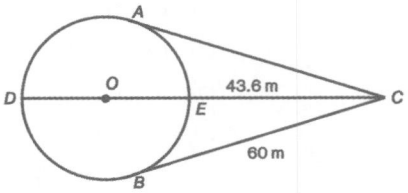

(a) Find the measure of the angle between walkways $\overline{CA}$ and $\overline{CB}$.
(b) Find the diameter of the circular garden, to the *nearest meter*.

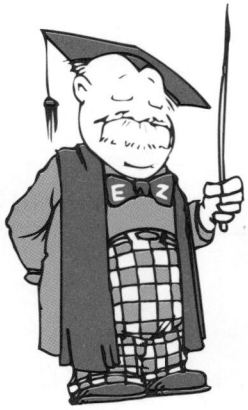

Some Geometric Relationships and Formulas Worth Remembering

POSTULATES OF CONGRUENCE AND EQUALITY

- Reflexive property: $\angle A \cong (=) \angle A$.
- Symmetric property: If $\angle A \cong (=) \angle B$, then $\angle B \cong (=) \angle A$
- Transitive property: If $\angle A \cong (=) \angle B$ and $\angle B \cong (=) \angle C$, then $\angle A \cong (=) \angle C$.

RELATIONSHIPS BETWEEN ANGLES FORMED BY PARALLEL LINES

Type of Angle Pair	Angle Relationship	Line $\ell \parallel$ Line m
• Alternate interior angles	• $\angle 3 \cong \angle 6$ and $\angle 4 \cong \angle 5$	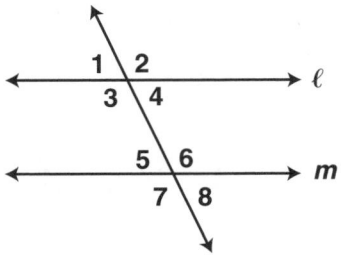
• Corresponding angles	• $\angle 1 \cong \angle 5$ and $\angle 3 \cong \angle 7$; $\angle 2 \cong \angle 6$ and $\angle 4 \cong \angle 8$	
• Interior angles on the same side of the transversal	• $m\angle 3 + m\angle 5 = 180$ and $m\angle 4 + m\angle 6 = 180$	

ANGLE-SUM RELATIONSHIPS

Figure	Sum
• Triangle	• Sum of interior angles = 180
• Quadrilateral	• Sum of interior angles = 360
• n-Sided polygon	• Sum of interior angles = $(n - 2) \times 180$
• n-Sided polygon	• Sum of exterior angles = 360

INEQUALITIES IN A TRIANGLE

- if $x > y$, then $a > b$
- if $a > b$, then $x > y$
- $d > a$ and $d > c$
- $z < x + y$, $x < y + z$, and $y < x + z$

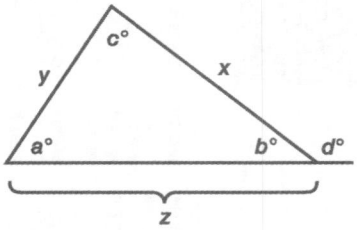

ANGLE MEASUREMENT IN CIRCLES

Location of Vertex of Angle	Measure of Angle Equals. . .
• At center of circle	• the measure of the intercepted arc
• On circle	• one-half of the measure of the intercepted arc
• Inside circle	• one-half of the sum of the measures of the intercepted arcs
• Outside circle	• one-half of the difference of the measures of the intercepted arcs

CHORD, TANGENT, AND SECANT-SEGMENT RELATIONSHIPS

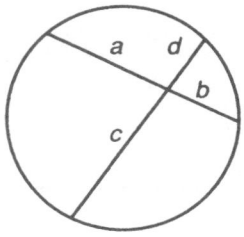

$a \times b = c \times d$

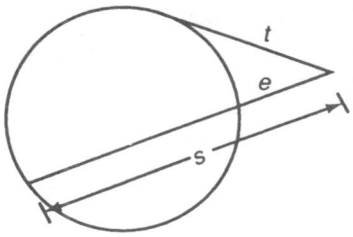

$t^2 = s \times e$

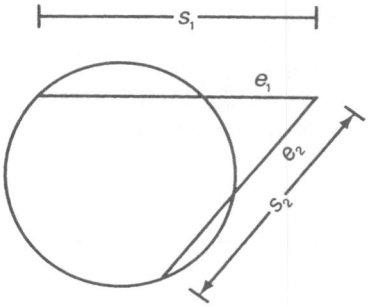

$s_1 \times e_1 = s_2 \times e_2$

AREAS OF POLYGONS

Figure	Area Formula
• Triangle	• $\frac{1}{2} \times \text{base} \times \text{height}$
• Equilateral triangle	• $\frac{(\text{side})^2}{4} \times \sqrt{3}$
• Rectangle	• length × width
• Square	• side × side or $\frac{1}{2} \times (\text{diagonal})^2$
• Parallelogram	• base × height
• Rhombus	• $\frac{1}{2} \times \text{diagonal}_1 \times \text{diagonal}_2$
• Trapezoid	• $\frac{1}{2} \times \text{height} \times \text{sum of bases}$
• Regular polygon	• $\frac{1}{2} \times \text{apothem} \times \text{perimeter}$

CIRCUMFERENCE AND AREA OF CIRCULAR REGIONS

Quantity	Formula
• Circumference of circle	• $2\pi \times \text{radius}$
• Length of arc formed by central angle of $n°$	• $\frac{n}{360} \times \text{circumference of circle}$
• Area of circle	• $\pi \times (\text{radius})^2$
• Area of sector formed by central angle of $n°$	• $\frac{n}{360} \times \text{area of circle}$

COORDINATE FORMULAS

Description	Formula
Let $A = (x_A, y_A)$ and $B = (x_B, y_B)$	
• Midpoint (x_m, y_m) of $\overline{AB}$.	• $x_m = \dfrac{x_A + x_B}{2}$; $y_m = \dfrac{y_A + y_B}{2}$
• Length of $\overline{AB}$.	• $AB = \sqrt{(x_B - x_A)^2 + (y_B - y_A)^2}$
• Slope m of $\overleftrightarrow{AB}$.	• $m = \dfrac{y_B - y_A}{x_B - x_A}$
• $\overleftrightarrow{AB}$ is a horizontal line.	• Slope of $\overleftrightarrow{AB} = 0$.
• $\overleftrightarrow{AB}$ is a vertical line.	• Slope of $\overleftrightarrow{AB}$ is undefined.
• $\overleftrightarrow{AB}$ is parallel to $\overline{CD}$.	• Slope of $\overleftrightarrow{AB}$ = slope of $\overleftrightarrow{CD}$.
• $\overleftrightarrow{AB}$ is perpendicular to $\overline{CD}$.	• Slope of $\overleftrightarrow{AB} = \dfrac{-1}{\text{slope of } CD}$.
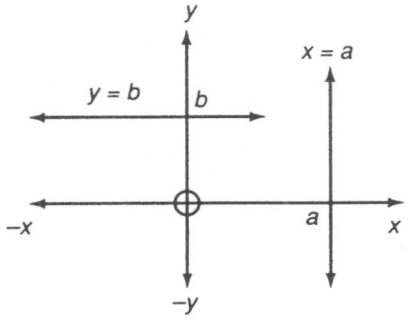	Equation of horizontal line: $y = b$ Equation of vertical line: $x = a$
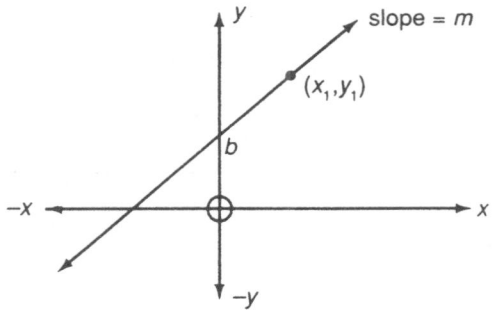	Equation of oblique line: $y = mx + b$ or $y - y_1 = m(x - x_1)$
Equation of circle having radius r and center:	
• at the origin	• $x^2 + y^2 = r^2$
• at (h, k)	• $(x - h)^2 + (y - k)^2 = r^2$

PROPERTIES OF TRANSFORMATIONS

Transformation	Properties Preserved	*Isometry	Coordinate Rules
Line Reflection	• collinearity • angle measure • distance	Opposite	• $r_{x\text{-axis}}(x, y) = (x, -y)$ • $r_{y\text{-axis}}(x, y) = (-x, y)$ • $r_{\text{origin}}(x, y) = (-x, -y)$ • $r_{y=x}(x, y) = (y, x)$ • $r_{y=-x}(x, y) = (-y, -x)$
Translation	• collinearity • angle measure • distance • orientation	Direct	$T_{h,k}(x, y) = (x + h, y + k)$
Rotation	• collinearity • angle measure • distance • orientation	Direct	• $R_{90°}(x, y) = (-y, x)$ • $R_{180°}(x, y) = (-x, -y)$ • $R_{270°}(x, y) = (y, -x)$
Dilation	• collinearity • angle measure • orientation	Image is similar to the original figure	$D_k(x, y) = (kx, ky)$
Glide Reflection	• collinearity • angle measure • distance	Opposite	

*Isometry is a transformation that produces a congruent image. A direct isometry preserves orientation and an opposite isometry reverses orientation.

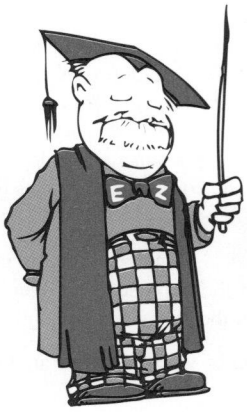

GLOSSARY

AA Theorem of Similarity — Two triangles are similar if two pairs of corresponding angles are congruent.

AAS Theorem of Congruence — Two triangles are congruent if two pairs of corresponding angles are congruent and a nonincluded pair of corresponding sides are congruent.

Abscissa — The x-coordinate of a point in the coordinate plane.

Absolute value — For a number x, denoted by $|x|$, its distance from 0 on the number line. Thus, $|x|$ always represents a nonnegative number.

Acute angle — An angle whose degree measure is greater than 0 and less than 90.

Acute triangle — A triangle with three acute angles.

Adjacent angles — Two angles that have the same vertex and share one side, but do not have any interior points in common.

Alternate interior angles — Pairs of angles formed when a transversal intersects two lines. The two angles in each pair are between the two lines, have different vertices, and lie on opposite sides of the transversal.

Altitude — A segment that is perpendicular to the side of the figure to which it is drawn.

Angle — The union of two rays that have the same end point.

Angle bisector — A line or any part of a line that contains the vertex of an angle and that divides the angle into two congruent angles. An angle has exactly one angle bisector.

Angle of depression — An angle formed by a horizontal ray of sight and the ray that is the line of sight to an object below the horizontal ray.

Angle of elevation — An angle formed by a horizontal ray of sight and the ray that is the line of sight to an object above the horizontal ray.

Apothem — For a regular polygon, the radius of its inscribed circle.

Arc of a chord — The minor arc of a circle whose end points are the end points of the chord. If the chord is a diameter, then either semicircle is an arc of the diameter.

Area — For a plane geometric figure, the number of square units it contains.

ASA Theorem of Congruence — Two triangles are congruent if two pairs of corresponding angles are congruent and the sides included by these angles are congruent.

Base angles of an isosceles triangle — The congruent angles that lie opposite the congruent sides of an isosceles triangle.

Base of an isosceles triangle — The noncongruent side of the isosceles triangle.

Bases of a trapezoid — The parallel sides of a trapezoid.

Betweenness of points — A term that refers to the order of three collinear points. If A, B, and C are three different collinear points, point C is between points A and B if $AC + CB = AB$.

Bisect — To divide into two equal parts.

Center of a regular polygon — The common center of the circles inscribed and circumscribed in the polygon.

Central angle of a circle — An angle whose vertex is at the center of a circle and whose sides are radii.

Central angle of a regular polygon — An angle whose vertex is the center of the regular polygon and whose sides terminate at consecutive vertices of the polygon.

Centroid of a triangle — The point at which the three medians of the triangle intersect.

Chord of a circle — A segment whose end points are on the circle.

Circle — The set of all points in a plane at a fixed distance from a given point called the *center*. The fixed distance is called the *radius* of the circle. An equation of a circle with center at point (h, k) and radius length r is $(x - h)^2 + (y - k)^2 = r^2$.

Circumference of a circle — The distance around the circle.

Circumscribed circle about a polygon — A circle that passes through each vertex of the polygon.

Circumscribed polygon about a circle — A polygon that has all of its sides tangent to the circle.

Collinear points — Points that lie on the same line.

Common external tangent to two circles — A line that is tangent to both circles and does not intersect the line segment whose end points are the centers of the two circles.

Common internal tangent to two circles — A line that is tangent to both circles and intersects the line segment whose end points are the centers of the two circles.

Complementary angles — Two angles whose measures add up to 90.

Composite transformation — A sequence of two or more transformations in which each transformation after the first is performed on the image of the transformation that was applied before it.

Concentric circles — Circles in the same plane that have the same center but have radii of different lengths.

Congruent angles — Angles that have the same measure.

Congruent circles — Circles with congruent radii.

Congruent line segments — Line segments that have the same length.

Congruent polygons — Polygons with the same number of sides that have the same size and the same shape. The symbol for congruence is $\cong$.

Congruent triangles — Triangles whose vertices can be paired so that any one of the following conditions is true: (1) the sides of one triangle are congruent to the corresponding sides of the other triangle ($SSS \cong SSS$); (2) two sides and the included angle of one triangle are congruent to the corresponding parts of the other triangle ($SAS \cong SAS$); (3) two angles and the included side of one triangle are congruent to the corresponding parts of the other triangle ($ASA \cong ASA$); (4) two angles and the side opposite one of these angles of one triangle are congruent to the corresponding parts of the other triangle ($AAS \cong AAS$). Two right triangles are congruent if the hypotenuse and a leg of one right triangle are congruent to the corresponding parts of the other triangle ($Hy - Leg \cong Hy - Leg$).

Converse of a conditional statement — Another conditional statement formed by interchanging the hypothesis ("Given") with the conclusion ("To Prove") of the original statement.

Convex polygon — A polygon each of whose interior angles measures less than 180.

Coordinate plane — A plane that is divided into four equal regions, called *quadrants*, by a horizontal number line and a vertical number line, called *axes*, intersecting at their zero points, called the *origin*. Each point in a coordinate plane is located by an ordered pair of numbers of the form (x, y). The first member, x, of the ordered pair gives the directed distance of the point from the zero point of the x-axis (horizontal). The second member, y, of the ordered pair gives the directed distance of the point from the zero point of the y-axis.

Corollary — A theorem that can be easily proved by means of a closely related theorem.

Corresponding angles — Pairs of angles formed when a transversal intersects two lines. The two angles in each pair lie on the same side of the transversal, but one angle is between the two lines and the other angle is exterior to the two lines.

Cosine of an acute angle of a right triangle — The ratio of the length of the leg that is adjacent to the acute angle to the length of the hypotenuse.

Decagon — A polygon with 10 sides.

Deductive reasoning — A step-by-step process by which a set of accepted facts is used to arrive at a conclusion.

Degree — A unit of angle measure. One degree is the measure of an angle formed by 1/360 of one complete rotation of a ray about its end point.

Diagonal of a polygon — A line segment whose end points are nonconsecutive vertices of the polygon.

Diameter of a circle — A chord of the circle that contains the center of the circle.

Dilation — A size transformation that produces an image similar to the original figure.

Direct isometry — An isometry that preserves orientation.

Distance formula — A formula used to find the length of the segment determined by two points in the coordinate plane. The distance d between two points $A(x_A, y_A)$ and $B(x_B, y_B)$ is given by the formula $d = \sqrt{(x_B - x_A)^2 + (y_B - y_A)^2}$.

Distance from a point to a line — The length of the perpendicular segment from the point to the line.

Dodecagon — A polygon with 12 sides.

Equiangular polygon — A polygon in which all of the angles have the same measure.

Equiangular triangle — A triangle in which all three angles have the same measure.

Equidistant — having the same distance.

Equilateral polygon — A polygon in which all of the sides have the same length.

Equilateral triangle — A triangle whose three sides have the same length.

Exterior angle of a polygon — An angle formed by a side of the polygon and the extension of an adjacent side of the polygon.

Externally tangent circles — Tangent circles that lie on opposite sides of the common tangent.

Extremes — The first and fourth terms of a proportion. In the proportion $\frac{a}{b} = \frac{c}{d}$, a and d are the extremes.

Geometric mean — See **Means**.

Glide reflection — The composition of a reflection in a line and a translation in the direction parallel to the reflecting line.

Half-plane — The set of points in a plane that lie on one side of a line.

Hexagon — A polygon with six sides.

Hypotenuse — The side of a right triangle that is opposite the right angle.

Incenter of a triangle — The point at which the three angle bisectors of the triangle intersect.

Indirect proof — A method of proof in which each possibility except the one that needs to be proved is eliminated by showing that it contradicts some known or given fact.

Inscribed angle of a circle — An angle whose vertex lies on the circle and whose sides are chords of the circle.

Inscribed circle of a polygon — A circle that is tangent to each side of the polygon.

Inscribed polygon — A polygon that has all of its vertices on a circle.

Internally tangent circles — Tangent circles that lie on the same side of the common tangent.

Isometry — A transformation that produces an image congruent to the original figure.

Isosceles trapezoid — A trapezoid whose nonparallel sides, called *legs*, have the same length.

Isosceles triangle — A triangle with two sides, called *legs*, that have the same length.

Leg of a right triangle — Either of the two sides of the right triangle that are not opposite the right angle.

Line — A term undefined in geometry; a line can be described as a continuous set of points forming a straight path that extends indefinitely in two opposite directions.

Line of centers of two circles — The line segment whose end points are the centers of the circles.

Line segment — Part of a line that consists of two different points on the line, called *end points*, and the set of all points on the line that are between them. The notation $\overline{AB}$ refers to the line segment with end points A and B, while AB refers to the length of $\overline{AB}$.

Line symmetry — When a line can be drawn that divides the figure into two parts that coincide when folded along the line.

Locus — (plural: *loci*) The set of all points, and only those points, that satisfy a given condition.

Major arc — An arc of a circle whose degree measure is greater than 180.

Means — The two middle terms of a proportion. In the proportion $\frac{a}{b} = \frac{c}{d}$, the terms b and c are the means. If $b = c$, then either b or c is called the *mean proportional* between a and d, which are called the *extremes*.

Median of a trapezoid — A line segment whose end points are the midpoints of the legs of the trapezoid.

Median of a triangle — A line segment whose end points are a vertex of the triangle and the midpoint of the side opposite that vertex.

Midpoint formula — A formula used to find the coordinates of the midpoint of a line segment in the coordinate plane. The midpoint of a line segment whose end points are $A(x_A, y_A)$ and $B(x_B, y_B)$ is $\left(\frac{x_A + x_B}{2}, \frac{y_A + y_B}{2}\right)$.

Midpoint of a line segment — The point on the line segment that divides the segment into two segments that have the same length.

Minor arc — An arc of a circle whose degree measure is less than 180.

Obtuse angle — An angle whose degree measure is greater than 90 and less than 180.

Obtuse triangle — A triangle that contains an obtuse angle.

Octagon — A polygon with eight sides.

Opposite isometry — An isometry that reverses orientation.

Opposite rays — Two rays that have the same end point and form a line.

Ordinate — The y-coordinate of a point in the coordinate plane.

Origin — The zero point on a number line.

Orthocenter of a triangle — The point at which the three altitudes of the triangle intersect.

Parallel lines — Lines in the same plane that do not intersect.

Parallelogram — A quadrilateral that has two pairs of parallel sides.

Pentagon — A polygon with five sides.

Perimeter of a polygon — The sum of the lengths of the sides of the polygon.

Perpendicular bisector of a segment — A line, ray, or line segment that is perpendicular to the segment at its midpoint.

Perpendicular lines — Two lines that intersect at 90° angles.

Plane — A term undefined in geometry; a plane can be described as a flat surface that extends indefinitely in all directions.

Point — A term undefined in geometry; a point can be described as a dot with no size that indicates location.

Point symmetry — A figure with 180° rotational symmetry.

Polygon — A closed figure in a plane whose sides are line segments that intersect at their end points.

Postulate — A statement whose truth is accepted without proof.

Proportion — An equation that states that two ratios are equal. In a proportion the product of the means equals the product of the extremes. Hence, if $\frac{a}{b} = \frac{c}{d}$, then $b \times c = a \times d$.

Pythagorean theorem — In a right triangle, the sum of the squares of the lengths of the legs is equal to the square of the length of the hypotenuse.

Quadrant of a coordinate plane — One of the four equal rectangular regions into which the coordinate plane is divided.

Quadrilateral — A polygon with four sides.

Radius (plural: *radii*) *of a circle* — A line segment whose end points are the center of the circle and any point on the circle.

Radius of a regular polygon — The radius of its circumscribed circle.

Ratio — A comparison of two numbers by division. The ratio of a to b can be represented by the fraction $\frac{a}{b}$, provided that $b \neq 0$.

Ratio of similitude of two similar polygons The constant ratio of the lengths of any two corresponding sides.

Ray — The part of a line that consists of a fixed point, called an *end point*, and the set of all points on one side of the end point.

Reciprocal of a nonzero number — The number that, when multiplied by the original number, gives 1. For example, the reciprocal of 5 is $\frac{1}{5}$ since $5 \times \frac{1}{5} = 1$.

Rectangle — A parallelogram with four right angles.

Reflection Over a Line — An isometry that "flips" a figure over a line while reversing orientation.

Rhombus — A parallelogram with four sides that have the same length.

Right angle — An angle whose degree measure is 90.

Right triangle — A triangle that contains a right angle.

Rotation — An isometry that "turns" a figure a specified number of degrees in a given direction (clockwise or counterclockwise) about some fixed point called the center of rotation.

Rotational Symmetry — A figure has rotational symmetry if it coincides with its image for some rotation of 180° or less.

SAS Theorem of Congruence — Two triangles are congruent if two pairs of corresponding sides are congruent and the angles formed by these sides are congruent.

Scalene triangle — A triangle in which no two sides have the same length.

Secant of a circle — A line that intersect the circle in two different points.

Semicircle — An arc whose end points are a diameter of the circle.

Similar polygons — Figures that have the same shape but may have different sizes. Two polygons with the same number of sides are similar if corresponding angles are congruent and the lengths of corresponding sides are in proportion.

Sine of an acute angle of a right triangle — The ratio of the length of the leg that is opposite the acute angle to the length of the hypotenuse.

Slope — A numerical measure of the steepness of a nonvertical line. The slope of a line is the difference of the y-coordinates of any two different points on the line divided by the difference of the corresponding x-coordinates of the two points. The slope of a horizontal line is 0, and the slope of a vertical line is not defined.

Slope formula — A formula used to calculate the slope of a nonvertical line when the coordinates of two points on the line are given. The slope m of a nonvertical line that contains points $A(x_A, y_A)$ and $B(x_B, y_B)$ is given by the formula $m = \frac{y_B - y_A}{x_B - x_A}$.

Slope-intercept form of an equation of a nonvertical line — An equation that has the form $y = mx + b$, where m is the slope of the line and b is the y-coordinate of the point at which the line crosses the y-axis.

Square — A rectangle all of whose sides have the same length.

SSS Theorem of Congruence — Two triangles are congruent if three pairs of corresponding sides are congruent.

Supplementary angles — Two angles whose measures add up to 180.

Tangent circles — Circles in the same plane that are tangent to the same line at the same point.

Tangent ratio of an acute angle of a right triangle — The ratio of the length of the leg that is opposite a given acute angle to the length of the leg that is adjacent to the same angle.

Tangent to a circle — A line that intersects the circle in exactly one point, called the *point of tangency*.

Theorem — A generalization that can be proved.

Transformation — A mapping of the elements of two sets where the elements are points such that each point of the object is mapped onto exactly one point called its *image* and each image point corresponds to exactly one point of the original object called the *preimage*.

Translation — An isometry that "slides" all points of a figure the same distance in the same direction.

Transversal — A line that intersects two lines at different points.

Trapezoid — A quadrilateral with exactly one pair of parallel sides.

Triangle — A polygon with three sides.

Undefined term — A term that can be described but is so basic that it cannot be defined. The terms *point, line*, and *plane* are undefined in geometry.

Vertex (plural: *vertices*) *of a polygon* — The point at which two sides of the polygon intersect.

Vertex angle of an isosceles triangle — The angle formed by the congruent sides of the isosceles triangle.

Vertical angles — Pairs of nonadjacent (opposite) angles formed by two intersecting lines.

Volume of a solid figure — The capacity of the solid figure as measured by the number of cubic units it contains.

x-axis — The horizontal number line in the coordinate plane.

x-coordinate — The first number in the ordered pair that represents the coordinates of a point in the coordinate plane. The x-coordinate gives the directed horizontal distance of the point from the origin.

y-axis — The vertical number line in the coordinate plane.

y-coordinate — The second number in the ordered pair that represents the coordinates of a point in the coordinate plane. The y-coordinate gives the directed vertical distance of the point from the origin.

y-intercept — The y-coordinate of the point at which a nonvertical line crosses the y-axis.

Answers to Chapter Exercises

Chapter 1

1. **(a)** $\overrightarrow{BA}, \overrightarrow{BE}, \overrightarrow{BC}, \overrightarrow{BD}$.
   (b) $\overleftrightarrow{BD}, \overleftrightarrow{BE}, \overleftrightarrow{DE}$
   (c) $\overleftrightarrow{AB}, \overleftrightarrow{BC}, \overleftrightarrow{AC}$.
   (d) Angles ABE, EBC, CBD, and DBA.
   (e) $\overrightarrow{BE}$ and $\overrightarrow{BD}$; $\overrightarrow{BA}$ and $\overrightarrow{BC}$.
2. **(a)** D
   (b) A
   (c) B
3. **(a)** $\measuredangle AEF$
   (b) $\measuredangle BCA$
   (c) $\measuredangle EFA$
4. B, N, R, and W.
5. $\overline{NB}$ and $\overline{NW}$.
6. Triangles ABN, NKR, TKR, TRW, NTR, and NTW.
7. Angles NRK, KRT, TRW, KRW, NRW, and NRT.
8. Angle ANT.
9. $\overrightarrow{NB}$ and $\overrightarrow{NW}$; $\overrightarrow{RB}$ and $\overrightarrow{RW}$.
10. $\overline{TR}$ is a side of $\triangle TKR$ and $\triangle TWR$. Also, $\overline{KR}$ is a side of $\triangle NKR$ and $\triangle TKR$.
11. $\overline{RN}$ and $\overline{RW}$.
12. **(a)** An angle whose measure is less than 90° is an acute angle.
    (b) A triangle having three sides equal in length is an equilateral triangle.
    (c) A ray (or segment) that divides an angle into two congruent angles is the bisector of the angle.
13. **(a)** Inductive.
    (b) Deductive.
    (c) Inductive.
    (d) Deductive.
14. The length of the median drawn to the hypotenuse of a right triangle is equal to one-half the length of the hypotenuse
15. The medians intersect at the same point.
16. **(a)** Henry has green eyes.
    (b) The measure of the third angle of the triangle is 80.
17. **(a)** The formula generates prime numbers for integer values of N from $n = 0$ to $n = 15$.
    (b) For $n = 16$, the formula produces the value 289, which is not a prime number since it is divisible by 17.

Chapter 2

1. **(a)** Acute.
   (b) Straight.
   (c) Right.
   (d) Acute.
   (e) Right.
   (f) Obtuse.
   (g) Acute.
   (h) Obtuse.
2. 8
3. Z.
4. $m\measuredangle RPL = 10$.
5. 12
6. 42
7. Right.
8. 20
9. $\overline{EY} \cong \overline{AS}$; $\measuredangle PYS \cong \measuredangle IAE$, $\measuredangle YEP \cong \measuredangle ASI$.
10. $\overline{AF} \cong \overline{CF}$.

11. $\measuredangle STP \cong \measuredangle OTP$.
12. $\overline{BE} \cong \overline{DE}$; $\measuredangle ADB \cong \measuredangle CDB$.
13. Transitive.
14. Reflexive.
15. Transitive.
16. Substitution.
17. Addition.
18. Subtraction.
19. Addition.
20. Addition.
21. Halves of equals are equal.
22. Subtraction.
23. Subtraction.
24. Addition.
25. 3. Reflexive.
    4. Addition.
26. 1. Given.
    2. Congruent angles are equal in measure.
    3. Subtraction.
    4. Substitution.
    5. Angles equal in measure are congruent.
27. 1. Given
    2. Reflexive.
    3. Subtraction.
    5. Given.
    6. Congruent segments are equal in measure.
    7. Reflexive.
    8. Subtraction.
    10. Segments equal in measure are congruent.
28. 2. Congruent angles are equal in measure.
    3. Given.
    4. Definition of angle bisector.
    5. Halves of equals are equal.
    6. Angles equal in measure are congruent.

Chapter 3

1. (1)
2. (3)
3. (2)
4. (2)
5. 75
6. 136
7. 45
8. The measure of the angle is 52 and its complement has measure 38.
9. 56
10. **(a)** 9
    (b) 40
    (c) 18
11. 143
12. True.
13. False. Consider triangle ABC in which $AC = BC$. Point C is equidistant from points A and B, but point C is not the midpoint of side $\overline{AB}$.
14. (1) $\measuredangle 1 \cong \measuredangle 4$ (Given);
    (2) $\measuredangle 2 \cong \measuredangle 1$ and $\measuredangle 3 \cong \measuredangle 4$ (Vertical angles are congruent);
    (3) $\measuredangle 2 \cong \measuredangle 3$ (Transitive property).
15. (1) $\overline{BD}$ bisects $\measuredangle ABC$ (Given);
    (2) $\measuredangle 3 \cong \measuredangle 4$ (Definition of angle bisector);
    (3) $\measuredangle 1 \cong \measuredangle 2$ (Supplements of congruent angles are congruent).
16. (1) $\measuredangle 3$ is complementary to $\measuredangle 1$, $\measuredangle 4$ is complementary to $\measuredangle 2$ (Given);
    (2) $\measuredangle 1 \cong \measuredangle 2$ (Vertical angles are congruent);
    (3) $\measuredangle 3 \cong \measuredangle 4$ (Complements of congruent angles are congruent)
17. (1) $\overline{AB} \perp \overline{BD}$, $\overline{CD} \perp \overline{BD}$, $\measuredangle 2 \cong \measuredangle 4$ (Given);
    (2) $\measuredangle 1$ is complementary to $\measuredangle 2$ and $\measuredangle 3$ is complementary to $\measuredangle 4$ (Adjacent angles whose exterior sides are perpendicular are complementary);
    (3) $\measuredangle 1 \cong \measuredangle 3$ (Complements of congruent angles are congruent).

18. (1) $\overline{KL} \perp \overline{JM}$, $\overline{KL}$ bisects $\measuredangle PLQ$ (Given);
    (2) $\measuredangle 1$ is complementary to $\measuredangle 2$ and $\measuredangle 4$ is complementary to $\measuredangle 3$ (Adjacent angles whose exterior sides are perpendicular are complementary);
    (3) $\measuredangle 2 \cong \measuredangle 3$ (Definition of angle bisector);
    (4) $\measuredangle 1 \cong \measuredangle 4$ (Complements of congruent angles are congruent).
19. (1) $\overline{NW} \perp \overline{WT}$, $\overline{WB} \perp \overline{NT}$, $\measuredangle 4 \cong \measuredangle 6$ (Given);
    (2) $\measuredangle 4 \cong \measuredangle 3$ (Vertical angles are congruent);
    (3) $\measuredangle 3 \cong \measuredangle 6$ (Transitive);
    (4) $\measuredangle 5$ is complementary to $\measuredangle 6$ and $\measuredangle 2$ is complementary to $\measuredangle 3$ (Adjacent angles whose exterior sides are perpendicular are complementary);
    (5) $\measuredangle 5 \cong \measuredangle 2$ (Complements of congruent angles are congruent).
20. (1) $\overline{MT}$ bisects $\measuredangle ETI$, $\overline{KI} \perp \overline{TI}$, $\overline{KE} \perp \overline{TE}$, $\measuredangle 3 \cong \measuredangle 1$, $\measuredangle 5 \cong \measuredangle 2$ (Given);
    (2) $\measuredangle 1 \cong \measuredangle 2$ (Definition of angle bisector);
    (3) $\measuredangle 3 \cong \measuredangle 5$ (Transitive);
    (4) $\measuredangle 4$ is complementary to $\measuredangle 3$ and $\measuredangle 6$ is complementary to $\measuredangle 5$ (Adjacent angles whose exterior sides are perpendicular are complementary);
    (5) $\measuredangle 4 \cong \measuredangle 6$ (Complements of congruent angles are congruent);

Chapter 4

1. (4)
2. (3)
3. (2)
4. **(a)** $x = 112, y = 68$.
   (b) $x = 104, y = 104$.
   (c) $x = 48, y = 80$.
   (d) $x = 45, y = 117$.
5. **(a)** 22
   (b) 110
   (c) 60
6. **(a)** 138 and 42
   (b) 144 and 36
7. **(a)** 55
   (b) 10
8. 30
9. **(a)** $m\measuredangle 2 = 80$ $m\measuredangle 8 = 100$
   $m\measuredangle 3 = 100$ $m\measuredangle 9 = 80$
   $m\measuredangle 4 = 80$ $m\measuredangle 10 = 130$
   $m\measuredangle 5 = 80$ $m\measuredangle 11 = 50$
   $m\measuredangle 6 = 50$ $m\measuredangle 12 = 130$
   $m\measuredangle 7 = 50$ $m\measuredangle 13 = 50$
   (b) $m\measuredangle 1 = 127$ $m\measuredangle 8 = 127$
   $m\measuredangle 2 = 53$ $m\measuredangle 9 = 53$
   $m\measuredangle 3 = 127$ $m\measuredangle 10 = 116.5$
   $m\measuredangle 4 = 53$ $m\measuredangle 11 = 63.5$
   $m\measuredangle 5 = 53$ $m\measuredangle 12 = 116.5$
   $m\measuredangle 6 = 63.5$ $m\measuredangle 13 = 63.5$
   $m\measuredangle 7 = 63.5$
10. **(a)** If I live in the United States, then I live in New York. (False.)
    (b) If two angles are equal in measure, then the angles are congruent. (True.)
    (c) If two angles are congruent, then they are vertical angles. (False.)
    (d) If the sum of the measures of two angles is 90, then the angles are complementary. (True.)
    (e) If two angles have the same vertex, then they are adjacent. (False.)
    (f) If two lines are parallel, then they are perpendicular to the same line. (False.)

11. (1) $\overline{BA} \parallel \overline{CF}, \overline{BC} \parallel \overline{ED}$ (Given);
    (2) $\measuredangle 1 \cong \measuredangle C$ (Postulate 4.1);
    (3) $\measuredangle C \cong \measuredangle 2$ (Theorem 4.2);
    (4) $\measuredangle 1 \cong \measuredangle 2$ (Transitive).
12. (1) $\overline{LT} \parallel \overline{WK} \parallel \overline{AP}, \overline{PL} \parallel \overline{AG}$ (Given);
    (2) $\measuredangle 1 \cong \measuredangle KWP$ (Theorem 4.2);
    (3) $\measuredangle KWP \cong \measuredangle WPA$ (Postulate 4.1);
    (4) $\measuredangle 1 \cong \measuredangle WPA$ (Transitive);
    (5) $\measuredangle WPA \cong \measuredangle 2$ (Postulate 4.1);
    (6) $\measuredangle 1 \cong \measuredangle 2$ (Transitive).
13. (2)
14. (1) $\overline{QD} \parallel \overline{UA}, \overline{QU} \parallel \overline{DA}$ (Given);
    (2) $m\measuredangle 1 = m\measuredangle 4, m\measuredangle 2 = m\measuredangle 3$ (Postulate 4 1);
    (3) $m\measuredangle 1 + m\measuredangle 2 = m\measuredangle 3 + m\measuredangle 4$ (Addition);
    (4) $m\measuredangle QUA = m\measuredangle ADQ$ (Substitution);
    (5) $\measuredangle QUA \cong \measuredangle ADQ$ (Angles equal in measure are congruent).
15. (1) $\overline{AT} \parallel \overline{MH}, \measuredangle M \cong \measuredangle H$ (Given);
    (2) $\measuredangle A$ is supplementary to $\measuredangle M$ and $\measuredangle T$ is supplementary to $\measuredangle H$ (Theorem 4.3);
    (3) $\measuredangle A \cong \measuredangle T$ (Supplements of congruent angles are congruent).
16. (1) $\overline{IB} \parallel \overline{ET}$, IS bisects $\measuredangle EIB$, $\overline{EC}$ bisects $\measuredangle TEI$ (Given);
    (2) $m\measuredangle BIS = \frac{1}{2} m\measuredangle EIB$ and $m\measuredangle TEC = \frac{1}{2} m\measuredangle TEI$ (Definition of angle bisector);
    (3) $m\measuredangle EIB = m\measuredangle TEI$ (Postulate 4.1);
    (4) $m\measuredangle BIS = m\measuredangle TEC$ (Halves of equals are equal);
    (5) $\measuredangle BIS \cong \measuredangle TEC$ (Angles equal in measure are congruent).
17. (1) $\measuredangle B \cong \measuredangle D, \overline{BA} \parallel \overline{DC}$ (Given);
    (2) $\measuredangle B \cong \measuredangle C$ (Postulate 4.1);
    (3) $\measuredangle C \cong \measuredangle D$ (Transitive);
    (4) $\overline{BC} \parallel \overline{DE}$ (Postulate 4.2).
18. (1) $k \parallel \ell, \measuredangle 5 \cong \measuredangle 8$ (Given);
    (2) $\measuredangle 5 \cong \measuredangle 1$ (Theorem 4.2);
    (3) $\measuredangle 1 \cong \measuredangle 8$ (Transitive);
    (4) $j \parallel \ell$ (Postulate 4.2).
19. (1) $\measuredangle K = \measuredangle P, m\measuredangle J + m\measuredangle P = 180$ (Given);
    (2) $m\measuredangle J + m\measuredangle K = 180$ (Substitution);
    (3) $\overline{KL} \parallel \overline{JP}$ (Theorem 4.6).
20. (1) $\overline{AB} \perp \overline{BC}$, $\measuredangle ACB$ is complementary to $\measuredangle ABE$ (Given);
    (2) $m\measuredangle ABE + m\measuredangle ABC + m\measuredangle CBD = 180$ (A straight angle has measure 180);
    (3) $m\measuredangle ABE + 90 + m\measuredangle CBD = 180$ (Substitution);
    (4) $m\measuredangle ABE + m\measuredangle CBD = 90$ (Subtraction);
    (5) $\measuredangle CBD$ is complementary to $\measuredangle ABE$ (Definition of complementary angles);
    (6) $\measuredangle ACB \cong \measuredangle CBD$ (Complements of the same angle are congruent);
    (7) $\overleftrightarrow{AC} \parallel \overleftrightarrow{EBD}$ (Postulate 4.2).
21. (1) $\overline{AG} \parallel \overline{BC}, \overline{KH} \parallel \overline{BC}, \measuredangle 1 \cong \measuredangle 2$ (Given);
    (2) $\overline{AG} \parallel \overline{KH}$ (Segments parallel to the same segment are parallel to each other—see Exercise 20);
    (3) Angles 1 and 2 are supplementary (Theorem 4.3);
    (4) Angles 1 and 2 are right angles (If two angles are supplementary and congruent, then each is a right angle);
    (5) $\overline{HK} \perp \overline{AB}$ (Segments that intersect to form a right angle are perpendicular).
22. GIVEN: $\ell \parallel p$ and $m \parallel p$.
    PROVE: $\ell \parallel m$
    PLAN: Show $\measuredangle 1 \cong \measuredangle 2$.
    $\measuredangle 1 \cong \measuredangle 3$ and $\measuredangle 2 \cong \measuredangle 3$. Since $\measuredangle 1 \cong \measuredangle 2$, $\ell \parallel m$ by Theorem 4.4.
23. GIVEN: $\overrightarrow{XL}$ bisects $\measuredangle AXY$, $\measuredangle YM$ bisects $\measuredangle DYX$, $\overrightarrow{XL} \parallel \overrightarrow{YM}$.
    PROVE: $\overleftrightarrow{AXB} \parallel \overleftrightarrow{CYD}$
    PLAN: Show $m\measuredangle AXY = m\measuredangle DYX$
    $m\measuredangle 1 = m\measuredangle 2$; $m\measuredangle AXY = 2m\measuredangle 1$ and $m\measuredangle DYX = 2m\measuredangle 2$. Since doubles of equals are equal, $m\measuredangle AXY = m\measuredangle DYX$ and $\overleftrightarrow{AXB} \parallel \overleftrightarrow{CYD}$ by Postulate 4.2.

24. GIVEN: $\overleftrightarrow{RAS} \parallel \overleftrightarrow{PCQ}$,
$\overrightarrow{AB}$ bisects $\measuredangle HAS$,
$\overrightarrow{CD}$ bisects $\measuredangle QCA$.
PROVE: $\overrightarrow{AB} \parallel \overrightarrow{CD}$
PLAN: Show $m\measuredangle 1 = m\measuredangle 2$.
$m\measuredangle HAS = m\measuredangle QCA$; $m\measuredangle 1 = \frac{1}{2} m\measuredangle HAS$ and $m\measuredangle 2 = \frac{1}{2} m\measuredangle QCA$. Hence, $m\measuredangle 1 = m\measuredangle 2$ and $\overrightarrow{AB} \parallel \overrightarrow{CD}$ by Theorem 4.4.

Chapter 5

1. (2)
2. (4)
3. (3)
4. (1)
5. (3)
6. **(a)** 52
   (b) 67
   (c) 47
   (d) 22
   (e) 127
7. 113
8. **(a)** 360
   (b) 720
   (c) 1,260
   (d) 1,980
9. **(a)** 12
   (b) 17
   (c) 5
   (d) 14
10. **(a)** 125
    (b) 104
    (c) 170
11. **(a)** 108
    (b) 165
    (c) 135
    (d) 156
12. **(a)** 20
    (b) 10
    (c) 9
    (d) 30
13. 10
14. **(a)** 8
    (b) 4
    (c) 15

15–17 PLAN: Show two angles of one triangle are congruent to two angles of the other triangle. Then apply Corollary 5.2.4

15. $\measuredangle B \cong \measuredangle D$ and $\measuredangle ACB \cong \measuredangle ECD$ so that $\measuredangle A \cong \measuredangle E$.
16. $\measuredangle C \cong \measuredangle DEB$ and $\measuredangle B \cong \measuredangle B$ so that $\measuredangle 1 \cong \measuredangle 2$.
17. $\measuredangle P \cong \measuredangle K$ and $\measuredangle KML \cong \measuredangle PRJ$ so that $\measuredangle J \cong \measuredangle L$. Hence, $\overline{KL} \parallel \overline{PJ}$ by Postulate 4.2.

18. Angles 1 and 2 are each complementary to $\measuredangle DCE$ and are therefore congruent to each other.

19. Angles CAX and ACY are supplementary Hence, $\frac{1}{2}\, m\measuredangle CAX + \frac{1}{2}\, m\measuredangle ACY = \frac{1}{2}(180) = 90$. By substitution, the sum of the measures of angles BAC and BCA is equal to 90. This implies that $m\measuredangle ABC = 90$ and $\measuredangle ABC$ is a right angle.

Chapter 6

1. (3)
2. (1)
3. Use SAS. $\overline{BM} \cong \overline{BM}$; right angle $BMA \cong$ right angle BMC; $\overline{AM} \cong \overline{CM}$.
4. Use ASA. $\measuredangle SRT \cong \measuredangle WRT$; $\overline{RT} \cong \overline{RT}$; $\measuredangle STR \cong \measuredangle WTR$.
5. Use ASA. $\measuredangle F \cong \measuredangle A$; by addition, $\overline{AC} \cong \overline{FD}$; $\measuredangle EDF \cong \measuredangle BCA$.
6. Use ASA. $\measuredangle R \cong \measuredangle T$; $\overline{SR} \cong \overline{ST}$; $\measuredangle S \cong \measuredangle S$.
7. Use ASA. Right angle $REW \cong$ right angle THW, $\overline{EW} \cong \overline{HW}$; $\measuredangle EWR \cong \measuredangle HWT$.
8. Use AAS. $\measuredangle UQX \cong \measuredangle DAX$; $\measuredangle UXQ \cong \measuredangle DXA$; $\overline{QU} \cong \overline{DA}$.
9. Use Hy-Leg. $\overline{JL} \cong \overline{EV}$ (Hyp) and by addition, $\overline{KL} \cong \overline{TV}$ (Leg).
10. Use ASA. Right angle $JKL \cong$ right angle ETV; $\overline{KL} \cong \overline{TV}$; by taking supplements of congruent angles, $\measuredangle EVT \cong \measuredangle JLK$.
11. Use AAS. $\measuredangle ARF \cong \measuredangle ARI$; by subtraction, $\measuredangle RFA \cong \measuredangle RIA$; $\overline{AR} \cong \overline{AR}$.
12. Use SAS. $\overline{TS} \cong \overline{RS}$; by subtraction, $\measuredangle TSW \cong \measuredangle RSP$, $\overline{SW} \cong \overline{SP}$.
13. Use AAS. Since $\overline{RP} \parallel \overline{SW}$, $\measuredangle TSW \cong \measuredangle SRP$ (Corresponding angles). Also, since $\overline{SP} \parallel \overline{TW}$, $\measuredangle STW \cong \measuredangle RSP$ $\overline{TW} \cong \overline{SP}$.
14. Use SSS. $\overline{AB} \cong \overline{DE}$; $\overline{BM} \cong \overline{EM}$; $\overline{AM} \cong \overline{DM}$ by transitivity.
15. Use AAS. $\measuredangle 1 \cong \measuredangle 2$. $\measuredangle A \cong \measuredangle D \cong \measuredangle 2$, so that $\measuredangle A \cong \measuredangle D$ (Angle); since complements of congruent angles are congruent, $\measuredangle AMB \cong \measuredangle DME$ (Angle); $\overline{BM} \cong \overline{EM}$ (Side).
16. Use ASA. Right angle AFC = right angle BDC, $\measuredangle C \cong \measuredangle C$, $\overline{FC} \cong \overline{DC}$.
17. Use SAS. $\overline{AD} \cong \overline{BF}$; $\measuredangle BAD \cong \measuredangle ABF$; $\overline{AB} \cong \overline{AB}$.
18. Use Hy-Leg. $\overline{AB} \cong \overline{DC}$ (Hyp); by subtraction, $\overline{AF} \cong \overline{CE}$ (Leg).
19. Use AAS. Right angle $KLG \cong$ right angle JRG; $\measuredangle G \cong \measuredangle G$; $\overline{KL} \cong \overline{JR}$.
20. Use AAS. Right angle KRO = right angle JLO; $\measuredangle ROK \cong \measuredangle LOJ$; by halves of equals are equal, $\overline{KR} \cong \overline{JL}$.
21. Use ASA. $\measuredangle AEB \cong \measuredangle DEC$; $\overline{EB} \cong \overline{EC}$; by halves of equals are equal, $\measuredangle 1 \cong \measuredangle 2$.
22. Use SSS (all other methods, except Hy-Leg, can also be used).

GIVEN: $\triangle ABC \cong \triangle XYZ$ and $\triangle RST \cong \triangle XYZ$.

PROVE: $\triangle ABC \cong \triangle RST$.

PLAN: Using the reverse of the definition of congruent triangles, $\overline{AB} \cong \overline{XY}$ and $\overline{RS} \cong \overline{XY}$ so that $\overline{AB} \cong \overline{RS}$. Similarly, show that each of the remaining pairs of corresponding sides are congruent.

Chapter 7

1. (1)
2. (3)
3. (2)
4. **(a)** 76
   (b) 20
   (c) 40
   (d) 32
   (e) 16
   (f) 36
5. Show $\triangle HGF \cong \triangle HJF$ by SAS. By CPCTC, $\measuredangle GFH \cong \measuredangle JFH$. $\measuredangle 1 \cong \measuredangle 2$ since supplements of congruent angles are congruent.
6. Show $\triangle ABC \cong \triangle DCB$ by Hy-Leg. $\measuredangle 1 \cong \measuredangle 2$ by CPCTC.
7. Show $\triangle JPY \cong \triangle KPX$ by ASA. $\overline{JY} \cong \overline{KX}$ by CPCTC.
8. $\overline{PJ} \cong \overline{PK}$ by addition. Show $\triangle KXP \cong \triangle JYP$ by SAS where $\overline{PK} \cong \overline{PJ}$, $\measuredangle P \cong \measuredangle P$, and $\measuredangle \overline{PX} \cong \overline{PY}$. By CPCTC, $\overline{KX} \cong \overline{JY}$.
9. Show $\triangle UTQ \cong \triangle DWA$ by SAS where $\overline{UT} \cong \overline{DW}$, $\measuredangle QTU \cong \measuredangle AWD$ (Alternate exterior angles); $\overline{QT} \cong \overline{AW}$ (Subtraction) By CPCTC, $\measuredangle UQT \cong \measuredangle DAW$ which implies $\overline{UQ} \parallel \overline{AD}$.
10. Show $\triangle SRT \cong \triangle HWT$ by AAS. By CPCTC, $\overline{RT} \cong \overline{WT}$ and T is the midpoint of $\overline{RW}$.
11. Show $\triangle ADB \cong \triangle CDB$ by SSS. By CPCTC, $\measuredangle ADB \cong \measuredangle CDB$ and $\overline{DB}$ bisects $\measuredangle ADC$.
12. Show $\triangle BAM \cong \triangle CDM$ by AAS. By CPCTC, $\overline{AM} \cong \overline{DM}$ and $\overline{BC}$ bisects $\overline{AD}$.
13. $\measuredangle RHK \cong \measuredangle NHK$ and $\measuredangle HRK \cong \measuredangle HNK$ (Supplements of congruent angles are congruent.) If two angles of one triangle are congruent to corresponding angles of another triangle, then the third pair of angles are congruent. Hence, $\measuredangle HKR \cong \measuredangle HKN$ which implies $\overline{HK} \perp \overline{RN}$.
14. Show $\triangle RTS \cong \triangle ACB$ by SSS. By CPCTC, $\measuredangle 1 \cong \measuredangle 2$. If two angles are supplementary (see Given) and congruent, then each is a right angle. Hence, $\overline{ST} \perp \overline{TR}$ and $\overline{BC} \perp \overline{AC}$.
15. (1) Show $\triangle RLS \cong \triangle RLT$ by Hy-Leg where $\overline{RL} \cong \overline{RL}$ (Hyp) and $\overline{RS} \cong \overline{RT}$ (Leg)
    (2) By CPCTC, $\overline{SL} \cong \overline{TL}$ and $RLS \cong \measuredangle RLT$. Taking supplements of congruent angles, $\measuredangle SLW \cong \measuredangle TLW$. $\triangle SLW \cong \triangle TLW$ by SAS: $\overline{SL} \cong \overline{TL}$ (Side), $\measuredangle SLW \cong \measuredangle TLW$ (Angle), and $\overline{LW} \cong \overline{LW}$ (Side). By CPCTC, $\measuredangle SWL \cong \measuredangle TWL$ so that $\overline{WL}$ bisects $\measuredangle SWT$.
16. Show $\triangle PMK \cong \triangle PML$ by Hy-Leg. By CPCTC, $\overline{KM} \cong \overline{LM}$ and M is the midpoint of $\overline{KL}$.
17. Show $\triangle PMK \cong \triangle PML$ by SSS. By CPCTC, $\measuredangle KMP \cong \measuredangle LMP$ which implies $\overline{PM} \perp \overline{KL}$.
18. Show $\triangle PSL \cong \triangle LTP$ by AAS. By CPCTC, $\overline{PS} \cong \overline{LT}$.
19. Show $\triangle XLR \cong \triangle XPS$ by Hy-Leg where $\overline{XR} \cong \overline{XS}$ (Hyp) and $\overline{XL} \cong \overline{XP}$ (Leg). By CPCTC, $\measuredangle R \cong \measuredangle S$ which implies $\triangle RTS$ is isosceles.
20. $\measuredangle 1 \cong \measuredangle 3$. Since parallel lines form congruent corresponding angles, $\measuredangle 1 \cong \measuredangle 2$ and $\measuredangle 3 \cong \measuredangle 4$. By transitivity, $\measuredangle 2 \cong \measuredangle 4$ which implies $\triangle PQR$ is isosceles.
21. Show $\triangle TWL \cong \triangle PXF$ by Hy-Leg where by addition $\overline{TL} \cong \overline{PF}$ (Hyp) and $\overline{WL} \cong \overline{XF}$ (Leg). By CPCTC, $\measuredangle 1 \cong \measuredangle 2$ which implies $\triangle FML$ is isosceles.
22. Show $\triangle KLE \cong \triangle ABW$ by AAS: $\measuredangle E \cong \measuredangle W$ (Theorem 7.1); $\measuredangle BAW \cong \measuredangle LKE$ (Supplements of congruent angles are congruent), $\overline{EL} \cong \overline{WB}$. By CPCTC, $\overline{KL} \cong \overline{AB}$.
23. By Theorem 7.1, $\measuredangle 1 \cong \measuredangle 2$ and $\measuredangle 3 \cong \measuredangle 4$ since supplements of congruent

angles are congruent. $\triangle VOK \cong \triangle VLZ$ by SAS. By CPCTC, $VK \cong VZ$, which implies that $\triangle KVZ$ is isosceles.

24. Show $\triangle BJH \cong \triangle KJL$ by ASA where right $\measuredangle BHJ$ is congruent to right $\measuredangle KLJ$, $\overline{HJ} \cong \overline{LJ}$ (Theorem 7.2), and $\measuredangle HJB \cong \measuredangle LJK$. By CPCTC, $\overline{JB} \cong \overline{JK}$.

25–29. These exercises involve double congruence proofs requiring that one pair of triangles be proved congruent in order to obtain congruent pairs of parts. These pairs may then be used to prove a second pair of triangles congruent.

25. First show $\triangle ABC \cong \triangle ADC$ by SAS in order to obtain $\overline{BC} \cong \overline{DC}$ and $\measuredangle ACB \cong \measuredangle ACD$. Taking supplements of congruent angles yields $\measuredangle BCE \cong \measuredangle DCE$. Since $\overline{CE} \cong \overline{CE}$, $\triangle BCE \cong \triangle DCE$ by SAS.

26. First show $\triangle BEC \cong \triangle DEC$ by SSS in order to obtain $\measuredangle BCE \cong \measuredangle DCE$ so that $\measuredangle BCA \cong \measuredangle DCA$. Show $\triangle ABC \cong \triangle ADC$ by SAS. By CPCTC, $\measuredangle BAC \cong \measuredangle DAC$ which implies that $\overline{CA}$ bisects $\measuredangle DAB$.

27. First show $\triangle ABL \cong \triangle CDM$ by SAS in order to obtain $\overline{BL} \cong \overline{DM}$ and $\measuredangle ALB \cong \measuredangle CMD$. Taking supplements of congruent angles gives $\measuredangle BLC \cong \measuredangle DMA$. Using the addition property, $\overline{AM} \cong \overline{CL}$. Hence, $\triangle CLB \cong \triangle AMD$ by SAS By CPCTC, $\measuredangle CBL \cong \measuredangle ADM$.

28. Show $\triangle AFD \cong \triangle CFE$ by ASA where right $\measuredangle ADF$ = right $\measuredangle CEF$, $\overline{AF} \cong \overline{CF}$ (Theorem 7.2), $\measuredangle AFD \cong \measuredangle CFE$ By CPCTC, $\overline{FD} \cong \overline{FE}$. $\triangle BDF \cong \triangle BEF$ by Hy-Leg. By CPCTC, $\measuredangle DBF \cong \measuredangle EBF$ which implies that $\overline{BF}$ bisects $\measuredangle DBE$.

29. Show $\triangle BDF \cong \triangle BEF$ by SSS so that $\measuredangle BDF \cong \measuredangle BEF$. Show $\triangle AFD \cong \triangle CFE$ by ASA where $\measuredangle DFA \cong \triangle EFC$; $\overline{DF} \cong \overline{EF}$; $\measuredangle ADF \cong \measuredangle CEF$. By CPCTC, $AF \cong CF$ which implies that $\triangle AFC$ is isosceles.

30. GIVEN: $\triangle ABC \cong \triangle RST$ and $\overline{BH}$ and $\overline{SK}$ are altitudes.
PROVE: $\overline{BH} \cong \overline{SK}$.
PLAN: Show $\triangle BAH \cong \triangle SRK$ by AAS where right $\measuredangle BHA$ = right $\measuredangle SKR$, and using the reverse of the definition of congruent triangles, $\measuredangle A \cong \measuredangle S$ and $AB \cong RS$. By CPCTC, $BH \cong SK$.

31. GIVEN: Equilateral $\triangle ABC$
PROVE: $\measuredangle A \cong \measuredangle B \cong \measuredangle C$.
PLAN: Since $\overline{AB} \cong \overline{BC}$, $\measuredangle A \cong \measuredangle C$. Since $\overline{AC} \cong \overline{BC}$, $\measuredangle A \cong \measuredangle B$. Hence, $\measuredangle B \cong \measuredangle C$ so that $\measuredangle A \cong \measuredangle B \cong \measuredangle C$.

32. GIVEN: $\triangle PEG$, $\overline{PE} \cong \overline{GE}$, altitudes $\overline{GH}$ and $\overline{PK}$ are drawn.
PROVE: $\overline{GH} \cong \overline{PK}$.
PLAN: Show $\triangle GHP \cong \triangle PKG$ by AAS where right $\measuredangle PHG \cong$ right $\measuredangle GKP$ $\measuredangle HPG \cong \measuredangle KGP$ (Base Angles Theorem); $\overline{PG} \cong \overline{PG}$. By CPCTC, $\overline{GH} \cong \overline{PK}$.

33. GIVEN: $\triangle ART$, $\overline{AR} \cong \overline{TR}$, medians $\overline{AY}$ and $\overline{TX}$ are drawn.
PROVE: $\overline{AY} \cong \overline{TX}$.
PLAN: Show $\triangle AXT \cong \measuredangle TYA$ by SAS where $\overline{AT} \cong \overline{AT}$; $\measuredangle XAT \cong \measuredangle YTA$ (Theorem 7.1); $\overline{AX} \cong \overline{TY}$ (Halves of equals are equal) By CPCTC, $\overline{AY} \cong \overline{TX}$.

34. GIVEN: M is the midpoint of $\overline{AB}$, $\overline{PM} \perp \overline{AB}$
PROVE: $\overline{PA} = \overline{PB}$
PLAN: Show $\triangle PMA \cong \triangle PMB$ by SAS since $\overline{PM} \cong \overline{PM}$, rt. $\measuredangle PMA \cong$ rt. $\measuredangle PMB$, and $\overline{AM} \cong \overline{BM}$.

35. GIVEN: $\overline{AB}$ and points P and Q such that $PA = PB$ and $QA = QB$. $\overleftrightarrow{PQ}$ intersects $\overline{AB}$ at point M.
PROVE: $\overline{AM} \cong \overline{BM}$ and $\overleftrightarrow{PQ} \perp \overline{AB}$
PLAN: $\triangle PAQ \cong \triangle PBQ$ by SSS so that $\measuredangle APM \cong \measuredangle BPM$. Therefore, $\triangle APM \cong \triangle BPM$ by SAS. By CPCTC, $\overline{AM} \cong \overline{BM}$ and $\measuredangle PMA \cong \measuredangle PMB$. Since angles PMA and PMB are adjacent, $\overleftrightarrow{PQ} \perp \overline{AB}$.

Chapter 8

1. (2)
2. (3)
3. (3)
4. (2)
5. (2)
6. $\overline{AB}$.
7. $\overline{AB}$.
8. $\overline{BC}$.
9. $\overline{AB}$.
10. $\overline{KL}$.
11. $\overline{BC}$.
12. $\measuredangle S$.
13. **(a)** True.
    (b) False
    (c) True.
    (d) True.
    (e) True.
    (f) False.
14. $\measuredangle B$.
15. **(a)** $\measuredangle R$.
    (b) $\measuredangle S$.
16. **(a)** Yes.
    (b) No.
    (c) No.
    (d) Yes.
17. $m\measuredangle 1 = m\measuredangle 2$ and $m\measuredangle 3 > m\measuredangle 2$. By substitution, $m\measuredangle 3 > m\measuredangle 1$ which implies $AB > BD$.
18. GIVEN: $\overline{AB} \not\cong \overline{BC}$, $\overline{BD} \perp \overline{AC}$.

    PROVE: $\overline{BD}$ does not bisect $\overline{AC}$.

    PROOF: Assume $\overline{BD}$ bisects $\overline{AC}$, so $\overline{AD} \cong \overline{CD}$ (side). As angles BDA and BDC are right angles, $\angle BDA \cong \angle BDC$. Since $\overline{BD} \cong \overline{BD}$, $\triangle BDA \cong \triangle BDC$ by SAS $\cong$ SAS. Then $\overline{AB} \cong \overline{BC}$ by CPCTC, but this contradicts the Given. Hence, $\overline{BD}$ does not bisect $\overline{AC}$.
19. $m\measuredangle 1 > m\measuredangle A$. By substitution, $m\measuredangle 2 > m\measuredangle A$ which implies $AD > ED$.
20. $m\measuredangle 4 > m\measuredangle 3$ and $m\measuredangle 3 > m\measuredangle AEC$. Hence, $m\measuredangle 4 > m\measuredangle AEC$.
21. Since $AC > BC$, $m\measuredangle 3 > m\measuredangle EAC$. Since $m\measuredangle 2 < m\measuredangle EAC$, $m\measuredangle 3 > m\measuredangle 2$ which implies $AD > BD$.
22. Since $AD > BD$, $m\measuredangle 3 > m\measuredangle 2$ m. $\measuredangle 1 = m\measuredangle 2$ and by substitution $m\measuredangle 3 > m\measuredangle 1$. But $m\measuredangle 4 > m\measuredangle 3$. Hence, $m\measuredangle 4 > m\measuredangle 1$ which implies $AC > DC$.
23. Assume $\measuredangle 1 \cong \measuredangle 2$. By transitivity, $\measuredangle 2 \cong \measuredangle 3$. But this contradicts Theorem 5.4. Hence, $\measuredangle 1 \not\cong \measuredangle 2$.
24. Assume $\overline{AB} \cong \overline{BC}$. Then $\measuredangle 2 \cong \measuredangle C$ and, by transitivity, $\measuredangle 1 \cong \measuredangle C$. But this contradicts Theorem 5.4. Hence, $\overline{AB} \not\cong \overline{BC}$.
25. Assume $RW = WL$ Then $m\measuredangle WRL = m\measuredangle WLR$ $m\measuredangle WLR > m\measuredangle T$. By substitution, $m\measuredangle WRL > m\measuredangle T$. Since $m\measuredangle SRT > m\measuredangle WRL$ $m\measuredangle SRT > m\measuredangle T$ which implies that $TS > RS$. But this contradicts the Given ($RS = TS$). Hence, $RW \not\cong WL$.
26. Assume $\triangle ADC$ is isosceles. Then $AD = CD$ $\triangle ADB \cong \triangle CDB$ by SAS by CPCTC, $\overline{AB} \cong \overline{CB}$, but this contradicts the Given ($\triangle ABC$ is *not* isosceles) Hence, $\triangle ADC$ is *not* isosceles.
27. Assume $\overline{AB} \parallel \overline{DE}$. Then $\measuredangle B \cong \measuredangle D$. Since $\overline{DE} \cong \overline{CE}$, $\measuredangle D \cong \measuredangle DCE$. By transitivity, $\measuredangle B \cong \measuredangle DCE$ Since $\measuredangle ACB \cong \measuredangle DCE$, $\measuredangle B \cong \measuredangle ACB$ which implies $\overline{AC} \cong \overline{AB}$. But this contradicts the Given ($AC > AB$). Hence $\overline{AB}$ is *not* parallel to $\overline{DE}$.
28. Assume $\overline{BD} \perp \overline{AC}$. $\triangle ABD \cong \triangle CBD$ by ASA. By CPCTC, $\overline{AB} \cong \overline{CB}$ which contradicts the assumption that $\triangle ABC$ is

scalene Hence, $\overline{BD}$ is *not* perpendicular to $\overline{AC}$.

29. Since $\overline{BE}$ bisects $\measuredangle ABD$, $m\measuredangle ABE = m\measuredangle DBE$. By substitution, $m\measuredangle E > m\measuredangle DBE$. Therefore, $BD > ED$. In $\triangle BCD$, since $\measuredangle BDC \cong \measuredangle C$, $BC = BD$. By substitution, $BC > ED$.

30 **(a)** By subtracting the measures of pairs of congruent base angles, $m\measuredangle CAD = m\measuredangle CBD$.

(b) In $\triangle EBC$, the measure of exterior angle AED is greater than the measure of angle CBD. By substitution, $m\measuredangle AED > m\measuredangle CAD$. Therefore, $AD > DE$.

31. GIVEN: Equilateral $\triangle RST$. Point X is any point on $\overline{RT}$ and $\overline{SX}$ is drawn.

PROVE: $SX < RS$.

PROOF: $m\measuredangle SXR > m\measuredangle T$. Since an equilateral triangle is also equiangular, $m\measuredangle R = m\measuredangle T$, then $m\measuredangle SXR > m\measuredangle R$, which implies that $RS > SX$ or, equivalently, $SX < RS$. Since $RS = ST = RT$. $SX < ST$ and $SX < RT$.

32. GIVEN: $\triangle BET$ with $\overline{BE} \cong \overline{TE}$. Angle E is obtuse.

PROVE: $BT > BE$.

PROOF: $m\measuredangle 1 < 90$ since a triangle may have almost one nonacute angle. Hence, $m\measuredangle E > m\measuredangle T$ which implies $BT > BE$. Since $BE = TE$, $BT > TE$.

33. Suppose point P is any point not on line ℓ. Draw a segment from point P perpendicular to ℓ, intersecting ℓ at point A. Choose any other point on ℓ, say point X, so that A and X are different points. Draw $\overline{PX}$. $\overline{PX}$ is the hypotenuse of right triangle PAX. Since $\overline{PX}$ lies opposite the greatest angle of the tri-angle, it must be the longest side of the triangle. Hence, $PA < PX$.

34. GIVEN: Acute scalene $\triangle ABC$. $\overline{BH}$ is the altitude drawn to side $\overline{AC}$.

PROVE: $\overline{BH}$ does not bisect $\measuredangle ABC$.

PROOF: Assume $\overline{BH}$ bisects $\measuredangle ABC$. Then $\triangle ABH \cong \triangle CBH$ by ASA so that $\overline{AB} \cong CB$ by CPCTC. This contradicts the assumption that $\triangle ABC$ is scalene. Hence, $\overline{BH}$ does not bisect $\measuredangle ABC$.

Chapter 9

1. (4)
2. (3)
3. (2)
4. (1)
5. 36
6. $m\measuredangle T = m\measuredangle M = 105$.
   $m\measuredangle H = m\measuredangle A = 75$.
7. $m\measuredangle R = m\measuredangle G = 55$.
   $m\measuredangle T = m\measuredangle I = 125$.
8. 23
9. $EF = 19$, $RT = 38$
10. **(a)** $YT = 28.5$ and $BE = 41.5$.
    (b) $YT = 35$, $LM = 41$, $BE = 47$.
11. 22
12. 45
13. In $\square ABCD$, $\overline{AB} \cong \overline{CD}$ and $\overline{AB} \perp \overline{CD}$ so that $\measuredangle BAE \cong \measuredangle DCF$. Show $\triangle AEB \cong \triangle CFD$ by SAS. By CPCTC, $\measuredangle ABE \cong \measuredangle CDF$.
14. In $\square ABCD$, $\overline{AD} \parallel \overline{BC}$ so that $\overline{BH} \parallel \overline{DE}$ and $\measuredangle H \cong \measuredangle E$. $\measuredangle BAD \cong \measuredangle BCD$. Taking supplements of congruent angles, $\measuredangle FAE \cong \measuredangle GCH$. $\triangle FAE \cong \triangle GCH$ by AAS. By CPCTC, $\overline{AF} \cong \overline{CG}$.

15. In $\square ABCD$, $\overline{AB} \cong \overline{CD}$ and $\overline{AB} \parallel \overline{DC}$. $\overline{AB} \cong \overline{BE}$ (Definition of midpoint). By transitivity, $\overline{CD} \cong \overline{BE}$; $\measuredangle EBF \cong \measuredangle DCF$, $\measuredangle EFB \cong \measuredangle DFC$; $\triangle BEF \cong \triangle CDF$ by AAS. By CPCTC, $\overline{EF} \cong \overline{FD}$.

16. Right $\triangle ABM \cong$ right $\triangle DCM$ by SAS. By CPCTC, $\overline{AM} \cong \overline{DM}$, which implies that $\triangle AMD$ is isosceles.

17. Since the diagonals of a rhombus are the perpendicular bisectors of each other, $\overline{AR} \cong \overline{CR}$ and angles ARS and CRS are right angles. Show $\triangle SRA \cong \triangle SRC$ by SAS. By CPCTC, $\overline{SA} \cong \overline{SC}$ which implies that $\triangle ASC$ is isosceles.

18. $\measuredangle EBC \cong \measuredangle ECB$. Since $\overline{AD} \parallel \overline{BC}$, $\measuredangle AFB \cong \measuredangle EBC$ and $\measuredangle DGC \cong \measuredangle ECB$. By transitivity, $\measuredangle AFB \cong \measuredangle DGC$. $\measuredangle A \cong \measuredangle D$ and $\overline{AB} \cong \overline{DC}$ so that $\triangle FAB \cong \triangle GDC$ by AAS. By CPCTC, $\overline{AF} \cong \overline{DG}$.

19. Since $\overline{AD} > \overline{DC}$, $m\measuredangle ACD > m\measuredangle DAC$. Since $\overline{AB} \parallel \overline{DC}$, $m\measuredangle BAC = m\measuredangle ACD$. By substitution, $m\measuredangle BAC > m\measuredangle DAC$.

20. Since $ABCD$ is a parallelogram, $m\measuredangle ABC = m\measuredangle ADC$. $m\measuredangle RBS = m\measuredangle SDR$ since halves of equals are equal. In triangles BAR and DCS, $\measuredangle A \cong \measuredangle C$ and $\measuredangle ABR \cong \measuredangle CDS$. This implies that angles CSD and ARB, the third pair of angles, are congruent. Since supplements of congruent angles are congruent, $\measuredangle BRD \cong \measuredangle DSB$. $BRDS$ is a parallelogram since both pairs of opposite angles are congruent.

21. $\overline{BL} \cong \overline{DM}$ and $\overline{BL} \parallel \overline{DM}$. Taking supplements of congruent angles, $\triangle ALB \cong \triangle CMD$. $\triangle ALB \cong \triangle CMD$ by SAS. By CPCTC, $AB \cong CD$ and $\measuredangle BAL \cong \measuredangle DCM$ so that $\overline{AB} \parallel \overline{CD}$. Hence, $ABCD$ is a parallelogram by Theorem 9.9.

22. $\overline{AB} \cong \overline{CD}$ and $\overline{AB} \parallel \overline{CD}$ so that $\measuredangle BAL \cong \measuredangle DCM$. $\triangle ALB \cong \triangle CMD$ by ASA, $\overline{BL} \cong \overline{DM}$ and $\measuredangle ALB \cong \measuredangle CMD$ by CPCTC. Taking supplements of congruent angles, $\measuredangle BLM \cong \measuredangle DML$ so that $\overline{BL} \parallel \overline{DM}$. Hence, $BLDM$ is a parallelogram by Theorem 9.9.

23. $\overline{AB} \cong \overline{BC}$ since a rhombus is equilateral. $\triangle ABL \cong \triangle BCM$ by SSS. By CPCTC, $\measuredangle B \cong \measuredangle C$. Since $\overline{AB} \parallel \overline{DC}$, angles B and C are supplementary. Hence, each is a right angle and $ABCD$ is a square.

24. Use an indirect proof. Assume $ABCD$ is a rectangle. Then $\triangle BAD \cong \triangle CDA$ by SAS. By CPCTC, $\measuredangle 1 \cong \measuredangle 2$ which contradicts the Given ($m\measuredangle 2 \cong m\measuredangle 1$).

25. $DE = \frac{1}{2}BC$ and $DF = \frac{1}{2}AB$. Since $DE = DF$, $\frac{1}{2}BC = \frac{1}{2}AB$ implies that $BC = AB$. Hence, $\triangle ABC$ is isosceles.

26. If $\overline{BC} \parallel \overline{WT}$, then $\overline{AC} \parallel \overline{WT}$ (Extensions of parallel segments are parallel.) $\overline{AW} \parallel \overline{TC}$ (Segments of parallel lines are parallel). Hence, $WACT$ is a parallelogram.

27. Show $\triangle RSW \cong \triangle WTR$ by SAS. By CPCTC, $\measuredangle TRW \cong \measuredangle SWR$ so that $\overline{RP} \cong \overline{WP}$ (Theorem 7.2).

28. By addition, $AG = DF$, $\overline{EF} \cong \overline{EG}$ (Given) and by Theorem 7.2 $\measuredangle EGF \cong \measuredangle EFG$. $\overline{BG} \cong \overline{CF}$ (Given). $\triangle BAG \cong \triangle CDF$ by SAS. By CPCTC, $\overline{AB} \cong \overline{DC}$. Hence, trapezoid $ABCD$ is isosceles.

29. By Theorem 7.1, $\measuredangle PLM \cong \measuredangle PML$. Since LM is a median, $\overline{LM} \parallel \overline{AD}$ so that $\measuredangle PLM \cong \measuredangle APL$ and $\measuredangle PML \cong \measuredangle DPM$. By transitivity, $\measuredangle APL \cong \measuredangle DPM$. Show $\triangle LAP \cong \triangle MDP$ by SAS. By CPCTC, $\measuredangle A \cong \measuredangle D$ which implies trapezoid $ABCD$ is isosceles.

30. Since $ABCD$ is a trapezoid, $\overline{BC} \parallel \overline{KD}$. Show $\overline{BK} \parallel \overline{DC}$ as follows: $\measuredangle BAK \cong \measuredangle CDA$ and by transitivity $\measuredangle BKA \cong \measuredangle CDA$; since corresponding angles are congruent, $\overline{BK} \parallel \overline{DC}$. $BKDC$ is a parallelogram since both pairs of opposite sides are parallel.

31. **(a)** Since $\measuredangle B \cong \measuredangle C$, $\overline{AB} \cong \overline{AC}$ (Theorem 7.2). $\overline{BD} \cong \overline{CF}$ (Halves of congruent segments are congruent.) Therefore, $\triangle DEB \cong \triangle FGC$ by SAS.

    (b) $\measuredangle 1 \cong \measuredangle 2$ (CPCTC). Since angles 1 and 2 are also supplementary (Given), $\overline{FG}$ and $\overline{DE}$ are perpendicular to $\overline{BC}$, $\overline{DE} \parallel \overline{FG}$ (lines perpendicular to the same line are parallel) and $\overline{DE} \cong \overline{FG}$ (by CPCTC). Hence, $DEGF$ is a parallelogram. Parallelogram $DEGF$ is a square since it contains a right angle (for example, $\angle DEG$ is a right angle since perpendicular lines meet to form right angles) and a pair of congruent adjacent sides ($\overline{FD} \cong \overline{FG}$ is given).

32. See the hint in the statement of the problem. Since the diagonals of a rhombus bisect its angles, $m\measuredangle EDA > m\measuredangle EAD$. This implies that $AE > DE$. Equivalently, $2AE > 2DE$. Since the diagonals of a rhombus bisect each other, AC may be substituted for $2AE$ and BD may be substituted for $2DE$ so that $AC > BD$.

33. See the hint in the statement of the problem. By dropping perpendiculars, a parallelogram is formed. Since the opposite sides of a parallelogram are equal in length, the perpendicular segments have the same length.

34. GIVEN: $\square ABCD$. Points X, Y, Z, and W are the midpoints of sides $\overline{AB}$, $\overline{BC}$, $\overline{CD}$, and $\overline{DA}$, respectively.

    PROVE: Quadrilateral $WXYZ$ is a parallelogram.

    PLAN: Draw diagonal $\overline{BD}$. In $\triangle BAD$, $\overline{XW}$ is parallel to $\overline{BD}$. In $\triangle BCD$, $\overline{YZ}$ is parallel to $\overline{BD}$. Hence, $\overline{XW} \parallel \overline{YZ}$. Similarly, by drawing diagonal $\overline{AC}$, show that $\overline{XY} \parallel \overline{WZ}$. Since both pairs of opposite sides are parallel, quadrilateral $WXYZ$ is a parallelogram.

35. GIVEN: Rectangle $ABCD$. Points E, F, G, and H are the midpoints of sides $\overline{AB}$, $\overline{BC}$, $\overline{CD}$, and $\overline{DA}$, respectively.

    PROVE: Quadrilateral $EFGH$ is a rhombus.

    PLAN: First prove $EFGH$ is a parallelogram. Next, show that an adjacent pair of sides are congruent. Show $\overline{HE} \cong \overline{HG}$ by proving $\triangle EAH \cong \triangle GDH$ by SAS.

36. In trapezoid $ABCD$ with congruent diagonals $\overline{BD}$ and $\overline{CA}$, altitudes $\overline{BE}$ and $\overline{CF}$ form right triangles BED and CFA which are congruent by Hy-Leg. Since $\measuredangle CAD \cong \measuredangle BDA$ by CPCTC, $\triangle ABD \cong \triangle DCA$ by SAS. By CPCTC, $\overline{AB} \cong \overline{DC}$.

Chapter 10

1. 84
2. 72
3. 75, 105, 75, 105
4. **(a)** 24
   (b) 10
   (c) $\frac{47}{8}$
   (d) 5
   (e) –3 or 7
5. **(a)** 8
   (b) $6e^2$
   (c) $\frac{1}{4}$
   (d) $\sqrt{54}$

6. **(a)** Yes.
   (b) Yes.
   (c) No.
   (d) No.
7. $\frac{AL}{AB} = \frac{AM}{AG}; \frac{LB}{AB} = \frac{MG}{AG}; \frac{AL}{LB} = \frac{AM}{MG}.$
8. **(a)** Yes.
   (b) Yes.
   (c) No.
   (d) No.
   (e) Yes.
9. **(a)** 15
   (b) 4
   (c) 7
   (d) 24
   (e) $KE = 9, KH = 3.$
10. $\measuredangle G \cong \measuredangle S, \measuredangle A \cong \measuredangle H, \measuredangle L \cong \measuredangle E$
    $\frac{GA}{SH} = \frac{AL}{HE} = \frac{GL}{SE}$
11. 40
12. $MY = 5, MX = 35.$
13. 44
14. 12 and 27
15. **(a)** $JB = 12$ and $RA = 8$
    (b) LK 17.5 = and $TS = 14$
    (c) $ST = 10$ and $KL = 16$
    (d) 5
16. 2. Parallel Postulate.
    3. Postulate 10.1.
    4. Substitution.
    5. If two lines are parallel, then the corresponding angles are congruent.
    6. Given.
    7. Transitive property.
    8. If two angles of one triangle are congruent to two angles of another triangle, then the third pair of angles are congruent.
    9. ASA Postulate.
    10. CPCTC.
    11. Substitution.

17–29. are based on the application of Theorem 10.3.

17. $\measuredangle W \cong \measuredangle Y$ and right $\measuredangle HAW \cong$ right $\measuredangle KBY$ implies $\triangle HWA \sim \triangle KYB$.
18. $\measuredangle XBC \cong \measuredangle YST$ and $\measuredangle C \cong \measuredangle T$ (Reverse of the definition of similar triangles.) Hence, $\triangle BXC \sim \triangle SYT$.
19. $\measuredangle FCD \cong \measuredangle AEF$ and $\measuredangle AFE \cong \measuredangle CFD$. Therefore, triangles EFA and CFD are similar by Theorem 10.3.
20. $\measuredangle AEF \cong \measuredangle AFE$ and $\measuredangle DFC \cong \measuredangle FCD$. Since $\measuredangle AFE \cong \measuredangle DFC$, by the transitive property, $\measuredangle AEF \cong \measuredangle FCD$. Therefore, $\triangle AFE \sim \triangle DFC$ by Similarity 10.3 and $\frac{AF}{EF} = \frac{DF}{FC}$.
21. $\measuredangle CTM \cong \measuredangle AWB$ (Reverse of the definition of similar triangles.) Since $\measuredangle AWB \cong \measuredangle CWB$ (Definition of angle bisector), $\measuredangle CTM \cong \measuredangle CWB$ by transitivity. $\measuredangle MCT \cong \measuredangle BCW$. Hence, $\triangle MCT \sim \triangle BCW$.
22. $\measuredangle S \cong \measuredangle MWB$ (Angle) and $\measuredangle A \cong \measuredangle WCA$ (Theorem 7.1). Since $\overline{AW} \parallel \overline{ST}$, $\measuredangle A \cong \measuredangle BTS$ and by transitivity, $\measuredangle BTS \cong \measuredangle WCA$ (Angle). Hence, $\triangle BCW \sim \triangle BTS$.
23. $\measuredangle WBC \cong \measuredangle WCB$ and taking supplements of these angles, $\measuredangle ABW \cong \measuredangle TCW$ (Angle). Since $\overline{AT}$ bisects $\measuredangle STW$, $\measuredangle STA \cong \measuredangle WTC$. $\measuredangle A \cong \measuredangle STA$ since $\overline{ST} \parallel \overline{AW}$. By transitivity, $\measuredangle A \cong \measuredangle WTC$ (Angle). Hence, $\triangle ABW \sim \triangle TCW$.
24. Show $\triangle TAX \sim \triangle WHY$, $\measuredangle W \cong \measuredangle ATX$ and $\measuredangle Y \cong \measuredangle AXT$ since $\parallel$ line segments have congruent corresponding angles.
25. Show $\triangle RMN \sim \triangle RAT$. $\measuredangle RMN \cong \measuredangle A$ and $\measuredangle RNM \cong \measuredangle T$. Write $\frac{MN}{AT} = \frac{RN}{RT}$. By Theorem 7.2 $NT = MN$ Substitute NT for MN in the proportion.
26. Show $\triangle SQP \sim \triangle WRP$, $\measuredangle SPQ \cong \measuredangle WPR$ (Angle). Since $\overline{SR} \cong \overline{SQ}$, $\measuredangle SRQ \cong \measuredangle SQR$. $\measuredangle SRQ \cong \measuredangle WRP$ (Definition of angle bisector). By transitivity, $\measuredangle SQR \cong \measuredangle WRP$ (Angle).
27. Show $\triangle PMQ \sim \triangle MKC$. Right $\measuredangle MCK \cong$ right $\measuredangle PMQ$. Since $\overline{TP} \cong \overline{TM}$, $\measuredangle TPM \cong \measuredangle TMP$.

28. Show $\triangle PMT \sim \triangle JKT$. Since $\overline{MP} \cong \overline{MQ}$, $\measuredangle MPQ \cong \measuredangle MQP$. Since $\overline{JK} \parallel \overline{MQ}$, $\measuredangle J \cong \measuredangle MQP$ so that by transitivity, $\measuredangle J \cong \measuredangle MPQ$ (Angle) $\measuredangle K \cong \measuredangle QMT$ (Congruent alternate interior angles.) $\measuredangle QMT \cong \measuredangle PMT$ (since $\triangle MTP \cong \triangle MTQ$ by SSS). By transitivity, $\measuredangle K \cong \measuredangle PMT$ (Angle). Write the proportion $\frac{PM}{JK} = \frac{PT}{JT}$. Substitute TQ for PT (see Given) in the proportion.

29. Show $\triangle EIF \sim \triangle HIG$. Since $\overline{EF}$ is a median, $\overline{EF} \parallel \overline{AD}$ so that $\measuredangle FEI \cong \measuredangle GHI$ and $\measuredangle EFI \cong \measuredangle HGI$.

30. Rewrite the product as $ST \times ST = TW \times RT$. Show $\triangle SWT \sim \triangle RST$. Right $\measuredangle RST$ = right $\measuredangle SWT$ and $\measuredangle T \cong \measuredangle T$.

31. Show $\triangle AEH \sim \triangle BEF$. $\measuredangle BEF \cong \measuredangle HEA$ and $\measuredangle EAH \cong \measuredangle EBF$ (Halves of equals are equal.)

32. Apply Postulate 10.1. Since $\overline{XY} \parallel \overline{LK}$. $\frac{JY}{YK} = \frac{JX}{XL}$. Since $\overline{XZ} \parallel \overline{JK}$, $\frac{JX}{XL} = \frac{KZ}{ZL}$. By transitivity, $\frac{JY}{YK} = \frac{KZ}{ZL}$. $YK = XZ$ since $YKZX$ is a parallelogram. Using substitution, $\frac{JY}{XZ} = \frac{KZ}{ZL}$. Cross-multiplying yields the desired product.

33. Rt. $\measuredangle B \cong$ Rt. $\measuredangle DCA$. Angles BAC and CDA are complementary to $\measuredangle CAD$ and are, therefore, congruent to each other. Triangles ABC and DCA are similar which means $\frac{AD}{AC} = \frac{AC}{BC}$, and $BC \times AD = (AC)^2$.

34. **(a)** $\overline{EF} \parallel \overline{AC}$, $\measuredangle ACF \cong \measuredangle GFE$ since parallel lines form congruent alternate interior angles. Similarly, since $\overline{DE} \parallel \overline{AB}$, $\measuredangle AFC \cong \measuredangle EGF$. Hence, $\triangle CAF \sim \triangle FEG$.

(b) $\triangle CAF \sim \triangle CDG$. By the transitive property, $\Delta CDG \sim \Delta FEG$ so that $\frac{DG}{EG} = \frac{GC}{GF}$ which implies $DG \times GF = EG \times GC$.

Chapter 11

1. $r = 8, s = 8\sqrt{3}, t = 4\sqrt{3}$.
2. $r = 25, s = 5\sqrt{5}, t = 10\sqrt{5}$.
3. $r = 16, s = 9, t = 15$.
4. 11.8
5. 6
6. $4\sqrt{6}$
7. $9\sqrt{5}$
8. 7
9. $4\sqrt{10}$
10. **(a)** 4 and 16
    (b) $8\sqrt{5}$
11. 8
12. $6\sqrt{2}$
13. 24
14. $100\sqrt{7}$
15. $\frac{20}{3}$
16. 80
17. 18
18. $5\sqrt{3}$
19. $\sqrt{74}$
20. **(a)** 4
    (b) $8\sqrt{3}$
21. $6\sqrt{2}$
22. $36\sqrt{2}$

23. **(a)** 4
    (b) $4\sqrt{3}$
    (c) $4\sqrt{2}$
24. **(a)** Altitude = $8/\sqrt{3}$, leg = $16/\sqrt{3}$.
    (b) Altitude = 8, leg = $8\sqrt{2}$.
    (c) Altitude = $8\sqrt{3}$, leg = 16.
25. $x = 4\sqrt{3}$, $y = 4$, $z = 2\sqrt{37}$.
26. $2\sqrt{37}$
27. $10\sqrt{3}$
28. $\frac{24}{25}$
29. $\cos R = \frac{40}{41}$, $\tan R = \frac{9}{40}$.
30. 32°
31. Base = 15, leg = 12.
32. **(a)** 19.1
    (b) 29.6
33. $x = 12.9$, $y = 3.6$.
34. $AD = 6.2$, $CD = 21.9$.
35. 5
36. 81.3
37. 6.7
38. **(a)** 56°
    (b) 12.4
39. 2.8
40. 32°

Chapter 12

1. 66
2. 52.5
3. 25
4. 66
5. 12.5
6. 38
7. 66
8. 61
9. 24
10. 75
11. **(a)** 36
    (b) 30
12. 26
13. 106
14. 25
15. **(a)** 74
    (d) 16
    (g) 16
    (b) 16
    (e) 32
    (h) 16
    (c) 74
    (f) 90
    (i) 74
16. 2:1
17. $x = 82$, $y = 59$, $z = 39$.
18. **(a)** 30
    (b) 65
    (c) 75
    (d) 55
    (e) 35
    (f) 100
19. **(a)** $m\widehat{WA} = 30$, $m\widehat{AT} = 90$, $m\widehat{ST} = 150$, $m\widehat{SW} = 90$.
    (b) 45.
    (c) 90.
    (d) 120.
    (e) 15.
20. **(a)** $m\widehat{KF} = 76$, $m\widehat{FM} = 104$, $m\widehat{JK} = 100$, $m\widehat{JM} = 80$, $m\widehat{HM} = 28$.
    (b) 76.
    (c) 76.
    (d) 12.
21. Show $\triangle AOM \cong \triangle BOM$ by Hy-Leg. By CPCTC, central $\measuredangle AOX \cong$ central $\measuredangle BOX$ which implies $\widehat{AX} \cong \widehat{BX}$.

22. By subtraction, $\widehat{RS} \cong \widehat{WT}$. $\measuredangle RTS \cong \measuredangle WST$ since inscribed angles that intercept congruent arcs are congruent.

23. GIVEN: $\overline{PA}$ tangent to circle O, B is between P and A.

PROVE: $\overline{OB}$ is *not* perpendicular to $\overline{PA}$.

PROOF: Draw $\overline{OA}$. Angle OAB is a right angle since a radius is perpendicular to a tangent at the point of tangency. Assume $\overline{OB}$ is perpendicular to $\overline{PA}$. Then $\angle OBA$ is a right angle. Because a triangle can contain, at most, one right angle, $\overline{OB}$ is *not* perpendicular to $\overline{PA}$.

24. Since $OA > AC$, $m\measuredangle OCA > m\measuredangle AOC$, $m\measuredangle BOC > m\measuredangle OCA$ so that $m\measuredangle BOC > m\measuredangle AOC$. Since $m\measuredangle BOC = m\widehat{BC}$ and $m\measuredangle AOC = m\widehat{AC}$, by substitution, $m\widehat{BC} > m\widehat{AC}$.

25. Since $m\widehat{BC} > m\widehat{AC}$, $m\measuredangle BOD > m\measuredangle AOD$. $m\measuredangle ADO > m\measuredangle BOD$ so that $m\measuredangle ADO > m\measuredangle AOD$. This implies that $OA > AD$.

26. Show $\triangle XPM \cong \triangle YQM$ by AAS where right $\measuredangle XPM \cong$ right $\measuredangle YQM$, $\measuredangle XMP \cong \measuredangle YMQ$, $\overline{XP} \cong \overline{YQ}$.

27. Since $\overline{FE} \cong \overline{FG}$, $m\measuredangle G = m\measuredangle E$. $m\widehat{FG} = 2m\measuredangle E$ so that $m\measuredangle BFG = m\measuredangle E$. By transitivity, $m\measuredangle G = m\measuredangle BFG$ so that $\overline{AB} \parallel \overline{EG}$.

28. Show $\triangle AMB \cong \triangle AMD$ by SAS. $\overline{BM} \cong \overline{DM}$; $\measuredangle M$ is inscribed in a semicircle so that right $\measuredangle AMB \cong$ right $\measuredangle AMD$; $\overline{AM} \cong \overline{AM}$. By CPCTC, $\measuredangle BAM = \measuredangle DAM$ which implies that $\widehat{BM} \cong \widehat{CM}$.

29. **(a)** $\measuredangle W \cong \measuredangle RSW$ since inscribed angles that intercept congruent arcs are congruent. $\measuredangle RSW \cong \measuredangle NSW$ (Definition of angle bisector.) By transitivity, $\measuredangle W \cong \measuredangle NSW$ which implies that $\triangle NSW$ is isosceles.

(b) $\widehat{WT} \cong \widehat{RW}$ since congruent inscribed angles intercept congruent arcs. Since $\widehat{RW} \cong SK$ (Given), $\widehat{WT} \cong \widehat{SK}$. $\measuredangle STK \cong \measuredangle TKW$ (Inscribed angles that intercept congruent arcs are congruent). Since the base angles of $\triangle NTK$ are congruent, the triangle is isosceles.

30. $m\widehat{LM} \cong m\widehat{KL}$ implies that $m\measuredangle MKL > m\measuredangle LMK$. Since $\overline{JK} \parallel \overline{LM}$, $m\measuredangle JKM = m\measuredangle LMK$. By substitution, $m\measuredangle MKL > m\measuredangle JKM$.

31. Since $\overline{MA} \cong \overline{MT}$, $\measuredangle MAT \cong \measuredangle ATM$. $\measuredangle SRA \cong \measuredangle ATM$ (Inscribed angles that intercept the same arc are congruent). By transitivity, $\measuredangle SRA \cong \measuredangle MAT$ which implies that $\overline{SR} \parallel \overline{ATW}$. Quadrilateral $RSTW$ is a parallelogram since $\overline{SR} \cong \overline{TW}$ (Given) and $\overline{SR} \cong \overline{TW}$.

Chapter 13

1. **(a)** 2 common internal tangents.
   2 common external tangents.
   (b) No common tangent.
   (c) 1 common external tangent.
2. **(a)** 11
   (b) 3
3. 6
4. 9
5. 14
6. 1 or 21
7. 18
8. 2
9. 20
10. 1

11. 5
12. 27
13. 5
14. 12
15. 12
16. **(a)** 13
    (b) $JW = 48, OA = 7$.
    (c) 12
17. $2\sqrt{5}$
18. 6
19. 9π
20. $9\pi\sqrt{2}$
21. $3\pi\sqrt{2}$
22. 16
23. 81
24. 6.7
25. 2.7
26. Triangles OAR and OBR are right triangles since angles S and T are right angles and $\overline{OA} \parallel \overline{MS}$ and $\overline{OB} \parallel \overline{MT}$ making corresponding angles congruent. $\triangle OAR \cong \triangle OBR$ by Hy-Leg. By CPCTC, $OA = OB$. $\overline{SR} \cong \overline{TR}$ by Theorem 13.2.
27. Because it is given that $\overline{BE} \parallel \overline{ADC}$, $\angle ABE \cong \angle CAB$. Angle E is inscribed in a semicircle, so it is a right angle. Angle ABC is a right angle, as it is formed by a diameter drawn to a tangent at its point of tangency. Hence, $\angle E \cong \angle ABC$. By Theorem 10.3, $\triangle ABE \sim \triangle CAB$.
28. Since a square contains 4 right angles and is equilateral, $\overline{OX} \perp \overline{QT}$, $\overline{OY} \perp \overline{PJ}$, and $OX = OY$. By Theorem 13.2, $\overline{QT} \cong \overline{PJ}$ so that $\widehat{QT} \cong \widehat{PJ}$. By arc subtraction, $\widehat{QP} \cong \widehat{JT}$.
29. Show $\triangle HBW \sim \triangle MBL$. $\measuredangle ABW \cong \measuredangle H$ since they are measured by one-half of the same arc measure. Since $ABLM$ is a parallelogram, $\overline{AB} \parallel \overline{ML}$ so that $\measuredangle ABW \cong \measuredangle BML$. By transitivity, $\measuredangle H \cong \measuredangle BML$ (Angle) $\measuredangle MBL \cong \measuredangle HBW$ (Angle).
30. Show $CD : BC = BC : AB$ by proving $\triangle ABC \sim \triangle BCD$. Right $\measuredangle ACB \cong$ right $\measuredangle CDB$. $\measuredangle CAB \cong \measuredangle CBD$ since they are measured by one-half the same arc measure.
31. Show $\triangle KLP \sim \triangle KJM$. $\measuredangle LKP \cong \measuredangle JKM$ and right $\measuredangle KLP \cong$ right $\measuredangle KJM$.
32. Show $\triangle KLP \sim \triangle KJM$. Right $\measuredangle KLP \cong$ right $\measuredangle KJM$ (Angle). Since $\overline{JP} \cong \overline{JM}$, $\measuredangle JPM \cong \measuredangle JMK$ $\measuredangle KPL \cong \measuredangle JPM$. By transitivity, $\measuredangle KPL \cong \measuredangle JMK$ (Angle).
33. **(a)** $\measuredangle NTK \cong \measuredangle WTK$ Since $\overline{KW} \cong \overline{TW}$. $\measuredangle WKT \cong \measuredangle WTK$. By transitivity, $\measuredangle NTK \cong \measuredangle WKT$ which implies that $\overleftrightarrow{NTP} \parallel \overleftrightarrow{KW}$
    (b) Rewrite product as $TW \cdot TW = JT \cdot TK$. Show $\triangle JTW \sim \triangle WTK$. $\measuredangle T \cong \measuredangle T$ (Angle). $\measuredangle JWT \cong \measuredangle NTK$ and $\measuredangle K \cong NTK$ so that $\measuredangle K \cong \measuredangle JWT$ (Angle).
34. GIVEN: Circles A and B are tangent externally at point P. $\overleftrightarrow{XY}$ is tangent to $\odot A$ at point X and tangent to $\odot B$ at point Y. $\overleftrightarrow{PQ}$ intersects $\overleftrightarrow{XY}$ at point Q.
    PROVE: $\overline{QX} \cong \overline{QY}$.
    PLAN: By Theorem 13.3, $\overline{QX} \cong \overline{QP}$ and $\overline{QP} \cong \overline{QY}$. By transitivity, $\overline{QX} \cong \overline{QY}$.
35. GIVEN: $\overleftrightarrow{XY}$ is tangent to $\odot A$ at point X and tangent to $\odot B$ at point Y. $\overleftrightarrow{PQ}$ is tangent to $\odot A$ at point P and tangent to $\odot B$ at point Q.
    PROVE: $\overline{XY} \cong \overline{PQ}$.
    PLAN: *Case 1*. Assume circles A and B are not congruent. $\overleftrightarrow{XY}$ cannot be parallel to $\overleftrightarrow{PQ}$. Therefore, $\overleftrightarrow{XY}$ and $\overleftrightarrow{PQ}$ will intersect, extended if necessary, say at point C. By Theorem 13.3, $\overline{CX} \cong \overline{CP}$ and $\overline{CY} \cong \overline{CQ}$. By subtraction, $\overline{XY} \cong \overline{PQ}$.
    Case 2. Assume circles A and B are congruent. Then $\overleftrightarrow{XY} \parallel \overleftrightarrow{PQ}$. Draw the line of centers $\overline{AB}$ and radii $\overline{AX}$, $\overline{BY}$, $\overline{AP}$, and $\overline{BQ}$. Quadrilateral $AXYB$ is a parallelogram since $\overline{AX} \cong \overline{BY}$ (Congruent circles have congruent radii) and $\overline{AX} \parallel \overline{BY}$ (Segments perpendicular to the

(Segments perpendicular to the same line are parallel.) Similarly, $APQB$ can be shown to be a parallelogram. Hence, $\overline{XY} \cong \overline{AB}$ and $\overline{AB} \cong \overline{PQ}$ so that $\overline{XY} \cong \overline{PQ}$.

36. GIVEN: Lines ℓ, m, and k are tangent to circle P at points C, D, and E, respectively, so that $\ell \parallel m$ and line k intersects line ℓ at point A and line m at point B.

PROVE: $\measuredangle APB$ is a right angle.

PLAN: Since $\ell \parallel m$, angles CAB and DBA are supplementary. $\overline{PA}$ bisects $\measuredangle CAB$ and PB bisects $\measuredangle DBA$ (refer to the discussion of the proof of Theorem 13.3). Hence, the sum of the measures of angles PAB and PBA is 90, which implies that the remaining angle of $\triangle APB$, namely, $\measuredangle APB$, must have a measure of $180 - 90$ or 90.

Chapter 14

1. (1) 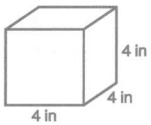

2. (4) 156
3. (3) 8
4. (4) 80
5. (3) 589
6. 60
7. $35\sqrt{3}$
8. $10\sqrt{2}$
9. $9\sqrt{3}$
10. 104
11. 52
12. 17.5
13. 162
14. $54 + 36\sqrt{3}$
15. **(a)** 8
    (b) $2\sqrt{11}$
    (c) $\frac{4}{\sqrt[4]{3}}$
16. **(a)** 12
    (b) 14
17. **(a)** 4
    (b) 5
    (c) 6
18. 240
19. **(a)** 42
    (b) 21
    (c) $21\sqrt{3}$
20. 5
21. **(a)** $\frac{9}{11}$
    (b) 55
22. 192
23. 45
24. 30
25. $150\sqrt{3}$
26. 12
27. $\frac{3}{4}$
28. **(a)** 30.8 cm
    (b) 3,080 cm^2
29. 270
30. 15 cm
31. 312.5 square units
32. 20π square units
33. 47
34. **(a)** 24π square units
    (b) $24\pi - 36\sqrt{3}$ square units
35. $8\pi - 16$ cm^2
36. $\frac{1}{2}$
37. $\frac{625}{8}\pi - 84$ square units
38. $256 - 64\pi$ square units
39. 80π cm^2
40. 2.6
41. 260
42. 169

43. 9 bags

44. **(a)** 103
(b) 40%

45. **(a)** 64π square units
(b) $96\sqrt{3}$ square units
(c) $\frac{32}{3}\pi - 16\sqrt{3}$ square units

46. **(a)** From point L draw the altitude to side $\overline{CV}$, intersecting $\overline{CV}$ at point H. Area $\triangle CLG = \frac{1}{2}CG \cdot LH$ and area $\triangle VLG = \frac{1}{2}VG \cdot LH$. Since $CG = VG$, area $\triangle CLG$ = area $\triangle VLG$. (*Note: In $\triangle CLV$, $\overline{LG}$ is a median. We have just proved that a median divides a triangle into two triangles having the same area.*)
(b) In $\triangle BCV$, $\overline{BG}$ is a median so that area $\triangle BCG$ = area $\triangle BVG$ (see part **a** of this exercise). Next, subtract:

$$\begin{array}{r} \text{Area } \triangle BCG = \text{area } \triangle BVG \\ -\text{Area } \triangle CLG = \text{area } \triangle VLG \\ \hline \text{Area } \triangle BLC = \text{area } \triangle BLV \end{array}$$

47. **(a)** 11
(b) 396 square units

48. **(a)** By subtraction, $\measuredangle ABC \cong \measuredangle EBD$ so that $\triangle ABC \cong \triangle EBD$ by SAS.
(b) Since congruent triangles have equal areas, by addition, area $\triangle ABC$ + area $\triangle BCD$ = area $\triangle EBD$ + area $\triangle BCD$. Therefore, area polygon $ABDC$ = area polygon $BEDC$.

49. The area of $\square ABCD$ is equal to the sum of the areas of triangles AEB, CEB and ADC. Since $\triangle CED \cong \triangle AEB$ and $\triangle AED \cong \triangle CEB$, the area of $\square ABCD$ is equal to the sum of the areas of triangles CED, AED, and ADC. $\overline{BE} \cong \overline{ED}$ so that $\overline{ED} \cong \overline{FD}$. Since $\overline{CD}$ and $\overline{AD}$ are medians of triangles ECF and EAF, respectively, area$\triangle CED$ = area $\triangle CDF$ and area $\triangle AED$ = area $\triangle FAD$. By substitution, the area of $\square ABCD$ is equal to the sum of the areas of triangles CDF, FAD, and ADC which is equivalent to $\triangle FAC$.

50. **(a)** Circle O is inscribed in equilateral $\triangle ABC$ where apothem $\overline{OX} \perp \overline{AC}$. In right $\triangle AOX$, $\overline{OA}$ represents the radius of the circumscribed circle, while OX is a radius of the inscribed circle. Since $m\measuredangle OAX = 30$ and $AX = \frac{1}{2}AC$, $OX = \frac{1}{\sqrt{3}}AX = \frac{1}{2\sqrt{3}}AC = \frac{\sqrt{3}}{6}AC$. The altitude to side $AC = \frac{\sqrt{3}}{2}AB = \frac{\sqrt{3}}{2}AC$. Hence, OX is one-third the length of the altitude.
(b) $AX = \frac{\sqrt{3}}{2}OA$ or $OA = \frac{2}{\sqrt{3}}AX = \frac{2}{\sqrt{3}}\left(\frac{1}{2}AC\right)$. OA is therefore equal to $\frac{1}{\sqrt{3}}$. Since OX may be represented by $\frac{1}{2\sqrt{3}}AC$ (see part **a**), $OX = \frac{1}{2}OA$. That is, the radius of the inscribed circle is one-half the length of the radius of the circumscribed circle. Since their areas have the same ratio as the square of the ratio of their radii, the ratio of the areas of the circle is 1:4.

51. **(a)** 150 square units
(b) 94 square units
(c) 64 cubic units

52. 1800π cm^3

53. 88 square units

54. $4\sqrt{3}$

Chapter 15

1. 20 square units
2. 12 square units
3. 17 square units
4. 24 square units
5. $(x-1)^2 + (y+2)^2 = 9$
6. **(a)** $O(-1, 1)$.
   (b) 5
   (c) Yes, since $OK = 5$.
7. $y = -1$.
8. $\sqrt{125}$
9. First show $JKLM$ is a parallelogram.

   Since slope of $JK = \frac{2}{3}$ and the slope of $KL = -\frac{3}{2}$, $\overline{JK} \perp \overline{KL}$. Since $\overline{JK} = \overline{KL} = \sqrt{13}$, $JKLM$ is a square.
10. **(a)** Slope of $JK = \frac{6}{5}$ and slope of $KL = -\frac{5}{6}$ so that $\overline{JK} \perp \overline{KL}$.
    (b) $JL = \sqrt{305}$, $KL = \sqrt{61}$, and $JK = \sqrt{244}$ so that $(JL)^2 = (KL)^2 + (JK)^2$.
    (c) 61 square units
11. **(a)** $\frac{8}{3}$
    (b) $-\frac{3}{2}$
    (c) $\frac{7}{6}$
12. **(a)** Collinear.
    (b) Noncollinear.
    (c) Collinear.
13. Slope of $\overline{ST} = \frac{8}{5}$, slope of $\overline{TA} = -\frac{5}{8}$, slope of $\overline{AR} = \frac{8}{5}$ and slope of $\overline{SR} = -\frac{5}{8}$.

    Hence, $\overline{ST} \parallel \overline{AR}$ and $\overline{TA} \parallel \overline{SR}$. Also, $\overline{ST} \perp \overline{TA}$ and so on.
14. $y = \frac{1}{2}x + 5$.
15. $y = \frac{3}{2}x - \frac{7}{2}$.
16. **(a)** $y = 7$.
    (b) $y = -x + 3$
    (c) $x = 6$.
    (d) $y = -4x + 1$.
17. **(a)** $y = 2$.
    (b) $x = -5$.
18. **(a)** $y = x + 1$.
    (b) $y = \frac{11}{9}x - \frac{25}{9}$.
    (c) $y = -\frac{2}{3}x + 12$.
19. $y = -\frac{1}{4}x - 2$.
20. $k = -25, t = -7$.
21. **(a)** $x^2 + y^2 = 16$.
    (b) 10
22. $x^2 + y^2 = 25$.
23. $k = \frac{5}{3}, h = \frac{36}{5}$.
24. $x^2 + (y-9)^2 = 25$.
25. Quadrilateral $TEAM$ is a parallelogram since opposite sides have the same slope and, as a result, are parallel: slope of $\overline{TE} = \text{slope of } \overline{AM} = \frac{7}{3}$ and slope of $\overline{EA} = \text{slope of } \overline{TM} = \frac{3}{7}$.

    Parallelogram $TEAM$ is a rhombus since a pair of adjacent sides have the same length: $TE = EA = \sqrt{58}$. Rhombus $TEAM$ is *not* a square since the slopes of adjacent sides are *not* negative reciprocals and, as a result, the sides do *not* intersect at right angles.

26. **(a)** 36.5 square units
    (b) 90 square units
    (c) 76 square units
    (d) 77 square units
27. **(a)** $k = 10$
    (b) $y = 2x + 4$
    (c) 24.5 square units
28. **(a)** $y = -x + 4$
    (c) $x = 0$
    (b) $y = 4$
    (d) $x^2 + y^2 = 16$
29. Show slope of $\overline{JA}$ = slope of $\overline{EK} = 0 \cong$ slope of $\overline{JE}$. Also show $JE = AK = a\sqrt{10}$.
30. Quadrilateral *ABCD* is a parallelogram since opposite sides have the same slope and, as a result, are parallel: slope of $\overline{AB} = \text{slope of } \overline{CD} = \frac{3}{8}$ and slope of $\overline{BC} = \text{slope of } \overline{AD} = \frac{5}{2}$.
    Parallelogram *ABCD* is *not* a rectangle since the slopes of adjacent sides are *not* negative reciprocals and, as a result, the sides do *not* intersect at right angles.
31. **(a)** Quadrilateral *KATE* is a trapezoid since only one pair of opposite sides have the same slope and, as result, are parallel: slope of $\overline{KA} = \text{slope of } \overline{TE} = \frac{2}{3}$ and slope of $\overline{AT} = -\frac{4}{3}$ while the slope of $\overline{KE}$ is undefined.
    (b) Since $AT = 5$ and $KE = 6$, the legs of trapezoid *KATE* are unequal so *KATE* is *not* an isosceles trapezoid.
32. Let the coordinates of ▱*ABCD* be represented by $A(0, 0)$, $B(s, t)$, $C(r + s, t)$, and $D(r, O)$. Use the distance formula to verify $AB = DC = \sqrt{s^2 + t^2}$ and $BC = AD = r$.
33. Let the coordinates of right $\triangle ABC$ be represented by $A(0, 0)$, $B(0, 2r)$, and $C(2s, 0)$. If M is the midpoint of $\overline{BC}$, then its coordinates are $M(s, r)$. $AM = \sqrt{s^2 + r^2}$ and $BC = 2\sqrt{s^2 + r^2}$ so that $AM = \frac{1}{2}BC$.
34. Let the coordinates of $\triangle ABC$ be represented by $A(0, 0)$, $B(2r, 2s)$, and $C(2t, 0)$. Let L and M be the midpoints of $\overline{AB}$ and $\overline{BC}$, respectively. Then $L(r, s)$ and $M(r + t, s)$. Show slope of $\overline{LM}$ = slope of $\overline{AC}$ and that $LM = \frac{1}{2}AC$.
35. Let the coordinates of isosceles trapezoid *ABCD* be represented by $A(0, 0)$, $B(r, s)$, $C(t, s)$, and $D(r + t, 0)$. $AC = \sqrt{t^2 + s^2} = BD$.

Chapter 16

1. (1)
2. (4)
3. (4)
4. (4)
5. (2)
6. Two circles concentric with the original circle having radii of 7 cm and 9 cm.
7. A line midway between the two parallel lines, parallel to each line, and 5 in. from each line.
8. A line parallel to the surface at a distance equal to the radius length.
9. A concentric circle having a radius of 9 cm.
10. The perpendicular bisector of the base.

11. Two points determined by the intersection of the perpendicular bisector of $\overline{AB}$ and the two lines that are parallel to and on either side of $\overleftrightarrow{AB}$ and 4 in. from $\overleftrightarrow{AB}$.
12. Four points determined by the angle bisector of each pair of vertical angles formed by the intersecting lines and a circle whose center is their point of intersection.
13. No points.
14. The two points determined by the intersection of the perpendicular bisector of $\overline{PQ}$ and the circle whose center is at point X and which has a radius of 5 cm.
15. **(a)** 2
    (b) 0
    (c) 1
16. **(a)** 0
    (b) 1
    (c) 2
17. **(a)** $y = 5, y = -5$.
    (b) $x = 4, x = -4$.
    (c) $y = 2x$.
    (d) $2x + 3y = 6$.
    (e) $y = x + 5$.
18. 2
19. 0
20. **(a)** 1
    (b) 2
    (c) 0
21. 1
22. **(a)** $x = 2$
    (b) $y = -5$
    (c) $y = 3x + 5$
23. 4
24. 2
25. **(a)** **1** A circle having point P as its center and a radius d units in length.
    2 The parallel lines $x = +1$ and $x = -1$.
    (b) **1** 1 **2** 3 **3** 4
26. (3)
27. Construction

Chapter 17

1. (1)
2. (1)
3. (3)
4. (4)
5. (2)
6. (4)
7. (3)
8. (1)
9. (3)
10. (1)
11. (4)
12. (3)
13. (1)
14. (4)
15. (2)
16. (4)
17. (1)
18. (3)
19. (2)
20. (3)
21. Reflection over $\overline{EM}$
22. $T_{-7,+9}$
23. Glide reflection over x-axis, $T_{+3,0}$
24. $R_{180°}$ about point M.
25. $r_{y\text{-axis}}$ and $T_{+3,-2}$
26. **(a)**

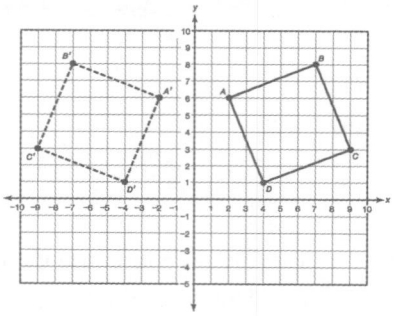

(b) 29

27. $y = 2x + 7$

28. **(a)** $A''(0,0), B''(-1,8), C''(-4,8)$
    (b) $R_{90°}$

29. **(a)** D
    (b) $\overline{DE}$
    (c) C

30. Center: $(-4,2)$; radius = 18

31. **(a)** A reflection in point E maps $\triangle AED$ onto $\triangle CEB$
    (b) Since the diagonals of a parallelogram bisect each other, $AE = EC$ and $DE = EB$ so points C and B are the images of the reflections of points A and D, respectively, in point E. Point E is its own image in a reflection in point E. Thus, under a reflection in point E, $A \rightarrow C, D \rightarrow B$, and $E \rightarrow E$. Because three non-collinear points determine a triangle, $\triangle CEB$ is the image of $\triangle AED$ under a reflection in point E.

Solutions to Cumulative Review Exercises

Chapters 1–7

1. 18
2. 130
3. 20
4. 144
5. 40
6. 40
7. Show a counterexample such as a 40°–40°–100° triangle.
8. 22 cm and 27 cm, or 24.5 cm and 24.5 cm.
9. $\triangle DAF \cong \triangle ECG$ by ASA since $\measuredangle A \cong \measuredangle C$, $\overline{AF} \cong \overline{CG}$ (halves of congruent segments are congruent) and $\angle AFD \cong \angle CGE$ (supplements of congruent angles are congruent). By CPCTC, $\overline{FD} \cong \overline{GE}$.
10. $\triangle DAF \cong \triangle ECG$ by AAS since $\measuredangle FDA \cong \measuredangle GEC$, $\measuredangle AFD \cong \measuredangle CGE$, and $\overline{AD} \cong \overline{CE}$ (using the subtraction property). By CPCTC, $\measuredangle A \cong \measuredangle C$. Therefore, $\triangle ABC$ is isosceles.
11. **(a)** $\measuredangle STR \cong \measuredangle BRT \cong \measuredangle TRS$. By the transitive property, $\measuredangle STR \cong \measuredangle TRS$. Therefore, $\overline{RS} \cong \overline{ST}$ (Theorem 7.2).
    (b) $\triangle RSM \cong \triangle TSM$ by SSS. By CPCTC, $\measuredangle RSM \cong \measuredangle TSM$. Therefore $\overline{SM}$ bisects $\measuredangle RST$.
12. $\overline{AM} \cong \overline{BM}$ (definition of median), $\measuredangle CMB \cong \measuredangle PMA$, and $\overline{CM} \cong \overline{PM}$ so$\triangle AMP \cong \triangle BMC$ by SAS. By CPCTC, $\measuredangle P \cong \measuredangle BCM$. Since alternate interior angles are congruent, $\overline{AP} \parallel \overline{CB}$.
13. **(a)** $\measuredangle A \cong \measuredangle B$, $\overline{AF} \cong \overline{BD}$ (addition property), $\measuredangle x \cong \measuredangle y$. Therefore, $\triangle AFG \cong \triangle BDE$ by ASA.
    (b) $\overline{AC} \cong \overline{BC}$ (Theorem 7.2), $\overline{AG} \cong \overline{BE}$ by CPCTC. Therefore, $\overline{GC} \cong \overline{EC}$ (subtraction property).
14. Show $\triangle KQM \cong \triangle LQN$ by SAS. By CPCTC, $\measuredangle K \cong \measuredangle L$ and $\overline{MK} \cong \overline{NL}$ so that $\triangle PMK \cong \triangle RNL$ by ASA. Therefore, $\overline{PM} \cong \overline{NR}$ by CPCTC.
15. By Theorem 7.1. $\measuredangle CAB \cong \measuredangle CBA$. By the subtraction property, $\measuredangle EAP \cong \measuredangle DBP$. In $\triangle APB$, $\overline{AP} \cong \overline{BP}$ by Theorem 7.2. Since $\measuredangle EPA \cong \measuredangle DPB$, $\Delta EPA \cong \triangle DPB$ by ASA. By CPCTC, $\overline{PE} \cong \overline{PD}$.
16. First show $\triangle CAE \cong \triangle CBD$ by AAS since $\measuredangle A \cong \measuredangle B$, $\measuredangle CEA \cong \measuredangle CDB$, and $\overline{AC} \cong \overline{BC}$. Next, show $\triangle GAE \cong \triangle FBD$ by SAS since $\overline{AE} \cong \overline{BD}$ (CPCTC), $\measuredangle A \cong \measuredangle B$, and $\overline{AG} \cong \overline{BF}$ (halves of congruent segments are congruent). By CPCTC, $\overline{EG} \cong \overline{DF}$.

Chapters 8–11

1. (2)
2. (4)
3. (3)
4. (4)
5. (4)
6. (1)
7. (1)
8. (1)
9. 24
10. 1
11. 40
12. $6\sqrt{2}$
13. 6
14. $2\sqrt{3}$
15. $\frac{5}{13}$
16. **(a)** 58.7°
    (b) 122
17. **(a)** 77°
    (b) 4.6
18. Use Theorem 10.3. $\measuredangle CMB \cong \measuredangle E$ (all right angles are congruent) and $\measuredangle B \cong \measuredangle D$ since complements of congruent (vertical) angles are congruent. Therefore, $\triangle CMB \sim \triangle DEA$.
19. Since the measure of an exterior angle of a triangle is greater than the measure of a non-adjacent interior angle of the triangle, $m\measuredangle ADC > m\measuredangle BCD$. Since $m\measuredangle BCD = m\measuredangle ACD$, by substitution $m\measuredangle ADC > m\measuredangle ACD$ which implies $CA > DA$.
20. Assume $\overline{AC}$ bisects $\measuredangle BAD$. Therefore, $\measuredangle BAC \cong \measuredangle DAC$. Since $\overline{BC} \parallel \overline{AD}$, $\measuredangle BCA \cong \measuredangle DAC$. By the transitive property, $\measuredangle BAC \cong \measuredangle BCA$, but this contradicts the Given, ($\triangle ABC$ is *not* isosceles). Therefore, the assumption $\overline{AC}$ bisects $\measuredangle BAD$ is false, and the statement $\overline{AC}$ does *not* bisect $\measuredangle BAD$ is true.
21. **(a)** $\angle CEF \cong \angle EDF$. $\angle CFE \cong \angle DEF$ since each angle is complementary to the same angle (angle DFE). $\triangle FEC \sim \triangle EDF$ by Theorem 10.3.
    (b) Since $\angle A \cong \angle C$ (Theorem 7.1) and $\angle DFA \cong \angle CEF$ (right angles are congruent), $\triangle DFA \sim \triangle FEC$. Since $\triangle FEC \sim \triangle EDF$, by the transitive property of similarity, $\triangle DFA \sim \triangle EDF$.
22. **(a)** $\overline{EB} \parallel \overline{CD}$ (extensions of parallel segments are parallel) and $\overline{CE} \parallel \overline{DB}$ (given) so that quadrilateral $BECD$ is a parallelogram.
    (b) Since the diagonals of the trapezoid are congruent, the trapezoid is isosceles so that triangles DAB and CBA are congruent. Thus, $\angle CAB \cong \angle DBA$. Since $\angle E \cong \angle DBA$ (parallel lines form congruent alternate interior angles), by the transitive property. $\angle CAB \cong \angle E$. By Theorem 7.2 $\overline{AC} \cong \overline{CE}$.
    (c) Since $\angle CAB \cong \angle DBA$ (see part **b**), ΔAFB is isosceles.
23. **(a)** Since $ABCD$ is a rectangle, $\overline{AD} \cong \overline{BC}$, $\angle D \cong \angle C$, and $\overline{DE} \cong \overline{CF}$ (by addition). Therefore, $\triangle ADE \cong \triangle BCF$ by SAS and $\angle 1 \cong \angle 2$ by CPCTC.
    (b) Angles A and B are right angles, $\angle 3 \cong \angle 4$ (complements of congruent angles are congruent).
    (c) $\overline{AG} \cong \overline{GB}$ by Theorem 7.2.
24. $\overline{BC} \parallel \overline{AD}$ since alternate interior angles are congruent. $\triangle CEB \cong \triangle AED$ by ASA since $\angle 1 \cong \angle 2$, $\overline{AE} \cong \overline{CE}$, and $\angle AED \cong \angle CEB$. Therefore, $\overline{BC} \cong \overline{AD}$ by CPCTC. Quadrilateral $ABCD$ is a parallelogram since the same pair of sides are both parallel and congruent.

25. First show $\triangle RWT \sim \triangle RVS$. Since $\overline{TW} \cong \overline{TV}$, $\measuredangle W \cong \measuredangle TVW \cong \measuredangle SVR$. By the transitive property, $\measuredangle W \cong \measuredangle SVR$. Since $\overline{RVW}$ bisects $\measuredangle SRT$, $\measuredangle TRW \cong \measuredangle SRV$, and $\triangle RWT \sim \triangle RVS$ by Theorem 10.3.

Therefore, $\dfrac{RW}{RV} = \dfrac{TW}{SV}$ so that $RW \times SV = RV \times TW$.

Chapters 12–17

1. **(a)** $-\dfrac{4}{3}$
   (b) 10
2. 10 square units
3. $y = -\dfrac{1}{3}x$
4. 120°
5. 30
6. 8
7. 2
8. **(a)** $x = 4$
   (b) $y = 2$
9. 18π
10. 72 square units
11. 2
12. $y = \dfrac{1}{2}x + 1$
13. 24π
14. $12\sqrt{3}$ square units
15. 12
16. (7, 17)
17. (1,–1)
18. $\dfrac{3}{2}$
19. 9π
20. $y = -\dfrac{2}{3}x$
21. 6
22. 2
23. $16 - 4\pi$ square units
24. 30
25. Show slope of $\overline{RA}$ = slope of $\overline{PT} = \dfrac{1}{3}$ and slope of $\overline{RT} \neq$ slope of $\overline{AP}$. $RT = AP = 5$ so the trapezoid is isosceles.
26. **(a)** 5
    (b) $x^2 + y^2 = 25$
    (c) $\dfrac{3}{4}$
    (d) $y = \dfrac{3}{4}x + \dfrac{25}{4}$
    (e) $(0, \dfrac{25}{4})$
27. Show $AB = BC = CD = DA = 5$ and slope $\overline{AB} \times$ slope $\overline{AD} = -1$.
28. **(a)** $y = 2$
    (b) $x = -4$
    (c) $(-4, 2)$
    (d) $(x - 2)^2 + (y + 2)^2 = 16$
    (e) 1
29. 6
30. **(a)** 60
    (b) 20
    (c) 60
    (d) 20
    (e) 110

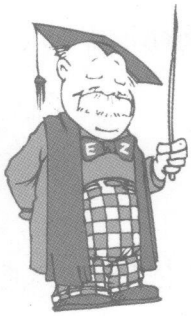

MOVE TO THE HEAD OF YOUR CLASS THE EASY WAY!

Barron's presents **THE E-Z SERIES** (formerly THE EASY WAY SERIES)—specially prepared by top educators, it maximizes effective learning while minimizing the time and effort it takes to raise your grades, brush up on the basics, and build your confidence. Comprehensive and full of clear review examples, **THE E-Z SERIES** is your best bet for better grades, quickly!

ISBN 978-0-7641-4256-7 **E-Z Accounting**—$16.99, *Can$19.99*
ISBN 978-0-7641-4257-4 **E-Z Algebra**—$16.99, *Can$19.99*
ISBN 978-1-4380-0039-8 **E-Z Algebra 2**—$16.99, *Can$19.50*
ISBN 978-0-7641-4258-1 **E-Z American History**—$18.99, *Can$22.99*
ISBN 978-0-7641-4458-5 **E-Z American Sign Language**—$16.99, *Can$19.99*
ISBN 978-0-7641-4468-4 **E-Z Anatomy and Physiology**—$16.99, *Can$19.99*
ISBN 978-0-7641-4466-0 **E-Z Arithmetic**—$16.99, *Can$19.99*
ISBN 978-0-7641-4134-8 **E-Z Biology**—$18.99, *Can$21.99*
ISBN 978-0-7641-4133-1 **E-Z Bookkeeping**—$14.99, *Can$17.99*
ISBN 978-0-7641-4259-8 **E-Z Business Math**—$14.99, *Can$17.99*
ISBN 978-0-7641-4461-5 **E-Z Calculus**—$16.99, *Can$19.99*
ISBN 978-0-7641-4128-7 **E-Z Chemistry**—$16.99, *Can$19.99*
ISBN 978-0-7641-2579-9 **Creative Writing the Easy Way**—$14.99, *Can$17.99*
ISBN 978-0-7641-4464-6 **E-Z Earth Science**—$16.99, *Can$19.99*
ISBN 978-0-7641-3736-5 **English for Foreign Language Speakers the Easy Way**—$18.99, *Can$21.99*
ISBN 978-0-7641-4260-4 **E-Z English**—$14.99, *Can$17.99*
ISBN 978-0-7641-3050-2 **Forensics the Easy Way**—$14.99, *Can$17.99*
ISBN 978-0-7641-4455-4 **E-Z French**—$16.99, *Can$19.99*
ISBN 978-0-7641-2435-8 **French Grammar the Easy Way**—$16.99, *Can$19.99*
ISBN 978-0-7641-3918-5 **E-Z Geometry**—$16.99, *Can$19.99*
ISBN 978-0-7641-4261-1 **E-Z Grammar**—$14.99, *Can$16.99*
ISBN 978-0-7641-4454-7 **E-Z Italian**—$14.99, *Can$17.99*
ISBN 978-0-8120-9627-9 **Japanese the Easy Way**—$18.99, *Can$22.99*
ISBN 978-0-7641-3237-7 **Macroeconomics the Easy Way**—$14.99, *Can$17.99*
ISBN 978-0-7641-9369-9 **Mandarin Chinese the Easy Way**—$21.99, *Can$26.50*
ISBN 978-0-7641-4132-4 **E-Z Math**—$14.99, *Can$17.99*
ISBN 978-0-7641-1871-5 **Math Word Problems the Easy Way**—$16.99, *Can$19.99*
ISBN 978-0-7641-4456-1 **E-Z Microbiology**—$18.99, *Can$22.99*
ISBN 978-0-8120-9601-9 **Microeconomics the Easy Way**—$16.99, *Can$19.99*
ISBN 978-0-7641-4467-7 **E-Z Organic Chemistry**—$16.99, *Can$19.99*
ISBN 978-0-7641-4126-3 **E-Z Physics**—$16.99, *Can$19.99*
ISBN 978-1-4380-0011-4 **E-Z Pre-Algebra**—$16.99, *Can$19.50*
ISBN 978-0-7641-4465-3 **E-Z Precalculus**—$16.99, *Can$19.99*
ISBN 978-0-7641-4462-2 **E-Z Psychology**—$16.99, *Can$19.99*
ISBN 978-0-7641-4129-4 **E-Z Spanish**—$16.99, *Can$19.99*
ISBN 978-0-7641-4249-9 **E-Z Spanish Grammar**—$16.99, *Can$19.99*
ISBN 978-0-7641-4459-2 **E-Z Spelling**—$14.99, *Can$17.99*
ISBN 978-0-7641-3978-9 **E-Z Statistics**—$16.99, *Can$19.99*
ISBN 978-0-7641-4251-2 **E-Z Trigonometry**—$16.99, *Can$19.99*
ISBN 978-0-8120-9765-8 **World History the Easy Way, Vol. One**—$18.99, *Can$22.99*
ISBN 978-0-8120-9766-5 **World History the Easy Way, Vol. Two**—$21.99, *Can$26.50*

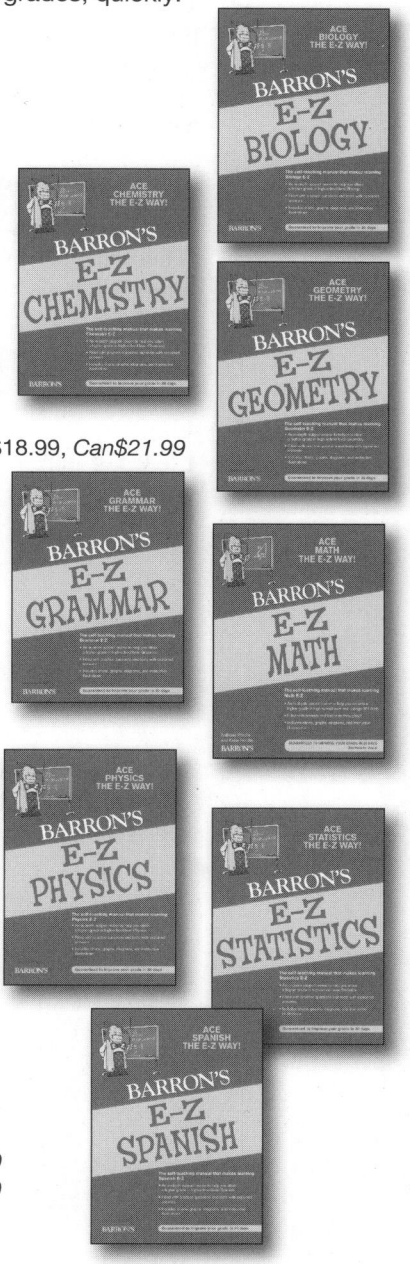

To order

Available at your local book store or visit **www.barronseduc.com**

BARRON'S

Barron's Educational Series, Inc.
250 Wireless Blvd.
Hauppauge, N.Y. 11788
Order toll-free: 1-800-645-3476
Order by fax: 1-631-434-3217

In Canada:
Georgetown Book Warehouse
34 Armstrong Ave.
Georgetown, Ontario L7G 4R9
Canadian orders: 1-800-247-7160
Order by fax: 1-800-887-1594

(#45) R4/12

Prices subject to change without notice.

31. **(a)** Show $AB = AC = \sqrt{40}$.
    (b) Midpoint of $\overline{BC}$ is $M(2, -1)$. Show slope $\overline{AM} \times$ slope $\overline{BC} = -1$.
32. **(a)** A pair of parallel lines whose equations are $x = 2 - d$ and $x = 2 + d$.
    (b) The circle whose equation is $x^2 + y^2 = 1$
    (c) **1** 2 **2** 1 **3** 0
33. First show $\triangle ANR \sim \triangle ABN$. $\measuredangle ANR \cong \measuredangle ABN$ (all right angles are congruent) and $\measuredangle A \cong \measuredangle A$. Therefore, $\dfrac{AN}{AB} = \dfrac{AR}{AN}$ so that $(AN)^2 = AR \times AB$.
34. Show $\triangle AEB \cong \triangle CFD$. $\measuredangle B \cong \measuredangle D$ (Theorem 7.1) so $ACD \cong CAB$. By arc subtraction, $AB \cong CD$ which implies $\overline{AB} \cong \overline{CD}$. Since $\measuredangle E \cong \measuredangle F$, $\triangle AEB \cong \triangle CFD$ by AAS, $\overline{AE} \cong \overline{CF}$ by CPCTC.
35. **(a)** 90
    (b) 4
    (c) **1** 63° **2** 63°
36. **(a)** 72
    (b) 36
    (c) 54
    (d) 18
    (e) 72
37. **(a)** Line parallel to and midway between the given lines.
    (b) Circle having P as its center and a radius of k units.
    (c) **1** 1 **2** 2 **3** 0
38. **(a)** Show $(AB)^2 + (BC)^2 = (AC)^2$
    (b) 12.5 square units
39. 17.5 square units
40. First show $\triangle BCD \sim \triangle ABE$. $\measuredangle DBC \cong \measuredangle A$ since the measure of each angle is equal to one-half the measure of the same arc (BF). $\measuredangle ABE$ is a right angle. Since $\overline{BA} \parallel \overline{CD}$ and interior angles on the same side of the transversal are supplementary, $\measuredangle DCB$ is also a right angle so $\measuredangle DCB \cong \measuredangle ABC$. Therefore, $\triangle BCD \sim \triangle ABE$ by Theorem 10.3, so $\dfrac{BD}{AE} = \dfrac{CD}{BE}$.
41. **(a)** 36°
    (b) 39